Lecture Notes in Computer Science 9126

Commenced Publication in 1973
Founding and Former Series Editors:
Gerhard Goos, Juris Hartmanis, and Jan van Leeuwen

More information about this series at http://www.springer.com/series/7412

Hans van Assen · Peter Bovendeerd
Tammo Delhaas (Eds.)

Functional Imaging and Modeling of the Heart

8th International Conference, FIMH 2015
Maastricht, The Netherlands, June 25–27, 2015
Proceedings

 Springer

Editors
Hans van Assen
Eindhoven University of Technology
Eindhoven
The Netherlands

Peter Bovendeerd
Eindhoven University of Technology
Eindhoven
The Netherlands

Tammo Delhaas
Maastricht University
Maastricht
The Netherlands

ISSN 0302-9743 ISSN 1611-3349 (electronic)
Lecture Notes in Computer Science
ISBN 978-3-319-20308-9 ISBN 978-3-319-20309-6 (eBook)
DOI 10.1007/978-3-319-20309-6

Library of Congress Control Number: 2015941125

LNCS Sublibrary: SL6 – Image Processing, Computer Vision, Pattern Recognition, and Graphics

Springer Cham Heidelberg New York Dordrecht London

Printed on acid-free paper

Springer International Publishing AG Switzerland is part of Springer Science+Business Media
(www.springer.com)

Preface

FIMH 2015 was the 8th International Conference on Functional Imaging and Modeling of the Heart. It was held in Maastricht, The Netherlands, during June 25–27, 2015. The previous editions of this conference were held in Helsinki (2001), Lyon (2003). Barcelona (2005), Salt Lake City (2007), Nice (2009), New York (2011), and London (2013). This biennial scientific event aims to integrate the research and development efforts in the fields of cardiovascular modeling and image analysis. For this particular edition, we especially welcomed contributions related to cardiac imaging through ultrasound. The main goal of FIMH is to encourage collaboration among scientists in signal and image processing, imaging, applied mathematics, biophysics, biomedical engineering, and experts in cardiology, radiology, biology, and physiology. The final objective is to contribute to the diagnosis and treatment of cardiovascular diseases.

These proceedings contain the 54 contributions that were selected out of 72 original submissions through a rigorous review process. Each paper, with a length of 8–11 pages in the proceedings format, was reviewed by two to four members of the International Program Committee, composed of approximately 50 prominent scientists in the field. The authors had to submit a point-by-point response to the reviewers' comments, and the final selection of the papers was made on the basis of the reviewers' opinion after this rebuttal. The proceedings, published in a dedicated volume of the Springer series *Lecture Notes in Computer Science* (LNCS), were available at the time of the event.

The conference was greatly enhanced by keynote lectures given by four invited speakers. The common topic of these lectures was the relation between measurement data, their interpretation, and the role of models in this interpretation. Dr. Mark K. Friedberg (Associate Professor of Paediatrics, Hospital for Sick Children and University of Toronto) revealed the harsh world of getting data, especially from kids. Prof. Darrel P. Francis, (Professor of Cardiology, Imperial College London) conveyed the reliability of input data for models in relation to the way doctors collect and interpret data. Prof. Michiel van den Broeke (Professor of Polar Meteorology, Utrecht University) reported on modeling in the seemingly completely different field of meteorology, with surprisingly similar issues to our field: sample resolution, sample quality, reliability of models, what do politicians want to hear, what is hearsay and what are facts? Prof. Antoon F.M. Moorman (Emeritus Professor of Anatomy and Embryology, Academic Medical Center, Amsterdam) disclosed how imaging techniques and analyses can and did help to understand cardiac development.

We want to express our special thanks to the members of the Program Committee, the additional reviewers, the organizing team, and the participants of the meeting, who contributed to the success of this event.

June 2015

Hans van Assen
Peter Bovendeerd
Tammo Delhaas

Organization

FIMH 2015 was organized by Maastricht University and Eindhoven University of Technology.

Organizing Committee

General Co-chairs

Hans van Assen	Eindhoven University of Technology, The Netherlands
Peter Bovendeerd	Eindhoven University of Technology, The Netherlands
Tammo Delhaas	Maastricht University, The Netherlands

Local Organization

Claire Meertens	Maastricht University, The Netherlands

Program Committee

Elsa Angelini	Columbia University, USA
Theo Arts	Maastricht University, The Netherlands
Leon Axel	NYU, USA
Olivier Bernard	Creatis INSA Lyon, France
Hans Bosch	Erasmus MC, The Netherlands
Oscar Cámara	Universitat Pompeu Fabra, Spain
Dominique Chapelle	Inria, France
Patrick Clarysse	University of Lyon, France
Pierro Colli Franzone	Pavia University, Italy
Dorin Comaniciu	Siemens, USA
Herve Delingette	Inria, France
Jan D'hooge	KU Leuven, Belgium
Olaf Doessel	Karlsruhe University, Germany
James Duncan	Yale University, USA
Alejandro Frangi	University of Sheffield, UK
Jean-Frederic Gerbeau	Inria, France
Arun Holden	Leeds University, UK
Robert Howe	Harvard University, USA
Ivana Isgum	Image Sciences Institute, The Netherlands
Peter Kohl	Imperial College, UK
Chris de Korte	Radboud University MC, The Netherlands
Pablo Lamata	King's College London, UK
Boudewijn Lelieveldt	Leiden University, The Netherlands
Cristian Lorenz	Philips Research, Germany
Lasse Løvstakken	NTNU, Norway
Isabelle Magnin	University of Lyon, France

Steven Niederer	King's College, UK
Alison Noble	Oxford University, UK
Sebastien Ourselin	University College London, UK
Terry Peters	Robarts Research Institute, Canada
Caroline Petitjean	University of Rouen, France
Mihaela Pop	Sunnybrook Research Institute, Canada
Kawal Rhode	King's College London, UK
Daniel Rueckert	Imperial College London, UK
Frank Sachse	University of Utah, USA
Gunnar Seemann	KIT, Germany
Maxime Sermesant	Inria, France
Pengcheng Shi	Rochester Institute of Technology, USA
Wenzhe Shi	Imperial College, UK
Nic Smith	University of Auckland, New Zealand
Larry Staib	Yale University, USA
Regis Vaillant	GE Healthcare, France
Andreas Wahle	University of Iowa, USA
Jurgen Weese	Phillips Research, Germany
Alistair Young	Auckland University, New Zealand
Xiahai Zhuang	Shangai Advanced Research Institute, China

Additional Reviewers

Luc Florack	Eindhoven University of Technology, The Netherlands
Andrea Fuster	Eindhoven University of Technology, The Netherlands
Simona Turco	Eindhoven University of Technology, The Netherlands

Contents

Function

Imaging

Models of Mechanics

Models of Electrophysiology

Function

Learning a Global Descriptor of Cardiac Motion from a Large Cohort of 1000+ Normal Subjects

Wenjia Bai[1]([✉]), Devis Peressutti[2], Ozan Oktay[1], Wenzhe Shi[1],
Declan P. O'Regan[3], Andrew P. King[2], and Daniel Rueckert[1]

[1] Biomedical Image Analysis Group, Department of Computing,
Imperial College London, London, UK
w.bai@imperial.ac.uk
[2] Division of Imaging Sciences and Biomedical Engineering,
King's College London, London, UK
[3] MRC Clinical Sciences Centre, Hammersmith Hospital,
Imperial College London, London, UK

Abstract. Motion, together with shape, reflect important aspects of cardiac function. In this work, a new method is proposed for learning of a cardiac motion descriptor from a data-driven perspective. The resulting descriptor can characterise the global motion pattern of the left ventricle with a much lower dimension than the original motion data. It has demonstrated its predictive power on two exemplar classification tasks on a large cohort of 1093 normal subjects.

1 Introduction

Motion, together with shape, reflect important aspects of cardiac function. Detecting abnormal motion of the cardiac ventricles is of clinical interest for both diagnosis and prognosis [1]. To characterise the dynamics of cardiac motion, many clinical features have been proposed, such as the ejection fraction, segmental velocity, strain, strain-rate, time to peak velocity, time to peak strain, wall motion score index etc. [1–3]. Apart from these, some statistical features have also been proposed, such as the cross correlation of intensity profiles across time [4] or the distance of the intensity distribution across time [5]. These indices are normally empirically defined and present an intuitive picture of the motion profile.

Apart from the empirical features, an alternative way is to describe the motion from a purely data-driven perspective and to learn a descriptor of the motion pattern from a large group of subjects. For example, in [6], a pixelwise Gaussian distribution of the velocity is learnt from a normal population and used for detection of motion abnormality. A challenge for cardiac motion analysis is that its data is normally of high dimension. Therefore, dimensionality reduction techniques are often used to discover the underlying structure of the data. For example, in [7–9], principal component analysis (PCA) and independent component analysis (ICA) have been proposed to learn the modes of myocardial shape variation at end-systole (ES). Whereas PCA assumes a Gaussian distribution

H. van Assen et al. (Eds.): FIMH 2015, LNCS 9126, pp. 3–11, 2015.
DOI: 10.1007/978-3-319-20309-6_1

and performs linear dimensionality reduction, manifold learning is a non-linear technique which aims to preserve the local structure of the data. It has recently gained a lot of attention in the medical imaging community [10,11].

In this work, we propose to learn a global descriptor of cardiac motion from segmental motion trajectories using dimensionality reduction techniques. The descriptor is not confined to a single voxel or a single segment, but instead it characterises the global motion pattern of the whole left ventricle (LV). We compare the performance of both PCA and Isomap manifold learning for dimensionality reduction. To demonstrate the value of the motion descriptor, we use it to predict the gender and age of a subject and evaluate the performance on a large data set of 1093 normal subjects.

2   Methods

Prior to the learning of a motion descriptor, we first estimate motion from cine cardiac MR images. Since the heart of each subject lies at different locations and

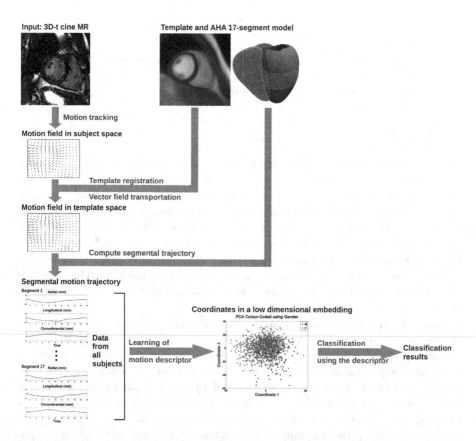

Fig. 1. The flowchart consists of motion tracking, spatial normalisation, descriptor learning and classification.

with different orientations, we perform spatial normalisation by registering and transporting all motion fields to a template space. The segmental motion trajectories are extracted and concatenated to form a high-dimensional feature vector. Dimensionality reduction is applied to the high-dimensional data leading to a global motion descriptor. Finally, we use the motion descriptor in exemplar classification tasks for gender classification and age prediction. Figure 1 illustrates the flowchart of the method and we will explain each step in the following.

2.1 Motion Tracking

In this work, we use cine MR for cardiac motion analysis. Other imaging modalities such as tagged MR or ultrasound (US) can also be used to capture the motion of the heart, which can provide different spatio-temporal resolution and image quality. The proposed motion descriptor is not confined to a specific imaging modality.

Motion tracking is performed for each subject using a 4D spatio-temporal B-spline image registration method with a sparseness regularisation term (TSFFD) [12]. The motion field estimate is represented by a displacement vector at each voxel and at each time frame t, which measures the displacement from the 0-th frame to the t-th frame. All the cine images in this work were acquired using the same imaging protocol, consisting of 20 time frames across a cardiac cycle with the 0-th frame representing the end-diastolic (ED) frame. Therefore, we do not perform temporal normalisation for the motion field.

2.2 Template Image and Spatial Normalisation

A template image is built by registering all the subject images at the ED frame and computing the average intensity image. In addition, the subject images are all segmented using a multi-atlas segmentation method [13]. The segmentation of the template image is then inferred by averaging all the subject segmentations. A template surface mesh is reconstructed from its segmentation and manually divided into 17 segments using the AHA model. The template and the segmental surface mesh are displayed at the top-right corner of Fig. 1.

The motion field estimate lies within the space of each subject. To enable inter-subject comparison and analysis, all the subject images are aligned to the template image by non-rigid B-spline image registration [14]. Using the transformation between the template space and subject space, we transport the motion field of each subject to the template space. Let $x' = T(x)$ denote the transformation from the template space to the subject space, where x and x' are respectively the coordinates in the template space and in the subject space. By considering the spatial transformation as a change of coordinates, we have,

$$d(x,t) = J_{T^{-1}}(x')d'(x',t) \tag{1}$$

where d' denotes an infinitesimal displacement in the subject space, d denotes the corresponding infinitesimal displacement in the template space and $J_{T^{-1}}(x') \equiv \frac{dx}{dx'}$ denotes the Jacobian matrix of the inverse transformation.

2.3 Segmental Motion Trajectory

To characterise cardiac motion both spatially and temporally, we empirically define a high-dimensional feature vector using the segmental motion trajectory. S denotes the number of left ventricular segments. Since we use the AHA 17-segment model, $S = 17$. T denotes the number of time frames, which is equal to 20 for our data set. d denotes the dimension of the displacement vector and $d = 3$, which consists of radial, longitudinal and circumferential components. We compute the mean displacement for each segment at each time frame. The displacements across time for all the segments are concatenated to form the feature vector, which has the dimension of $S \times T \times d$ and contains information about the cardiac motion both spatially and temporally.

In principle, we can increase the spatial segments S so the feature vector describes more detailed motion at a higher spatial resolution. For example, we can compute the displacement for all the vertices of the myocardial mesh and concatenate them. However, we have found that it becomes computationally prohibitive to perform dimensionality reduction for vertex-wise motion data using techniques such as PCA. Also, since the cardiac motion is estimated using B-splines, displacements at neighbouring vertices are very similar and we may not need all the vertices to represent the motion data. Therefore, we adopt segment-wise motion data in this work.

2.4 Learning of a Motion Descriptor

Given the high-dimensional feature vector, we perform dimensionality reduction in order to find a descriptor which can characterise the motion with a low dimension. We compare two techniques, PCA and Isomap manifold learning [15]. The resulting low-dimensional coordinates are used as a motion descriptor.

PCA looks for a low-dimensional embedding of the data points that best preserves the variance. In the new coordinate system, the greatest variance of the data lies on the first coordinate, the second variance of the data on the second coordinate and so on. It is accomplished by eigen-decomposition of the data covariance matrix. In contrast to this, Isomap looks for a low-dimensional embedding that best preserves the geodesic distances between pairs of data points, i.e. the local data structure. It analyses the data structure as a graph, where each node denotes a data point and it is connected with K neighbours. The geodesic distances in the neighbourhood are preserved in the new coordinate system.

2.5 Application to Classification Tasks

To demonstrate the abundant information contained in the motion descriptor, we use the motion descriptor for two exemplar classification tasks, training SVM classifiers namely for gender classification and age prediction.

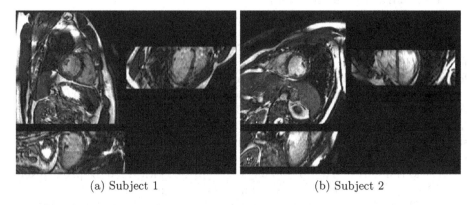

(a) Subject 1   (b) Subject 2

Fig. 2. Two exemplar cardiac MR images. Three orthogonal views are shown for each subject.

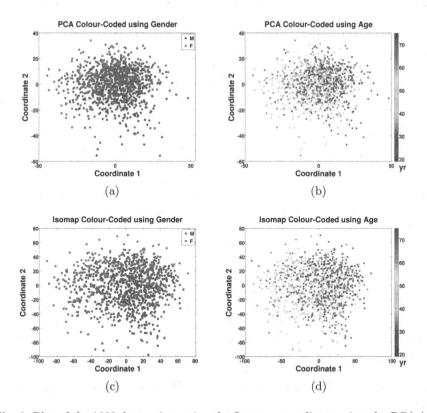

Fig. 3. Plot of the 1093 data points using the first two coordinates given by PCA (top row) or Isomap (bottom row). The data points are colour-coded using gender or age.

3   Experiments and Results

The data set used in this work consists of cardiac MR images of 1093 normal subjects (493 males, 600 females; age range 19–75 yr, mean 40.1 yr), which forms part of the UK 1000 Cardiac Phenomes project Cardiac MR was performed on a 1.5T Philips Achieva system (Best, Netherlands). The maximum gradient strength was 33 mT/m and the maximum slew rate 160 mT/m/ms. A 32 element cardiac phased-array coil was used for signal reception. Scout images were obtained and used to plan a single breath-hold 3D cine balanced steady-state free precession (b-SSFP) images in the left ventricular short axis (LVSA) plane from base to apex using the following parameters: repetition time msec/echo time msec, 3.0/1.5; flip angle, 50°; bandwidth, 1250 Hz/pixel; pixel size 2.0 × 2.0 mm; section thickness 2 mm overlapping; reconstructed voxel size, 1.25 × 1.25 × 2 mm; number of sections, 50–60; cardiac phases, 20; sensitivity encoding (SENSE) factor, 2.0 anterior-posterior and 2.0 right-left direction. Two exemplar images are displayed at Fig. 2.

We extracted motion descriptors using PCA or Isomap from this data set. Figure 3 shows the first coordinates given by PCA and Isomap. The data points are colour-coded using gender or age. Figure 3(a) shows that the male subjects are more likely to be distributed at the bottom-right corner using the PCA coordinates, whereas the female subjects are more likely to be at the top-left corner. Figure 3(b) shows that the age of the subjects follows a right-to-left trend, gradually changing from young to old. The Isomap coordinates in Fig. 3(c) and (d) reflect a similar trend as the PCA coordinates. This means that the motion descriptor, though in a low dimension, contains abundant information for motion data analysis.

SVM classifiers were trained using the segmental trajectories or the motion descriptors as input. We used the radial basis function (RBF) as the kernel and the default parameter settings in R. Ten-fold cross-validation was performed for performance evaluation. The performance was evaluated using the percentage of correct gender classification and the age prediction error. There is one parameter for Isomap, which is the number of neighbours K. We tuned the parameter and found that $K = 20$ achieved the best performance.

Table 1 lists the classification performance using the original segmental trajectories and using the motion descriptors with different dimensions. The original dimension of the feature is $S \times T \times d = 1020$. It shows that the best accuracy is achieved using the original high-dimensional feature vector, which is over 89 % accuracy rate for gender classification and only –0.13 yr error for age prediction. However, with a much lower dimension of only 100, PCA can also achieve very high accuracy, with over 87 % accuracy rate for gender classification and –0.36 yr error for age prediction. More dimensions in PCA do not necessarily improve the performance, since the first few coordinates of PCA have already encoded most of the data variance.

For Isomap manifold learning, it achieves worse gender accuracy rate of 80.15 % and age prediction error of –0.47 yr, when the same dimension of 100 is used. This may hint that the high-dimensional feature vectors are located

Table 1. Classification performance using the original segmental trajectories and using the motion descriptors with different dimensions. Gender classification is evaluated using the percentage of correct classification and age prediction is evaluated using the prediction error.

	Dim	Gender	Age (yr)		Dim	Gender	Age (yr)
Original	1020	89.94 %	-0.13 ± 7.09				
PCA	10	76.21 %	-0.65 ± 9.27	Isomap	10	76.30 %	-1.03 ± 9.82
	50	84.63 %	-0.50 ± 8.12		50	80.60 %	-0.60 ± 9.36
	100	87.38 %	-0.36 ± 7.77		100	80.15 %	-0.47 ± 9.47
	150	86.74 %	-0.30 ± 8.03		150	78.78 %	-0.47 ± 9.69
	200	86.46 %	-0.16 ± 8.42		200	77.68 %	-0.50 ± 9.94

in a relatively flat manifold and therefore do not need non-linear techniques for dimensionality reduction. We have also tested two other manifold learning methods, namely locally linear embedding (LLE) and Laplacian eigenmaps. They achieve similar or slightly worse performance than Isomap.

4   Discussion and Conclusions

In this work, we learn a motion descriptor completely from a data-driven perspective on a large population of subjects by looking for a low-dimensional embedding which can explain either the data variance (PCA) or the local data structure (Isomap). We have demonstrated the resulting motion descriptor can be useful for both data visualisation and classification tasks.

There are mainly two reasons that motivate us to use the dimensionality reduction techniques. First, it allows convenient visualisation of high-dimensional data so that we can appreciate the data distribution in a better way. Second, it avoids the curse of dimensionality that may occur during data analysis. However, the SVM classifier seems to be well adapted to high-dimensional data. This explains why classification using the original high-dimensional data also yields a very good performance, as shown in Table 1.

In this work, segmental displacement trajectory is used as a representation of cardiac motion. However, other representations such as velocity, strain and diastolic filling rate can characterise motion in different ways. Our future work includes estimation of myocardial strain and potentially segmental strain can also be concatenated into the feature vector and used for learning a motion descriptor.

Registration is a key step to normalise the motion field of each subject into a template space. Two measures are taken to reduce the impact of potential registration errors on subsequent motion analysis. First, six landmarks are used to initialise the image registration so that severe registration error is less likely to happen. Second, the mean motion trajectory is computed for each segment, which is more robust than vertex-wise motion trajectory.

Although we only show exemplar applications for gender and age prediction on a data set of normal subjects, potentially we can also apply the motion descriptor to other tasks which require motion as input. In the future, we plan to explore its application on cardiac patient classification, for example, to classify between CRT respondents and non-respondents by combining the motion descriptor with other information such as QRS duration, myocardial scar amount etc., which may together reveal interesting findings about cardiac diseases.

References

1. Mor-Avi, V., Lang, R.M., Badano, L.P., Belohlavek, M., et al.: Current and evolving echocardiographic techniques for the quantitative evaluation of cardiac mechanics. Eur. J. Echocardiogr. **12**(3), 167–205 (2011)
2. Lang, R.M., Bierig, M., Devereux, R.B., Flachskampf, F.A., et al.: Recommendations for chamber quantification. Eur. J. Echocardiogr. **7**(2), 79–108 (2006)
3. Garcia-Barnés, J., Gil, D., Badiella, L., Hernandez-Sabate, A., et al.: A normalized framework for the design of feature spaces assessing the left ventricular function. IEEE Trans. Med. Imaging **29**(3), 733–745 (2010)
4. Lu, Y., Radau, P., Connelly, K., Dick, A., Wright, G.: Pattern recognition of abnormal left ventricle wall motion in cardiac MR. In: Yang, G.-Z., Hawkes, D., Rueckert, D., Noble, A., Taylor, C. (eds.) MICCAI 2009, Part II. LNCS, vol. 5762, pp. 750–758. Springer, Heidelberg (2009)
5. Afshin, M., Ben Ayed, I., Punithakumar, K., Law, M., et al.: Regional assessment of cardiac left ventricular myocardial function via MRI statistical features. IEEE Trans. Med. Imaging **33**(2), 481–494 (2014)
6. Duchateau, N., De Craene, M., Piella, G., Silva, E., et al.: A spatiotemporal statistical atlas of motion for the quantification of abnormal myocardial tissue velocities. Med. Image Anal. **15**(3), 316–328 (2011)
7. Remme, E., Young, A.A., Augenstein, K.F., Cowan, B., Hunter, P.J.: Extraction and quantification of left ventricular deformation modes. IEEE Trans. Biomed. Eng. **51**(11), 1923–1931 (2004)
8. Leung, K.Y.E., Bosch, J.G.: Local wall-motion classification in echocardiograms using shape models and orthomax rotations. In: Sachse, F.B., Seemann, G. (eds.) FIHM 2007. LNCS, vol. 4466, pp. 1–11. Springer, Heidelberg (2007)
9. Suinesiaputra, A., Frangi, A.F., Kaandorp, T., Lamb, H.J., et al.: Automated detection of regional wall motion abnormalities based on a statistical model applied to multislice short-axis cardiac MR images. IEEE Trans. Med. Imaging **28**(4), 595–607 (2009)
10. Bhatia, K.K., Rao, A., Price, A., Wolz, R., Hajnal, J., Rueckert, D.: Hierarchical manifold learning for regional image analysis. IEEE Trans. Med. Imaging **33**(2), 444–461 (2014)
11. Ye, D.H., Desjardins, B., Hamm, J., Litt, H., Pohl, K.: Regional manifold learning for disease classification. IEEE Trans. Med. Imaging **33**(6), 1236–1247 (2014)
12. Shi, W., Jantsch, M., Aljabar, P., Pizarro, L., Bai, W., et al.: Temporal sparse free-form deformations. Med. Image Anal. **17**(7), 779–789 (2013)
13. Bai, W., Shi, W., O'Regan, D.P., Tong, T., et al.: A probabilistic patch-based label fusion model for multi-atlas segmentation with registration refinement: application to cardiac MR images. IEEE Trans. Med. Imaging **32**(7), 1302–1315 (2013)

14. Rueckert, D., Sonoda, L.I., Hayes, C., Hill, D.L.G., Leach, M.O., Hawkes, D.J.: Nonrigid registration using free-form deformations: application to breast MR images. IEEE Trans. Med. Imaging **18**(8), 712–721 (1999)
15. Tenenbaum, J.B., De Silva, V., Langford, J.C.: A global geometric framework for nonlinear dimensionality reduction. Science **290**(5500), 2319–2323 (2000)

Steps Towards Quantification
of the Cardiological Stress Exam

R. Chabiniok[1,2]($\boxtimes$), E. Sammut[2], M. Hadjicharalambous[2], L. Asner[2],
D. Nordsletten[2], R. Razavi[2], and N. Smith[2]

[1] Inria Saclay Ile-de-France, MƎDISIM Team, Palaiseau, France
radomir.chabiniok@inria.fr
[2] Division of Imaging Sciences and Biomedical Engineering,
St Thomas' Hospital, King's College London, London, UK

Abstract. In this work we aim to advance the translation of model-based myocardial contractility estimation to the clinical problem of quantitative assessment of the dobutamine stress exam. In particular, we address the question of limited spatial resolution of the observations obtained from cine MRI during the stress test, in which typically only a small number of cine MRI slices are acquired. Due to the relative risk during the dobutamine infusion, a safe acquisition protocol with a healthy volunteer under the infusion of a beta-blocker is applied in order to get a better insight into the contractility estimation using such a type of clinical data. The estimator is compared for three types of observations, namely the processed short axis cine stack contiguously covering the ventricles, the short axis stack limited to only 3 slices and the combination of 3 short and 3 long axis slices. A decrease of contractilities in AHA regions under the beta-blocker infusion was estimated for each observation. The corrected model (by using the estimated parameters) was then compared with the displacements extracted from 3D tagged MRI.

1 Introduction

A dobutamine stress test is used clinically to identify regional wall motion abnormalities to guide management on therapeutic options, such as revascularisation [7]. It is performed at increasing doses of dobutamine, an inotropic drug, with simultaneous imaging by echocardiography or magnetic resonance imaging (MRI) at each dosage level. Significant risks such as arrhythmia and cardiac arrest at higher doses, and the unpleasant sensation of the pharmacologically induced stress requires a rapid data acquisition which should not exceed three minutes at each dobutamine dosage. This limits the spatio-temporal resolution of the obtained image data. In particular in cardiac MRI – the modality used in this paper – a typical clinical protocol includes an acquisition of 3 slices in short axis placed in basal, mid- and apical third of left ventricle (LV), see Fig. 1 (middle), and 3 standard long axis images of LV (the so-called 2-, 3- and 4-chamber views). In addition to the limited spatial coverage, the demanding breath-holding under the dobutamine infusion often compromises the quality of image data.

© Springer International Publishing Switzerland 2015
H. van Assen et al. (Eds.): FIMH 2015, LNCS 9126, pp. 12–20, 2015.
DOI: 10.1007/978-3-319-20309-6_2

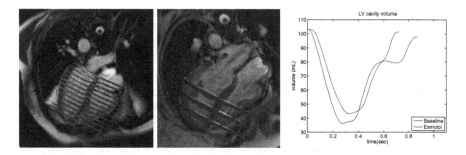

Fig. 1. Usual coverage of whole ventricles by short axis *full cine stack* (left) as oppose to the *3 short axis* slices typically used in the stress exam (middle). The LV volume plot on the right demonstrates a decreased ejection during infusion of beta-blocker esmolol.

A positive test is defined by a regional wall motion abnormality during stress and is currently clinically assessed entirely qualitatively. An estimation of some constitutive parameters in a biomechanical model is a way of assessing mechanical properties of myocardium [4,12,13], and in [3] quantitative values directly targeting regional myocardial contractility were estimated using cine MRI as observations. Such an approach has a potential to provide a higher reproducibility of the exam, lower inter-observer variability and possibly may even allow a decrease in the dose of dobutamine as the quantitative measure may increase the sensitivity. However, in comparison to [3] – where a full short axis cine stack covering whole ventricles was used for contractility estimation – in the stress test we need to deal with the limited spatio-temporal coverage. The presented study focuses on exploring the feasibility of translating the methodology of [3] into the clinical setup of the dobutamine stress exam with respect to the challenge of limited coverage.

Due to the associated risks in using dobutamine, it would be impossible to perform trial acquisitions in a healthy volunteer or significantly prolong the patient exam to acquire additional short axis slices under the dobutamine stimulation. Therefore, to get an insight into various image sampling modes, we estimate a change of contractility in a healthy volunteer under the safe pharmacological stimulus of the beta-blocker esmolol. This allows safe acquisition of high quality image data of a *full cine stack*, which can be then downsampled in space or time, in accordance to the real clinical data of the dobutamine stress test. As opposed to the positively inotropic drug dobutamine, the negatively inotropic beta-blocker globally reduces the contractility. This is reflected in the decreased stroke volume (see Fig. 1, right) and we expect that the estimated contractility values should follow a similar trend.

After describing the experimental data together with image processing in Sect. 2, and the modeling and parameter estimation framework (Sect. 3), we present in Sect. 4 the results of contractility estimation at baseline and under beta-blocker stimulation by using processed cine MRI of various resolution. We discuss our results and give some perspective remarks in Sect. 5 and conclude the paper in Sect. 6.

2   Clinical Data

2.1   Data Acquistion

Cardiac MRI was performed on 1.5T Philips Achieva system. The acquisition of a healthy volunteer dataset was performed at baseline and under the infusion of beta-blocker esmolol (50–200 μg/kg/min to achieve 10–20 % decrease in heart rate) and the following data were acquired in each part of the study:

- Cine bSSFP sequence in retrospective ECG gating with spatio-temporal resolution 2 × 2 × 8 mm and 40 time frames/cardiac cycle, FOV 350 × 350 mm. The short axis cine stack covering contiguously whole ventricles and standard long axis views of left ventricle (2-, 3-, and 4-chamber view) were taken.
- 3D tagged MRI of LV in prospective ECG triggering with acquired spatial resolution 3.4 × 7.7 × 7.7 mm (for 3 orientations of tag planes) reconstructed into one resulting 3D tagged image interpolated to 1 × 1 × 1 mm voxel size and temporal resolution ∼33 ms for both baseline and esmolol scan.

2.2   Data Processing

Image processing consists first of spatial registration of all the image sequences (rigid registration using the Image Registration Toolkit IRTK[1]). Then, we select 3 slices from the short axis stack such that the inter-slice spacing corresponds to that one in real stress exam. Non-rigid image registration based motion tracking [11] is consequently employed to track the LV endo- and epicardial surfaces in cine MRI data. By motion tracking of various types (and combinations) of cine images, we generate three types of *observations* (deforming LV endo- and epicardial surfaces) which are used in sequel for the contractility estimation. These will be denoted as:

1. *full s.a. stack:* the short axis cine stack covering whole ventricles with the temporal resolution downsampled to 20 time frames/cardiac cycle;
2. *3-s.a.+3-l.a.:* the selected 3 slices in short axis with 3 long axis views in original temporal sampling 40 time frames/cycle (standard clinical protocol);
3. *3-slice s.a.:* the selected 3 slices of the full stack (40 time frames/cycle).

Remark 1. The temporal resolution for *full s.a. stack* used in the parameter estimation is such that the MRI acquisition time corresponds to the standard clinical protocol *3-s.a.+3-l.a.*

Additionally, we extract full tissue displacements from 3D tagged MRI by using the tracking algorithm [11] tuned to the specific spatio-temporal resolution of the tagged data. The full displacements will not be used in this study as observations, however the error between the simulated and extracted displacement will be assessed. More details about image processing of such particular type of MRI dataset can be found in [2].

---

[1] http://www.doc.ic.ac.uk/~dr/software.

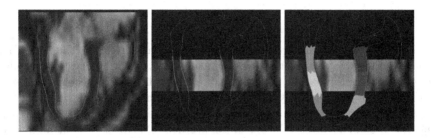

Fig. 2. From left to right: "pseudo 4-chamber" view reconstructed from full short axis cine stack consisting of 8 mm-slices without gap with the contours of geometrical model; p4-ch view from 3-s.a. slices (inter-slice spacing 16 mm); standard AHA segments. The red parts of LV endo- and epicaridal surfaces are used in the estimation to measure the data-model discrepancy (Color figure online).

3 Modeling and Parameter Estimation Framework

The model and data-asimilation procedure used in this work are described in [3,5], respectively, with more particular details of the estimator setup in [2]. In a nutshell, we use a continuum mechanics based model of biventricular anatomy. The model consists of an active contraction component (Bestel-Clément-Sorine model) and the passive tissue is represented by the visco-elastic material with the hyperelastic potential given by $W_e = \kappa_1(J_1 - 3) + \kappa_2(J_2 - 3) + \kappa(J - 1 - \ln J)$ (Ciarlet-Geymonat). The model is activated by an analytically-prescribed physiological electrical activation wave.

The data-assimilation is based on reduced-order unscented Kalman filtering [8]. The discrepancy between the model and processed image data is evaluated by means of signed distances between the corresponding LV endo- and epicardial surfaces [10]. The parts of the cardiac surfaces on which the model-data discrepancy is evaluated is given by the coverage of the heart by the observations (see red surfaces in Fig. 2). Anatomically-created AHA regions are encompassed by the *full s.a.* and *3-s.a.+3-l.a.* observation surfaces. In the *3-slice s.a.* case the mid-cavity AHA segments are nearly fully encompassed by those surfaces, however the top half of the basal segments and the apical segments are not included in the imaging data (Fig. 2, right).

The model is pre-calibrated manually at the baseline so that the global volume indicators correspond to the clinical data. The Windkessel models representing the circulations are calibrated so that the simulated end-systolic and end-diastolic aortic pressures correspond to the measured values at baseline, and physiological values for the pulmonary circulation are assumed. This pre-calibration is used as an initialization for the contractility estimation both for the baseline and esmolol cases. The estimation of the contractility parameter is performed on the standard AHA subdivision for all 3 types of observations.

4   Results

Figure 3 (top) shows consistent estimates of myocardium contractility in AHA regions of a healthy volunteer at baseline using all types of observations. Under the beta-blocker stimulation, a decrease of contractility was detected in all segments when the *full s.a. stack* was used, and we can notice that the contractility decrease is lowest in the septal wall (particularly the apical septal part). When using the *3-s.a.+3-l.a.* standard clinical protocol the drop of contractility is smaller but from a qualitative point of view it is similar to the *full s.a. stack* case – relatively homogeneous except for the septal wall. Finally, the *3-s.a. slices* observations detect the lowest contractility drop and in particular "do not see" any contractility change in the regions not covered by the observation surfaces, i.e. the apical regions.

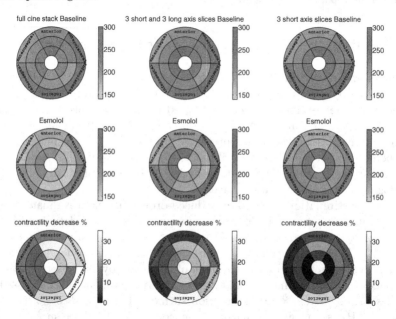

Fig. 3. Contractility (in kPa) estimated using various type of observation surfaces at baseline (top) and during esmolol infusion (mid row); relative drop of contractility under the infusion of beta-blocker for each type of observations (bottom).

A global decrease of contractility is expected under the beta-blocker infusion in a healthy volunteer. However, the results from Fig. 3 cannot be compared to any ground truth as this is not available in vivo. To support our result, we employ the displacements extracted from the 3D tagged MRI and assess the error of radial component of displacement in each node of the tetrahedral mesh within the AHA regions over the systole and average it over the AHA region i:

$$error_i = \frac{1}{|n \in AHA_i||t \in systole|} \sum_{t \in systole} \sum_{node\ n \in AHA_i} |(d^t_{n,model} - d^t_{n,3Dtag}) \cdot u^r_n|,$$

with d denoting the incremental displacement between two consecutive time frames of tagged MRI and u^r_n the vector of radial direction. Figure 4 shows the

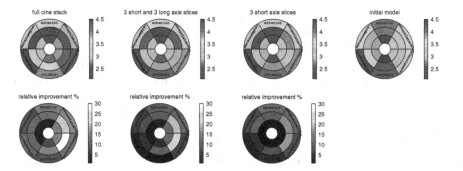

Fig. 4. Evaluation of radial error in the model (the initial parametrization and after the contractility estimation) with respect to radial displacements extracted from 3D tagged MRI: absolute error (top); relative drop of error per segment (bottom).

improvement of this radial error for the esmolol case after inputting the estimated contractility parameters into the heart model with respect to the initial model parameterization.

5 Discussion

The presented paper addresses a specific issue of applying model-based contractility estimation [3] in the quantification of dobutamine stress tests, used routinely to assess cardiac patients for consideration of revascularisation, namely the decreased spatial sampling of cine MRI. In this work we used esmolol, a negatively inotropic agent, to evaluate the optimal coverage and resolution of image data, and tested the whole data acquisition-processing-modeling and parameter estimation framework using a high quality volunteer dataset with pharmacologically modulated contractility.

As shown in Fig. 3 (top, estimation at baseline), if the model is sufficiently pre-calibrated by means of a low discrepancy with the observations, the estimator does not need to do much correction and the final regionally estimated parameters are consistent in between all types of observations. However, if the initialization is further apart from the data – as in our esmolol case – the estimator needs to correct more. As expected, the estimated quantities better correct the part of the model encompassed by the observation surfaces, and the regions further from those surfaces (e.g. apical region in the *3-s.a. observations*) will be less sensitive and the estimator will perform worse. Evaluating Grammian matrix [9] would provide an insight into the sensitivity of parameters in each region, and will enrich our work in future.

In this study we have focused on detecting the decrease of regional contractility and on the associated correction of the model displacements. In the absence of ground truth, we evaluated the error with respect to the displacements extracted from the 3D tagged MRI – the image data not used in the modeling-parameter estimation framework applied in this paper.

Visually, a simple qualitative assessment of the patterns of relative contractility decrease and the relative error correction for the esmolol case (Figs. 3 and 4) supports the idea that the model is being corrected by decreasing the contractility. These figures show that in the septal segments (particularly the apical septal part) the estimator was not able to detect a contractility drop sufficiently and the error in the displacements in the model was not corrected. The apical septal regions may suffer either from a low quality of observations (suboptimal quality of segmentation) or from a more significant modeling error. Both may be caused by the complex geometry and fibre directions at the insertion place of RV. Adding the observations obtained by segmenting the RV surfaces might improve this issue.

In our tests, both the *full cine stack* and the *3-s.a.+3-l.a.* protocol showed qualitatively similar decrease of contractility with esmolol. This suggests that the usual clinical protocol of *3-s.a.+3-l.a.* could be as suitable for estimating the relative contractility change as the *full s.a. cine stack*. The *3-s.a.+3-l.a.* protocol may however have two drawbacks: Firstly, the quantitative values of the detected contractility drop are significantly lower than if the full cine stack is used. Secondly, an important prerequisite for the motion tracking combining *s.a.* and *l.a.* views [11] is that those images are very well spatially aligned (to correct the non-reproducibility of patient's breath-holding). This spatial registration was efficient for our volunteer scan in which the full short axis cine stack provided sufficient information from the whole heart. When using patient data the spatial registration of the 3-s.a. slices with 3-l.a. slices would be a very challenging task.

The acquisition time for the full short axis cine stack in a spatio-temporal resolution as was used in our estimation trials is comparable to the acquisition of *3-s.a.+3-l.a. slices* in standard temporal resolution. The presented work shows that a full-coverage short axis cine stack with a reduced temporal resolution may be an interesting alternative if quantification of results becomes a priority. We are aware that this hypothesis would need to be assessed by a study including a higher number of subjects, and it is one of our current targets.

We have calibrated the aortic pressure to the measured diastolic and systolic values at baseline, and we kept the pressure for the case with drug. It is reasoned by the fact that esmolol acts as a cardiac specific beta-blocker and therefore has a lower effect on the periphery circulation. Even if the pressure was not taken at each level of drug, the estimation would still provide a relative change in contractilities between the LV regions. Although we would be missing the scaling factor for the contractilities, the heterogeneity between the regions might provide the information to diagnose regional defects in contractility.

Long running times of the estimation (convergence of parameters) could be addressed by the preconditioning for instance by a reduced model [1] or statistical methods [2]. The state estimator [9] was not used in this work to allow for a higher sensitivity to the parameters, but would most likely accelerate the convergence. The optimal setup of the estimator is our ongoing work. A coarser subdivision e.g. into 6 radial regions may be considered when only a limited spatial coverage by images is available (e.g. the *3-s.a. slices* in our trial). This result of estimation might be of a similar clinical value for a significantly lower computational cost.

We did not consider tagged MRI as observations, although it was shown that they are superior to cine MRI for estimating the tissue contractility [6]. In a very limited image acquisition time and the current practice of qualitative assessment of the dobutamine stress exam, it is unlikely that the tagged data would be acquired in near future in such an exam.

6   Conclusion

In this paper, we have investigated the types of data acquisition able to estimate a change of myocardial contractility in the clinical setup of dobutamine stress test. By carefully designing experimental data acquisition and adjusting the modeling-parameter estimation framework, the approach can provide an insight into the optimal imaging protocol to most accurately answer the clinical question. Simultaneously to the "computer-assisted diagnosis" we are therefore targeting the topic of "model-assisted optimal imaging".

The problem of estimating the contractilities in dobutamine stress tests has a number of unresolved subtasks and this work is providing initial steps to some of them, in particular the issue of coverage of myocardium by the image data, and the level of spatial discretization for the estimated parameters depending on the image data. Further development of the modeling and data-assimilation components, extending the number of subjects, challenging the image processing part by the image quality realistic in the dobutamine stress examinations and starting the estimation in the real dobutamine cases – with an accordingly modified scanning protocol – is a natural continuation of the presented work.

Acknowledgments. The authors acknowledge the support of Engineering and Physical Sciences Research Council EP/H046410/1, British heart foundation grant NH/11/5/29058 and Cardiovascular Healthcare Technology Cooperative. In addition, the author are thankful to P. Moireau and D. Chapelle (Inria, France) for providing the HeartLab software library, used in this work for all modeling and estimation computations, and to D. Rueckert and W. Shi (Imperial College London, Ixico) for providing the IRTK based motion tracking and valuable discussions. This research was supported by the National Institute for Health Research (NIHR) Biomedical Research Centre at Guy's and St. Thomas' NHS Foundation Trust and King's College London. The views expressed are those of the author(s) and not necessarily those of the NHS, the NIHR or the Department of Health.

References

1. Caruel, M., Chabiniok, R., Moireau, P., Lecarpentier, Y., Chapelle, D.: Dimensional reductions of a cardiac model for effective validation and calibration. Biomech. Model. Mechanobiol. **13**(4), 897–914 (2014)
2. Chabiniok, R., Bhatia, K.K., King, A.P., Rueckert, D., Smith, N.: Manifold learning for cardiac modeling and estimation framework. In: Camara, O., Mansi, T., Pop, M., Rhode, K., Sermesant, M., Young, A. (eds.) STACOM 2014. LNCS, vol. 8896, pp. 284–294. Springer, Heidelberg (2015)

3. Chabiniok, R., Moireau, P., Lesault, P.-F., Rahmouni, A., Deux, J.-F., Chapelle, D.: Estimation of tissue contractility from cardiac cine-MRI using a biomechanical heart model. Biomech. Model. Mechanobiol. **11**(5), 609–630 (2012)

4. Chapelle, D., Fragu, M., Mallet, V., Moireau, P.: Fundamental principles of data assimilation underlying the Verdandi library: applications to biophysical model personalization within euHeart. Med. Biol. Eng. Comput. **51**(11), 1221–1233 (2013)

5. Chapelle, D., Le Tallec, P., Moireau, P., Sorine, M.: An energy-preserving muscle tissue model: formulation and compatible discretizations. Int. J. Multiscale Comput. Eng. **10**(2), 189–211 (2012)

6. Imperiale, A., Chabiniok, R., Moireau, P., Chapelle, D.: Constitutive parameter estimation methodology using tagged-MRI data. In: Metaxas, D.N., Axel, L. (eds.) FIMH 2011. LNCS, vol. 6666, pp. 409–417. Springer, Heidelberg (2011)

7. Jahnke, C., Nagel, E., Gebker, R., Kokocinski, T., Kelle, S., Manka, R., Fleck, E., Paetsch, I.: Prognostic value of cardiac magnetic resonance stress tests: adenosine stress perfusion and dobutamine stress wall motion imaging. Circulation **115**, 1769–1776 (2007)

8. Moireau, P., Chapelle, D.: Reduced-order unscented Kalman filtering with application to parameter identification in large-dimensional systems. ESAIM Control Optimisation Calc. Var. **17**, 380–405 (2011)

9. Moireau, P., Chapelle, D., Le Tallec, P.: Joint state and parameter estimation for distributed mechanical systems. Comput. Methods Appl. Mech. Eng. **197**, 659–677 (2008)

10. Moireau, P., Chapelle, D., Le Tallec, P.: Filtering for distributed mechanical systems using position measurements: perspectives in medical imaging. Inverse Problems, 25(3):035010, p. 25 (2009)

11. Shi, W., Zhuang, X., Wang, H., Duckett, S., Luong, D.V.N., Tobon-Gomez, C., Tung, K., Edwards, P., Rhode, K., Razavi, R., Ourselin, S., Rueckert, D.: A comprehensive cardiac motion estimation framework using both untagged and 3D tagged MR images based on non-rigid registration. IEEE Trans. Med. Imaging **31**(6), 1263–1275 (2012)

12. Wang, V.Y., Lam, H.I., Ennis, D.B., Cowan, B.R., Young, A.A., Nash, M.P.: Modelling passive diastolic mechanics with quantitative MRI of cardiac structure and function. Med. Image Anal. **13**(5), 773–784 (2009)

13. Xi, J., Lamata, P., Niederer, S., Land, S., Shi, W., Zhuang, X., Ourselin, S., Duckett, S., Shetty, A., Rinaldi, C., Rueckert, D., Razavi, R., Smith, N.: The estimation of patient-specific cardiac diastolic functions from clinical measurements. Med. Image Anal. **17**(2), 133–146 (2013)

Personalization of Atrial Electrophysiology Models from Decapolar Catheter Measurements

Cesare Corrado[✉], Steven Williams, Henry Chubb, Mark O'Neill,
and Steven A. Niederer

Division of Imaging Sciences and Biomedical Engineering,
King's College London, London, UK
cesare.corrado@kcl.ac.uk

Abstract. A novel method to characterize biophysical atria regional ionic models from multi-electrode catheter measurements and tailored pacing protocols is presented. Local atria electrophysiology was described by the Mitchell and Schaeffer 2003 action potential model. The pacing protocol was evaluated using simulated bipolar signals from a decapolar catheter in a model of atrial tissue. The protocol was developed to adhere to the constraints of the clinical stimulator and extract the maximum information about local electro-physiological properties solely from the time the activation wave reaches each electrode. Parameters were fitted by finding the closest parameter set to a data base of 3125 pre computed solutions each with different parameter values. This fitting method was evaluated using 243 randomly generated in silico data sets and yielded a mean error of ±10.46 % error in estimating model parameters.

Keywords: Electrograms · Computational models personalization · Multi-electrode catheters

1 Introduction

Computational models represent a novel framework for studying pathologies of the human atria and offer a pathway for selecting patients and personalizing treatment, [1,2]. Of particular interest is the study of atrial fibrillation (AF), a pathology where the underpinning mechanisms triggering and maintaining the arrhythmia are not known. The use of increasingly sophisticated electro anatomical mapping systems, high fidelity imaging techniques and inverse ECG methods has significantly improved patient outcomes [3]. These improved diagnostic modalities are still only able to provide information on the current state of the patient and are unable to provide predictions of the outcome of treatments.

Biophysical modeling provides a formal framework that combines our understanding of atria physiology, physical constraints and patient measurements to make quantitative predictions of patient response to treatment. These models have provided fundamental insight into the mechanisms that underpin arrhythmia's in the ventricle and the atria, [4] but their potential to inform clinical procedures had been limited by their inability to capture the significant variability in physiology inherent both between and within AF patients.

© Springer International Publishing Switzerland 2015
H. van Assen et al. (Eds.): FIMH 2015, LNCS 9126, pp. 21–28, 2015.
DOI: 10.1007/978-3-319-20309-6_3

The aim of this study is to develop and test a robust and rapid pacing protocol and a model fitting method that allow for local characterization of cellular biophysical parameters and electrophysiology restitution within patients atria. Validation of the pacing protocol will be presented in Sect. 3. Validation of the method proposed here is purely based on synthetic data. Application to experimental data will be a topic of a future work.

2   Method

2.1   Computational Model

The first step of the method proposed is to build a data set of restitution curves for each permutation of a set of parameters known a priori. A numerical model describing the action potential propagation across the left atrium tissue is implemented. The model is then used to reproduce electrogram signals from the poles of a decapolar catheter when an ectopic pacing stimulus is applied to the central poles. The procedure for building the restitution curves is described in Sect. 2.2.

Atria tissue electrophysiology was modeled by the mono-domain equations [5], a simplification of the bi-domain model, [6] when intra- and extra-cellular conductivities are considered proportionals up to a constant, λ. Due to assumed local symmetry and negligible thickness in the atrium the model was reduced to a 1D fiber model as follows:

$$\frac{\partial v_m}{\partial t} - \sigma_m \frac{d^2 v_m}{dx^2} = I_{ion} + I_{app} \tag{1}$$

$$\sigma_m = \frac{\lambda}{1+\lambda} \tag{2}$$

where I_{ion} is an external applied stimulus, while I_{ion} characterize the ionic currents defined by the ionic model, [7].

Substituting the mono-domain approximation into the extra-cellular potential equilibrium equation, and simplifying the conductivity it follows that:

$$\sigma_i \frac{d^2 v_m}{dx^2} + (\sigma_i + \sigma_e) \frac{d^2 \phi_e}{dx^2} = 0 \qquad \qquad \sigma_e = \lambda \sigma_i$$

$$\sigma_i \frac{d^2}{dx^2} \left(v_m + (1+\lambda) \phi_e \right) = 0 \tag{3}$$

$$\frac{d}{dx} \left(v_m + (1+\lambda) \phi_e \right) = const$$

where ϕ_e represents the extra-cellular potential, v_m the trans-membrane potential, σ_i the intra-cellular conductivity and σ_e the extra-cellular conductivity. The constant term appearing on the right-hand side of (3) is fixed by imposing a zero spatial mean on the extra-cellular potential:

$$\phi_e = -\frac{1}{1+\lambda} \left(v_m - \bar{v}_m \right)$$

where $\bar{v}_m$, denotes the spatial mean of the trans-membrane potential.

The catheter configuration and the catheter dimensions are depicted in Fig. 1; bipolar signals were evaluated as the difference of the extra-cellular potentials between a pair of electrodes spaced by a distance d_1.

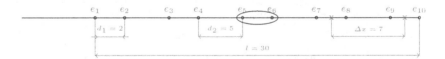

Fig. 1. Decapolar catheter configuration and dimensions. Dimensions are expressed in mm. The pacing stimulus is applied to the central poles, highlighted by the gray ellipse. Bipolar poles are determined by the pairs: (e_1,e_2), (e_3,e_4), (e_5,e_6), (e_7,e_8), (e_9,e_{10})

Ionic currents were described using the Mitchell Schaefer 2003 (MS), [7] ionic current model. Despite its simplicity, the MS model captures the Effective Refractory Period (ERP) and the Conduction Velocity (CV) restitutions with a minimal number of parameters. The MS model has four parameters that describe the opening (τ_{open}, τ_{in}) and closing (τ_{close}, τ_{out}) rate of the fast activation and slow repolarization currents respectively.

2.2 Pacing Protocol and Restitution Curves

Atria were paced from the central poles (e_5,e_6) of the decapolar catheter depicted in Fig. 1; activations were measured at proximal and distal poles, (e_1,e_2), (e_3,e_4), (e_7,e_8), (e_9,e_{10}).

The initial (maximum) inter-beat interval was fixed at $s^0 = 700$ ms and the minimum inter-beat interval $s_{min} = 200$ ms, for each decrement step ΔT_{stim} the inter-beat delay s^{i+1} is defined recursively as:

$$
\begin{aligned}
s^{i+1} &= s^i - \Delta T_{stim}, &\quad i &= 1, \ldots N \\
\left(s^N - s_{min}\right) &= \min\left(s^i - s_{min}\right), \quad \left(s^i - s_{min}\right) \geq 0
\end{aligned}
\tag{4}
$$

and depicted graphically in Fig. 2.

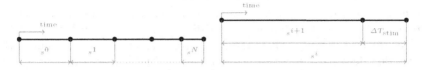

Fig. 2. Inter-beat sequence. Left: the sequence generated by the recursion defined in (4). Right: definition of the $(i + 1)$-th pacing interval as a function of the previous one.

In this work, values of $\Delta T_{stim} = (100, 80, 60, 40, 20, 10)$ ms were chosen.

For each bipolar pair the Non-Linear Energy Operator (NLEO), [8], was evaluated between two subsequent stimulations. In Fig. 3 (left) the output of a bipolar electrode is depicted; in the same Figure (right) activations are showed.

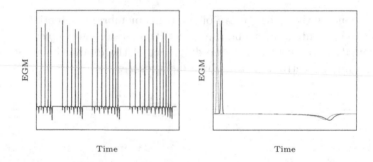

Fig. 3. Example of output of a bipolar electrode (left) and particular for two subsequent electrodes. Peaks corresponds to activations.

The time of maximal NLEO was defined as the activation time. We have assumed that we do not have access to repolarization times as these can not be observed reliably in an atria electrogram. Repolarization times were available in [9], when a mono phasic action potential catheter was used to make recordings. In the absence of repolarization times we estimate the ERP by identifying when activation fails to propagate for a given inter-beat interval.

CV was evaluated as the ratio between the distance Δx defined in Fig. 1 and the time elapsed during which the activation front propagates from one electrode to the subsequent one.

2.3 Parameter Fitting

A data base of simulation results for 3125 combinations of model parameters was created for the pacing protocol described above. The data base was created once on an HPC in approximately one hour and a half. The range of parameter values explored in the data set are described in Table 1. Restitutions were then evaluated for each member of the data base.

The parameter fitting is performed in two steps. First, for each member of the data set a cost C_1 is evaluated as the sum of the mean square difference of the "measured" and the data base CV on each electrode and for each pacing decrement. A monotone C_1 cost function is then obtained by re-ordering the data set. A subset of N_1 solutions is chosen that have a C_1 cost below a cut off value. The cut off is the minimum of a defined cost function value (Fig. 4 left panel) or the cost function where the derivative is less than the modulus of the cost function derivative, to avoid plateau regions (Fig. 4 right panel).

Table 1. Parameter values used for building the data set

	Conductivity (S/cm)	τ_{in} (ms)	τ_{out} (ms)	τ_{open} (ms)	τ_{close} (ms)
Min	0.001	0.225	4.5	75	120
Max	0.003	0.375	7.5	125	180
Step	0.0005	0.0375	0.75	12.5	15

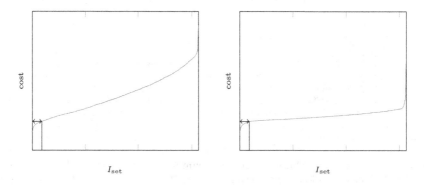

Fig. 4. Example of cost function and choice of candidate subset. Left panel: cost criteria determined by a maximum cost value; this happens when the cost as a function of I_{set} is enough steeper. If this is not the case, the selection of I_{set} is performed using the derivative of the cost function (right panel)

On the set of N_1 candidates a new cost C_2 is evaluated as the sum of the mean square difference between the "measured" and data base ERP on each electrode and with each pacing decrement.

A new sub-set of N_2 candidates is then chosen as the sub-set yielding an ERP-related cost smaller than a fixed tolerance. From these available parameter sets, the one that best fits the CV was chosen as the CV measurement had the highest fidelity. The application of the estimation procedure was performed in less than one minute on 2.66 Ghz Xeon desktop machine.

3   Results

To evaluate the error properties and robustness of our approach a set of 243 combination models was generated by choosing parameter randomly in the [min, max] interval reported in Table 1. For each combination, the parameter set determined from the fitting process was compared with the known true solution and the L_2 (mean error on the 5 parameters) and the L_∞ error (maximum error between the 5 parameters) were then evaluated as a percentage of the known true solution. Figure 5 shows the L_2 and L_∞ error distributions and the corresponding cumulative distribution function (CDF). For the L_2 error, a maximum value of 29.45 % was found with a mean error of 10.46 % and a standard deviation on the error of 5.4 %. As depicted by the CDF, 95 % of the estimated parameters analyzed here have a L_2 error not greater than 20 %.

Figure 6 shows the error distribution for each parameter together with its CDF. Higher relative errors occur when the value of the parameter to estimate is close to the minimum value adopted for building the data base of Table 1. The best performances are obtained in estimating the conductivity and τ_{in} parameters.

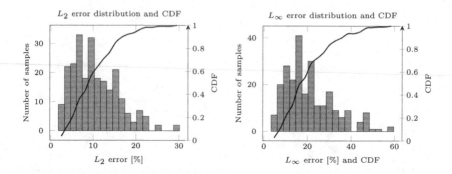

Fig. 5. Left: L_2 error distribution (bars) and cumulative distribution (thick line) evaluated for 243 set of randomly chosen parameters. Right: L_∞ error distribution (bars) and cumulative distribution (thick line) evaluated for 243 set of randomly chosen parameters.

According to [11], τ_{open}, τ_{close} parameters characterize ERP restitution: a poorer approximation of these two parameters was expected, since the available data is best able to constrain parameters affecting CV restitutions.

The parameter τ_{out} is constrained by both CV and ERP restitution. This parameter characterizes the outward repolarization current, [7]. The inability to accurately measure ERP through activation times alone means that this parameter is often poorly constrained leading to it often being the worst fit parameter (see Fig. 6, bottom right).

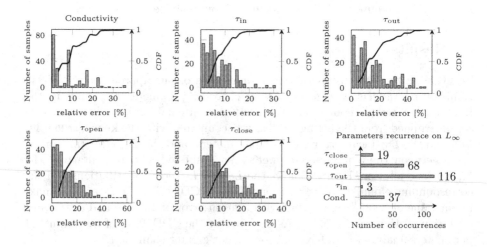

Fig. 6. Error distribution and CDF for each parameter; Recurrence of the maximum error for each parameter

Figure 6 (bottom, right) shows the number of occurrences each parameter defines the L_∞ error. The τ_{out} and τ_{open} parameters appear to be the least well constrained (116 occurrences for τ_{out}, 68 for τ_{open}).

Robustness with respect to noise was evaluated by adding a white noise signal to each of the bipolar electrograms with an intensity equal to the 10 % of the maximum absolute value of the electrode output. For the L_2 error, a maximum value of 29.45 % was found with a mean error of 10.72 % and a standard deviation on the error of 5.5 %. This result is not surprising since the proposed method fits the model parameters using activation time values, i.e. where the signal to noise ratio is maximum.

4   Discussion

A pacing protocol designed to constrain the biophysical parameters of a cellular ionic model from a multi-electrodes catheter bipolar electrograms was derived and tested. The simplified model used in this study reflects the level of complexity available from clinical data where only endocardial surface activation times can be recorded. Differently from [10], the same catheter is used to both pacing and measuring the activation times, reducing uncertainty in the relative orientation of the wave propagation and the catheter. The CV and ERP restitutions with respect to the pacing decrement were determined only using the activation times. The robustness of the method was tested on a set of 243 combination of randomly chosen parameters.

In the test performed, higher relative errors occurred when the value of parameter to estimate was close to the minimum value adopted for building the data base. The best performances were obtained in the estimation of the conductivity and τ_{in} parameters. These parameters mainly characterize the CV restitution, the measurements which are well represented by the data set. Conversely, τ_{open} and τ_{close} characterize ERP restitution. The accuracy in determining ERP depends on ΔT_{stim}, thus a poorer approximation on ERP could lead to a poorer approximation on τ_{open} and τ_{close}.

The τ_{out} parameter appears to be the least well constrained (116 occurrences in determining the L_∞ error); and poorly approximated. This parameter characterize the outward current, and thus the repolarization: a possible reason of the poor performances in constraining τ_{out} could be ascribed to a greater influence of this parameter on the ERP restitution than the CV restitution.

Another limitation on the accuracy of the proposed methods depends on the number of values adopted for each parameter in building the data set.

The proposed method represents a first step in personalizing atrial electrophysiology models to individual patient physiology and pathology on clinical time scales.

5   Conclusions

In this work were introduced a robust and potentially clinically tractable protocol and fitting algorithm for characterizing local tissue electro-physiology properties by biophysical ionic cell models.

References

1. Kneller, J., Zou, R., Vigmond, E., et al.: Cholinergic atrial fibrillation in a computer model of two-dimensional sheet of canine atrial cells with realistic ionic properties. Circ. Res. **90**(9), e73–e87 (2002)
2. Aslandi, O.V., Colman, M.A., Stott, J., et al.: 3D virtual human atria: a computational platform for studying clinical atrial fibrillation. Prog. Biophys. Mol. Biol. **107**(1), 156–168 (2011)
3. Kistler, P.M., Ho, S.Y., Rajappan, K., et al.: Electrophysiologic and anatomic characterization of sites resistant to electrical isolation during circumferential pulmonary vein ablation for atrial fibrillation: a prospective study. Circ. Res. **90**(9), e73–e87 (2002)
4. Colli Franzone, P., Pavarino, L.F., Savaré, G.: Computational electrocardiology: mathematical and numerical modeling. In: Quarteroni, A., Formaggia, L., Veneziani, A. (eds.) Complex Systems in Biomedicine, pp. 187–241. Springer, Milan (2006)
5. Potse, M., Dubé, B., Richer, J., et al.: A comparison of monodomain and bidomain reaction-diffusion models for action potential propagation in the human heart. IEEE Trans. Biomed. Eng. **53**, 2425–2435 (2006)
6. Tung, L.: A bi-domain model for describing ischemic myocardial D-C potentials. Ph.D. thesis, MIT (1978)
7. Mitchell, C.C., Schaeffer, D.G.: A two-current model for the dynamics of cardiac membrane. Bull. Math. Bio. **65**, 767–793 (2003)
8. Mukhopadhyay, S., Ray, G.C.: A new interpretation of nonlinear energy operator and its efficacy in spike detection. IEEE Trans. Biomed. Eng. **45**(2), 180–187 (1998)
9. Franz, M.R., Karasik, P.L., Li, C., et al.: Electrical remodeling of the human atrium: similar effects in patients with chronic atrial fibrillation and atrial flutter. JACC **30**(7), 1785–1792 (1997)
10. Weber, F.M., Luik, A., Schilling, C., et al.: Conduction velocity restitution of the human atrium-an efficient measurement protocol for clinical electrophysiological studies. IEEE Trans. Biomed. Eng. **58**(9), 2648–2655 (2011)
11. Relan, J.: Personalised electrophysiological models of ventricular tachycardia for radio frequency ablation therapy planning. Ph.D. Thesis (2012)

Automatic LV Feature Detection and Blood-Pool Tracking from Multi-plane TEE Time Series

Shusil Dangi[1], Yehuda K. Ben-Zikri[1], Yechiel Lamash[4], Karl Q. Schwarz[2,3], and Cristian A. Linte[1,2(✉)]

[1] Chester F. Carlson Center for Imaging Science, Rochester, NY, USA
{sxd7257,calbme}@rit.edu
[2] Biomedical Engineering, Rochester, NY, USA
[3] Division of Echocardiography, University of Rochester Medical Center, Rochester, NY, USA
[4] Technion - Israel Institute of Technology, Haifa, Israel

Abstract. Multi-plane, 2D TEE images constitute the clinical standard of care for assessment of left ventricle function, as well as for guiding various minimally invasive procedure that rely on intra-operative imaging for real-time visualization. We propose a framework that enables automatic, rapid and accurate endocardial left ventricle feature identification and blood-pool segmentation using a combination of image filtering, graph cut, non-rigid registration-based motion extraction, and 3D LV geometry reconstruction techniques applied to the TEE image series. We evaluate our proposed framework using several retrospective patient tri-plane TEE image sequences and demonstrate comparable results to those achieved by expert manual segmentation using clinical software.

1 Introduction

Over the past two - three decades, ultrasound (US) imaging has evolved as the preferred, standard-of-care imaging modality for the diagnosis, screening, monitoring, and real-time guidance of several conditions. Specifically, thanks to its real-time capabilities, relatively inexpensive cost (compared to other modalities), and lack of exposure, US imaging has become the "first-line" modality for patient screening, diagnosis, and cardiac function assessment.

Trans-esophageal echocardiography (TEE) enables heart imaging while minimizing signal attenuation and optimizing field-of-view. As such, TEE is not only used for screening and diagnosis, but also for intra-operative therapy monitoring and/or image-guided cardiac interventions. Since the mid-2000s, TEE technology has accommodated 3D image acquisition and visualization of the cardiac anatomy in lieu of simple 2D renderings. However, despite the added bonus of 3D and 4D (3D + time) displays, the inherent trade-off between frame rate, and extent of anatomy covered, has determined clinicians to resort to the acquisition and visualization of multi-planar (orthogonal bi-plane or tri-plane) images to estimate the required parameters to assess cardiac function (i.e., ejection fraction) or identify critical features for image-guided therapy.

© Springer International Publishing Switzerland 2015
H. van Assen et al. (Eds.): FIMH 2015, LNCS 9126, pp. 29–39, 2015.
DOI: 10.1007/978-3-319-20309-6_4

Despite their high frame rate, 2D US images are hampered by several well known limitations: challenging interpretation and uncertainty in identifying structures of interest due to inherent specular appearance. Several approaches for LV segmentation in echocardiography [1] have been popularly formulated as a contour finding problem, with the active contour method [2,3] being extensively used. Given its edge-based energy approach, the active contour method often produced many local minima and is also sensitive to the initialization. Inspired by the active contours, the level set method [4,5] uses both edge- and region-based energy, making it more robust and less sensitive to initialization.

Active shape [6] and active appearance models [7] incorporate knowledge of the LV shape and appearance from manually segmented training sets, but assume a Gaussian distribution of the shape and appearance derived from the training sets, requiring an initial approximation close to the final solution. On the other hand, database-guided segmentation [8] overcome the initialization problem by implicitly encoding prior knowledge from the expert-annotated databases, yet at the expense of a highly complex search process. Other supervised learning techniques, such as artificial neural networks [9], have been used to detect endocardial border pixels using expert annotated training sets, but require large training sets and are unable to handle cases well outside of the training set.

In this work we propose the implementation and clinical validation of an automatic workflow that encompasses well-evaluated filtering, segmentation, registration, and volume reconstruction techniques as a means to provide a rapid, robust and accurate framework for feature tracking from multi-plane ultrasound image sequences. The proposed computational framework was developed in close collaboration with our echocardiography colleagues, motivated by the need to reduce user-dependent and user-induced bias and reduce the uncertainty associated with the process of manually identifying features from US image sequences. The impact and contribution of the proposed work is the integration of several image processing techniques (i.e., phase-based filtering, segmentation, registration and volume reconstruction) into a streamlined workflow that utilizes traditional standard of care images and fits seamlessly within the current workflows associated with both cardiac function assessment and intra-operative cardiac intervention guidance and monitoring.

2   Methodology

Speckle noise and signal dropouts inherent in US images render intensity based approaches unreliable; rather, local-phase based approaches [10], theoretically invariant to the intensity magnitude, have been preferred for detecting endocardium. Here we exploit the robustness of phase-based feature detection and combine it with the power of graph cut-based techniques [11] that use both region and boundary regularization, to obtain a rapid, automatic piecewise smooth segmentation of the LV blood pool and muscle regions. In addition, we conducted a preliminary study using retrospective clinical patient data consisting of tri-plane (60° to one another) TEE image sequences through the cardiac cycle to validate the proposed tools and demonstrate their clinical utility and performance against commercial, clinical-grade, clinician-operated software.

The proposed methodology encompasses three steps: (1) endocardial left ventricle (LV) feature extraction and blood-pool segmentation from the raw 2D multi-plane image sequences, (2) frame-to-frame feature tracking and propagation through the cardiac cycle using non-rigid image registration, and (3) 3D reconstruction of the LV blood pool geometry at the desired cardiac phases using spline-based interpolation and convex hull fitting.

2.1 LV Feature Extraction and Blood-Pool Segmentation

Image Pre-processing via Monogenic Filtering: Unlike intensity-based edge detection algorithms are inefficient in identifying features from US images, intensity invariant local phase-based techniques have shown promising results [10], where a local phase of $\pm\pi/2$ signifies high symmetry, while a local phase of 0 or π signifies high asymmetry [12]. The local phase computation of a 1D signal uses a complex analytic signal comprised of the original signal as the real part and its corresponding Hilbert transform as the imaginary part. However, since the Hilbert transform is mathematically restricted to 1D with no straightforward extension to 2D and 3D, we used the method described in [13] to extend the concept of the analytic signal to higher dimensions using a monogenic signal. The higher dimension monogenic signal is generated by combining a bandpass Gaussian-derivative filter with a vector-valued odd filter (i.e., a Reisz filter). The low frequency variations in the local phase are extracted using a high spread (σ) Gaussian-derivative filter, while the high frequency components are extracted using a low spread (σ) Gaussian-derivative filter. The described monogenic filtering sequence is used to transform each of the three tri-plane 2D US images into corresponding "cartoon" images in which the blood pool and myocardial wall appear enhanced, facilitating their segmentation in the subsequent step.

Graph Cut-Based Segmentation: The resulting "cartoon" image is used to construct a four neighborhood graph structure in which each pixel is connected to its east, west, north and south neighbors. Three special nodes called terminals are added, which represent three classes (labels): background, blood pool and myocardium. The segmentation can be formulated as an energy minimization problem to find the labeling f, such that it minimizes the energy:

$$E(f) = \sum_{\{p,q\}\in\mathcal{N}} V_{p,q}(f_p, f_q) + \sum_{p\in\mathcal{P}} D_p(i_p, f_p), \tag{1}$$

where the first term represents smoothness energy, which forces pixels p and q defined by a set of interacting pair $\mathcal{N}$, towards the same label. The second term represents the data energy that reduces the disagreement between the labeling f and the observed data i_p. The links between each pixel and the terminals (i.e., t-links) are formulated as the negative logarithm of the normal distribution [14]:

$$D_p(i_p, f_p) = -\ln\left(\frac{1}{\sigma\sqrt{2\pi}}exp\left(-\frac{(i_p - \mu)^2}{2\sigma^2}\right)\right), \tag{2}$$

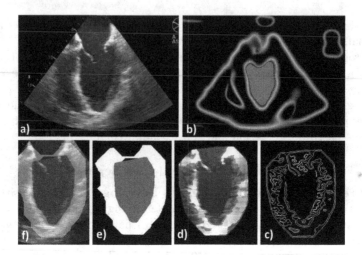

Fig. 1. Segmentation Workflow: (a) original US image, (b) high spread (σ) low frequency monogenic filter applied to the "2D + time" image dataset shown with the high confidence blood pool mask, (c) low spread (σ) high frequency monogenic filter output with blood pool removed, (d) "cartoon" image with enhanced regions, and (e) graph cut segmentation output (f) superimposed onto the original image.

where μ and σ are the mean and standard deviation for the three classes obtained from the image. The links between neighboring pixels, called n-links, are weighted according to their similarity to formulate the smoothness energy:

$$V_{p,q}(f_p, f_q) = \begin{cases} 2K \cdot T(f_p \neq f_q) & \text{if } |I_p - I_q| \leq C \\ K \cdot T(f_p \neq f_q) & \text{if } |I_p - I_q| > C \end{cases} \tag{3}$$

where $T(\cdot)$ is 1 if its argument is true, and otherwise 0, K is a constant, and C is a intensity threshold that forces the neighboring pixels within the threshold towards the same label. The minimum cut equivalent to the maximum flow is obtained via the expansion algorithm in [11] yielding optimal segmentation of background, blood-pool, and myocardium (Fig. 1e).

2.2   Frame-to-frame Feature Tracking and Propagation

Image Pre-processing: Once a single-phase image is segmented using the procedure outlined in Sect. 2.1, the extracted features are tracked and propagated throughout the cardiac cycle using non-rigid registration (Fig. 2). Prior to registration, each "2D + time" image sequence corresponding to each of the triplane views is first "prepared" by identifying a region of interest-based "bounding box" centered on the features that belong to the LV. To ensure the chosen "bounding box" spans the entire LV including blood-pool, myocardium, and surrounding region, this window is selected based on the high confidence blood pool mask obtained after the application of the high spread Gaussian-derivative filter

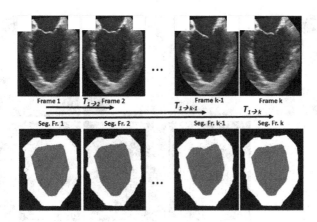

Fig. 2. The frame-to-frame motion transforms $(T_{(k-1)\rightarrow k})$ are estimated by non-rigidly registering adjacent images in the sequence, then concatenated $(T_{1\rightarrow k} = T_{1\rightarrow 2} \cdot \ldots \cdot T_{(k-1)\rightarrow k})$ and applied to the segmented end-diastolic (ED) frame $(F_k = T_{1\rightarrow k} \cdot F_1 = T_{(k-1)\rightarrow k} \cdot \ldots \cdot T_{1\rightarrow 2} \cdot F_1)$.

employed in Sect. 2.1 to the entire image sequence, followed by an isotropic dilation to ensure full coverage beyond the LV myocardial boundary. Moreover, the mitral valve region is "trimmed" using a straight line joining the leaflet hinges.

Non-rigid Registration Algorithm: The employed registration algorithm is a modified version of the biomechanics-based algorithm proposed by Lamash *et al.* [15]. The LV anatomy is modeled as a two compartment model consisting of muscle — linear elastic, isotropic, and incompressible, and blood-pool, with prescribed smoothness constraints to allow rapid motion of the endocardial contour. We initialize the algorithm by first discretizing the endocardial and epicardial contours, then constructing a mesh of the blood-pool and myocardium. Rather than resorting to a rectangular grid, we account for the local curvature of the endocardial border using a finite-element like mesh defined via linear shape functions. The algorithm deforms the mesh by estimating the required deforming forces that minimize the sum of the squared difference between the initial and target images (Fig. 3). To avoid large deformations and ensure a smooth displacement field, a linear elastic regularization approach [16] is utilized.

2.3 3D LV Volume Reconstruction

Following the segmentation of each of the tri-plane views at end-diastole using the technique in Sect. 2.1 and their propagation throughout the cardiac cycle, the resulting images are re-inserted into a pseudo-3D image volume along the same orientation at which they were originally acquired (i.e., 60° apart) corresponding to each cardiac phase. The boundary points of each segmented contour at the same elevation are then fitted using the parametric variational cubic spline

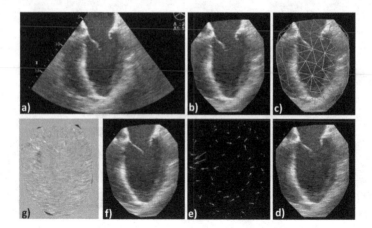

Fig. 3. Registration workflow: (a) the original image is "prepared" by automatically identifying an LV-centered ROI (b) onto which the mesh is applied (c), then registered to the target image (d); the resulting displacement field (e) is applied to the pre-registered image (b) to obtain the registered image (f), which can be compared to the target image (d) by visualizing the digitally subtracted image (g).

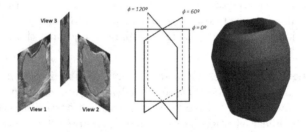

Fig. 4. Schematic illustration of the 3D LV reconstruction: the tri-plane views at 60° (a) are inserted at their appropriate orientation (b), followed by spline interpolation and convex hull generation (c).

technique in [17]. The spline interpolated data is used to generate a convex hull using the algorithm proposed in [18] (Fig. 4).

3   Evaluation and Results

We conducted a preliminary study using retrospective tri-plane time series data spanning multiple cardiac cycles from patients who underwent TEE imaging for cardiac function assessment. Since the proposed framework encompasses three different components — automatic extraction of endocardial features, registration-based feature tracking and propagation, and volume reconstruction — we assessed the performance of each component against the ground truth, which consists of the blood-pool representation annotated manually by the expert clinician, using the EchpPac PC clinical software. In addition, we also evaluated the performance at each stages of our application running in MATLAB on an Intel® Xenon® 3.60 GHz 32 GB RAM PC.

Table 1. Comparison between the blood-pool area measurements (Mean ± Std. Dev. [cm^2]) annotated by the expert (Ground Truth) and the area obtained via A — automatic feature detection from individual frames; B — single phase automated feature detection + registration-based propagation; and C — single phase expert manual annotation + registration-based propagation. Measurements are evaluated at two cardiac phases — end-diastole (ED) and end-systole (ES) — and averaged across all views and cardiac cycles spanned by the acquired data.

Blood-pool area [cm^2] vs. Method	ED	ES
Ground truth: multi phase expert manual seg	52.1 ± 3.2	50.4 ± 4.6
Method A: multi phase auto seg	51.2 ± 3.5	48.9 ± 4.3
Method B: single phase auto seg + Reg	50.1 ± 4.0	48.3 ± 4.6
Method C: single phase manual expert seg + Reg	49.8 ± 4.6	48.2 ± 5.1

Automatic Direct Frame Endocardial Feature Extraction Evaluation: We first evaluated the accuracy of our automatic, direct frame endocardial feature extraction component against expert manual annotation of the same features from the same frames performed by a cardiologist using the GE EchoPac PC clinical software. Table 1 summarizes the blood-pool area measurements annotated by the expert (Ground Truth) and the area obtained via A — automatic feature detection from individual frames; B — single phase automated feature detection + registration-based propagation; and C — single phase expert manual annotation + registration-based propagation. Measurements are evaluated at two cardiac phases — end-diastole (ED) and end-systole (ES) — and averaged across all views and multiple cardiac cycles spanned by the acquired sequences. Our automatic blood-pool extraction technique required 26.5 s to segment a "2D + time" 15 frame TEE tri-plane sequence.

Registration-Based Blood-Pool Tracking and Propagation Evaluation: To evaluate the accuracy with which the non-rigid registration algorithm propagates the extracted features throughout the cardiac cycle, we employed several metrics, including the DICE correlation, Hausdorff distance, mean absolute distance error and endocardial target registration error (TRE) computed between the ground truth blood-pool manually annotated by the expert and the blood-pool depicted via three other methods under consideration (Table 2).

Figure 5 visually compares the ground truth blood-pool annotation performed by the expert clinician to that extracted via direct frame feature identification, as well as registration-based propagation of the single-frame blood-pool annotated either manually by the expert or automatically using the first component of our proposed framework. The segmentation propagation technique required 162 s to run through a 15 frame tri-plane TEE sequence.

3D Volume Reconstruction and Ejection Fraction Evaluation: Lastly, we assessed the accuracy of the 3D LV reconstruction procedure by comparing

Table 2. Mean ± Std. Dev. of several metrics — DICE Coefficient [%], Hausdorff Distance [mm], Mean Absolute Distance (MAD) Error [mm], and Endocardial TRE [mm] — used to compare the expert clinicians' blood-pool annotations (Ground Truth) with the blood-pool annotation obtained via A — automatic feature detection from individual frames; B — single phase automated feature detection + registration-based propagation; and C — single phase expert manual annotation + registration-based propagation. Measurements are evaluated at two cardiac phases — end-diastole (ED) and end-systole (ES).

Comparison metrics	Expert vs. A		Expert vs. B		Expert vs. C	
	ED	ES	ED	ES	ED	ES
DICE coeff [%]	94.9 ± 0.7	94.7 ± 1.4	93.8 ± 0.9	94.6 ± 1.0	95.1 ± 1.0	95.2 ± 1.8
Haussdorf dist [mm]	4.7 ± 0.9	5.2 ± 1.3	7.9 ± 3.5	5.9 ± 1.3	6.4 ± 1.7	5.4 ± 2.1
MAD error [mm]	1.5 ± 0.3	1.6 ± 0.6	1.9 ± 0.4	1.7 ± 0.5	1.7 ± 0.2	1.8 ± 0.7
Endocardial TRE [mm]	1.9 ± 0.2	2.0 ± 0.7	2.6 ± 0.7	2.1 ± 0.5	2.2 ± 0.2	2.2 ± 0.8

Table 3. Comparison between the LV blood-pool volume and Ejection Fraction (EF) between expert manual annotations (Ground Truth) and A — automatic feature detection from individual frames; B — single phase automated feature detection + registration-based propagation; and C — single phase expert manual annotation + registration-based propagation. Measurements were evaluated at two cardiac phases — end-systole (ES) and end-diastole (ED).

LV assessment metric	EchoPac		Auto A		Manual + Reg C		Auto + Reg B	
	ED	ES	ED	ES	ED	ES	ED	ES
Mean vol [mL]	249.0	223.0	247.6	220.8	232.0	209.6	242.0	217.7
Std dev vol [mL]	3.5	10.8	3.5	3.8	10.4	9.8	2.0	1.5
Mean LV EF (%)	10.4 ± 5.6		10.9 ± 2.0		9.6 ± 0.4		10.0 ± 0.8	
Std dev LV EF (%)	5.6		2.0		0.4		0.8	

the reconstructed LV volume to that estimated by the GE EchoPac PC clinical software following expert manual segmentation. The end-diastolic and systolic volume measurements are summarized in Table 3, along with the corresponding ejection fraction measurements. Performance-wise, the LV volume reconstruction from a tri-plane sequence requires 11.6 s.

4   Discussion

We described the implementation and clinical data evaluation of a rapid, automatic framework that encompasses well-evaluated filtering, segmentation, registration, and volume reconstruction techniques as a means to provide a rapid, robust and accurate framework for feature tracking from multi-plane ultrasound image sequences. All components of the proposed technique — segmentation, registration-based feature tracking and propagation, and 3D blood-pool volume

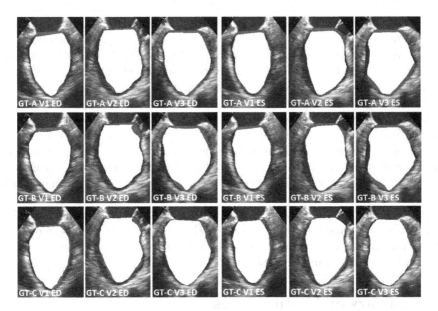

Fig. 5. Visual comparison of the blood-pool annotations achieved via *A — automatic feature detection from individual frames; B — single phase automated feature detection + registration-based propagation; and C — single phase expert manual annotation + registration-based propagation* vs. the ground truth expert manual blood-pool annotation (GT) quantified at end-diastole (ED) and end-systole (ES) for the three tri-plane views (V1, V2 and V3). White regions are common between the GT and each of the three A, B and C blood-pool estimates, red regions belong to the expert annotated blood-pool (GT), while the blue regions belong to the blood-pool area depicted by each of the three annotation methods A, B or C under comparison. Panels are named according to the same convention — i.e., the panel labeled *GT-B V2 ES compares the ground truth expert-annotated blood-pool (GT) to the blood-pool annotated using Method B displayed in View 2 at end-systole* (Color figure online).

reconstruction — were assessed against expert manual segmentation at both the systolic and diastolic cardiac phases and demonstrated accurate and consistent performance, while significantly minimizing user-induced variability. Furthermore, unlike other techniques that operate on 3D datasets, this technique enables rapid and consistent analysis of multi-plane, 2D US image sequences — the standard format for acquisition, interpretation, and analysis of cardiac US images.

As the proposed workflow integrates multiple algorithms, the influence of different parameters in the segmentation result is an important consideration. The frequency specific to the monogenic filter operates over a wide range of values and yields a good quality "cartoon image" for further segmentation. Similarly, for the graph cut algorithm, the mean and standard deviations for the blood pool, muscle and background regions are adaptively extracted from the image content, while the threshold 'C' that constraints the pixels towards same label can span a sufficiently wide range without significantly effecting the segmentation result. Furthermore, Lamash *et al.* [15] have thoroughly studied the effects of various regularization parameters in the biomechanics-based registration; for our purpose we selected the optimal parameters as suggested by the paper [15]. In summary, the proposed workflow yields a consistent segmentation result over a wide range of parameter values.

Unlike expert manual segmentation that is highly sensitive to intra- and inter-observer variability, the proposed technique provides a consistent result for each dataset, which can be reviewed and improved, if needed, by expert clinicians. The single-phase feature extraction, followed by tracking and propagation via registration further reduces uncertainty, avoiding the need to segment each frame independently by using the *a priori* frame information along with the image sequence to achieve optimal segmentation. Hence, should the expert clinician choose to perform any adjustments to the single-phase segmentation, their precise tracking and propagation throughout the cardiac cycle is guaranteed by the registration-based implementation.

5 Summary and Future Work

The impact and contribution of the proposed work is the integration of several image processing techniques (i.e., phase-based filtering, segmentation, registration and volume reconstruction) into a streamlined workflow that utilizes traditional standard of care images and fits seamlessly within the current workflows associated with both cardiac function assessment and intra-operative cardiac intervention guidance and monitoring.

Ongoing and future efforts include further evaluation and demonstration of how the proposed technique can cater to dynamically reconstructing 3D endocardial LV representations that facilitate computer-assisted assessment of stroke volume and ejection fraction, as well as employing intra-operative multiplane 2D TEE data to dynamically update and animate CT and/or MRI anatomy depicted pre-operatively to better represent the intra-operative conditions. Lastly, although we believe the most meaningful assessment is still against the expert clinicians analysis of the same input data, we acknowledge the importance of assessing the output of our proposed framework against the output of other techniques and extend the analysis to a large dataset of multi-plane image sequences acquired across multiple cardiac cycles.

Besides its direct application to computer-aided cardiac function assessment, the proposed framework is readily adaptable to the guidance and monitoring of image-guided cardiac interventions, most of which involve the use of real-time ultrasound imaging the clinical standard of care for cardiac procedures.

Acknowledgments. The authors would like to acknowledge Dr. Nathan Cahill for sharing his technical expertise and Aditya Daryanani for his help with image segmentation. In addition, we acknowledge funding support from the Kate Gleason Research Fund and the RIT College of Engineering Faculty Development Grant.

References

1. Noble, J., Boukerroui, D.: Ultrasound image segmentation: a survey. IEEE Trans. Med. Imaging **25**(8), 987–1010 (2006)
2. Mishra, A., Dutta, P., Ghosh, M.: A GA based approach for boundary detection of left ventricle with echocardiographic image sequences. Image Vis. Comput. **21**(11), 967–976 (2003)
3. Mignotte, M., Meunier, J.: A multiscale optimization approach for the dynamic contour-based boundary detection issue. Comput. Med. Imaging Graph. **25**(3), 265–275 (2001)
4. Bernard, O., Friboulet, D., Thevenaz, P., Unser, M.: Variational B-spline level-set: a linear filtering approach for fast deformable model evolution. IEEE Trans. Image Process. **18**(6), 1179–1191 (2009)
5. Cremers, D., Osher, S.J., Soatto, S.: Kernel density estimation and intrinsic alignment for shape priors in level set segmentation. Int. J. Comput. Vis. **69**(3), 335–351 (2006)
6. Cootes, T.F., Taylor, C.J., Cooper, D.H., Graham, J.: Active shape models-their training and application. Comput. Vis. Image Underst. **61**(1), 38–59 (1995)
7. Bosch, J., Mitchell, S., Lelieveldt, B., Nijland, F., Kamp, O., Sonka, M., Reiber, J.: Automatic segmentation of echocardiographic sequences by active appearance motion models. IEEE Trans. Med. Imaging **21**(11), 1374–1383 (2002)
8. Georgescu, B., Zhou, X., Comaniciu, D., Gupta, A.: Database-guided segmentation of anatomical structures with complex appearance. In: IEEE Computer Society Conference on Computer Vision and Pattern Recognition, CVPR 2005, vol. 2, pp. 429–436 (2005)
9. Jyh Herng Wu, E., De Andrade, M.L., Nicolosi, D.E., Pontes Jr., S.C.: Artificial neural network: border detection in echocardiography. Med. Biol. Eng. Comput. **46**(9), 841–848 (2008)
10. Mulet-Parada, M., Noble, J.A.: 2D + T acoustic boundary detection in echocardiography. Med. Image Anal. **4**, 21–30 (2000)
11. Boykov, Y., Veksler, O., Zabih, R.: Fast approximate energy minimization via graph cuts. IEEE Trans. PAMI **23**, 1222–1239 (2001)
12. Kovesi, P.: Symmetry and asymmetry from local phase. In: Proceedings of 10th Australian Joint Conference Artificial Intelligence, pp. 2–4 (1997)
13. Rajpoot, K., Grau, V., Noble, J.A.: Local-phase based 3D boundary detection using monogenic signal and its application to real-time 3-D echocardiography images. In: Proceedings of IEEE International Symposium Biomedical Imaging, pp. 783–786 (2009)
14. Uzkent, B., Hoffman, M.J., Cherry, E., Cahill, N.: Processing IEEE western NY image signal process workshop, pp. 47–51 (2014)
15. Lamash, Y., Fischer, A., Carasso, S., Lessick, J.: Strain analysis from 4D cardiac CT image data. IEEE Trans. Biomed. Eng. **62**, 511–521 (2015)
16. Zitova, B., Flusser, J.: Image registration methods: a survey. Image Vis. Comput. **21**, 977–1000 (2003)
17. Lee, E.T.Y.: Choosing nodes in parametric curve interpolation. Comput. Aided Des. **21**, 363–370 (1989)
18. Barber, C.B., Dobkin, D.P., Huhdanpaa, H.: The quickhull algorithm for convex hulls. ACM Trans. Math. Softw. **22**, 469–483 (1996)

Assessment of Septal Motion Abnormalities in Left Bundle Branch Block Patients Using Computer Simulations

Peter R. Huntjens[1,2(✉)], John Walmsley[1], Vincent Wu[3], Tammo Delhaas[1], Leon Axel[3], and Joost Lumens[1,2]

[1] Department of Biomedical Engineering, Cardiovascular Research Institute Maastricht (CARIM), Maastricht University, Maastricht, The Netherlands
[2] L'Institut de Rythmologie et Modélisation Cardiaque (IHU-LIRYC), Université de Bordeaux, Bordeaux, France
`p.huntjens@maastrichtuniversity.nl`
[3] Department of Radiology and Medicine, NYU School of Medicine, New York, USA

Abstract. Septal Flash (SF) is a rapid leftward – rightward motion of the septal wall during the isovolumic contraction phase that is frequently but not always observed in heart failure patients with left bundle branch block (LBBB). The goal of the present study is to evaluate the feasibility of detecting SF by assessing septal curvature both in patients with LBBB using MRI and in simulations using the CircAdapt model of the heart and circulation. In both patients and simulations, SF was characterized by a decrease of septal wall curvature and septum to lateral wall distance, followed by a rapid increase prior to aortic valve opening. Additionally, computer simulations revealed that SF can be explained by an intra-left ventricular (septal-to-lateral wall) activation delay. Reducing contractility in the left ventricular free wall abolished the rightward SF motion in LBBB. This finding suggests that lack of SF may indicate co-morbidities that can result in non-response to cardiac resynchronization therapy.

1 Introduction

Septal Flash (SF) is an abnormal motion of the septum that is frequently observed during isovolumic contraction phase in left bundle branch block (LBBB) patients with heart failure (HF) [1]. SF is a leftward motion of the early activated septum, rapidly followed by a rightward movement, which coincides with lateral wall contraction. Abnormal septal motion, assessed by magnetic resonance imaging (MRI) or echocardiography, has been associated with cardiac resynchronization therapy (CRT) response [2, 3]. However it remains unclear why presence of SF is a predictor of response to CRT. In addition, causes of SF absence in LBBB patients are not entirely clear. Differences in electrical activation patterns have been proposed as a possible explanation for SF absence in LBBB patients [4]. Furthermore, it was shown in an animal model that SF was not present in hearts with septal myocardial scar [5]. Simulations using a computational model of the cardiovascular system (CircAdapt) and patient recordings revealed that local differences in contractility could affect septal longitudinal deformation

H. van Assen et al. (Eds.): FIMH 2015, LNCS 9126, pp. 40–47, 2015.
DOI: 10.1007/978-3-319-20309-6_5

patterns [6]. The goal of the present study is to evaluate the feasibility of detecting SF by assessing septal curvature both in patients with LBBB using MRI and in simulations using the CircAdapt model. Using CircAdapt simulations, we tested our hypothesis that septal motion is possibly affected by both intra-left-ventricular activation delay and left ventricular free wall (LVFW) contractility differences.

2   Methods

2.1   Magnetic Resonance Imaging

Two patients with QRS duration > 120 ms and LBBB ECG morphology, one of them with SF, were retrospectively selected. Presence of SF was based on visually inspection of four chamber MRI images, obtained by a Siemens 1.5 T system (Avanto, Siemens Medical Systems, Germany). No scar tissue was present based on late gadolinium enhancement images in either patient. Phase flow quantification images were acquired to determine aortic valve opening time. An MRI set of short axis images at the level of the papillary muscles was identified and used for septal curvature (C_{sept}) and displacement measurements. The time delay between the moment of onset QRS and the acquisition of the first image was ignored. Matlab (MathWorks Inc., MA, USA) was used to select the superior and inferior RV-LV junction points at the endocardial site in the first frame (Fig. 1A: J1 and J2). The image was rotated so that the line through the junction points was perpendicular to the y-axis. Two endocardial points on the perpendicular bisector of the junction points were selected: one on the septum and one on the LVFW (Fig. 1A: S and L). Using these reference endocardial points, C_{sept} was calculated as previously described by Dellegrottaglie et al. [7] (Fig. 1B). Briefly, two chords,

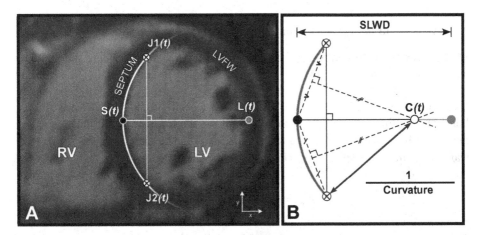

Fig. 1. Septal curvature measurements: **(A)** Points tracked in short-axis images for measurement of septal curvature: midpoints of septum (S) and LVFW (L), and anterior (J1) and inferior (J2) LV/RV junction points. **(B)** Geometric method for calculation of location of center and radius of circle through three points. Distance between endocardial septal mid-point (S) and the endocardial LVFW mid-point (L) was defined as the septum-lateral wall distance (SLWD).

connecting the mid-wall point to the superior and inferior RV-LV junction points, were defined. The intersection point of the two lines which bisects these chords perpendicularly was defined as the centroid of the circular arc segment intersecting the defined midwall and two junction points. C_{sept} was defined as the inverse of the distance of the line drawn from the calculated centroid C to any point on the arc segment. The distance from mid-point of the septum to the mid-point of the LVFW (both endocardial site) along the x-axis was defined as septal-to-lateral wall distance (SLWD).

The four endocardial reference points were tracked over time using an automated-feature-tracking algorithm based on normalized cross-correlation [8]. Briefly, a 20 pixel square area surrounding the selected point in the initial view was compared to a 40 pixel square search area in the next phase. The pixel which correlated the most to the selected point in the previous frame was selected as the location of the new point. This procedure was performed for all frames, to provide measurements of septal curvature and displacement throughout the cardiac cycle.

2.2   CircAdapt Model

The effects of dyssynchronous activation and LVFW hypocontractility on septal motion were tested using the CircAdapt computational model of the human heart and circulation [9, 10]. CircAdapt consists of modules, representing cardiac walls, cardiac valves, large blood vessels, systemic and pulmonary peripheral vasculature, the pericardium and local cardiac tissue mechanics. The geometrically simplified LV and RV cavities of the heart in the CircAdapt model are surrounded by three spherical caps: the LVFW, RVFW and interventricular septum. The cardiac walls are surrounded by a passive elastic sheet, the pericardium. The ventricular walls are mechanically coupled at the septal junction through equilibrium of tensile forces (Fig. 2) [10].

2.3   Assessment of Septal Motion Using CircAdapt

SF was assessed in the CircAdapt model by calculating traces of curvature and SLWD, and by generating a pseudo-MMode image. Transmural pressure calculations take place on the midwall surface dividing the volume of the wall in two (Fig. 2B) [10]. The thickness of tissue on either side of the mid-wall surface depends on mid-wall curvature (C_m). For each cap, the displacement of the mid-wall surface from the septal junction (x_m), the mid-wall surface area (A_m), the wall volume (V_w) and C_m are known (Fig. 2B). The wall area of the entire sphere $A_{m,s}$ (Fig. 2C) can be calculated by:

$$A_{m,s} = 4\pi \left(\frac{1}{C_m}\right)^2 \tag{1}$$

The cap therefore takes up a proportion of the whole sphere:

$$\frac{A_m}{A_{m,s}} = \frac{A_m C_m^2}{4\pi} \tag{2}$$

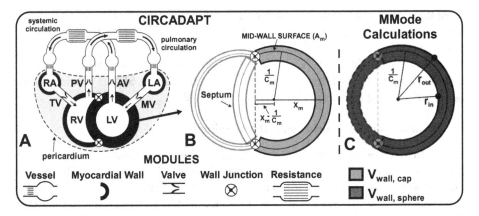

Fig. 2. (**A**) CircAdapt model designed as a network of modules representing atrial and ventricular cardiac walls, blood vessels, cardiac valves: AV: aortic; MV: mitral; PV: pulmonary; TV: tricuspid valve, and the pulmonary and systemic peripheral vasculatures. (**B**) Cross-sectional view of ventricular walls. The walls are mechanically coupled at the junction points. C_m: mid-wall curvature; x_m: distance from junction plane to mid-wall. (**C**) r_{out}: radius of outer wall; r_{in}: radius of inner wall. Solid lines denote the shaded cap in panel B, dashed lines are constructed in the calculation. Panels A and B are adapted from Lumens et al. [10]

The corresponding wall volume of the entire sphere is:

$$V_{w,s} = \frac{V_w 4\pi}{A_m C_m^2} \qquad (3)$$

Solving the following equations then gives the inner and outer radii of this sphere:

$$\frac{4}{3}\pi r_{in}^3 = V_{m,s} - \frac{1}{2}V_{w,s}, \quad \text{and} \quad \frac{4}{3}\pi r_{out}^3 = V_{m,s} + \frac{1}{2}V_{w,s} \qquad (4)$$

In which $V_{w,s}$ is the wall volume of the entire sphere, while r_{in} and r_{out} are the distances of the epicardial and endocardial sides of the wall, respectively, from the center of the sphere on which the cap lies, which differs between caps. We calculate the locations of inner (x_{in}) and outer (x_{out}) surfaces of the wall relative to the septal junction (Fig. 2A) as follows:

$$x_{in} = x_m - \frac{1}{C_m} + r_{in}, \quad \text{and} \quad x_{out} = x_m + \frac{1}{C_m} + r_{in} \qquad (5)$$

It is possible for the septum to become flat, *i.e.* the C_m becomes zero and hence Eq. 1 is undefined. When the septum is flat, mid-wall surface divides the width of the wall in half and the inner and outer surface radii are calculated as follows:

$$r_{in} = +\frac{1}{2}\frac{V_w}{A_m}, \quad \text{and} \quad r_{out} = -\frac{1}{2}\frac{V_w}{A_m} \qquad (6)$$

When the outer and inner radii of all ventricular walls are known, the positions of the mid-wall, endocardial and epicardial boundaries of the ventricular walls are plotted in time. The position of the LV epicardial surface is taken to be the zero reference position in our pseudo-MMode figures. C_m of the septal wall was used to compare curvature of model simulations to MRI-derived curvatures, C_{sept}. In addition, SLWD was defined as the sum of the r_{in} of the septal wall and r_{in} of the LVFW.

2.4 Simulation of Dyssynchronous Heart Failure

Starting from a reference simulation representing the normal adult heart and circulation [6], a failing heart without LBBB, with LBBB and with LBBB and LVFW hypocon-tractility were simulated. In all simulations heart rate was set to 70 bpm. Circulating blood volume and peripheral vascular resistance of the systemic circulation were adjusted so that cardiac output and mean arterial pressure equaled 3.8 L/min and 84 mmHg [10]. HF was induced by a reduction of contractility of all ventricular walls

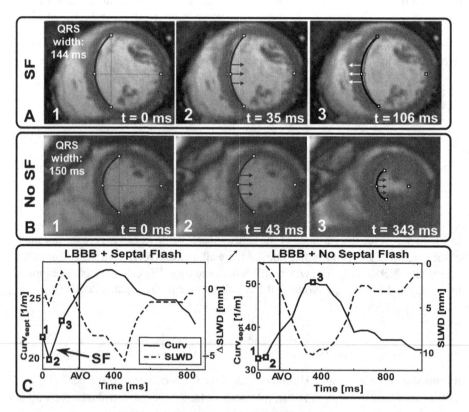

Fig. 3. Septal deformation patterns in an LBBB patient with SF and an LBBB patient without SF. Short axis MRI images of an LBBB patient with SF (Panel A) and an LBBB patient without SF (Panel B). Measured septal curvature changes (solid, onset QRS: t = 0 ms) and septal to lateral wall distance (SLWD, dashed) of SF patient (left) and no SF patient (right) are shown in lower panel (Panel C). Numbers correspond to frames shown in Panel A and B.

to 60 % of its normal value. In the LBBB simulations, dyssynchronous activation was simulated by inducing a delay in onset of mechanical activation of 20 ms in the septum and 60 ms in the LVFW, relative to the RVFW. To assess the effect of local differences in contractile myofibre function on SF, the contractility in the LVFW was further reduced to 20 % of its normal value.

3 Results

In the patient with SF, SF was associated with a decrease in SLWD, followed by an increase. In addition, septal flattening could be observed as a decrease in curvature, i.e. increase in septal wall radius. An increase in curvature was caused by both the rightward movement of the septal wall and inward movement of the junction points (Fig. 3, Panels A & C). In the LBBB patient without SF, the

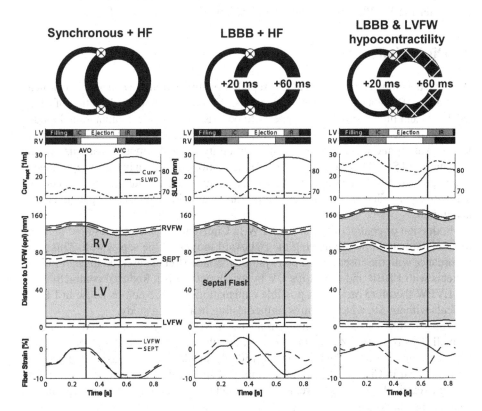

Fig. 4. Overview of CircAdapt simulations. Left: HF with synchronous mechanical activation; middle: HF with LBBB; right: HF with LBBB and decrease LVFW contractility. Septal curvature (curv$_{sept}$; solid) and SLWD (dashed) are shown. Distance of cardiac walls to epicardial LVFW mid-point is shown in MMode figures. The upper wall represents the RVFW, middle wall: septum, lower wall LVFW. Fiber strain is shown in lower panel: dashed line: septal strain; solid line: LVFW strain. Vertical solid lines indicate aortic valve opening (AVO: left) and closure (AVC; right).

absence of a rightward motion of the septal wall resulted in a decrease of SLWD. However, curvature increased throughout the cardiac cycle due to inward motion of the junction points (Fig. 3, Panels B & C). In the simulation of a dyssynchronous heart an abrupt leftward – rightward motion during the isovolumic contraction phase could be observed in either SLWD or pseudo-MMode (Fig. 4). During ejection SLWD was decreasing, while curvature of the septum was increasing. SF became apparent when intra-ventricular dyssynchrony of mechanical activation was present, as seen in both curvature and SLWD, and in the MMode images (Fig. 4, middle column). Fibers in the septum shortened prior to ejection, stretching the LVFW, as shown by changes in fiber strain (Fig. 4, lower panel). When the LVFW became activated, LVFW fibers started to shorten, while septal fibers lengthened (= rebound stretch), coinciding with the SF motion pattern. When the contractility of the LVFW fibers was reduced (Fig. 4 right column), rebound stretch of the septal fibers was absent. No SF pattern could be observed in neither the SLWD and curvature traces, nor in the pseudo-MMode.

4   Discussion

We have developed a method to assess septal curvature and SLWD changes in clinical short-axis MRI data and in model simulations. With this method the detection of SF in MRI data and model simulations appears to be feasible. Using a computational model of the cardiovascular system, we have evaluated the effect of dyssynchrony of mechanical activation and local differences in contractility on septal motion patterns. Septal curvature and SLWD dynamics of an LBBB activation simulation in CircAdapt showed good agreement with measured SF patient data. However, the number of patients and simulations included in our study are limited and validation with a larger patient cohort is required.

Reduction of contractility in the LVFW in CircAdapt simulations resulted in a 'no SF' motion pattern by eliminating the rightward motion of the septal wall. This is in accordance with the absence of septal systolic rebound stretch observed in heart failure patients with LBBB and coexisting LV hypocontractility [6]. Reduced contractility of the LVFW therefore provides a possible explanation for CRT non-response in LBBB patients without SF, which requires further testing in a patient study.

The absolute LV diameter change reflected in MMode figures and SLWD patterns is small. This is a result of inducing HF in the simulations by decreasing global contractility, causing dilation of the LV (EDV > 150 ml). In addition, short-axis diameters are overestimated in CircAdapt due to the simplified spherical geometry of the ventricles. At high ventricular volumes, a relatively small change in radius is required to cause a large change in volume.

Despite the fact that SLWD and curvature both reflect the evolving equilibrium between septal tension, trans-septal pressure gradient and cavity volumes, both features can behave differently. This is the case in the patient without SF, in which a leftward movement of the septum accompanied with an increase in curvature can be observed. However, in our CircAdapt simulation with reduced LVFW contractility, both curvature and SLWD patterns appeared to be comparable. The relation between curvature and displacement could be valuable to assess inter-ventricular mechanics and tissue properties, such as contractility reserve.

Simulating SF motion patterns in CircAdapt could give us the possibility to further investigate the underlying mechanisms of SF. Combining model simulations with patient data could provide us with a better understanding of mechanisms contributing to SF and its relation to CRT response.

References

1. Dillon, J.C., Chang, S., Feigenbaum, H.: Echocardiographic manifestations of left bundle branch block. Circulation **49**, 876–880 (1974)
2. Sohal, M., Amraoui, S., Chen, Z., Sammut, E., Jackson, T., Wright, M., O'Neill, M., Gill, J., Carr-White, G., Rinaldi, C.A., Razavi, R.: Combined identification of septal flash and absence of myocardial scar by cardiac magnetic resonance imaging improves prediction of response to cardiac resynchronization therapy. J. Interv. Card. Electrophysiol. **40**, 179–190 (2014)
3. Leenders, G.E., Cramer, M.J., Bogaard, M.D., Meine, M., Doevendans, P.A., De Boeck, B.W.: Echocardiographic prediction of outcome after cardiac resynchronization therapy: conventional methods and recent developments. Heart Fail. Rev. **16**, 235–250 (2011)
4. Duckett, S.G., Camara, O., Ginks, M.R., Bostock, J., Chinchapatnam, P., Sermesant, M., Pashaei, A., Lambiase, P.D., Gill, J.S., Carr-White, G.S., Frangi, A.F., Razavi, R., Bijnens, B.H., Rinaldi, C.A.: Relationship between endocardial activation sequences defined by high-density mapping to early septal contraction (septal flash) in patients with left bundle branch block undergoing cardiac resynchronization therapy. Europace **14**, 99–106 (2012)
5. Duchateau, N., Sitges, M., Doltra, A., Fernandez-Armenta, J., Solanes, N., Rigol, M., Gabrielli, L., Silva, E., Barcelo, A., Berruezo, A., Mont, L., Brugada, J., Bijnens, B.: Myocardial motion and deformation patterns in an experimental swine model of acute LBBB/CRT and chronic infarct. Int. J. Cardiovasc. Imaging **30**, 875–887 (2014)
6. Leenders, G.E., Lumens, J., Cramer, M.J., De Boeck, B.W., Doevendans, P.A., Delhaas, T., Prinzen, F.W.: Septal deformation patterns delineate mechanical dyssynchrony and regional differences in contractility: analysis of patient data using a computer model. Circ. Heart Fail. **5**, 87–96 (2012)
7. Dellegrottaglie, S., Sanz, J., Poon, M., Viles-Gonzalez, J.F., Sulica, R., Goyenechea, M., Macaluso, F., Fuster, V., Rajagopalan, S.: Pulmonary hypertension: accuracy of detection with left ventricular septal-to-free wall curvature ratio measured at cardiac MR. Radiology **243**, 63–69 (2007)
8. Wu, V., Chyou, J.Y., Chung, S., Bhagavatula, S., Axel, L.: Evaluation of diastolic function by three-dimensional volume tracking of the mitral annulus with cardiovascular magnetic resonance: comparison with tissue Doppler imaging. J. Cardiovasc. Magn. Reson. **16**, 71 (2014)
9. Arts, T., Delhaas, T., Bovendeerd, P., Verbeek, X., Prinzen, F.W.: Adaptation to mechanical load determines shape and properties of heart and circulation: the CircAdapt model. Am. J. Physiol. Heart Circ. Physiol. **288**, H1943–H1954 (2005)
10. Lumens, J., Delhaas, T., Kirn, B., Arts, T.: Three-wall segment (TriSeg) model describing mechanics and hemodynamics of ventricular interaction. Ann. Biomed. Eng. **37**, 2234–2255 (2009)

Quantifying Structural and Functional Differences Between Normal and Fibrotic Ventricles

Prashanna Khwaounjoo[1(✉)], Ian J. LeGrice[2], Mark L. Trew[1], and Bruce H. Smaill[1,2]

[1] Auckland Bioengineering Institute, The University of Auckland, Auckland, New Zealand
pkhw002@aucklanduni.ac.nz

[2] Department of Physiology, The University of Auckland, Auckland, New Zealand

Abstract. Fibrosis is a significant component of cardiac remodeling in heart failure. However, such remodeling has not been fully quantified across a range of scales and the functional impacts on arrhythmogenesis are still poorly understood. Transmural ventricular tissue samples from WKY and SHR rats are imaged and analyzed structurally at the scale of myocardial laminae. New imaging protocols and immunohistochemical labeling are investigated for 3D reconstructions of cell distributions and interconnectivity. At larger scales, there are obvious structural differences between WKY and SHR tissue in fiber rotation and tissue connectivity. Electrical activation models show less significant differences in functional behavior between the two tissue types. Imaging extended volume 3D cell connectivity provides promising insights and will be used in the future to inform modeling at larger scales.

Keywords: Cardiac remodeling · Cardiomyopathy · Heart failure · Biophysical modelling · Imaging techniques

1   Introduction

Sudden cardiac death (SCD) due to ventricular arrhythmia is a major cause of mortality in heart failure (HF) [1]. Pathological remodeling in HF tissue includes changes in the 3D tissue structure, cell geometry and cellular proteins [2], as well as increased expression of the collagen network, or fibrosis [3]. It is well known that the 3D organization of myocardium is a major determinant of the spread of electrical activation in the heart [4], and that arrhythmic substrates such as delayed and discontinuous activation are linked to ventricular fibrosis [5]. Computer modeling studies have shown that the risk of wave-break is intensified by structural discontinuities across a range of spatial scales [6]. Such studies demonstrate that fibrosis is likely to give rise to re-entry and fibrillation [7, 8]. While it is believed that SCD is closely associated with the character and extent of fibrosis, the exact mechanisms responsible and the spatial scales at which they operate remain unclear.

Here these issues are investigated through structural and functional analysis of left ventricular (LV) myocardium from the spontaneous hypertensive rat (SHR), a genetic animal model that replicates the progression of human hypertensive heart disease to HF [9, 10]. We extend tools and techniques [11, 12] that have previously been used to

© Springer International Publishing Switzerland 2015
H. van Assen et al. (Eds.): FIMH 2015, LNCS 9126, pp. 48–56, 2015.
DOI: 10.1007/978-3-319-20309-6_6

characterize 3D cardiac structure at a range of scales and outline their application to relating fibrosis to heart rhythm disturbance.

2   Materials and Methods

2.1   3D Tissue Imaging

All tissue collections were approved by the Animal Ethics Committee of The University of Auckland and conform to the Guide for the Care and Use of Laboratory Animals. Hearts from 12 month-old SHR rats and age-matched Wistar Kyoto (WKY) controls were perfused with fixative and picrosirius red stain. Transmural LV freewall segments were embedded in resin and imaged in 3D (2 μm voxels) using extended volume confocal microscopy. Reconstructed image volumes from two such hearts are shown in Fig. 1.

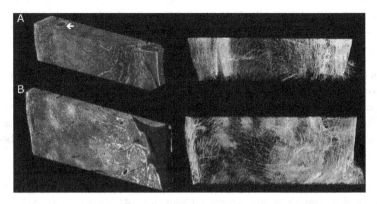

Fig. 1. Reconstructed transmural views **A.** WKY tissue (**left**) and the collagen structure (**right**). Arrow indicates large vessel **B.** SHR tissue (**left**) and the collagen structure (**right**). Image volumes for A and B are 3.50 × 0.92 × 0.73 mm and 4.04 × 1.50 × 0.30 mm, respectively.

2.2   Tissue Modeling

Connectivity and Volume Distributions. Quantitative comparisons of the WKY and SHR tissues were made at a supra-cellular scale. Viable myocytes, collagen and extracellular space were differentiated on the basis of image intensity using custom ranges in ImageJ (www.imagej.nih.gov). To calculate volume and connectivity measures, spherical and shell summation weighted filters (radii 5 voxels or 0.01 mm) were applied to the 3D segmented images using FFTs. A binominal filter was used to smooth the tissue volume and surface area to 8 μm resolution. Connectivity is the relative measure of the surface area of segmented myocytes in physical contact with neighboring myocytes.

Structural Orientations. The principal structural directions of the image volume were estimated using structure tensor analysis [13]. A 1D FFT approach with optimal gradient operators provided an estimate of voxel intensity gradients throughout the volume and these were used to construct structure tensors. Eigen-analysis of the structure tensors defined fiber orientations (smallest eigenvalue) and the cleavage planes that separate layers of myocytes (largest eigenvalue).

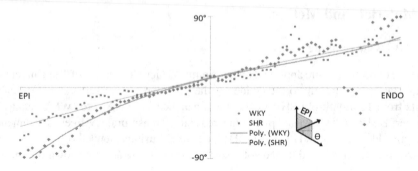

Fig. 2. Myofiber organisation of WKY (blue) and SHR (red) tissue. Fiber angle distribution from the epicardium and endocardium are presented along with a fitted cubic trendline (Color figure online).

Network Descriptions of Tissue. The topology of a segmented image of viable myocardium was described by a 3D network of node volume and edges. To create a network, the image volume was tessellated into a mesh of conforming voxel units, the discrete volumes and representative nodal locations of surviving myocardium within each unit were determined, and the connection areas and representative edge locations between units were identified. Multiple discrete fiber tracts through a voxel unit were treated as independent network components. The complete topologies of a network description of both the tissue sets were determined from the independent components using an efficient algorithm based on multiple applications of Quicksort [14]. Tissue fiber and sheet orientations were projected onto the network description.

Activation Modeling. Activation was simulated through the tissue network description using a finite volume discretization of a monodomain reaction-diffusion model. Cell membrane currents were solved using a Luo-Rudy dynamic model [15]. The parameters used were as follows: time step 0.005 ms, cell membrane surface area to volume ratio 200 mm^{-1}, membrane capacitance 0.0112 µF/mm^2, and $(\sigma_f, \sigma_s, \sigma_n)$ (0.1368,0.0583,0.0583) ms/mm. Stimuli were located in the midwall of each tissue sample. Activation and repolarization times were used to determine conduction velocities and action potential distributions.

2.3   3D Cellular Imaging

In addition to the myocardial sheet scale tissue comparisons, analysis at the cellular coupling scale is important for understanding fibrotic tissue remodeling in SHR vs WKY rats. Adult rat hearts were excised and immediately perfused using a Langendorff system with cold saline and fixed with Paraformaldehyde (PFA) 4 %. The tissue was cut transversely into 3 mm thick rings and placed overnight in 4 % PFA. To label for intercalated disks and viable myocytes the tissue was then placed overnight into a Phosphate-buffered saline (PBS) triton 0.3 % solution with anti Cx43 (Sigma) and phalloidin conjugated to 488 nm (Sigma) at dilutions of 1:1000 and 1:100 respectively. A secondary antibody 568 nm label was used to mark for the Cx43. An alternate labeling for cell membranes was wheat germ agglutinin (WGA) conjugated with a 488 nm secondary antibody (LifeTech) at 1:50 dilution. The tissue was thoroughly washed with PBS to remove unattached immunolabels. Specimens were dehydrated in a graded ethanol series and optically cleared with BABB solution until transparent (~3 days). Specimens were imaged using a Nikon TE2000 confocal microscope with 20x dry and 40x oil immersion objectives.

3   Results

Key structural similarities and differences between two WKY and SHR tissue samples are shown in Fig. 1. Additional tissue samples will be analyzed in future work.

A typical transmural myofiber rotation occurs in both the WKY and SHR samples through most of the heart wall (Fig. 2). However, there is some difference in the sub epicardial region.

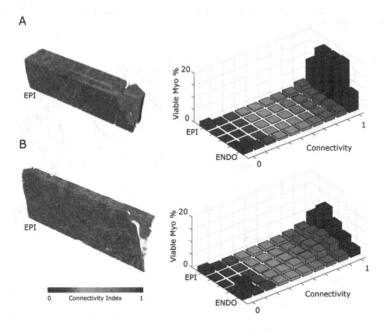

Fig. 3. Myocyte connectivity and volume fraction measures. **A.** Preserved myocytes rendered with normalised connectivity field for WKY tissue. The frequency histograms display variation in viable myocyte and connectivity changes in subvolumes from epi to endo. **B.** SHR tissue measurements as in A.

Figure 3 shows the relative surface area in physical contact (connectivity) for the segmented myocardium. The WKY tissue is more connected compared to the SHR. The right-hand panels show the proportion of viable myocardium across the heart wall, for a given connectivity. Connectivity is more symmetrically distributed in the WKY tissue. The majority of the tissue is well connected through the midwall, however marked reductions are observed in the both the epicardial and endocardial regions. In contrast, in the SHR sample, the distribution of highly connected tissue is asymmetric; biased toward the epicardium. There are more instances of reduced connectivity through the midwall than in the WKY sample.

The functional impacts of structural remodeling were investigated by simulating electrical activity in both the WKY and SHR models. Figure 4 shows that for these samples the progress of activation is similar (Fig. 4A and B), although activation is slower moving toward the epicardium in the case of the SHR model compared to the WKY (Fig. 4D). The opposite is the case moving toward the endocardium (Fig. 4D). The coupling (and hence loading differences) between the WKY and SHR models is illustrated by the different potential profiles relative to the stimulus site (Fig. 4D). The endocardial delayed activation in the WKY model arises from the narrow coupling tract between the main bulk of tissue and the endocardial trabeculation. Both the conduction velocities (CV) and the action potential durations (APD) are similar for both models (Fig. 5). The CV distributions (Fig. 5A and B, insets) indicate that overall the CV is slower for the SHR model.

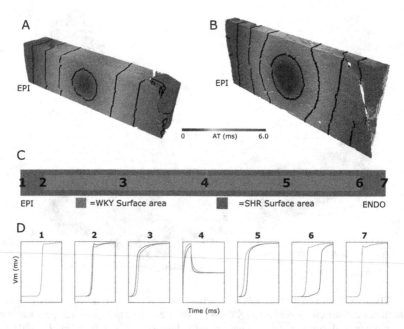

Fig. 4. Predicted activation times (AT) and potentials. **A.** Activation in WKY tissue. Contours at 1 ms. **B.** Activation in SHR tissue. **C.** Relative widths of tissue models. Locations of potential traces are indicated (1–7). **D.** WKY (blue) and SHR (red) potential (Color figure online).

4   Discussion and Conclusions

This study quantified structural and functional differences between age-matched (12 months) normal (WKY) and fibrotic (SHR) rat ventricular tissue samples. It also outlined the development of imaging techniques with potential for examining 3D cell coupling differences between WKY and SHR rat ventricular tissue. With ongoing refinement and expansion, these structural and functional data will ultimately improve our understanding of the impact of tissue remodeling on cardiac function in a number of disease states and senescence.

The most obvious structural differences are that the SHR tissue does not exhibit the same sub-epicardial fiber rotation as the WKY tissue and that the tissue coupling is reduced and asymmetrically biased toward the epicardium in the SHR compared to the WKY tissue. Functionally, there is some evidence of slowing as activation moves toward the epicardium compared to the endocardium in the SHR tissue, and the conduction velocity is also slightly slower in the SHR case compared to the WKY. A limitation of this study to date is that only one sample of each tissue type has been analyzed in detail and only WKY tissue has been imaged in 3D at the cellular level. However, more tissue samples have been imaged (4 image sets at various ages from 6–36 months) and are in the process of being analyzed, and additional tissue samples are being labeled and imaged at the cellular scale.

Figure 1A (left) shows that there is a significant vessel in the sub-epicardium of the WKY rat tissue. However, this is not sufficient to explain the trend shown in Fig. 2 where the SHR fiber angle diverges from WKY almost from the midwall. The sub- epicardial

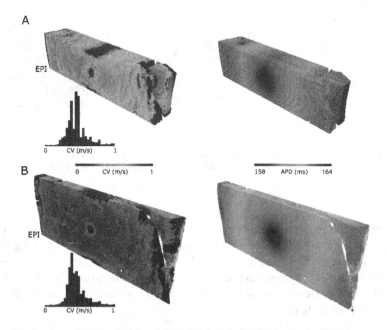

Fig. 5.   Conduction velocities (**left**) and APD (**right**): **A.** WKY tissue. **B.** SHR tissue.

collagen structure is clearly different between the WKY and SHR tissue. The collagen structure is strongly correlated with the collagen orientation [16].

Variation in connectivity (Fig. 3) is expected due to the increase in collagen found throughout the diseased tissue. Studies have shown that SHR tissue has substantially increased endomysial collagen and collagen I density [16, 17]. The connectivity shown in Fig. 3 is determined only for volumes containing myocytes, the coupling reduction in SHR tissue is attributed to decreased lateral coupling from the increased volume of collagen. The structural remodeling captured by this coupling metric is consistent with other analysis of SHR tissue [16].

Functional analysis of the WKY and SHR tissues is challenging. Modeling predicts some slowing and asymmetry of SHR vs WKY activation (Figs. 4 and 5). The changes in activation time may be attributed to thick patches of fibrosis which act as barriers to propagation [18]. There is compelling evidence from in vivo experiments that structural anisotropy provides a substrate for slow propagation, block and electrical instability [19]. However, these differences may be muted by the small tissue volumes with the model boundaries masking the true impact of the increased fibrosis in the SHR tissue. A new extended volume confocal imaging rig currently under development will enable larger volume tissue samples to be considered.

To improve mechanistic analysis it is important to develop a 3D understanding of myocyte coupling at the cellular scale. Our investigations confirm that a combination of labels using WGA to highlight cell membranes and a Cx43 label to indicate the locations of the intercalated disks are effective (Fig. 6) [20]. With high resolution imaging it is difficult to obtain a comprehensive 3D representation of the overall cell- group structures [21] and sectioning tissue can lead to destruction and lateralization of tissue structure [22].

Fig. 6. 3D reconstructions of BABB tissue volumes. **A.** Labeled for Cx43 (Anti Cx43 – blue) and actin (phalloidin - orange). **B.** Labeled for cell membranes (WGA - green) **C.** High resolution image of tissue labeled for cell membranes (WGA – green) and Cx43 (Anti Cx43 - blue) (Color figure online).

To enable 3D reconstructions over multiple cell depths, we applied an optical clearing step using BABB to increase the imaging depth [23]. The resulting image volumes (Fig. 6) can be utilized in functional models by applying appropriate segmentation techniques [24] to determine cell coupling patterns in normal WKY tissue and fibrotic SHR tissue. Healthy tissue has been shown to have an average of 11 myocytes connected to 1 myocyte [25].

The preliminary data presented in this paper suggest that fibrosis in SHR rat tissue alters the electrical activity. However, both the image structural analysis and the model functional analysis will require additional refinements. There is further scope to test the functional predictions using experimental optical mapping techniques [26].

References

1. Zipes, D.P., Wellens, H.J.J.: Sudden cardiac death. Circulation **98**, 2334–2351 (1998)
2. Schaper, J., Kostin, S., Hein, S., Elsässer, A., Arnon, E., Zimmermann, R.: Structural remodelling in heart failure. Exp. Clin. Cardiol. **7**, 64–68 (2002)
3. Houser, S.R., Margulies, K.B., Murphy, A.M., Spinale, F.G., Francis, G.S., Prabhu, S.D., Rockman, H.A., Kass, D.A., Molkentin, J.D., Sussman, M.A., Koch, W.J.: Animal models of heart failure: a scientific statement from the American heart association. Am. Heart Assoc. Counc. Basic Cardiovasc. Sci. Counc. Clin. Cardiol. Counc. Funct. Genomics Transl. Biol. **111**, 131–150 (2012)
4. Smaill, B.H., Zhao, J., Trew, M.L.: Three-dimensional impulse propagation in myocardium: arrhythmogenic mechanisms at the tissue level. Circ. Res. **112**, 834–848 (2013)
5. Kawara, T., Derksen, R., de Groot, J.R., Coronel, R., Tasseron, S., Linnenbank, A.C., Hauer, R.N.W., Kirkels, H., Janse, M.J., de Bakker, J.M.T.: Activation delay after premature stimulation in chronically diseased human myocardium relates to the architecture of interstitial fibrosis. Circulation **104**, 3069–3075 (2001)
6. Pertsov, A.M.: Scale of geometric structures responsible for discontinuous propagation in myocardial tissue. In: Spooner, P.M., Joyner, R.W., Jalife, J. (eds.) Discontinuous Conduction in the Heart, pp. 273–293. Futura Press, Armonk (1997)
7. Tanaka, K., Zlochiver, S., Vikstrom, K.L., Yamazaki, M., Moreno, J., Klos, M., Zaitsev, A.V., Vaidyanathan, R., Auerbach, D.S., Landas, S., Guiraudon, G., Jalife, J., Berenfeld, O., Kalifa, J.: Spatial distribution of fibrosis governs fibrillation wave dynamics in the posterior left atrium during heart failure. Circ. Res. **101**, 839–847 (2007)
8. TenTusscher, K.H.W.J., Panfilov, A.V.: Influence of diffuse fibrosis on wave propagation in human ventricular tissue. Europace **9**, vi38–vi45 (2007)
9. Cingolani, O.H., Yang, X.-P., Cavasin, M.A., Carretero, O.A.: Increased systolic performance with diastolic dysfunction in adult spontaneously hypertensive rats. Hypertension **41**, 249–254 (2003)
10. Slama, M., Ahn, J., Varagic, J., Susic, D., Frohlich, E.D.: Long-term left ventricular echocardiographic follow-up of SHR and WKY rats: effects of hypertension and age. Am. J. Physiol. Heart Circ. Physiol. **286**, H181–H185 (2004)
11. Rutherford, S.L., Trew, M.L., Sands, G.B., LeGrice, I.J., Smaill, B.H.: High-resolution 3-dimensional reconstruction of the infarct border zone: impact of structural remodeling on electrical activation. Circ. Res. **111**, 301–311 (2012)
12. Seidel, T., Draebing, T., Seemann, G., Sachse, F.B.: A semi-automatic approach for segmentation of three-dimensional microscopic image stacks. Funct. Imaging Model. Heart **7945**, 300–307 (2013)
13. Jähne, B.: Digital Image Processing. Springer, Heidelberg (2005)
14. Hoare, C.A.R.: Quicksort. Comput. J. **5**, 10–16 (1962)
15. Faber, G.M., Rudy, Y.: Action potential and contractility changes in [Na(+)](i) overloaded cardiac myocytes: a simulation study. Biophys. J. **78**, 2392–2404 (2000)

16. LeGrice, I.J., Pope, A.J., Sands, G.B., Whalley, G., Doughty, R.N., Smaill, B.H.: Progression of myocardial remodeling and mechanical dysfunction in the spontaneously hypertensive rat. Am. J. Physiol. Heart Circ. Physiol. **303**, H1353–H1365 (2012)

17. Boluyt, M.: Matrix gene expression and decompensated heart failure: the aged SHR model. Cardiovasc. Res. **46**, 239–249 (2000)

18. Jansen, J.A., van Veen, T.A.B., de Jong, S., van der Nagel, R., van Stuijvenberg, L., Driessen, H., Labzowski, R., Oefner, C.M., Bosch, A.A., Nguyen, T.Q., Goldschmeding, R., Vos, M.A., de Bakker, J.M.T., van Rijen, H.V.M.: Reduced Cx43 expression triggers increased fibrosis due to enhanced fibroblast activity. Circ. Arrhythmia Electrophysiol. **5**, 380–390 (2012)

19. Van Rijen, H.V.M., Eckardt, D., Degen, J., Theis, M., Ott, T., Willecke, K., Jongsma, H.J., Opthof, T., de Bakker, J.M.T.: Slow conduction and enhanced anisotropy increase the propensity for ventricular tachyarrhythmias in adult mice with induced deletion of connexin43. Circulation **109**, 1048–1055 (2004)

20. Schwab, B.C., Seemann, G., Lasher, R.A., Torres, N.S., Wulfers, E.M., Arp, M., Carruth, E.D., Bridge, J.H.B., Sachse, F.B.: Quantitative analysis of cardiac tissue including fibroblasts using three-dimensional confocal microscopy and image reconstruction: towards a basis for electrophysiological modeling. IEEE Trans. Med. Imaging **32**, 862–872 (2013)

21. Dodt, H.-U., Leischner, U., Schierloh, A., Jährling, N., Mauch, C.P., Deininger, K., Deussing, J.M., Eder, M., Zieglgänsberger, W., Becker, K.: Ultramicroscopy: three-dimensional visualization of neuronal networks in the whole mouse brain. Nat. Methods **4**, 331–336 (2007)

22. Young, A.A., Legrice, I.J., Young, M.A., Smaill, B.H.: Extended confocal microscopy of myocardial laminae and collagen network. J. Microsc. **192**, 139–150 (1998)

23. Dickie, R., Bachoo, R.M., Rupnick, M.A., Dallabrida, S.M., Deloid, G.M., Lai, J., Depinho, R.A., Rogers, R.A.: Three-dimensional visualization of microvessel architecture of whole-mount tissue by confocal microscopy. Microvasc. Res. **72**, 20–26 (2006)

24. Seidel, T., Draebing, T., Seemann, G., Sachse, F.B.: A semi-automatic approach for segmentation of three-dimensional microscopic image stacks of cardiac tissue. In: Ourselin, S., Rueckert, D., Smith, N. (eds.) FIMH 2013. LNCS, vol. 7945, pp. 300–307. Springer, Heidelberg (2013)

25. Saffitz, J.E., Kanter, H.L., Green, K.G., Tolley, T.K., Beyer, E.C.: Tissue-specific determinants of anisotropic conduction velocity in canine atrial and ventricular myocardium. Circ. Res. **74**, 1065–1070 (1994)

26. Khwaounjoo, P., Rutherford, S.L., Scrcek, Ma., LeGrice, I.J., Trew, M.L., Smaill, B.H.: Image-based motion correction for optical mapping of cardiac electrical activity. Ann. Biomed. Eng., 1–12 (2014)

Sparsity and Biomechanics Inspired Integration of Shape and Speckle Tracking for Cardiac Deformation Analysis

Nripesh Parajuli[1](✉), Colin B. Compas[5], Ben A. Lin[3], Smita Sampath[6], Matthew O'Donnell[7], Albert J. Sinusas[3,4], and James S. Duncan[1,2,4]

[1] Departments of Electrical Engineering, Yale University, New Haven, CT, USA
nripesh.parajuli@yale.edu
[2] Departments of Biomedical Engineering, Yale University, New Haven, CT, USA
[3] Departments of Internal Medicine, Yale University, New Haven, CT, USA
[4] Departments of Diagnostic Radiology, Yale University, New Haven, CT, USA
[5] IBM Research-Almaden, San Jose, CA, USA
[6] Merck Sharp and Dohme, Singapore, Republic of Singapore
[7] Department of Bioengineering, University of Washington, Seattle, WA, USA

Abstract. Cardiac motion analysis, particularly of the left ventricle (LV), can provide valuable information regarding the functional state of the heart. We propose a strategy of combining shape tracking and speckle tracking based displacements to calculate the dense deformation field of the myocardium.

We introduce the use and effects of l_1 regularization, which induces sparsity, in our integration method. We also introduce regularization to make the dense fields more adhering to cardiac biomechanics. Finally, we motivate the necessity of temporal coherence in the dense fields and demonstrate a way of doing so.

We test our method on ultrasound (US) images acquired from six open-chested canine hearts. Baseline and post-occlusion strain results are presented for an animal, where we were able to detect significant change in the ischemic region. Six sets of strain results were also compared to strains obtained from tagged magnetic resonance (MR) data. Median correlation (with MR-tagging) coefficients of 0.73 and 0.82 were obtained for radial and circumferential strains respectively.

Keywords: Echocardiography · Motion · Shape tracking · Speckle tracking · Radial basis functions · Regularization

1 Introduction

A reliable quantitative and qualitative assessment of the motion behavior can assist in the detection, localization, control and treatment of myocardial injuries. Within this context, echocardiography is a popular and important tool because it is relatively inexpensive and highly portable. It also provides better temporal resolution in comparison to other modalities.

© Springer International Publishing Switzerland 2015
H. van Assen et al. (Eds.): FIMH 2015, LNCS 9126, pp. 57–64, 2015.
DOI: 10.1007/978-3-319-20309-6_7

Over the years several shape based tracking methods and speckle based tracking methods have been proposed to calculate dense motion field for the LV and ultimately calculate regional strains. Shape based methods track myocardial boundaries or automatically selected shape and intensity based features which predominantly arise from myocardial boundaries [1,2]. Hence, mid-wall displacements are usually missing or not reliable.

Popular speckle tracking methods can be broadly classified into block-matching algorithms such as Lubinski et al. [3] and non-rigid registration based methods such as Heyde et al. [4]. Speckle tracking methods are better at estimating mid-wall displacements but do not always successfully track the boundaries due to inconsistent speckle patterns at the blood tissue interface.

Compas et al. [5] proposed a method to calculate dense field across the myocardium by integrating displacements from shape and speckle tracking methods using radial basis function (RBF) interpolation. The use of RBFs provided a flexible and simple framework by eliminating the need of complicated grids and meshes to model cardiac geometry. Hence, it provided a simpler mechanism of modeling complex, and sometimes incomplete, cardiac geometries.

In this work, we pose the problem in a Bayesian setting and then present it in a convex optimization format. We use l_1 regularization to tackle noise and also, indirectly, to select features. Drawing inspiration from cardiac biomechanics, we add penalties on divergences and derivatives. We also motivate and demonstrate the need to make motion fields temporally coherent. We test our methods on 2D canine heart images and present appropriate results.

2   Methods

We will first briefly discuss the elementary methods to obtain initial shape and speckle tracking based estimation. Then we will discuss in detail the integration framework where our contribution lies.

2.1   Elementary Motion Estimation Methods

Shape Tracking. We use boundary features derived from an automated level-set segmentation method [6]. Once endocardial and epicardial boundary points are found, a modified Generalized Robust Point Matching (GRPM) algorithm is used to obtain displacement estimates across frames [1].

The algorithm matches two point sets X and Y by solving for a fuzzy correspondence matrix m. The original form was solved using an iterative setup. However, since the objective function was convex, we determined that we could solve this directly using a convex optimization solver [7]. The GRPM formulation allowed us to control the relative weight between minimizing distance between the point sets and minimizing the difference in their shape characteristics such as curvatures. It also allowed us to control the fuzziness (or uniqueness) of the correspondence.

Speckle Tracking. Speckle patterns are inherent to US and, while they seem random in nature, they exhibit some consistency in a small time window. This can be leveraged and a correlation based tracking of local image patches is done.

For a pixel in a given image, a kernel of one speckle length around the pixel, in the complex signal, is correlated with neighboring kernels in the next frame [3]. The peak correlation value is used to determine the matching kernel in the next frame and calculate displacement and confidence measure. The velocity in the echo beam direction is further refined using zero-crossings of the phase of the complex correlation function.

2.2   Integration Framework

Bayesian Viewpoint. For an image frame, the dense displacement is U and the initial shape and speckle tracking displacements are U^{sh} and U^{sp} respectively. Conditioning using Bayes' theorem, and assuming independence between U^{sh} and U^{sp}, we get

$$P(U \mid U^{\mathrm{sh}}, U^{\mathrm{sp}}) \propto P(U^{\mathrm{sh}} \mid U)P(U^{\mathrm{sp}} \mid U)P(U) \qquad (1)$$

$P(U)$ captures the assumptions we make about our displacement field as a prior. We assume that U^{sp} and U^{sh} are both normally distributed with respect to U. Taking log on both sides, we get:

$$\log(P(U \mid U^{\mathrm{sh}}, U^{\mathrm{sp}})) \propto k_1 - \frac{1}{\sigma_{\mathrm{sh}}^2}||U^{\mathrm{sh}} - U||_2^2 - \frac{1}{\sigma_{\mathrm{sp}}^2}||U^{\mathrm{sp}} - U||_2^2 + L(P(U)) \quad (2)$$

Here, the σ's are the standard deviations of the shape and speckle displacements. $L(P(U))$ is the log of the prior distribution on U. We obtain the final dense field by maximizing the log posterior.

RBF Interpolation. An RBF is parametrized by its location $c \in \mathbb{R}^2$ and its width $\sigma \in \mathbb{R}^2$ (in 2D). Hence, during an RBF interpolation process, different resolutions can be captured by altering the width and the number of RBFs. In our case, we place RBFs around the myocardium and use a multi-level scheme, where in each iteration RBFs of different widths and varying quantity are used [8].

For each iteration, first we evaluate the value of all the RBFs for consideration and consolidate them as a design matrix (or a dictionary of bases) H. Then, the displacement values are expressed as a linear combination of these bases as $U = (U^{\mathrm{x}}, U^{\mathrm{y}}) = (Hw_{\mathrm{x}}, Hw_{\mathrm{y}})$. Therefore, our goal is to solve for the linear weights $w = (w_{\mathrm{x}}, w_{\mathrm{y}})$, which we accomplish using an optimization setup:

$$\hat{w}_{\mathrm{x}}, \hat{w}_{\mathrm{y}} = \operatorname*{argmin}_{w_{\mathrm{x}}, w_{\mathrm{y}}} \sum_{i \in A_{\mathrm{sh}}} c_i^{\mathrm{sh}} \left([Hw]_i - U_i^{\mathrm{sh}}\right)^2 + \sum_{i \in A_{\mathrm{sp}}} c_i^{\mathrm{sp}} \left([Hw]_i - U_i^{\mathrm{sp}}\right)^2 \quad (3)$$

A_{sh} and A_{sp} refer to shape and speckle areas in the myocardium. The confidences of each shape and speckle displacements are c_i^{sp} and c_i^{sh}. The above equation is the RBF parametrized form of Eq. 2, except for the prior part, which we discuss next.

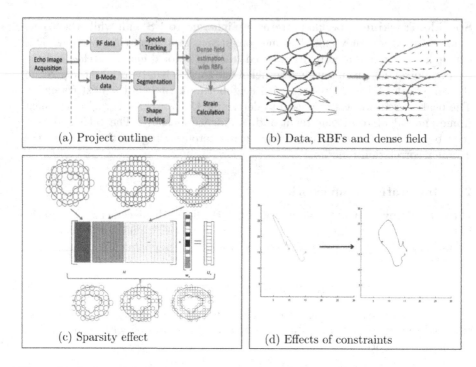

Fig. 1. (a) Outline with our contribution area highlighted, (b) Speckle (blue) and Shape (red) tracking displacements, RBFs and dense field, (c) Basis selection with sparsity, (d) Path of a point in myocardium - with biomechanical constraints and temporal smoothing (blue, on right) and without (magenta, on left) (Color figure online).

Regularization and Biomechanics. We add the l_1 norm of our linear weights w's to our objective in Eq. 3 as a prior. This regularization step captures our belief that the interpolation should be smooth and should not overfit our data.

Furthermore, the l_1 norm also induces sparsity on the weights w's. A large number of w's will end up becoming zero. This, in essence, is a form of feature selection. Of all the RBFs we consider for interpolation, only a handful will influence our final values. This ultimately allows us to capture varying and appropriate levels of details at different areas of the myocardium.

Next, we add the norm of the divergence of the motion field as a penalty to our in Eq. 3. We do so under the assumption that the cardiac tissue is roughly incompressible and therefore, motion vector field should be roughly divergence free [9]. Finally, we also mildly penalize the norm of the spatial derivatives to discourage jumps and discontinuities in the motion fields. Therefore, we have the following form:

$$\hat{w} = \underset{w}{\mathrm{argmin}}\, f_{\mathrm{adh}}(Hw, U^{\mathrm{sh}}, U^{\mathrm{sp}}) + \lambda_1 ||w||_1 + \lambda_2 f_{\mathrm{biom}}(U)$$
$$f_{\mathrm{biom}}(U) = ||\nabla \cdot U||_2^2 + \alpha ||\nabla U||_2^2$$

(4)

Here, $U = Hw$, f_{adh} refers to the data adherence terms in Eq. 3. Since we do not want derivatives to be penalized as much as divergences, α is set to be 0.25.

Cross Validation for Parameter Selection. Since regularization is key to our approach, appropriate selection of the regularization parameters $\lambda = [\lambda_1, \lambda_2, \lambda_3]$ is crucial. We use K-fold cross validation (CV) to select our parameters.

Assuming $K = 5$ (without loss of generality), if the value of function f (displacements in our case) is known at n points, then the data is partitioned randomly into 5 sets, each of size $n/5$. From these 5 groups, 4 are chosen randomly and are assigned as the training set X_{train}^k and the remaining one is the test set X_{test}^k. This is done 5 times, hence we have 5 sets of training and test sets. Corresponding function values are also labeled f_{train}^k and f_{test}^k, where $k \in \{1, 2, \ldots, 5\}$.

For a choice of λ, for each k, first solve for w^k, only using the training set:

$$
\hat{w}^k = \underset{w^k}{\text{argmin}} \sum_{i=1}^{|X_{\text{train}}|} (f_{\text{train}}^k(x_i) - [Hw^k]_i)^2 + \lambda_1 ||w^k||
$$
$$
+ \lambda_2 f_{\text{biom}}(Hw^k) \tag{5}
$$
$$
\hat{f}^k(x) = H\hat{w}^k
$$

where, $x_i \in X_{\text{train}}$.

Next, for all $x \in X_{\text{test}}$, we calculate and accumulate the squared difference between $\hat{f}^k(x)$ and $f_{\text{test}}^k(x)$ as our error in estimation. The set of λ's yielding the lowest total error value is chosen as the appropriate set of parameters.

Temporal Smoothing. Since strain calculation involves the accumulation of displacement values across time frames, temporal consistency and smoothness is necessary. Once we carry out all our frame-to-frame calculations, we obtain temporal smoothness by doing a spatio-temporal interpolation of our results.

3   Results

Philips iE33 ultrasound system, with the X7-2 probe and a hardware attachment that provided RF data, was used to acquire six open chest canine heart images. For all six canines, baseline and post-occlusion (induced by the occlusion of left anterior descending (LAD) coronary artery) ultrasound images were analyzed (in 2D). However, only three baseline (of three canines) and three post-occlusion (of the three remaining canines) tagged MR images, acquired using 1.5T Sonata MR scanner, were available and analyzed using HARP MR software [10]. The MR derived strains are used as validation of our method. All experiments were conducted in compliance with the Institutional Animal Care and Use Committee policies.

We first present how end-systolic (ES) radial strains have changed from baseline to occlusion condition for one animal in Fig. 2. The median strain value for

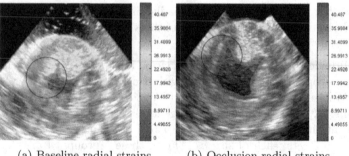

(a) Baseline radial strains   (b) Occlusion radial strains

Fig. 2. Baseline and post-occlusion radial strain values shown across the myocardium for corresponding end systolic (ES) frames (with the anterior region circled).

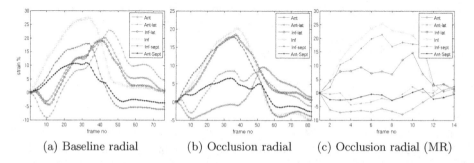

(a) Baseline radial   (b) Occlusion radial   (c) Occlusion radial (MR)

Fig. 3. Radial strain values for: Anterior, Anterolateral, Inferolateral, Inferior, Inferoseptal and Anteroseptal regions for ED-ED in (a) Baseline (b) Occlusion (c) Occlusion MR

the entire myocardium has dropped from 14.12 to 9.70. We can also see that the change in strain is most significant in the anterior region, which is expected since the LAD artery was occluded.

In Fig. 3, we compare average radial strain values for each segment, for the entire cardiac cycle (end diastole - end diastole (ED)), between baseline and occlusion conditions of one animal. Confirming with our qualitative analysis for Fig. 2, the anterior region, and to some extent the neighboring regions, goes through the most significant change. We can also see how our post occlusion strains correspond well with our validation MR results.

Next, we compare our strains for the six image sets (for which tagged MR data was available) with tagged MR strains. We compare correlation values of segmental strain plots for the entire cardiac cycle (ED-ED). We also expand our analysis by comparing our results with results from Compas et al. in [5], only shape tracking, only speckle tracking and B-spline non-rigid registration (code obtained from [11]). Box plots illustrating the distribution of correlation values for all the methods, are given in Fig. 4.

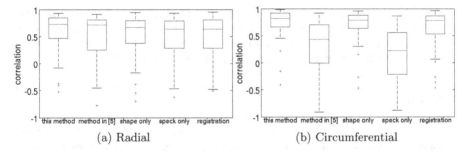

(a) Radial   (b) Circumferential

Fig. 4. Box plots illustrating the median value and the distribution of correlation (with MR) of segmental strains obtained from different methods.

Table 1. Overall median correlations

	This method	Method in [5]	Shape tracking	Speckle tracking	Registration
Radial	**.73**	.72	.67	.64	.64
Circumferential	**.82**	.43	.79	.22	.78

The median correlation values for all segmental radial and circumferential strains, obtained from six datasets are displayed in Table 1. We can see that our method performs better than the other methods considered for comparison. Correlations for radial strains are roughly similar for different methods and not very high. However, results for circumferential strains are significantly different for different methods, and high median correlation with MR-tagging was obtained for our method.

4   Conclusions

The use of l_1 regularization, biomechanical constraints and temporal smoothing has allowed us to develop a robust method of estimating dense displacement fields, which we tested on 2D data. Our bayesian inspired framework, which is complemented by the data driven parameter selection for optimization, has added to the stability of our method. The strains generated were in good agreement with MR-tagging based strains.

Our next step is to test the method in 3D. Since 3D data has lower temporal and spatial resolution, and hence, the elementary shape and speckle tracking displacements will possibly be noisier, the use of robust regularization strategy will potentially be of more value. We will also work to expand upon the regularization strategies to make the motion fields biomechanically and temporally more consistent.

Acknowledgment. Several members of Dr. Albert Sinusas's lab, including, but not limited to, Christi Hawley, were involved in the image acquisitions. Members of

Dr. Matthew O'Donnell's lab developed the RF speckle tracking algorithms and code. Dr. Xiaojie Huang provided us the segmentation code. We would like to sincerely thank everyone for their contributions. This work was supported in part by the National Institute of Health (NIH) grant numbers R01HL121226 and T32HL098069.

References

1. Lin, N., Duncan, J.S.: Generalized robust point matching using an extended free-form deformation model: application to cardiac images. In: IEEE International Symposium on Biomedical Imaging: Nano to Macro, pp. 320–323. IEEE (2004)
2. Shi, P., Sinusas, A.J., Constable, R.T., Ritman, E., Duncan, J.S.: Point-tracked quantitative analysis of left ventricular surface motion from 3-D image sequences. IEEE Trans. Med. Imaging **19**, 36–50 (2000)
3. Lubinski, M.A., Emelianov, S.Y., O'Donnell, M.: Speckle tracking methods for ultrasonic elasticity imaging using short-time correlation. IEEE Trans. Ultrason. Ferroelectr. Freq. Control **46**, 82–96 (1999)
4. Heyde, B., Bouchez, S., Thieren, S., Vandenheuvel, M., Jasaityte, R., Barbosa, D., Claus, P., Maes, F., Wouters, P., D'hooge, J.: Elastic image registration to quantify 3-D regional myocardial deformation from volumetric ultrasound: experimental validation in an animal model. Ultrasound Med. Biol. **39**, 1688–1697 (2013)
5. Compas, C.B., Lin, B.A., Sampath, S., Jia, C., Wei, Q., Sinusas, A.J., Duncan, J.S.: Multi-frame radial basis functions to combine shape and speckle tracking for cardiac deformation analysis in echocardiography. In: Metaxas, D.N., Axel, L. (eds.) FIMH 2011. LNCS, vol. 6666, pp. 113–120. Springer, Heidelberg (2011)
6. Huang, X., Dione, D.P., Compas, C.B., Papademetris, X., Lin, B.A., Bregasi, A., Sinusas, A.J., Staib, L.H., Duncan, J.S.: Contour tracking in echocardiographic sequences via sparse representation and dictionary learning. Med. Image Anal. **18**, 253–271 (2014)
7. CVX Research, I.: CVX: Matlab software for disciplined convex programming, version 2.0 (2012). http://cvxr.com/cvx
8. Floater, M.S., Iske, A.: Multistep scattered data interpolation using compactly supported radial basis functions. J. Comput. Appl. Math. **73**, 65–78 (1996)
9. Song, S.M., Leahy, R.M.: Computation of 3-D velocity fields from 3-D cine CT images of a human heart. IEEE Trans. Med. Imaging **10**, 295–306 (1991)
10. Osman, N.F., Prince, J.L.: Visualizing myocardial function using HARP MRI. Phys. Med. Biol. **45**, 1665 (2000)
11. Kroon, D.J.: B-spline grid, image and point based registration (2008). http://www.mathworks.com/matlabcentral/fileexchange/20057-b-spline-grid-image-and-point-based-registration

Characterization of Myocardial Velocities by Multiple Kernel Learning: Application to Heart Failure with Preserved Ejection Fraction

Sergio Sanchez-Martinez[1]([✉]), Nicolas Duchateau[3], Bart Bijnens[1,2],
Tamás Erdei[4], Alan Fraser[4], and Gemma Piella[1]

[1] DTIC, Universitat Pompeu Fabra, Barcelona, Spain
sergio.sanchezm@upf.edu
[2] ICREA, Barcelona, Spain
[3] Asclepios Research Project, INRIA Sophia Antipolis, Sophia Antipolis, France
[4] Wales Heart Research Institute, Cardiff University, Cardiff, UK

Abstract. The present study aims at improving the characterization of myocardial velocities in the context of heart failure with preserved ejection fraction (HFPEF) by combining multiple descriptors. It builds upon a recent extension of manifold learning known as multiple kernel learning (MKL), which allows the combination of data of different natures towards the learning. Such learning is kept unsupervised, thus benefiting from all the inherent explanatory power of the data without being conditioned by a given class. The methodology was applied to 2D sequences from a stress echocardiography protocol from 33 subjects (21 healthy controls and 12 HFPEF subjects). Our method provides a novel way to tackle the understanding of the HFPEF syndrome, in contrast with the diagnostic issues surrounding it in the current clinical practice. Notably, our results confirm that the characterization of the myocardial functional response to stress in this syndrome is improved by the joint analysis of multiple relevant features.

1 Introduction

Heart failure with preserved ejection fraction (HFPEF) is recognized as a public health problem of growing concern. This syndrome presents signs of heart failure but still maintains the ejection fraction into a normal range. Left ventricular diastolic dysfunction is a leading mechanism causing the syndrome [9] but recent studies suggest that HFPEF is a rather heterogeneous condition consisting of several pathophysiological subtypes [7]. Due to this heterogeneity, there is still an incomplete understanding of the syndrome, which results in a low statistical power of current diagnostic methods [10].

Previous attempts to characterize HFPEF [9] were mainly based on relatively simple measurements taken at rest, such as ejection fraction or E/e' (ratio of the early transmitral flow velocity by pulsed Doppler and the early mitral annular velocity by myocardial velocity imaging). Myocardial velocity imaging (MVI) provides a direct measurement of the velocities of the myocardial tissue with

© Springer International Publishing Switzerland 2015
H. van Assen et al. (Eds.): FIMH 2015, LNCS 9126, pp. 65–73, 2015.
DOI: 10.1007/978-3-319-20309-6_8

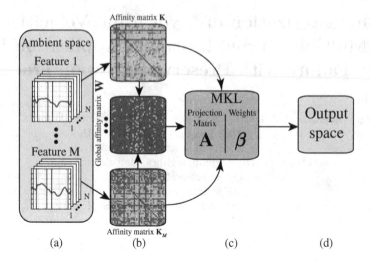

Fig. 1. Pipeline of the proposed method: (a) ambient space with M input features for each subject, (b) feature-specific and global affinity matrices used as input to the MKL optimization, (c) MKL optimization (d) Output space of reduced dimensionality.

high temporal resolution. However, they fail at summarizing the complexity of the observable abnormalities. In order not to discard useful information, further analysis on the data would be helpful.

We propose to take advantage of the full myocardial velocity traces (along the whole cardiac cycle), from two stages of a stress echocardiography protocol, to determine the diagnostic features of the disease. Functional atlases [2] provide a common spatiotemporal system of coordinates where such patterns can be compared. However, their analysis is often limited to linear statistical comparisons, or to voxel-wise observations. Non-linear dimensionality reduction techniques such as graph embedding [11] have been investigated to overcome these limitations and perform pattern-wise comparisons [3]. Nonetheless, such analyses only focus on single descriptors to characterize a given population in an unsupervised way (e.g. myocardial velocity in [3]). Depending on the complexity of the disease, a single descriptor strategy may not be sufficient to accomplish a proper characterization. In the present paper, we build upon a recently proposed framework, known as multiple kernel learning (MKL), which allows to optimally fuse heterogeneous information. A supervised learning formulation of MKL was given in [8].

Our main contribution is to adapt this formulation to an unsupervised setting suitable for our problem. We build upon the assumption that symptoms related to HFPEF will become apparent during exercise [4] and will be reflected in the myocardial velocity patterns especially of left ventricular long-axis function. Accordingly, the inputs for the MKL algorithm come from MVI acquisitions of healthy and suspected HFPEF subjects, at different stages of a stress protocol and in different regions of the left ventricular wall.

We demonstrate the relevance of jointly analyzing these multivariate data within a MKL framework. Notably, clusters can be intuited from the velocity traces and agree with clinical diagnosis. Additionally, we show that the analysis is effectively improved by the inclusion of multiple features. These results suggest that the characterization of the HPFEF syndrome can be improved using our representation.

2 Methods: Unsupervised MKL for Dimensionality Reduction

In the following, we describe how the MKL-based algorithm of [8] can be adapted to an unsupervised setting. Figure 1 outlines the pipeline of the proposed algorithm.

The starting point is the ambient space containing the input data, organized in a set of M features $\mathbf{x}_m$, $m \in [1, M]$. In the first step of MKL, a kernel-based affinity matrix $\mathbf{K}_m$ associated to each feature $\mathbf{x}_m$ is computed. We choose Gaussian kernels of the form: $\mathbf{K}_m(\mathbf{x}_{mi}, \mathbf{x}_{mj}) = \exp(-\|\mathbf{x}_{mi} - \mathbf{x}_{mj}\|^2 / 2\sigma_m^2)$, where $\mathbf{x}_{mi}$ stands for the feature m of subject i ($i \in [1, N]$) and σ_m for the kernel bandwidth, whose choice is discussed in Sect. 3.2. Additionally, a sparse global affinity matrix $\mathbf{W}$ (each of its elements W_{ij} codifying neighborhood membership) is calculated as $\mathbf{W} = \sum_{m=1}^{M} \hat{\mathbf{K}}_m$, where $\hat{\mathbf{K}}_m$ corresponds to the kernel $\mathbf{K}_m$ normalized by its variance. This serves to avoid that higher variability features have a larger impact in the neighborhood decision.

$\mathbf{W}$ and $\mathbf{K}_m$ are the inputs to the MKL block, which is the core of the algorithm. It is based on a Laplacian formulation [1] that ensures that subjects with similar characteristics (neighbors) remain close in the output space. In univariate graph embedding [11], the optimal embedding is obtained through the minimization of:

$$\min_{\mathbf{v}} \sum_{i,j=1}^{N} \|\mathbf{v}^\top \mathbf{x}_i - \mathbf{v}^\top \mathbf{x}_j\|^2 \mathbf{W}_{i,j}, \quad \text{s. t.} \quad \sum_{i=1}^{N} \|\mathbf{v}^\top \mathbf{x}_i\|^2 \mathbf{D}_{i,i} = 1, \tag{1}$$

where $\mathbf{D}$ is a diagonal matrix, whose values are the result of a row-wise summation of $\mathbf{W}$, N is the number of subjects, $\mathbf{x}_i$ is the value of the one and only descriptor (univariate setting) associated to subject i and $\mathbf{v}$ is the matrix that maps the input data into the output space. Notice that the values W_{ij} for neighbors are high, thus forcing their proximity in the embedding.

We adapt this formulation to combine multiple features in an unsupervised MKL framework [8], through the minimization of:

$$\min_{\mathbf{A},\boldsymbol{\beta}} \sum_{i,j=1}^{N} \|\mathbf{A}^\top \mathbb{K}^{(i)} \boldsymbol{\beta} - \mathbf{A}^\top \mathbb{K}^{(j)} \boldsymbol{\beta}\|^2 \mathbf{W}_{i,j}$$

$$\text{s.t.} \quad \sum_{i=1}^{N} \|\mathbf{A}^\top \mathbb{K}^{(i)} \boldsymbol{\beta}\|^2 \mathbf{D}_{i,i} = 1, \quad \beta_m \geq 0 \,, \forall m \in [1, M], \tag{2}$$

where $\mathbb{K}^{(i)}$ is defined for subject i as $\mathbb{K}^{(i)} = [\mathbf{K}_m(n,i)]_{(n,m)\in[1,N]\times[1,M]}$.

The unknowns in Eq. 2 are the matrix $\mathbf{A}$, which performs the final mapping to the output space, and the weights given to the different features $\boldsymbol{\beta} = [\beta_1 \ldots \beta_M]^\top$.

These values are calculated by means of a two-stage optimization strategy. The first stage of the optimization aims at optimizing $\mathbf{A}$, while $\boldsymbol{\beta}$ is fixed. This step is initialized by fixing $\beta_m = 1$, $\forall m \in [1, M]$. The problem turns out to be a generalized eigenvalue problem, with an explicit solution. The second stage of the optimization aims at optimizing $\boldsymbol{\beta}$, while fixing the previously calculated $\mathbf{A}$. This problem can be solved by quadratically constrained quadratic programming (QCQP), which is not straightforward. Nevertheless, it can be relaxed to a semi-definite programming problem, which can be solved more efficiently. In practice, this is addressed by the use of CVX, a package for solving convex programs [6]. Further details about the optimization can be found in [8].

The input samples are mapped to the output space by $\mathbf{y} = \mathbf{A}^\top \sum_{m=1}^{M} \mathbf{K}_m \beta_m$, where $\mathbf{y} = [\mathbf{y}_1; \mathbf{y}_2; \ldots; \mathbf{y}_N]$ gathers (row-wise) the coordinates of each input sample in the output space.

3   Experiments and Results

3.1   Echocardiographic Data

The method was applied to 2D sequences in a 4-chamber view from a stress echocardiography protocol using a semi-supine bicycle. Commercial software (EchoPAC, v.113, GE Healthcare, Milwaukee, WI) was used for information extraction. The database consisted of 33 subjects (21 healthy controls and 12 HFPEF subjects, diagnosed by Paulus' criteria [9]), with age 68 ± 6 years. The sequences were acquired at rest and during submaximal bicycle exercise (at a heart rate of 100–110 beats per minute, before fusion of the early and late diastolic (atrial) velocities of mitral inflow) [4]. Velocity patterns were extracted from myocardial velocity acquisitions, using a fixed sample (size 1×10 mm, located $\approx$ 10 mm above the mitral annulus at end-systole) at the basal septum and basal lateral wall of the left ventricle (LV). We consider that these regions are sufficient to account for the global longitudinal changes possibly present in the ventricles of the studied HFPEF subjects. The samples were kept fixed to minimize the variability that may appear when tracking the measured regions along the heart cycle, thus maintaining the acquisition as simple as possible. Case-per-case examination was performed to check that the sample area remained within the myocardial wall. The total number of extracted features was 4 (septal/lateral at rest/submaximal), referred to as *global analysis*, extendible to 16 for a *local analysis*, where different cardiac phases are treated independently (systole, isovolumic relaxation, early and late diastole). An example of the data extracted for a given subject is illustrated in Fig. 2. Class labels based on clinical diagnosis were provided together with the database. However, the subsequent learning was performed in an unsupervised way. Such labels were only used to compare the characterization of the learnt representation.

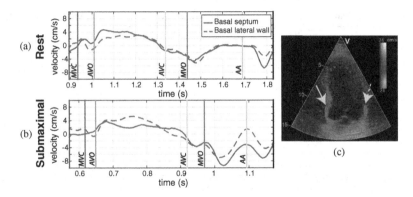

Fig. 2. Velocity patterns at rest (a) and submaximal (b) stages, corresponding to the snapshot (c) of a myocardial velocity acquisition from a sample subject. The patterns correspond to the basal septum and lateral wall regions, indicated by arrows in (c). Depicted physiological events: MVC (mitral valve closure), AVO (aortic valve opening), AVC (aortic valve closure), MVO (mitral valve opening) and AA (atrial activation).

Temporal Normalization. Subjects differ in terms of intrinsic heart rate and timing of cardiac phases (e.g. systole, diastole). This is even more pronounced between the rest and submaximal sequences of a given subject. In order to consistently compare subjects, their measurements need to match temporally. To this end, a two-stage normalization was applied. The first stage consisted in a piece-wise linear warping of the timescale, using physiological events normalized to the heart cycle. These events were: start and end of the heart cycle (onset of QRS on the ECG), mitral/aortic valves opening and closure (identified from Doppler flows, acquired separately but at similar heart rates) and atrial activation (onset of P-wave on the ECG). The timescales were redefined towards a common reference. This reference was selected from the healthy controls, as the one with the most central velocity patterns across a range of tests where each subject was considered successively as reference. The most central pattern is the one for which the sum of Euclidean distances to the remaining patterns is minimized. Then, after achieving a common temporal reference, the second stage consisted in resampling the velocity data to the new common temporal reference, through spline interpolation [2].

3.2 Parameters Tuning

The bandwidth σ_m of the kernel $\mathbf{K}_m$ was calculated feature-wise as the average of the pairwise Euclidean distances between each sample and its k-th nearest neighbor (looking at the corresponding feature). In this case $k = 11$. Then, the number of neighbors used to define the affinity matrix was fixed to 4. These two values were established heuristically, aiming at maximizing the spread of each class in the output space. Last, the number of iterations of the optimization was determined according to the convergence rate. In our implementation, it

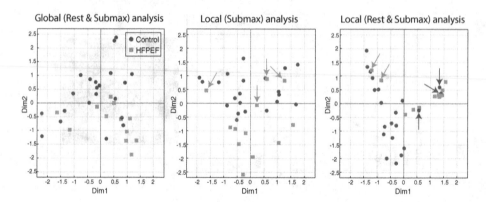

Fig. 3. Three examples of output space: (a) is the result of a *global* analysis, including rest and submaximal stages, (b) corresponds to a *local* analysis at submaximal stage and (c) is obtained by a *local* analysis at rest and submaximal stages jointly. When clusters are intuited, "misclassified" subjects are pointed with arrows.

was fixed to 3 iterations, since at this point the algorithm had converged to a final solution (noticed by the asymptotic decrease of the difference between the optimized variables among consecutive iterations). Unlike approaches using large databases of very different images [8], we think that the relative similarity of the studied features (physiological data in a coherent population) may contribute to this fast convergence.

3.3   Algorithm Output and Discussion of the Results

Learnt Representation. From the N-dimensional output space provided by the MKL algorithm, the two first dimensions (associated to the largest eigenvalues) are kept to form a representation that reduces the dimensionality of the complex ambient space and that is consistent with the phenomenon under study. In this particular application, we search to characterize the HFPEF syndrome, which mainly presents diastolic abnormalities in the LV. In an ideal representation, subjects would be mapped class-wise and clear clinically-defined clusters would stand out. Previous works [5] specifically target the recovery of a clustered space by integrating support vector machine techniques to the MKL framework. In contrast, in our application, the uncertainty about the definition of clear clinical clusters is high. Thus, we prefer to keep the problem unsupervised. This also leaves the door open for further exploitation of the data distribution in the output space, such as variability analysis or metrics computation.

Features Combination. We evaluated the contribution of using multiple descriptors consistent with the disease, against a single descriptor approach. *Global* and *local* temporal analyses were performed, taking into account the entire pattern or temporal windows corresponding to different cardiac phases, respectively. Figure 3 provides an example of three output spaces obtained by: (a) *global* analysis including rest and submaximal stages, (b) *local* analysis at submaximal stage, and (c) *local* analysis fusing rest and submaximal stages. In these experiments, septal and lateral regions were jointly analyzed, corresponding to rows 11, 10 and 12 in Table 1, respectively. This table summarizes some tested configurations with the available data.

Suggested Labeling. The number of "misclassified" subjects (second-to-last column in Table 1) serves as an indicator of the accuracy of the representation characterizing the HFPEF syndrome. It is qualitatively determined by visual inspection of the learnt representation by: first, recognizing class-related clusters from the output space; then, identifying subjects lying in a "wrong" cluster based on the clinical labels (indicated by arrows in Fig. 3). Empty values in some experiments indicate the absence of distinguishable regions within the representation. To support the qualitative interpretation, sensitivity and specificity were calculated (last column in Table 1), considering clinical decision as the gold standard. Notice that the highest sensitivity corresponds to the most exhaustive analysis.

The results of Table 1 suggest that a *local* analysis is more accurate than a *global* analysis in all cases. By splitting the pattern into physiological temporal windows (systole, iso-volumic relaxation, early and late diastole), the weights estimated by the algorithm point out their relative importance (early diastole the most prominent), which coincides with the physiological features of interest discussed in the literature [4]. Besides, by analyzing jointly different regions (basal septum and basal lateral wall) and different stages (rest and submaximal) the obtained representation is more explanatory. Better defined regions emerge, consistently with the class provided with the database. Finally, we observed that analyzing basal lateral wall alone was not sufficient to differentiate healthy from diseased subjects (for this reason these experiments are not shown in Table 1).

Clinical Potential. According to the requirements of clinical practice, the algorithm organizes the data such that a grading appears between healthy and diseased subjects. Referring to the best representation (Fig. 3c and last row in Table 1), the "misclassified" subjects pointed with green arrows were confirmed to be outliers by a clinically experienced observer. These subjects presented clinical conditions different from HFPEF, which were found to be: left bundle branch block (LBBB), right ventricular dysfunction (that affects the LV). These conditions may or may not be reflected in the velocity pattern, thus to characterize them properly, meaningful complementary descriptors should be added to the study. It is worth mentioning that although experiment 10 (Table 1) seems to be the best in terms of "misclassified" subjects, two of them do not lie in their corresponding region (being false negatives). On the contrary, no significant class-wise differences based on E/e' were reported in the literature [4].

Table 1. Different launches tested on the available data. Symbol "-" indicates the absence of distinguishable regions within the representation. "Misc." stands for misclassified and "Sens./Spec." stand for sensitivity and specificity.

Test	Region	Stage	Analysis	Misc	Sens./Spec. (%)
1	Septal	Rest	*Global*	-	-
2	Septal	Rest	*Local*	-	-
3	Septal	Submax	*Global*	-	-
4	Septal	Submax	*Local*	-	-
5	Septal	Rest + Submax	*Global*	7	50.0/95.2
6	Septal	Rest + Submax	*Local*	6	58.3/95.2
7	Septal + Lateral	Rest	*Global*	-	-
8	Septal + Lateral	Rest	*Local*	6	66.6/90.5
9	Septal + Lateral	Submax	*Global*	-	-
10	Septal + Lateral	Submax	*Local*	4	66.6/100.0
11	Septal + Lateral	Rest + Submax	*Global*	-	-
12	Septal + Lateral	Rest + Submax	*Local*	5	83.3/85.7

4   Conclusions

We presented a method for quantitatively characterizing myocardial velocities in a population. Its major strength lies in its ability to jointly analyze information, which is demonstrated to improve its explanatory capability. In the context of HFPEF, the method succeeded in differentiating healthy from diseased subjects, which contrasts with the poor performance of simpler analyses, such as those currently led in the clinical practice involving E/e' ratio. By generating hypotheses for discriminative variables, the method has high potential to reach a better understanding of the HFPEF syndrome, which is highly challenging due to its heterogeneity. Perspectives for the method consist of taking advantage of the lower-dimensional representation to quantitatively support clinical diagnosis on a broader dataset.

Acknowledgements. The authors acknowledge European Union 7th Framework Programme (VP2HF FP7-2013-611823) and the Spanish Ministry of Economy and Competitiveness (TIN2012-35874). The work of S. Sanchez-Martinez was supported by a fellowship from "la Caixa" Banking Foundation.

References

1. Belkin, M., Niyogi, P.: Laplacian eigenmaps for dimensionality reduction and data representation. Neural Comput. **15**, 1373–1396 (2003). MIT Press
2. Duchateau, N., De Craene, M., Piella, G., et al.: A spatiotemporal statistical atlas of motion for the quantification of abnormal myocardial tissue velocities. Med. Image Anal. **15**, 316–328 (2011)

3. Duchateau, N., De Craene, M., Piella, G., et al.: Constrained manifold learning for the characterization of pathological deviations from normality. Med. Image Anal. **16**, 1532–1549 (2012)

4. Erdei, T., Smiseth, O., Marino, P., et al.: A systematic review of diastolic stress tests in heart failure with preserved ejection fraction, with proposals from the EU-FP7 MEDIA study group. Eur. J. Heart Fail. **16**, 1345–1361 (2014)

5. Gönen, M.: Supervised multiple kernel embedding for learning predictive subspaces. IEEE Trans. Knowl. Data. Eng. **25**, 2381–2389 (2013)

6. Grant, M., Boyd, S.: CVX: matlab software for disciplined convex programming, version 2.0 beta, September 2013. http://cvxr.com/cvx

7. Komajda, M., Lam, C.: Heart failure with preserved ejection fraction: a clinical dilemma. Eur. Heart J. **35**, 1022–1032 (2014)

8. Lin, Y., Liu, T., Fuh, C.: Multiple kernel learning for dimensionality reduction. IEEE Trans. Pattern Anal. Mach. Intell. **33**, 1–14 (2011)

9. Paulus, W., Tschöpe, C., Sanderson, J., et al.: How to diagnose diastolic heart failure: a consensus statement on the diagnosis of heart failure with normal left ventricular ejection fraction by the heart failure and echocardiography associations of the european society of cardiology. Eur. Heart J. **28**, 2539–2550 (2007)

10. Senni, M., Paulus, W., Gavazzi, A., et al.: New strategies for heart failure with preserved ejection fraction: the importance of targeted therapies for heart failure phenotypes. Eur. Heart J. **35**(40), 2797–2811 (2014)

11. Yan, S., Xu, D., Zhang, B., et al.: Graph embedding and extensions: a general framework for dimensionality reduction. IEEE Trans. Pattern Anal. Mach. Intell. **29**, 40–51 (2007)

Quantitative Analysis of Lead Position vs. Correction of Electrical Dyssynchrony in an Experimental Model of LBBB/CRT

David Soto-Iglesias[1]([✉]), Nicolas Duchateau[3], Constantine Butakoff[1],
David Andreu[2], Juan Fernández-Armenta[2], Bart Bijnens[1], Antonio Berruezo[2],
Marta Sitges[2], and Oscar Camara[1]

[1] PhySense, DTIC, Universitat Pompeu Fabra, Barcelona, Spain
david.soto@upf.edu
[2] Cardiology Department, Thorax Institute, University of Barcelona, Hospital Clinic,
Barcelona, Spain
[3] Asclepios Research Project, Inria Sophia Antipolis, Sophia Antipolis, France

Abstract. Cardiac resynchronization therapy (CRT) is a recommended treatment in patients with electrical dyssynchrony such as left bundle branch block (LBBB). The determination of the optimal leads position, and the quantification of the changes in electrical activation are two current major challenges. In this paper, we investigate these aspects through electroanatomical data from a controlled experimental protocol, which studied pigs with no structural disease under LBBB and CRT conditions. We propose to use a quasi-conformal mapping technique to standardize electroanatomical maps of endo- and epi-cardial walls of both ventricles to a common reference geometry, in which simple quantitative indices can be computed. Then, we investigate the relation between leads and simple surrogates of the recovery of the electrical activation based either on total activation times or on the spatial distribution of the patterns. Our methodology allows a better understanding of the complex electrical activation patterns in LBBB and CRT, and confirms hypotheses about the optimal leads position from previous studies.

Keywords: Electro-anatomical mapping · LBBB · CRT

1 Introduction

Cardiac resynchronization therapy (CRT) is a treatment used to improve cardiac pump function in patients with heart failure, by correcting the electrical activation of the heart and then making it contract synchronously. However, the high rate of non-responders, around 30 %, and the associated cost are still issues of primary concern [1]. For CRT optimization some studies aimed at identifying the best placement of the CRT leads, but results are contradictory: while one study suggested to avoid apical locations [2], others reported no significant differences [3,4]. A recent study [5] evaluated the optimal location for the left

© Springer International Publishing Switzerland 2015
H. van Assen et al. (Eds.): FIMH 2015, LNCS 9126, pp. 74–82, 2015.
DOI: 10.1007/978-3-319-20309-6_9

ventricle (LV) lead through in-silico experiments. Its results suggest that the LV lead should be placed distant from both the septum and scar regions. Further comparisons [6] between in-silico, experimental models and real data also confirmed that biventricular pacing is useful for recovering the electrical dyssynchrony between RV and LV ventricles. Despite these advances, the optimal lead placement is still highly controversial.

Electro-anatomical maps (EAMs) acquired with electrophysiological mapping systems are well established into the daily clinical routine for guiding radio-frequency ablation therapies e.g., in the treatment of different arrhythmias. EAMs provide local electrical activation information fused with a 3D representation of the anatomical area that is mapped. This information helps improving the understanding of the electrical activation patterns in different cardiac pathologies and subsequent treatment, such as left bundle branch block (LBBB) and CRT pacing, through subject-specific observations. Nevertheless EAMs are difficult to interpret for non-experienced observers. Moreover, it is not straighforward to compare, quantify and perform population-based statistics with data corresponding to different subjects or at different stages of the disease.

Several researchers have recently proposed mapping techniques to construct 2D reference systems for different organs such as the left ventricle [7], the atria [8], vertebral bones [9] or faces [10]. In this paper, we use a quasi-conformal mapping technique that allows creating a 2D reference map for EAMs of the endocardial and epicardial walls of the heart, including both ventricles. The approach is inspired from previous works [11] that used it in an experimental swine model of LBBB-induced electrical dyssynchrony, implanted with a CRT device. However, such works only considered the LV endocardium. In the present study, the joint analysis of the endo- and epi-cardial walls allows understanding aspects of the transmurality of the electrical activation. Furthermore, in order to quantitatively interpret these data, we propose a set of simple indices directly measured from the activation maps. Through these indices, we provide insights into the links between lead location and the correction of electrical dyssynchrony, which confirm previous hypotheses about the optimal leads location.

2   Materials

2.1   Protocol

A total of 9 pigs (median weight of 34 kg) were studied from a dataset described in [12]. Animal handling was approved by the Ethics Committee at Anonymous hospital and conformed to international guidelines [12].

LBBB was induced in all animals using radio-frequency ablation, assisted by high rate pacing (160 bpm) during the burning process to prevent from ventricular fibrillation. Further details about the protocol can be found in [12]. A CRT device (bipolar right ventricle (RV) lead, model Beflex RF45; Sorin Group) was then positioned in the animal model: the RV lead was placed at the apex while the LV lead was positioned through a lateral coronary vein or through a lateral or anterolateral position of the LV epicardial surface, depending on the accessibility of the animal's anatomy.

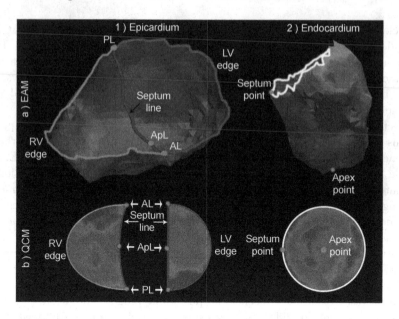

Fig. 1. (1) Epicardial EAM and reference 2D space. Septum line, anterior, (AL), apical (ApL) and posterior (PL) landmarks, respectively. (2) Endocardial EAM and reference 2D space.

2.2   Acquisition of Electroanatomical Maps (EAMs)

Contact mapping data was collected with an electroanatomical mapping device (CARTO-XP, Biosense Webster, Diamond Bar, CA) after introducing a 3.5-mm tip catheter (Thermo-Cool Navi-Star, Biosense Webster) through the femoral artery until reaching the endocardium of the LV by the retrograde aortic access. Then, using the same catheter and system, an epicardial map of the LV and RV was performed.

EAMs were subsequently generated by reconstructing information from the set of sparse acquisition points, where the clinician placed the catheter to measure the electrical activity (around 2500 ms at 1 kHz). These measurements were processed and rendered in 3D to display the Local Activation Time (LAT), which is the parameter retained in this study to investigate electrical activation patterns. All maps were acquired on the day of the experiment, at baseline, post-ablation (LBBB) and with CRT pacing. The average number of acquired points were 171 ± 66 and 255 ± 76, for endocardium and epicardium, respectively.

3   Methods

We introduce a two-step approach to jointly analyze different EAMs looking for relations between the status of the heart, electrical dyssynchrony and CRT leads position. First, we construct a common reference 2D frame using a quasi-conformal mapping. This has the advantage of standardizing both endo- and epi-cardial anatomies, thus facilitating the interpretation of corresponding EAMs. Then, we develop a set of both local and global quantitative measures to study standardized EAM data.

3.1 Quasi-conformal Mapping of Endo- and Epi-cardial Walls

In [11] we developed a quasi-conformal mapping (QCM) technique that takes advantage of the existence of an homeomorphism between the LV endocardial surface and a 2D disk. It was used to construct a reference common space for analyzing LV endocardial EAMs from different subjects. Here, we extend this technique to also incorporate information about the epicardium, i.e. including the external wall of the left and right ventricles of the heart. As can be seen in Fig. 1, epicardial EAMs only provide information of the LV lateral wall and RV free wall. Thus data on the septal wall are missing.

Since, we want to compare endo- with epicardial walls, we should not map the epicardium to a disk as in the LV endocardium data. Instead, we map the LV epicardium to the right half of a 2D disk, respecting the 17 AHA segment definition, as shown in Fig. 1. For the RV epicardium, we present a half-moon shape reference space as previously defined in [13].

The QCM of both heart walls are then computed by forcing every vertex coordinate to have a vanishing Laplacian, as shown in [11]. For the epicardium we only need to set the landmarks that provide correspondences between the EAM data and the 2D reference space. First, we define the septum line that divides LV and RV over the epicardial EAM mesh. To do so, three landmarks are manually selected: one over the anterior edge (AL) of the mesh (Fig. 1): one over the posterior (PL) one; and finally one in an apical (ApL) position. Then, we define the septum line in the 3D EAM mesh as the geodesic line over the mesh that joins these landmarks. In the 2D reference space, the anterior landmark is placed at the $(1,0)$ point and the posterior to the $(-1,0)$ one (see top and bottom points, respectively, in Fig. 1). Then the septum line is defined as a geodesic line linking these two landmarks, being the LV free wall mapped onto the contour of half disk. Finally, a similar approach is used for the RV epicardial mapping disk. It has the advantage of preserving relative distances between the boundary points at the original 3D EAM mesh.

3.2 Local and Global Quantitative Measures

EAMs provide rich information of electrical patterns of the heart, such as the time when every region of the heart is activated (LAT). Nevertheless EAMs are often analyzed globally by computing the total activation time (time difference between the start and end of activation). There is a need for more local measures of EAMs to better characterize the underlying electrical patterns with different therapy conditions. Here, we propose to derive indices from histograms of the isochrones in EAMs to better quantify electrical dyssynchrony of the endo-cardial LV free wall and the epicardium of both ventricles, as well as getting insights into the transmurality of the activation. Such indices are then compared to total activation times and to an apicality index representing lead position.

Total Activation Times (TATs). Since EAM data from both endo- and epi-cardial walls are already mapped onto the common reference space, it is straightforward to compute regional total activation times (for the LV, for the RV, for both ventricles) and to derive some indices comparing TATs at different stages of the experimental model. Therefore we define the following TAT-derived indices: the delay induced by LBBB is captured as the difference of the TATs between LBBB and baseline ($\Delta_{base} = TAT_{LBBB} - TAT_{baseline}$); the effect of the CRT therapy is quantified as the difference of the TAT between LBBB and CRT ($\Delta_{CRT} = TAT_{LBBB} - TAT_{CRT}$); and a recovery index expressing how close to the TAT after CRT is from its baseline value ($Recovery = (\Delta_{base} - \Delta_{CRT}) * 100$).

Indices from Histograms of Isochrones. The isochrones are defined for each EAM as the set of points with LATs within a certain range of the total TAT. The analysis of these isochrones can help on the interpretation of electrical patterns by studying the percentages of activated areas at a given timepoint. LAT values were previously normalized between 0 and 100 % of the total TAT. Ranges of 5 % are considered for the isochrones, corresponding to a total of 20 bins. An example of such histograms is shown in Figs. 2 and 3.

Apicality Index. For each lead, the apicality index is defined as: $Apicality = 1 - r$, where r is the distance between the lead and the center of the disk in polar coordinates. Leads position are identified from the CRT activation map using the k-means algorithm [14]. This served to identify the centers of two distinguishable clouds of activation within the initial 10ms isochrones of the epicardium EAM, already mapped into the 2D common reference space. With this formulation, an apicality index closer to 1 or zero would mean an apical or basal lead, respectively.

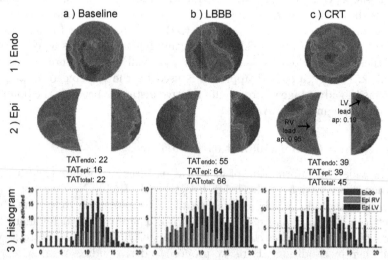

Fig. 2. Single case (#2) from group 1 (basal LV lead + apical RV lead). Rows: QCM of endocardial and epicardial EAMs, and histogram of isochrones, at each stage.

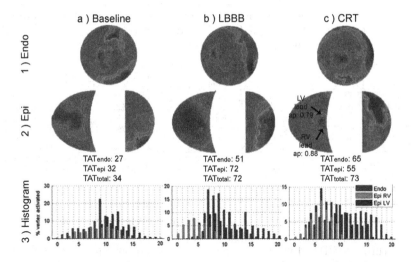

Fig. 3. Single case (#9) from group 4 (basal LV lead + apical RV lead). Rows: QCM of endocardial and epicardial EAMs, and histogram of isochrones, at each stage.

4   Results

Between baseline and LBBB, the QRS width (measured on the ECG signals of the electro-anatomical system) significantly increased (median value from 55 ms to 84 ms, respectively; p = 0.005).

The TAT values computed from the EAMs are summarized for each case in Table 1. Results are ordered according to four subgroups (G1-G4, different colours in Table 1). Such groups were created based on the apicality index, as follows: G1 (purple), *basal LV lead and apical RV lead*; G2 (red), *mid LV lead and apical RV lead*; G3 (green), *both mid LV and RV leads*; G4 (yellow), *both apical LV and RV leads*. The behaviour of a couple of cases belonging to groups G1 and G4 is illustrated in Figs. 2 and 3, respectively.

The *Recovery* index (Sect. 3.2), which is computed from TATs, was used to compare the outcome of these subgroups. Average recoveries for subgroups 1, 2 and 3 were > 23 %, 17 %, < 16 %, respectively. The two animals in G4 had a

Table 1. Endo- and epi-cardial TATs (ms) at each stage of the protocol. Leads configuration subgroups are highlighted by the color code: G1, purple; G2, red; G3, green; G4, yellow. Corresponding electrical recovery is also indicated.

Pig	Baseline					LBBB					CRT					Recovery
	LV		RV	Total	Total	LV		RV	Total	Total	LV		RV	Total	Total	
	Endo	Epi	Epi	Epi	Heart	Endo	Epi	Epi	Epi	Heart	Endo	Epi	Epi	Epi	Heart	%
1	43	20	21	24	43	52	67	53	82	82	43	66	40	67	67	38.46
2	22	16	8	16	22	55	45	51	64	66	39	39	32	39	45	47.73
3	35	25	18	25	36	98	49	42	69	98	55	63	70	70	70	45.16
4	47	23	24	24	47	63	72	58	94	94	67	83	78	83	83	23.40
5	30	X	X	X	X	52	58	30	58	65	48	57	60	61	61	17.39
6	37	18	19	23	37	61	36	32	49	63	46	56	48	59	59	15.38
7	42	33	23	33	42	81	48	42	70	82	79	58	37	58	79	7.5
8	32	27	25	30	43	38	54	35	64	64	53	46	44	47	54	47.62
9	27	22	22	32	34	51	55	31	72	72	65	46	47	55	73	-2.63

Table 2. Bin number corresponding to the isochrone within which the activation starts/ends, at baseline, LBBB and CRT. The two last columns indicate the differences between the initial and final number of 5 % isochrones.

Pig	Baseline			Difference		LBBB			Difference		CRT			Difference	
	LV		RV	LV	Epi	LV		RV	LV	Epi	LV		RV	LV	Epi
	Endo	Epi	Epi	Endo-Epi	LV-RV	Endo	Epi	Epi	Endo-Epi	LV-RV	Endo	Epi	Epi	Endo-Epi	LV-RV
1	0/20	3/13	2/11	3/-7	1/2	5/18	3/20	0/13	-2/2	3/7	2/16	0/20	0/12	-2/4	0/8
2	0/20	6/19	7/14	6/-1	-1/5	3/20	6/20	0/16	3/0	6/4	2/20	0/17	1/16	-2/-3	-1/1
3	1/20	1/14	0/11	0/-4	1/3	0/20	6/16	1/9	6/-4	5/7	5/20	1/20	0/20	-4/0	1/0
4	0/20	6/16	6/16	6/-4	0/0	3/17	5/20	0/11	2/3	5/9	3/19	0/20	0/18	-3/1	0/2
5	X	X	X	X	X	4/20	0/18	0/9	-4/-2	0/9	2/20	0/19	0/20	-2/-1	0/-1
6	0/20	10/18	8/16	10/-2	2/2	1/20	5/16	0/11	4/-4	5/5	5/19	0/20	5/20	-5/1	-5/0
7	0/18	3/20	3/12	3/2	0/8	0/20	9/20	3/13	9/0	6/7	0/20	2/13	3/17	2/-7	-1/-4
8	0/15	7/19	9/20	7/4	-2/-1	7/18	3/20	0/11	-4/2	3/7	0/20	0/17	0/16	0/-3	0/1
9	0/15	9/20	2/14	9/5	7/6	3/16	5/20	0/9	2/4	5/11	2/20	2/16	0/13	0/-4	2/3

recovery of -2.63 % and 47.62 %. The substantial difference in the recovery of G4 subjects, and the fact that the TAT of one of these animals lies within the range for G1, point out the limits of a global index such as TAT to quantify electrical recovery through CRT. In contrast, the simple isochrones indexes we propose partially exploit the spatial distribution of the activation patterns, which may help improving the assessment of electrical recovery.

An histogram of isochrones is computed for each endo- and epicardial EAM. Table 2 lists the bin number within which the electrical activation starts and finalizes for each EAMs at baseline, LBBB and CRT. Additionally, transmural and inter-ventricular differences between these bins are also shown. These *difference* columns provide insights into the restoration of the synchronization of RV and LV after CRT. Three subjects in G1 achieved a quasi-complete recovery in the initial and final bins, but the #1 did not recover the final one. G3, we can observe that the LV and RV are not completely synchronized after CRT. G4, subject number #8 improves its electrical synchronization, while subject #9 does not completely. Figure 3 provides elements of explanation for these electrical behaviours: both LV and RV leads are at close locations at the apex of the RV, inducing that the RV epicardium is activated two bins before the LV one, and also finalizing before the activation. Differences with the electrical pattern of a positive responder from G1 are visible in Fig. 2.3.c.

5   Discussion and Conclusions

The presented apicality index allows characterizing CRT leads positioning from EAM data. Our findings confirm the optimal positions suggested by previous studies [5], namely that optimal pacing sites should target one lead at the apex and the other lead distant from it. Global indices such as TATs seem limited for the assessment of electrical recovery, while simple indices that integrate the spatial distribution of the activation complement our understanding.

The histogram of isochrones introduced in this work allows a qualitative analysis of the electrical activity in both ventricles. Complementary information in Figs. 2 and 3 help identifying the LBBB-induced dyssynchrony, which is presented both transmurally (between LV endo- and epi-cardium) and intercavity (LV and RV epicardium) due to the LBBB. The correction resulting from CRT can also be assessed. By looking at large changes in the slope of the histograms, the entrance onto the Cardiac Conduction System (CCS) via the Purkinje network can also be identified (e.g. in Fig. 3 the entrance onto the CCS is produced at the bin #7 of the histogram).

One needs to be conscious of the sensibility of the method to the location of the landmarks used for the mapping, and the difficulty of acquiring EAMs in real patients with such a distributed sampling of the acquired points. However, the value of our study resides in allowing a better understanding of the complex mechanisms of electrical activation, in a LBBB/CRT protocol where external factors are controlled. The effect of damaged regions (e.g. regional infarct) on both the activation patterns and the optimal lead placement will be studied in further work.

Acknowledgments. This study was partially founded by the Spanish Ministry of Science and Innovation (TIN2011-28067) and the Spanish Industrial and Technological Development Center (cvREMOD-CEN-20091044). The authors gratefully acknowledge the support of N Solanes, M Rigol, E Silva, A Doltra, L Gabrielli, L Mont, J Brugada (Hospital Clnic, Barcelona, Spain) and A Barcelo (Sorin Group, Barcelona, Spain) on the experimental protocol.

References

1. Bleeker, G., et al.: Clinical versus echocardiographic parameters to assess response to cardiac resynchronization therapy. Am. J. Cardiol. **97**, 260–263 (2006)
2. Singh, J., et al.: Left ventricular lead position and clinical outcome in the multicenter automatic defibrillator implantation trial-cardiac resynchronization therapy (MADIT-CRT) trial. Circulation **123**, 1159–1166 (2011)
3. Gold, M., et al.: Comparison of stimulation sites within left ventricular veins on the acute hemodynamic effects of cardiac resynchronization therapy. Heart Rhythm **2**, 376–381 (2005)
4. Saxon, L., et al.: Influence of left ventricular lead location on outcomes in the COMPANION study. J. Cardiovasc. Electrophysiol. **20**, 764–768 (2009)
5. Huntjens, P., et al.: Influence of left ventricular lead position relative to scar location on response to cardiac resynchronization therapy: a model study. Europace **16**, 62–68 (2014)
6. Lumens, J., et al.: Comparative electromechanical and hemodynamic effects of left ventricular and biventricular pacing in dyssynchronous heart failure. JACC **62**, 2395–2403 (2013)
7. Soto-Iglesias, D., Butakoff, C., Andreu, D., Fernández-Armenta, J., Berruezo, A., Camara, O.: Evaluation of different mapping techniques for the integration of electro-anatomical voltage and imaging data of the left ventricle. In: Ourselin, S., Rueckert, D., Smith, N. (eds.) FIMH 2013. LNCS, vol. 7945, pp. 391–399. Springer, Heidelberg (2013)
8. Karim, R., et al.: Surface flattening of the human left atrium and proof-of-concept clinical applications. Comput. Med. Imaging Graph. **38**(4), 251–266 (2014)
9. Lam, K.C., et al.: Genus-one surface registration via teichmüller extremal mapping. MICCAI **3**, 25–32 (2014)

10. Zeng, W., et al.: Ricci flow for 3D shape analysis. IEEE PAMI **32**, 662–677 (2010)
11. Soto-Iglesias, D., et al.: Analyzing electrical patterns in an experimental swine model of dyssynchrony and CRT. In: IEEE Computing in Cardiology (CINC 2013), vol. 40, pp. 623–626 (2013)
12. Rigol, M., et al.: Development of a swine model of left bundle branch block for experimental studies of cardiac resynchronization therapy. J. Cardiovasc. Transl. Res. **6**(4), 616–622 (2013)
13. Jing, L., et al.: Patients with repaired tetralogy of fallot suffer from intra- and inter-ventricular cardiac dyssynchrony: a cardiac magnetic resonance study. J. Cardiovasc. Imaging **12**, 1333–1343 (2014)
14. MacQueen, J.: Some methods for classification and analysis of multivariate observations. In: Proceedings of the Fifth BSMSP, Statistics, vol. 1, pp. 281–297 (1967)

Principal Component Analysis for the Classification of Cardiac Motion Abnormalities Based on Echocardiographic Strain and Strain Rate Imaging

Mahdi Tabassian[1,2（✉）], Martino Alessandrini[2], Luca De Marchi[1],
Guido Masetti[1], Nicholas Cauwenberghs[3], Tatiana Kouznetsova[3],
and Jan D'hooge[2]

[1] Department of Electrical, Electronic and Information Engineering,
University of Bologna, Bologna, Italy
mahdi.tabassian2@unibo.it
[2] Department of Cardiovascular Sciences, Cardiovascular Imaging and Dynamics
Group, KU Leuven, Leuven, Belgium
[3] Department of Cardiovascular Sciences, Research Unit of Hypertension
and Cardiovascular Epidemiology, KU Leuven, Leuven, Belgium

Abstract. Clinical value of the quantitative assessment of regional myocardial function through segmental strain and strain rate has already been demonstrated. Traditional methods for diagnosing heart diseases are based on values extracted at specific time points during the cardiac cycle, known as 'techno-markers', and as a consequence they may fail to provide an appropriate description of the strain (rate) characteristics. This study concerns the statistical analysis of the whole cardiac cycle by the Principal Component Analysis (PCA) method and modeling the major patterns of the strain (rate) curves. Experimental outcomes show that the PCA features can outperform their traditional counterparts in categorizing healthy and infarcted myocardial segments and are able to drive considerable benefit to a classification system by properly modeling the complex structure of the strain rate traces.

Keywords: Strain/strain rate classification · Principal Component Analysis · Feature extraction

1 Introduction

Echocardiography is the modality of choice in clinical diagnostics and for the noninvasive assessment of heart function. In daily clinical practice, visual evaluation is widely used to determine regional abnormalities in myocardial wall motion. Although this qualitative assessment can be done easily, it suffers from the inter-observer variability which reduces its clinical value.

Tissue Doppler imaging and speckle tracking are two promising echocardiographic techniques that have been developed for the noninvasive study of

© Springer International Publishing Switzerland 2015
H. van Assen et al. (Eds.): FIMH 2015, LNCS 9126, pp. 83–90, 2015.
DOI: 10.1007/978-3-319-20309-6_10

myocardial function. Based on these techniques, strain (rate) imaging has been introduced to provide an effective approach for the assessment of changes in the regional myocardial wall motion deformation [6]. End-systolic strain and peak-systolic strain rate values are two traditional techno-markers that have been extensively used by clinicians to describe the strain (rate) profiles and to classify different heart diseases [9]. These traditional features, however, ignore the diastolic period of the cardiac cycle. They also represent the value of the strain (rate) profile at only one time point and as a result, cannot capture the temporal information available in the deformation curves.

Despite several studies for the classification of regional myocardial function based on the traditional features of the strain (rate) curves [7,9], only a few investigations have been carried out for the detection of heart abnormalities by taking into account the whole temporal behavior of the strain (rate) curves. The idea of analyzing the whole strain profiles, derived from tagged magnetic resonance imaging, by PCA was initially proposed in [3]. The authors showed that the statistical reference model that achieved by employing PCA and normal strain curves can properly detect abnormal strain patterns. Inspired by [3], PCA has been used in [1] to model ultrasonic strain and strain rate traces of the healthy subjects and it has been discussed that the PCA attributes can provide more information about strain (rate) curves than the traditional features. In [10], the artificial neural network (ANN) was used to classify strain profiles obtained at baseline and during experimentally induced acute ischemia using animal data. In a pre-processing step, each strain curve was represented by 70 equidistant samples and normalized in amplitude. The obtained profiles were then given to an ANN for categorization.

Following the results presented in [1], this paper addresses the subsequent issues: (i) building two PCA models by making use of the normal and acute infarcted strain and strain rate curves, (ii) incorporating the PCA and traditional features in a classification system and examining their capabilities for the categorization of normal and acute infarcted strain and strain rate traces and, (iii) comparing the PCA features extracted from the strain and strain rate traces in terms of the amount of the discriminatory information that they provide for a classification system.

The rest of the paper is organized as follows. Section 2 presents the details of the data acquisition procedure, reviews the PCA method and introduces the employed classifiers. In Sect. 3, the outcomes of the PCA implementation and the classification phase are presented. Discussion about the obtained results are given in Sect. 4. Finally, Sect. 5 draws conclusions and summarizes the paper.

2   Materials and Methods

2.1   Data Acquisition and Preprocessing

A group of 27 normal subjects and 54 subjects with acute myocardial infarction was used in this study. For the patients, myocardial segments were categorized into infarct, border and remote based on MRI-delayed enhancement and the perfusion territory of the infarct-related vessel [9]. Data acquisition was performed at high frame rate ($>180s^{-1}$) with a GE VingMed Vivid7 equipped with a 2.5 MHz transducer. For each subject, data were acquired in the apical 2-, 3- and 4-chamber views with optimization of the pulse repetition frequency in order to avoid aliasing. An event-driven graphical user interface called SPEQLE [4] was used for the post-processing of the data to extract longitudinal strain (rate) traces in an 18-segment model of the left ventricle [2]. Since the number of samples of the extracted curves could be different due to the differences in the heart rates of the subjects, a linear interpolation procedure was adopted to have the same number of samples in all traces. To avoid unwanted changes of the curves due to the interpolation procedure, each of the six cardiac phases (i.e. electromechanical coupling, isovolumetric contraction, ejection, isovolumetric relaxation, early filling and late filling [6]) was interpolated separately and then merged to have the whole heart cycle. The interpolated curves were then used in the PCA implementation and classification phases.

Table 1 lists the number of subjects that were selected randomly from the healthy and pathological groups for building training, validation and test sets. This random selection was repeated 10 times and the results presented in Sect. 3 are the average of running the classifier on these 10 different sets of data. Since for a pathological subject only the subset of acutely infarcted segments was used for the classification task, the number of utilized pathological subjects in Table 1 was more than the healthy ones so that both groups had roughly the same number of curves in the training, validation and test sets. Note that, the segmental strain and strain rate curves were sorted in two different groups of training, validation and test data to study their clinical relevance for discriminating normal and infarcted traces.

Table 1. Number of subjects taken from the healthy and pathological groups for the training, validation and test sets

	#Training	#Validation	#Test
Healthy subjects	12	5	10
Pathological subjects	25	9	20

2.2 Principal Component Analysis

PCA is a popular statistical approach for feature extraction, dimensionality reduction and data visualization [8,12]. Given a data set of random vectors $X = \{x_1 \ldots x_N\}$ where $x_i \in \mathbb{R}^n$, the PCA algorithm gives a representation of the data in $\mathbb{R}^m$ ($m < n$) such that the new variables are less redundant compared to the original ones.

The first step in building the PCA model is *centering* the data that can be done by first computing the mean vector of X and then subtracting every data vector from it. In the second step, the covariance matrix of the centralized data set is computed. The third step is to find the $n \times n$ eigenvector matrix $\Phi = \{\phi_1 \ldots \phi_n\}$ and diagonal eigenvalue matrix $\Lambda = diag\{\lambda_1 \ldots \lambda_n\}$ of the covariance matrix. The final step is to insert the first m eigenvectors of Φ with the largest eigenvalues, which are known as Principal Components (PCs), into a new matrix Ψ and project the data onto the space spanned by the eigenvectors of Ψ,

$$Y = \Psi^T \overline{X} \tag{1}$$

where the variance of the new low-dimensional data Y is maximized which means that the first axis has the largest variance, the second axis has the second largest variance and so on.

2.3 Classifier

Performance of a classification system depends on both the features extracted from the data and the classification technique. In order to differentiate between the effects of the employed features and the classification strategy on the final classification outcomes, two different classifiers are used in our experiments. The first classifier is called the *locan-mean based* (LMB) method [11] and considers the information of classes around a test sample for its classification. The second classifier is *support vector machine* (SVM) [5]. It works based on the idea of the maximum margin solution and finds a hyper-plane that has the greatest distance to the training samples in the boundaries of a binary classification problem.

3 Results

3.1 PCA Outcomes

The segmental strain and strain rate traces of the training sets were used to build two separate PCA models. Figures 1 and 2 illustrate the first three PCs of the strain and strain rate traces, respectively. The variance percentages accounted for the PCs and the results of adding and subtracting the PCs to and from the mean curves are also showed. In order to investigate the PCs' structures in terms of timing of the six mechanical phases of the cardiac cycle, the timing of the onset of each phase is shown with a red vertical line.

3.2 Classification Outcomes

Average classification results on the test data for the LMB and SVM are shown in Fig. 3 and Table 3, respectively. Table 2 gives the best average classification accuracies and their corresponding sensitivity and specificity values obtained by the LMB and selecting K nearest neighbors (KNNs) from [1 100] interval. The number of PCs was set to 10 based on the favorable results gained with the validation data.

4 Discussion

PCA can be considered as a linear model for approximating a set of data in which the main structures of the data are captured by the PCs and their eigenvalues

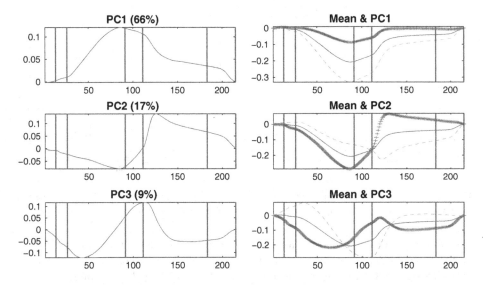

Fig. 1. The first three PCs of the strain curves, their contributions in the covariance matrix and the results of adding (red ++) and subtracting (green - -) them to and from the mean strain profile (Color figure online)

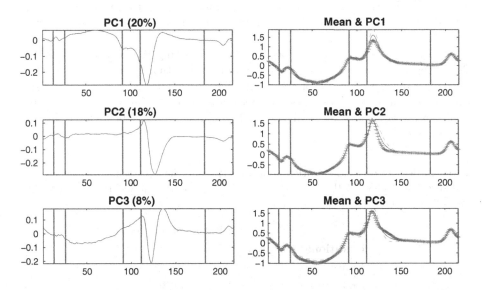

Fig. 2. The first three PCs of the strain rate curves, their contributions in the covariance matrix and the results of adding (red ++) and subtracting (green - -) them to and from the mean strain rate profile (Color figure online)

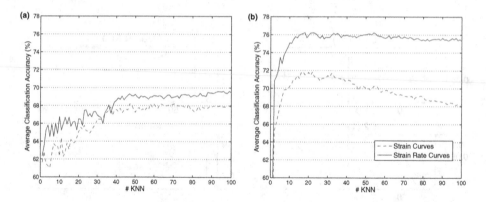

Fig. 3. Average classification accuracies (%) on the test strain and strain rate curves with the LMB method using (a) the traditional features and, (b) the PCA features

[12]. Figures 1 and 2 show that for both the strain and strain rate curves, PC1s and the mean curves are very similar in timing and shape except that they have opposite signs which is trivial in linear modeling and can be compensated by multiplying the weights of those PCs by –1. This similarity implies that the first PCs, which have noticeable contributions in data modeling, capture the average patterns of the strain and strain rate curves. The second and third PCs of the strain and strain rate curves, however, reflect important variations during early diastole and model more complicated variations around the mean. As an instance, while the first PC of the strain traces follows the mean curve, the second and third PCs yield to one and two intersections with the mean curve, respectively.

A key observation that can be made by comparing the percentages of the PCs belonging to the strain and strain rate curves is that in the first case, 92 % of the variances accounted for the whole eigenvectors are captured by the first three PCs while in the latter case this percentage is 46 %. This indicates that the strain rate traces have more diverse and complex temporal behavior than the strain curves.

Table 2. Best average classification accuracies (%) and their corresponding sensitivity and specificity values of the LMB method with the traditional and PCA features on the test strain and strain rate curves

	Accuracy	Sensitivity	Specificity
Traditional features			
Strain	68.22	67.51	68.86
Strain rate	69.63	73.35	66.35
PCA features			
Strain	71.86	64.98	78.90
Strain rate	76.26	69.93	83.44

Table 3. Average classification accuracies (%) and their sensitivity and specificity values of the SVM method with the traditional and PCA features on the test strain and strain rate curves

	Accuracy	Sensitivity	Specificity
Traditional features			
Strain	69.61	55.14	83.39
Strain rate	71.17	67.23	75.56
PCA features			
Strain	70.06	73.41	67.07
Strain rate	74.32	77.03	72.28

Tables 2 and 3 show that regardless of the classification methodology, the PCA features yielded to more satisfactory accuracy rates for both the strain and strain rate traces compared to the traditional features. This means that by exploring the whole cardiac cycle instead of only the systolic period, considerably more information about the characteristics of the strain and strain rate curves can be obtained. Another key point that is highlighted in Tables 2 and 3 is the difference between the classification accuracies obtained with features extracted from the strain and strain rate traces. While by using the PCA features, the classification accuracies on the strain rate curves are remarkably better than those of the strain traces, there is not much benefit to be gained by employing the traditional features and the accuracies obtained for the strain rate traces are slightly better than the strain curves. This observation suggests that the strain rate traces provide more discriminatory information than the strain curves and these extra information could be captured by an efficient statistical model like PCA.

Tables 2 and 3 also show that for both the PCA and traditional features, the sensitivity and specificity rates achieved by the LMB and SVM varied considerably. It means that the classification strategy has a direct effect on the final outcomes and suggests that different classification methodologies should be examined to find a suitable setup. These results also imply that by combining several classifiers that address the classification problem from different points of view, a good compromise between the sensitivity and specificity values could be obtained that is a topic for the future research.

5   Conclusion

In this study, it was hypothesized that the temporal behavior of the segmental strain (rate) curves contains valuable diagnostic information which can be captured by a rigorous statistical approach. The PCA method was then employed to statistically analyze the strain (rate) traces. In order to evaluate the usefulness of the PCA features, they were compared with the end-systolic strain and peak-systolic strain rate values as the traditional features. Experiments with a data set of strain (rate) curves of healthy and pathological subjects demonstrated that the PCs can provide more discriminatory information for the classification system than the traditional features. Our experiments also showed that analyzing the strain rate traces with PCA would lead to better results than the strain curves. However, more thorough analysis with a larger set of data is needed to improve the classification performance and to determine the role of different parameters that affect the obtained outcomes.

References

1. Aoued, F., Eroglu, E., Herbots, L., Rademakers, F., D'hooge, J.: A statistical model-based approach for the detection of abnormal cardiac deformation. In: Ultrasonics Symposium, vol. 1, pp. 512–515. IEEE (2005)
2. Cerqueira, M.D., Weissman, N.J., Dilsizian, V., Jacobs, A.K., Kaul, S., Laskey, W.K., Pennell, D.J., Rumberger, J.A., Ryan, T., Verani, M.S.: Standardized myocardial segmentation and nomenclature for tomographic imaging of the heart: a statement for healthcare professionals from the Cardiac Imaging Committee of the Council on Clinical Cardiology of the American Heart Association. Circulation 105, 539–542 (2002)
3. Clarysse, P., Han, M., Croisille, P., Magnin, I.: Exploratory analysis of the spatio-temporal deformation of the myocardium during systole from tagged MRI. IEEE Trans. Biomed. Eng. 11, 1328–1339 (2002)
4. Claus, P., D'hooge, J., Langeland, T.M., Bijnens, B., Sutherland, G.R.: SPEQLE (Software Package for Echocardiographic Quantification LEuven) an integrated approach to ultrasound-based cardiac deformation quantification. In: Computers in Cardiology, vol. 29, pp. 69–72. IEEE (2002)
5. Cristianini, N., Shawe-Taylore, J.: An Introduction to Support Vector Machines. Cambridge University Press, Cambridge (2000)
6. D'hooge, J., Bijnens, B., Thoen, J., Van de Werf, F., Sutherland, G., Suetens, P.: Echocardiographic strain and strain-rate imaging: a new tool to study regional myocardial function. IEEE Trans. Med. Imaging 21(9), 1022–1030 (2002)
7. Jamal, F., Kukulski, T., Sutherland, G.R., Weidemann, F., D'hooge, J., Bijnens, B., Derumeaux, G.: Can changes in systolic longitudinal deformation quantify regional myocardial function after an acute infarction? an ultrasonic strain rate and strain study. J. Am. Soc. Echocardiogr. 15(7), 723–730 (2002)
8. Jolliffe, I.T.: Principal Component Analysis, 2nd edn. Springer, New York (2002)
9. Herbots, L., D'hooge, J., Eroglu, E., Thijs, D., Ganame, J., Claus, P., Dubois, C., Theunissen, K., Bogaert, J., Dens, J., Kalantzi, M., Dymarkowski, S., Bijnens, B., Belmans, A., Boogaerts, M., Sutherland, G., Van de Werf, F., Rademakers, F., Janssens, S.: Improved regional function after autologous bone marrow-derived stem cell transfer in patients with acute myocardial infarction: a randomized, double-blind strain rate imaging study. Eur. Heart J. 30, 662–670 (2009)
10. McMahona, E.M., Korinekb, J., Yoshifukub, S., Senguptaa, P.P., Manducab, A., Belohlaveka, M.: Classification of acute myocardial ischemia by artificial neural network using echocardiographic strain waveforms. Comput. Biol. Med. 38, 416–424 (2008)
11. Mitani, Y., Hamamoto, Y.: A local mean-based nonparametric classifier. Pattern Recogn. Lett. 27(10), 1151–1159 (2006)
12. Wold, S., Esbensen, K., Geladi, P.: Principal Component Analysis. Chemometr. Intell. Lab. Syst. 2, 37–52 (1987)

Prediction of Clinical Information from Cardiac MRI Using Manifold Learning

Haiyan Wang[1], Wenzhe Shi[1], Wenjia Bai[1],
Antonio M. Simoes Monteiro de Marvao[2], Timothy J.W. Dawes[2],
Declan P. O'Regan[2], Philip Edwards[1], Stuart Cook[2], and Daniel Rueckert[1](✉)

[1] Biomedical Image Analysis Group, Imperial College London, London, UK
D.Rueckert@imperial.ac.uk
[2] Hammersmith Hospital, Imperial College London, London, UK

Abstract. Cardiac MR imaging contains rich information that can be used to investigate the anatomy and function of the heart. In this paper, we demonstrate that it is possible to learn anatomical and functional information from cardiac MR imaging without explicit segmentation in order to predict clinical variables such as blood pressure with high accuracy. To learn the anatomical variations, we build manifolds of different time points across different subjects. In addition, we investigate two different approaches to incorporate motion information into a manifold, and compare these manifolds to a manifold learned from a single time point. Combining both inter- and intra-subject variation, we are able to construct accurate and reliable classifiers to predict clinical variables. Our proposed method does not require any explicit image segmentation and motion estimation and is able to predict clinical variables with good accuracy.

1 Introduction

Cardiac MR images contain rich information that can be used to reveal important information about the anatomy and function of the heart. Given a large number of subjects, the shape, appearance and motion of the heart can vary significantly across subjects due to age, gender, blood pressure or disease progression. Finding an appropriate representation of a large dataset of cardiac MRI images in order to extract meaningful features has gained increasing attention and importance in recent years. The most popular approaches include atlasing techniques [6–9]. However, these atlasing techniques require image segmentation and motion estimation in order to extract the desired information. The process is usually computationally intensive and depends heavily on the accuracy and robustness of the image analysis techniques.

Dimensionality reduction techniques such as manifold learning simplify the data so that it can be efficiently processed and interpreted. The Euclidean distance between images is more meaningful in the low-dimensional manifold space [1]. Manifold learning has seen increasing use in the analysis of medical images

© Springer International Publishing Switzerland 2015
H. van Assen et al. (Eds.): FIMH 2015, LNCS 9126, pp. 91–98, 2015.
DOI: 10.1007/978-3-319-20309-6_11

ranging from the segmentation of brain MR images [15], the detection of respiratory motion in ultrasound [12], to study of regional appearance pattern in cardiac and brain MRI images [3] and classification of abnormal myocardial motion [5].

In this work, we propose to learn anatomical and functional information directly from cardiac MR images using manifold learning. Learning cardiac function directly from images is increasingly attractive due to its efficiency, which has been applied to estimate the cardiac ventricular volumes [14,16]. In general, manifold learning techniques can be divided into local and global methods: Local linear embedding (LLE) [10] and Laplacian Eigenmaps [2] embedding aim to preserve local distances between data points. Global methods like Isomap consider the distance between all pairs of data points in its objective function, hence preserves global distances. For our study, we are more interested in preserving the inter-subject and intra-subject variations across the data at the same time. Laplacian Eigenmaps have the advantage of generating feature coordinates that are not only dependent upon the original pairwise similarity between them, but implicitly emphasizes the natural clusters in the dataset. This makes the obtained embedding more robust [2].

We hypothesize that the anatomical information can be embedded in the differences between images of the same phase in the cardiac cycle across different subjects and that the functional information can be embedded in the differences between different cardiac phases of the same subject. To exploit both anatomical and functional information from cardiac MR images, we first learn a low dimensional manifold of a large number of cardiac MR images at a single time point to describe inter-subject variation. Then we consider two alternative approaches to incorporate motion information into a manifold. Both are compared to the manifold learned from a single time point. Finally, we demonstrate that by using different combinations of information, we are able to predict different clinical variables with good accuracy.

The remainder of the paper is organised as follows: Sect. 2 explains the methods used for derive different forms of manifold for incorporating motion information. Section 3 presents the results. Finally, Sect. 4 concludes the work.

2   Methods and Materials

2.1   Materials and Image Pre-processing

The dataset used in this paper consists of 209 subjects, of which 95 are male and 114 are female; 35 subjects with high blood pressure (systolic/diastolic pressure >140/90 mmHg on separate occasions) and 174 with normal blood pressure. Images were obtained on a 1.5T Philips Achieva system with voxel sizes of 2.0×2.0 mm; slice thickness 2 mm and 20 cardiac phases.

A region of interest (ROI) is defined by five manually placed landmarks in the ED phase of the MR image sequence. These five landmarks are carefully placed by clinicians and include two insertion points of the right ventricle, the apex point and two longitudinal LV axis points (one in the most basal slice and one in the most apical slice). Each image is then transformed by aligning the landmarks to a randomly chosen reference image using rigid registration. This spatial normalisation is performed to correct for variations in position and orientation across the subjects but to preserve variations due to size differences. The pre-processing is kept as simple as possible in order to avoid the loss of any useful information.

2.2   Manifold Learning

Manifold learning is used for dimensionality reduction: in the original, high-dimensional space each set of images $I = \{I_1, I_2, ..., I_n\} \in \mathbb{R}^M$ represents a data point of dimension M, where M is the number of voxels in each image. Assuming there exists a lower dimension manifold which can well represent the local geometric relationship between images, the manifold learning seeks to find a low dimensional representation $Y = \{y_1, y_2, ..., y_n\} \in \mathbb{R}^m$ of the input images I where $m \ll M$. Laplacian Eigenmaps [2] build a sparsely connected graph from a pairwise weight matrix W computed from the data set. The connectivity between each pair of data points in the graph is defined by a similarity measure S. In our work we use the sum of squared difference (SSD) as similarity measure.

We investigate three different ways of constructing manifold: a single manifold, a pooled manifold and an aligned manifold.

- To preserve the inter-subject variation and extract anatomical information, first a low dimensional manifold is derived from a single time-point. In this work, we use the ED phase as this single time point but other choices are possible. We refer to this as *single manifold*.
- To preserve the inter- and intra-subject variation, which captures functional and motion information in cardiac MR images, we perform the manifold learning on the cardiac MR image sequence. For this purpose, we propose two different approaches:
  - *Pooled manifold.* All time frames from the same subject are simultaneously embedded together with their ED phase images.
  - *Aligned manifold.* Previous studies revealed that for the pooled manifold important intra-subject variation can be lost in the embedding while the inter-subject variation dominates the manifold [4]. Therefore, we derive an individual manifold for each individual cardiac MR image sequences. These manifold are then aligned into the final manifold using Procrustes analysis [13].

In order to investigate that how many frames we need from the cardiac sequence to capture functional information, we derive the aligned manifold from images with different temporal resolutions. The single manifold can be seen as the manifold with a temporal resolution of one image during a cardiac cycle. We then increase the temporal resolution to two frames (the ED and ES frames) and so on. In this work, we gradually increase the temporal resolution from one frame to twenty frames per cardiac cycle and build aligned manifold for each different temporal resolution respectively.

2.3   Classification

The representation obtained through the manifold learning allows the embedding of inter- and intra-subject variation of cardiac anatomy and function into a lower dimensional space. This makes it possible to use the embedding coordinates y_i as the feature vector to analyze the population in terms of cardiac function. We investigate the capability of the extracted manifold coordinates by testing two classification scenarios: (a) male vs. female and (b) high blood pressure vs. normal. A support vector machine (SVM) approach [11] is used for the classification.

3   Experiments and Results

3.1   Impact of the Parameter Setting

Selecting Value for Neighbourhood Size and Feature Dimension. The neighbourhood size (k) and feature dimension (d) are two important parameters in the Laplacian eigenmap embedding. The correct choice of these two parameters is important to obtain meaningful representations of the data. In this work, these parameters are selected empirically using an exhaustive 2D grid search.

Figure 1 shows classification accuracies computed based on a single manifold within the parameter grid for classifying subjects according to gender and high vs normal blood pressure respectively. We show the average classification accuracies obtained after 100 runs of seven-fold cross-validation. As can be seen for the gender classification task, the accuracy initially increases sharply as k and d increase and reaches its peak at $k = 45, d = 7$. The classification accuracy decays significantly for large d. By contrast, for the blood pressure classification task, the accuracy increases steadily and remains stable for $k > 10$ and $d > 15$.

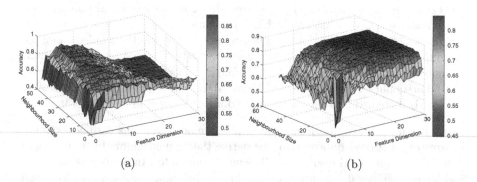

(a)   (b)

Fig. 1. Average classification accuracies computed based on the single manifold with the varying neighbourhood size and feature dimension for discriminating between (a) gender and (b) subject with high vs. normal blood pressure.

Selecting Temporal Resolution for Aligned-Manifold. In order to achieve the best classification rate and investigate the impact of neighbourhood size, feature dimension and temporal resolution for the pooled and aligned manifold, the optimal parameter settings are searched using grid search as well. The results for the classification according to blood pressure using the aligned manifold is reported in Fig. 2. 100 runs of seven-fold cross-validation are performed for each of the settings: neighbourhood size $k \in [2..50]$, feature dimension $d \in [1..20]$ and number of frames $f \in [1..20]$. The search range chosen for the neighbourhood size and feature dimension are based on the results from the previous section. Clear improvements in the classification accuracy can be observed in both figures when increasing any of the parameters. Thereafter the accuracy stabilizes with the best performance obtained at $k = 29, d = 4, f = 16$.

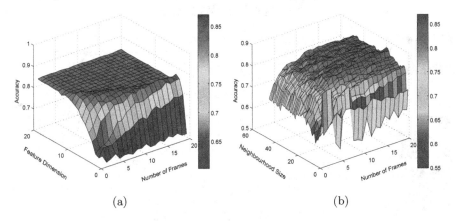

(a)   (b)

Fig. 2. The impact of (a) the temporal resolution f and feature dimension d, (b) neighbourhood size k and the temporal resolution f for classifying subjects with high and normal blood pressure using the aligned manifold.

3.2   Results

Visualising the Low Dimensional Embedding. The first two dimensions of the manifold embedding coordinates are selected for visualisation in Fig. 3. The single manifold illustrates a clear separation of subjects according to gender. In the pooled manifold, most frames from the same subject overlap with each other. This is due to the fact that the inter-subject variation in the dataset is much larger than the intra-subject [4]. The aligned manifold only shows 50 subjects for illustration purpose and the overlaid images show similar anatomical appearance with its neighbours.

Classification Results. The single manifold, pooled manifold and aligned manifold proposed in this work are used to select features for classification between the gender and between subjects with high and normal blood-pressure. The single manifold is learned from the ED images, while the pooled and aligned

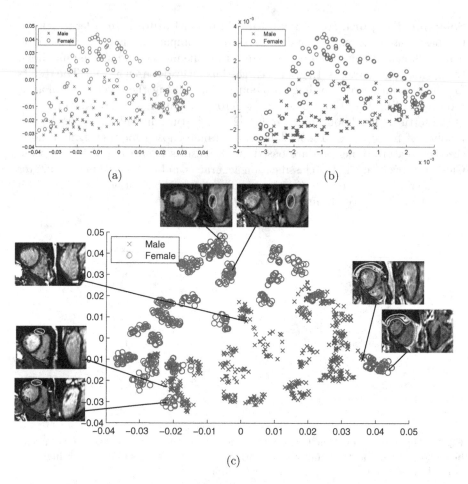

Fig. 3. 2D visualizations of manifolds labeled by genders, (a) single manifold, (b) pooled manifold with all frames for each subject and (c) aligned manifold with all frames for each subject. It can be observed that the frames from the same subject cluster together. The overlaid images show similar anatomical appearance with its neighbours (as annotated in yellow) (Color figure online).

manifolds are learned from different numbers of temporal frames. A SVM with a Gaussian radial basis function kernel is used for classification.

Table 1 reports the classification results based on the three types of manifolds for both classification tasks. For comparison, the classification rates obtained using clinical information such as LV volume, stroke volume and LV mass are also shown. It can be seen that the classifications accuracy using manifold learning coordinates as features outperforms the use of clinical information. Among the three types of manifolds, the aligned manifold has a distinct advantage over the other two manifold models for classification according to blood pressure. For gender discrimination, all the three manifold models reach the same accuracy.

Table 1. Comparison of classification results based on the different types of manifolds. For comparison, we have also investigated how well clinical measurements can be used to separate subjects.

	Gender	Blood pressure
single-manifold	0.89 ± 0.02	0.83 ± 0.01
pooled-manifold	0.89 ± 0.01	0.84 ± 0.01
aligned-manifold	0.89 ± 0.02	0.87 ± 0.01
LV volume (ED)	0.77 ± 0.01	0.57 ± 0.01
LV volume (ES)	0.75 ± 0.01	0.54 ± 0.02
Stroke volume	0.72 ± 0.02	0.58 ± 0.02
LV mass	0.85 ± 0.01	0.63 ± 0.01

4   Discussion and Conclusion

We successfully used manifold learning as a feature selection approach for a SVM-based classifier to analyse 209 cardiac MR image sequences. The SVM-based classifier directly operates on the manifold coordinates of the MR images without the need for any explicit image analysis. We also demonstrated that, by considering both inter- and intra-subject variation in the manifold learning, we are able to extract both anatomical and functional information. This can be used to construct powerful and reliable classifiers that are more predictive than global indices like LV volume and mass. The aligned manifold allows for investigating how much temporal information we need to help improving the classification performance.

Our experiments show that the neighbourhood size, feature dimension and amount of motion information have significant impact on the prediction results: for gender prediction additional temporal information and features beyond a certain point do not add useful information. This can be explained by the fact that gender is highly correlated to the volume of the heart but not to other properties of the myocardium. By contrast, blood pressure is closely related to both anatomy and function of the heart, and although including further additional manifold dimensions beyond a certain level has less of an influence on the classification result, temporal information is essential for accurate prediction. It is also interesting to see that the best classification performance is reached before using the full available temporal resolution.

The proposed approach is computationally efficient as no non-rigid registration or segmentation is required. The presented work so far has focused on healthy volunteers. Future work will investigate the prognostic capabilities of the proposed manifold learning method in diseases such as dilated cardiomyopathy and pulmonary arterial hypertension.

References

1. Aljabar, P., Wolz, R., Rueckert, D.: Manifold learning for medical image registration, segmentation, and classification. In: Machine Learning in Computer-aided Diagnosis: Medical Imaging Intelligence and Analysis. IGI Global (2012)
2. Belkin, M., Niyogi, P.: Laplacian eigenmaps for dimensionality reduction and data representation. Neural Comput. **15**(6), 1373–1396 (2003)
3. Bhatia, K., Rao, A., Price, A., Wolz, R., Hajnal, J., Rueckert, D.: Hierarchical manifold learning for regional image analysis. IEEE Trans. Med. Imaging **33**(2), 444–461 (2014)
4. Chang, W.-Y., Chen, C.-S., Hung, Y.-P.: Analyzing facial expression by fusing manifolds. In: Yagi, Y., Kang, S.B., Kweon, I.S., Zha, H. (eds.) ACCV 2007, Part II. LNCS, vol. 4844, pp. 621–630. Springer, Heidelberg (2007)
5. Duchateau, N., Craene, M.D., Piella, G., Frangi, A.F.: Constrained manifold learning for the characterization of pathological deviations from normality. Med. Image Anal. **16**(8), 1532–1549 (2012)
6. Duchateau, N., De Craene, M., Piella, G., Silva, E., Doltra, A., Sitges, M., Bijnens, B.H., Frangi, A.F.: A spatiotemporal statistical atlas of motion for the quantification of abnormal myocardial tissue velocities. Med. Image Anal. **15**(3), 316–328 (2011)
7. Fonseca, C.G., Backhaus, M., Bluemke, D.A., Britten, R.D., Do Chung, J., Cowan, B.R., Dinov, I.D., Finn, J.P., Hunter, P.J., Kadish, A.H.: The cardiac atlas projectan imaging database for computational modeling and statistical atlases of the heart. Bioinf. **27**(16), 2288–2295 (2011)
8. Hoogendoorn, C.: A Statistical Dynamic Cardiac Atlas for the Virtual Physiological Human: Construction and Application. Universitat Pompeu Fabra, Barcelona (2014)
9. Lombaert, H., Peyrat, J., Croisille, P., Rapacchi, S., Fanton, L., Cheriet, F., Clarysse, P., Magnin, I., Delingette, H., Ayache, N.: Human atlas of the cardiac fiber architecture: study on a healthy population. IEEE Trans. Med. Imaging **31**(7), 1436–1447 (2012)
10. Roweis, S.T., Saul, L.K.: Nonlinear dimensionality reduction by locally linear embedding. Science **290**(5500), 2323–2326 (2000)
11. Vapnik, V.: Statistical Learning Theory, 1st edn. Wiley, New York (1998)
12. Wachinger, C., Yigitsoy, M., Navab, N.: Manifold learning for image-based breathing gating with application to 4D ultrasound. In: Jiang, T., Navab, N., Pluim, J.P.W., Viergever, M.A. (eds.) MICCAI 2010, Part II. LNCS, vol. 6362, pp. 26–33. Springer, Heidelberg (2010)
13. Wang, C., Mahadevan, S.: Manifold alignment using procrustes analysis. In: Proceedings of the 25th international conference on Machine learning, pp. 1120–1127. ACM (2008)
14. Wang, Z., Ben Salah, M., Gu, B., Islam, A., Goela, A., Li, S.: Direct estimation of cardiac biventricular volumes with an adapted bayesian formulation. IEEE Trans. Biomed. Eng. **61**(4), 1251–1260 (2014)
15. Wolz, R., Aljabar, P., Hajnal, J., Hammers, A., Rueckert, D., Initi, A.D.N.: Leap: learning embeddings for atlas propagation. Neuroimage **49**(2), 1316–1325 (2010)
16. Zhen, X., Wang, Z., Islam, A., Bhaduri, M., Chan, I., Li, S.: Direct estimation of cardiac bi-ventricular volumes with regression forests. In: Golland, P., Hata, N., Barillot, C., Hornegger, J., Howe, R. (eds.) MICCAI 2014, Part II. LNCS, vol. 8674, pp. 586–593. Springer, Heidelberg (2014)

Revealing Differences in Anatomical Remodelling of the Systemic Right Ventricle

Ernesto Zacur[1]([⊠]), James Wong[2], Reza Razavi[2], Tal Geva[3],
Gerald Greil[2], and Pablo Lamata[1]

[1] Department of Biomedical Engineering, King's College London, London, UK
ernesto.zacur@kcl.ac.uk
[2] Department of Imaging Sciences, King's College London, London, UK
[3] Boston Children's Hospital, Harvard Medical School, Boston, MA, USA

Abstract. Cardiac remodelling, which refers to the change of the shape and size of the myocardium, is an adaptive response to developmental, disease and surgical processes. Traditional metrics of length, volume, aspect ratio or wall thickness are used in the clinic and in medical research, but have limited capabilities to describe complex structures such as the shape of cardiac ventricles. In this work we present an example of how computational analysis of cardiac anatomy can reveal more detailed description of developmental and remodelling patterns. The clinical problem is the analysis of the impact of two different surgical palliation techniques for hypoplastic left heart syndrome. Construction of a computational atlas and the statistical description of its variability are performed from the short axis stack of 128 subjects. Results unveil, for the first time in the literature, the differences in remodelling of the systemic right ventricle depending on the surgical palliation technique.

Keywords: Computational anatomy · Statistical shape analysis · Systemic right ventricle · Discriminative analysis

1 Introduction

Hypoplastic left heart syndrome (HLHS) is a congenital heart defect where a baby is born with the left ventricle severely underdeveloped. This condition requires a three-stage surgical palliation procedure that creates a circulation based on the two chambers of the right side of the heart that is compatible with life. The first stage (called Norwood procedure) is performed within the first weeks after birth. This stage aims to improve systemic perfusion by reconstructing the native aorta and arch, maintain pulmonary perfusion via insertion of a shunt and promote mixing of blood with an atrial septectomy. A subsequent stage of surgery (Stage II or Glenn procedure) is performed at ages 4 to 6 months, and seeks to form a superior cavo-pulmonary anastomosis in order to provide a more permanent supply of blood to the lungs. The final procedure (Stage III or Fontan procedure), performed at 2 to 4 years of age, involves a total cavo-pulmonary anastomosis in order to separate oxygenated and deoxygenated circulation.

One of the following two techniques can be used in the Norwood procedure (Stage I) as systemic-to-pulmonary artery shunt to supply blood to the lungs:

© Springer International Publishing Switzerland 2015
H. van Assen et al. (Eds.): FIMH 2015, LNCS 9126, pp. 99–107, 2015.
DOI: 10.1007/978-3-319-20309-6_12

- The modified Blalock-Taussig (MBT) shunt creates a connection between the innominate or subclavian artery and one of the branch pulmonary arteries. Some studies have associated the MBT shunt with haemodynamic instability and sudden unexpected death as continuous flow of blood into the low resistance pulmonary circulation during diastole may lead to under-perfusion of the coronary arteries.
- The right ventricle-to-pulmonary artery (RVPA) shunt (a.k.a. Sano shunt) involves the placement of a conduit between the pulmonary artery and the right ventricle. RVPA reduces the risk of under-perfusion as the circulation is primarily in systole. However, this technique results in a scar on the wall of the systemic ventricle and increases the risk of arrhythmias or aneurysmal dilatation of the outflow tract.

Recent evidence showed an association of the RVPA shunt with a higher transplantation free survival at a year of age, although no differences in measures of RV ejection fraction (EF) were found [5]. At 32 months of age, RV EF had deteriorated significantly in the RVPA shunt group and transplant free survival rates were indistinguishable between the both shunt strategies. At present, and despite these evidences, the direct impact of the ventriculotomy scar on the growth and motion of systemic RVs remains incompletely understood.

Cardiac magnetic resonance (CMR) provides detailed 3D descriptions of the RV [1], but applying traditional scalar measurements such as volumes, lengths, thickness or diameters may be insufficient to capture complex variations in the geometry of 3D shapes. However, tools from the Computational Anatomy discipline, in particular shape analysis from CMR images, allow the analysis, detection and location of small differences in ventricular geometries. These tools are now applied in the medical research field, revealing detailed anatomical patterns in diseases and several findings in relationships between anatomy and organ function [3, 10, 18].

In this study we examine differences in the anatomical remodelling of the systemic RV in patients with HLHS who have undergone MBT or RVPA shunt techniques. Patient anatomies were expressed as the geometrical shape of the RV, extracted from dynamic CMR images. All shapes of RVs are considered to share the same geometrical features, and the differences between these features can describe the population variability. We describe shape variability by means of spatial coordinates of homologous points among geometries. Statistical techniques of dimensionality reduction enable the identification of differences in the cardiac remodelling between the two shunt techniques.

2   Materials and Methods

2.1   Cohort and Image Acquisition

In order to investigate the anatomical cardiac remodelling regarding the surgical shunt technique, a cohort of 128 patients of HLHS were analysed. Patient population presents differences in the Norwood technique and in the stage of the procedure. Demographic information is presented in Table 1.

Datasets were acquired at 2 congenital cardiac centres; MBT from Evelina London Children's Hospital, UK (ethical approval 08/H0810/058), while RVPA patients' data were provided by Boston Children's Hospital, USA (ethical approval IRB-P00012488). Datasets consist of balanced steady state free precession (bSSFP) cine imaging in short axis orientations acquired on a 1.5-T Philips Achieva MRI scanner according to recent guidelines [4]. Images were acquired with pixel size varying from 0.85 to 1.4 mm in the short axis plane and from 5 to 10 mm inter-slice separation. Typically, the myocardial occupies from 6 to 13 slices depending on patient's heart size and inter-slice gap. CMR scans were acquired after Stage I at age 70 to 160 days, after Stage II at 20 to 35 months of age, and after Stage III at 3 to 14 years of age. Babies were scanned under anaesthetic.

Table 1. Demographic data.

	MBT shunt	RVPA shunt
Stage I	30	20
Stage II	29	14
Stage III	32	3
Total	91	37
Age at Norwood	7.1 ± 6.7 days	5.3 ± 2.8 days

2.2   Image Segmentation and Mesh Personalization

The systemic RV was assumed to have an elliptical shape, similarly to a healthy left ventricle (LV), and therefore it includes the septal wall. RV myocardium of all short-axis cine stacks was manually contoured by an expert at end-diastole. Trabeculations were excluded from the segmentation.

Segmented myocardial anatomies were fitted to a template mesh formed by cubic Hermite bricks [17] using the methods described in [8, 9]. The mesh fitting algorithm uses all slices with anatomical information of the systemic RV. Truncation of the basal ventricular anatomy was made at the image acquisition set, when the expert operator decides the plane that captures the most basal slice of the ventricles. A limited flexibility is provided to the template by means of the allowed degrees of freedom of the template. This results in a regularized fitting of the manual segmentation, reducing segmentation errors from human operation or from its discretized nature. All 128 scans were personalized reaching an average distance no larger than 1.5 mm to the manual segmentation. After the fitting step, the surfaces of the Hermite elements were tessellated into a triangular mesh. As a result of this process, the systemic RV anatomy was described by a set of 2834 descriptive points and each of these points is assumed to be in anatomical correspondence among all anatomies.

Differences in postural information were removed by pre-aligning all cases: all geometries were rotated to set the apical-base direction in the vertical Z axis and the right to left direction (identified by the centre of the LV blood pool manually located) in the horizontal X axis. Translational offset was removed by aligning the centre of RV blood pool. Finally, size is normalized by uniform scaling such that all segmented myocardia have a volume equal to the geometric mean of the original volumes. Notice that we are including geometries of hearts of ages from 3 months to 14 years old and without the volume normalization the most important geometrical feature results to be the size. An alternative size normalization could be to consider clinical information such as subject's weight, height or age as was done in [11]. Nevertheless, in a preliminary study, we found that size is not a relevant remodelling pattern in this population. This also explains why traditional shape indexes (volume, mass) have not been able to identify differences between surgery procedures.

Even if CMR scans are from the same patient at different stages, the analyses performed in this work considered all the data independent as in a transversal study.

2.3   Statistical Description of the Data

A mean geometry was constructed from the 128 geometrical descriptors by averaging along the population the spatial coordinates of each descriptive point. This mean geometry (also called *anatomical atlas*) comprises the geometrical characteristics shared among all patients, see Fig. 1.

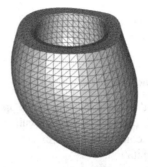

Fig. 1. Anatomical atlas (average shape of 128 cases) of the systemic right ventricle in HLHS. Shape coordinates characterise the differences in anatomy with respect to this mean shape.

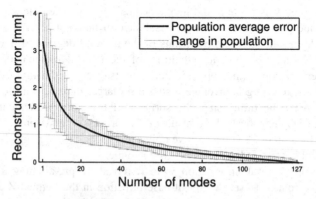

Fig. 2. Reconstruction error in terms of the number of PCA components. 16 PCA components are enough to reconstruct all instances within 1.5 mm of accuracy.

The residual geometrical information between the patients and the atlas is statistically analysed by means of principal components analysis (PCA) technique [6]. PCA performs an orthogonal linear transformation formed by the sequence of the *most expressive features* [14]. It is important to note that in the PCA space, the more coordinates are considered to reconstruct the data, the more accurate the reconstruction is. Figure 2 shows reconstruction errors among 128 descriptors and among the 2834 descriptive points. It can be noted that 127 PCA coordinates are required to represent the whole population (instead of the 2834×3 coordinates required in the Cartesian description). By considering only 16 PCA coordinates, all reconstructed instances are within 1.5 mm of accuracy (same as the mesh personalization process described before). The truncation of PCA coordinates can be also considered as a filtering of low-SNR additive noise in the data [12]. Figure 3 illustrates these concepts through a gradual reconstruction of a contour with increasing number of PCA components.

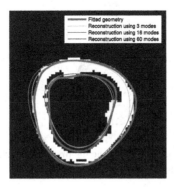

Fig. 3. Illustration of the mesh fitting and PCA reconstruction accuracy. It compares a segmented scan, subject to user and discretization/quantization/pixilation errors; the fitted geometry, stated to the regularization given in the fitting process; and corresponding reconstructed instances by taking into account different number of PCA components.

3   Qualitative Inspection and Quantitative Analysis

The principal modes in shape variation of the HLHS cohort, obtained by PCA, are illustrated in Fig. 4.

In order to perform a quantitative analysis of the discriminative properties of each mode, we performed a 'two-sample Student's t-test' [16] along each principal component. The Student's t-test has an associated p-value. This p-value indicates the degree of confidence that the instances of both groups correspond to random draws from two populations with different means. Without making strong assumptions on the underlying distribution of the data, the p-value can be empirically assessed with a randomized permutation test [7]. In a binary group discrimination task, the smaller the resulting p-value, the more distinguishable the groups are.

Specifically, we compute p-values between MBT *vs.* RVPA patient groups along the first 20 PCA coordinates. For each direction, the resulting p-value was computed by means of permutation test performed with 10^5 random permutations of the group label. The obtained p-value was used as an indication of the separability of the groups.

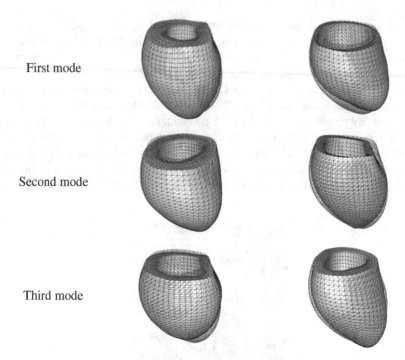

Fig. 4. Extremal shapes (*atlas* ± 2 std dev) along the first, second and third principal components directions. The atlas is shown in red to compare.

Figure 5 shows the *p*-values corresponding to *t*-test between subgroups MBT-Stage I *vs.* RVPA-Stage I and between subgroups MBT-Stage II *vs.* RVPA-Stage II. These results illustrate that groups are more similar between them at Stage I than at Stages II.

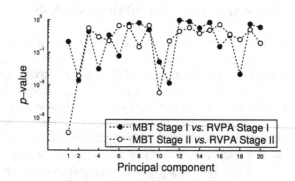

Fig. 5. *p*-values of the Student's *t*-test along each principal component. A significant difference at Stage II in mode 1 ($p < 0.001$) is revealed, without any other mode showing this level of significance at any of the two stages.

4 Revealing Anatomical Differences Between Surgical Techniques

It is expected that the impact of cardiac remodelling after the Norwood procedure increases along time, which is supported by results in Fig. 5. In order to illustrate the location of the most prominent remodelling differences between surgical techniques, a multivariate Hotelling's T2 test was performed for each descriptive point including its x,y,z coordinates [13]. Non-parametric permutation tests (with 10^4 random permutations) estimate the p-value of the T2 statistics. Subsequently, *false discovery rate* (FDR) correction [2] were used as control approach on simultaneous multiple comparisons. Figure 6 shows the corrected p-values obtained in the comparison between all-Stage I *vs.* MBT or RVPA at Stage II. This figure shows that the number of points with significant differences is much larger when Norwood procedure is performed by the RVPA than by the MBP shunt.

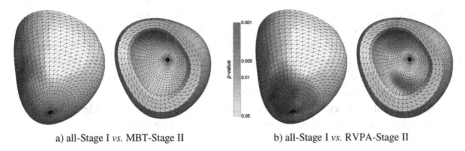

a) all-Stage I *vs.* MBT-Stage II   b) all-Stage I *vs.* RVPA-Stage II

Fig. 6. Anatomical location of remodelling differences (FDR corrected p-value of Hotelling's T2 test for each descriptive point): BMT sub-group shows an almost empty map of differences, suggesting a simple growth from Stage I to Stage II, but RVPA subgroup shows a large region with statistically significant changes, indicating a different remodelling pattern.

5 Discussion and Conclusions

The computational analysis of the systemic RV anatomy has revealed remodelling differences between the two shunt surgical techniques. This finding has never been reported using traditional geometrical indexes, and illustrates the added value of the proposed methodology.

RVPA shunt is considered a *more aggressive* surgical technique than MBP shunt. The necessary ventriculotomy affects directly to the myocardium muscle and larger cardiac remodelling effects are expected. Our results are in agreement with this consideration: the differences at Stage II between the RVPA and MBP subgroups are mainly originated by a different remodelling pattern of RVPA subgroup compared to a simple growth of MBP subjects, see Fig. 6.

Besides known findings obtained from correlations between traditional scalar measurements, such as volumes, lengths, thickness or diameters, [1] the use of high dimensional shape descriptions may improve our understanding of the cardiac remodelling. However, the use of such novel shape descriptions poses important challenges for its interpretation: a shape coefficient is a much more abstract and intangible number compared to traditional scalar measurements. We have dealt with this limitation by the illustration of the anatomical regions where changes are more significant, helping thus to extract valuable clinical conclusions from the data.

There are several limitations in this study. The construction of the computational atlas is limited to the available image resolution (a short axis stack). Full 3D resolution images would have revealed a higher level of anatomical detail. The automatic 3D mesh reconstruction of the systemic RV achieved a lower, although tolerable, accuracy compared to the adult healthy LV (average fitting error divided by average length: 1.43 mm/78 mm versus 1.28 mm/95 mm [10]), indicating the greater difficulty for manual segmentation and automatic reconstruction of a structure with thinner walls and larger shape variability.

HLHS is a condition with a very limited number of cases in the world (with a prevalence rate of two to three cases per 10,000 live births), and reaching statistical power from small study cohorts requires accurate biomarkers. Our results demonstrate the capability to quantify ventricular remodelling, which is a signature of the extra burden imposed to the systemic right ventricle. Moreover, in a recent parallel study we have identified the relationship between shape and function, where an RVPA shunt is also associated with an impaired contraction [15]. Shape metrics may thus help in understanding of the cardiac remodelling after the Norwood procedure. From a clinical viewpoint, shape metrics could improve the stratification and the planning of surgical procedures in subjects with HLHS, enable an early detection of maladaptive remodelling patterns, and guide the decisions for an optimal treatment.

Acknowledgements. This study has received funding by the Department of Health through the NIHR comprehensive Biomedical Research Centre award to Guy's & St Thomas' NHS Foundation Trust in partnership with King's College London and King's College Hospital NHS Foundation Trust, the Centre of Excellence in Medical Engineering (funded by the Wellcome Trust and EPSRC; grant number WT 088641/Z/09/Z) as well as the BHF Centre of Excellence (British Heart Foundation award RE/08/03). PL holds a Sir Henry Dale Fellowship funded jointly by the Wellcome Trust and the Royal Society (grant no. 099973/Z/12/Z).

References

1. Bellsham-Revell, H.R., Tibby, S.M., Bell, A.J., Witter, T., Simpson, J., Beerbaum, P., Anderson, D., Austin, C.B., Greil, G.F., Razavi, R.: Serial magnetic resonance imaging in hypoplastic left Heart syndrome gives valuable insight into ventricular and vascular adaptation. J. Am. Coll. Cardiol. **61**(5), 561–570 (2013)
2. Benjamini, Y., Hochberg, Y.: Controlling the false discovery rate: a practical and powerful approach to multiple testing. J. Roy. Stat. Soc.: Ser. B (Methodol.) **57**, 289–300 (1995)
3. Fonseca, C.G., Backhaus, M., Bluemke, D.A., Britten, R.D., Do Chung, J., Cowan, B.R., et al.: The Cardiac Atlas Project—an imaging database for computational modeling and statistical atlases of the heart. Bioinformatics **27**(16), 2288–2295 (2011)

4. Fratz, S., Chung, T., Greil, G.F., Samyn, M.M., Taylor, A.M., Valsangiacomo Buechel, E.R., Yoo, S.-J., Powell, A.J.: Guidelines and protocols for cardiovascular magnetic resonance in Children and Adults with congenital heart disease: SCMR expert consensus group on congenital Heart disease. J. Cardiovasc. Magn. Reson. **15**(1), 51 (2013)

5. Frommelt, P.C., Gerstenberger, E., Cnota, J.F., Cohen, M.S., Gorentz, J., Hill, K.D., et al.: Impact of initial shunt type on cardiac size and function in children with single right ventricle anomalies before the Fontan procedure: the single ventricle reconstruction extension trial. J. Am. Coll. Cardiol. **64**(19), 2026–2035 (2014)

6. Fukunaga, K.: Introduction to Statistical Pattern Recognition. Academic Press, San Diego (2013)

7. Good, P.I.: Permutation Tests: A Practical Guide to Resampling Methods for Testing Hypotheses. Springer, New York (2000). Springer Series in Statistics

8. Lamata, P., Niederer, S., Nordsletten, D., Barber, D.C., Roy, I., Hose, D.R., Smith, N.: An accurate, fast and robust method to generate patient-specific cubic Hermite meshes. Med. Image Anal. **15**(6), 801–813 (2011)

9. Lamata, P., Sinclair, M., Kerfoot, E., Lee, A., Crozier, A., Blazevic, B., et al.: An automatic service for the personalization of ventricular cardiac meshes. J. R. Soc. Interface **11**(91), 20131023 (2014)

10. Lewandowski, A.J., Augustine, D., Lamata, P., Davis, E.F., Lazdam, M., Francis, J., McCormick, K., et al.: Preterm Heart in adult life: cardiovascular magnetic resonance reveals distinct differences in left ventricular mass, geometry, and function. Circulation **127** (2), 197–206 (2013)

11. Medrano-Gracia, P., Cowan, B.R., Ambale-Venkatesh, B., Bluemke, D.A., Eng, J., et al.: Left ventricular shape variation in asymptomatic populations: the multi-ethnic study of atherosclerosis. J. Cardiovasc. Magn. Reson. **16**, 56 (2014)

12. Olmos, S., Garcia, J., Jané, R., Laguna, P.: ECG signal compression plus noise filtering with truncated orthogonal expansions. Sig. Process. **79**(1), 97–115 (1999)

13. Styner, M., Oguz, I., Xu, S., Pantazis, D., Gerig, G.: Statistical group differences in anatomical shape analysis using Hotelling T2 metric. In: Medical Imaging, International Society for Optics and Photonics, vol. 6512 (2007)

14. Swets, D., Weng, J.: Using discriminant Eigenfeatures for image retrieval. IEEE Trans. Pattern Anal. Mach. Intell. **18**, 831–836 (1996)

15. Wong, J., Lamata, P., Rathod, R.H., Bertaud, S., Dedieu, N., et al.: Using cardiac magnetic resonance and computational modelling to assess the systemic right ventricle following different Norwood procedures: a dual centre study. J. Cardiovasc. Magn. Reson. **17**(1), M12 (2015)

16. Woolson, R.F., Clarke, W.R.: Statistical Methods for the Analysis of Biomedical Data. Wiley, New Jersey (2011)

17. Young, A.A., Cowan, B.R., Thrupp, S.F., Hedley, W.J., Dell'Italia, L.J.: Left ventricular mass and volume: fast calculation with guide-point modeling on mr images. Radiology **216** (2), 597–602 (2000)

18. Zhang, X., Cowan, B.R., Bluemke, D.A., Finn, J.P., Fonseca, C.G., Kadish, A.H., et al.: Atlas-based quantification of cardiac remodeling due to myocardial infarction. PLoS ONE **9** (10), e110243 (2014)

Imaging

Assessment of Myofiber Orientation in High Resolution Phase-Contrast CT Images

V. Baličević[1]([✉]), S. Lončarić[1], R. Cárdenes[2], A. Gonzalez-Tendero[3,4],
B. Paun[2], F. Crispi[3,4], C. Butakoff[2], and B. Bijnens[2,5]

[1] Faculty of Electrical Engineering and Computing, University of Zagreb,
Zagreb, Croatia
vedrana.balicevic@fer.hr
[2] PhySense, N-RAS, Universitat Pompeu Fabra, Barcelona, Spain
[3] Fetal i+D Fetal Medicine Research Center, BCNatal - Barcelona Center
for Maternal-Fetal and Neonatal Medicine (Hospital Clinic and Hospital
Sant Joan de Deu), IDIBAPS, University of Barcelona, Barcelona, Spain
[4] Centre for Biomedical Research on Rare Diseases (CIBER-ER), Barcelona, Spain
[5] Institució Catalana de Recerca i Estudis Avançats (ICREA), Barcelona, Spain

Abstract. Complex helical organization of cardiac fibers is one of the
key factors for efficient beat-to-beat contraction and electrical impulse
propagation. Complete understanding of this (inter-individual) configuration is limited by image acquisition and analysis constraints.
Consequently, extensive quantification of myofiber orientation and
remodeling within diverse cases is still lacking. With its high resolution and contrast, synchrotron-based phase-contrast X-ray imaging offers
potential for assessing this information. Although it recently gained
increased attention for biomedical purposes, only few cardiac applications were presented to this date. In this paper, we used synchrotron-
based acquisitions of a healthy fetal rabbit heart and implemented a
structure tensor method for estimating fiber orientation. For comparison, we generated the common rule-based model, simulating fiber angles
distribution for the given geometry. Although we find similar fiber angle
transmural courses compared to the theoretical, high-resolution imaging
and analysis show that the myocardium in an individual is more complex
than often assumed.

Keywords: Myocytes arrangement · Fiber angle · Synchrotron imaging · Structure tensor · Streeter model

1 Introduction and Background

The efficiency of electrical and mechanical performance of a heart is substantially determined by its geometry and structure. The first discoveries of specific helical and locally laminar cardiac myofiber configuration were obtained
from dissections and histological measurements [1,2] and are still considered
the gold-standard. Thorough knowledge of the cardiac architecture opens up a

© Springer International Publishing Switzerland 2015
H. van Assen et al. (Eds.): FIMH 2015, LNCS 9126, pp. 111–119, 2015.
DOI: 10.1007/978-3-319-20309-6_13

series of applications: from detailed anatomical atlases of myofiber architecture, simulations of electrical wave propagation in cardiac muscle, towards patient-personalized applications [3,4]. For this, it is important to have individual information or at least know the variability in the architecture between hearts.

Nowadays, attempts towards assessing the individual myocardial fiber architecture are based on imaging of the heart. Most commonly it is the diffusion tensor magnetic resonance imaging (DT-MRI) [5–8], which employs the anisotropic water diffusion property of the tissue to determine its orientation. Approaches that similarly exploit other physical properties of the tissue involve ultrasound elastic tensor [9] or backscatter tensor imaging [10]. Aside from these, alternative methods for more directly extracting fiber orientation from the fibers appearance are applicable to high-resolution MRI [11], polarized light microscopy [12], confocal microscopy [13] and optical coherence tomography [14]. Within these methods, a good alternative to the most accepted diffusion tensor method as implemented in DT-MRI is found in deriving the structure tensor data from the fibers appearance. The equivalence of the two approaches comes from the property of the cardiac tissue that the largest diffusion coefficient corresponds to the direction with the smallest structural variation, and viceversa.

The limitation of existing acquisition techniques is in the trade-off between the resolution, the corresponding sample size and the noise sensitivity. Third generation synchrotron-based phase-contrast X-ray imaging offers a potential superior solution for imaging of cardiac fibers. We suggest that the structure tensor concept can be applied to phase-contrast acquisitions, obtaining accurate and reliable results for cardiac applications due to its resolution which is comparable to the cardiac myofiber dimensions. Similar approaches for the purpose of cardiac fiber orientation identification have recently been explored in acquisitions of fiber patches [15,16].

This paper is focused on analysing whole-heart high resolution phase-contrast images to quantify fiber distributions, based on the gradient structure tensor method and comparing it to the Streeter model of the expected cardiac fiber arrangement. The workflow of the approach is shown in Fig. 1.

2   Materials and Methods

2.1   Data Acquisition

An *ex vivo* healthy fetal rabbit heart was used. Preparation of the heart for imaging included formalin perfusion fixation and dehydration in ethanol followed by agarose embedding. Phase-contrast X-ray images were acquired at the European Synchrotron Radiation Facility (ESRF, beamline ID19; 19Kev). After reconstructing the projections, the images have an isotropic resolution of 7.43 μm and capture the whole volume of the isolated heart (approx. $1 \times 1 \times 2$ cm^3).

2.2   Data Preprocessing

Slicewise segmentation and labeling was performed by visual inspection, discriminating the left and right ventricle and excluding the surrounding container volume, non-tissue artifacts and cavities from the analysis.

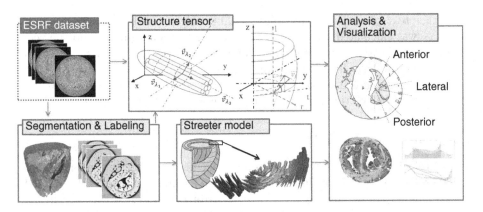

Fig. 1. Schematic display of the analysis performed on the phase-contrast dataset. The structure tensor box includes a scheme of a fiber system defined with eigenvectors $\mathbf{v}_\lambda$ obtained by tensor eigenanalysis, and the assumed cardiac cylindrical coordinate system defined with r-radial, t-tangential and l-longitudinal component. The Streeter model box displays the expected fiber orientation in a whole heart and a transmural patch. The evaluation box shows the sectors selected for quantification of results.

2.3 Gradient Structure Tensor

For calculating the myofiber orientation, a gradient structure tensor method was applied. For each voxel $\mathbf{x} = [x, y, z]$ in the image $I(\mathbf{x})$, the oriented gradient magnitudes in x, y and z-directions were obtained using a central difference algorithm, resulting in a gradient vector $\mathbf{g}(\mathbf{x}) := \nabla I(\mathbf{x}) = [g_x(\mathbf{x}), g_y(\mathbf{x}), g_z(\mathbf{x})]$. The structure tensor in a voxel is defined as $\mathbf{T}(\mathbf{x}) := \nabla I(\mathbf{x}) \nabla I(\mathbf{x})^T$, however, here information is gathered over a voxel's cubical neighborhood $N(\mathbf{x})$ as a linear combination of the belonging structure tensors, resulting in an local structure tensor estimation $\hat{\mathbf{T}}(\mathbf{x}) = \sum_{\mathbf{p} \in N(\mathbf{p})} \mathbf{T}(\mathbf{p})$, which is in a matrix notation:

$$\hat{\mathbf{T}} = \begin{bmatrix} \sum g_x{}^2 & \sum g_x g_y & \sum g_x g_z \\ \sum g_y g_x & \sum g_y{}^2 & \sum g_y g_z \\ \sum g_z g_x & \sum g_z g_y & \sum g_z{}^2 \end{bmatrix} \tag{1}$$

Eigen-decomposition of the structure tensor (Eq. 2) transforms the given gradient space into a space defined with orthogonal vectors $\mathbf{v}_i$ ($i \in 1, 2, 3$), encoding the appearance of a tubular structure, i.e. fiber (Fig. 1):

$$\hat{\mathbf{T}}\mathbf{v} = \lambda \mathbf{v} \tag{2}$$

The smallest eigenvalue λ_i corresponds to the vector $\mathbf{v}_i$ pointing in the fiber direction, while the largest eigenvalue, as opposed to MRI, is assumed to correspond to the vector pointing in the direction of the sheet normal [6]. For the smallest eigenvalue vector, the inclination angle α_F was calculated as an angle between the transverse plane and the vector projection to the local tangential plane of the cylindrical coordinate system of the heart, as illustrated in Fig. 1.

In further text, we refer to this angle as fiber angle. Similarly, the transverse angle α_T was calculated as the angle between the tangential plane and the vector projection to the transverse plane. These two angles for each voxel location $\mathbf{x}$ determine the local fiber vector orientation.

The application of the structure tensor method to the fetal cardiac dataset resulted in a fiber vector field and corresponding angle maps.

2.4   Rule-Based Fiber Model

The histological studies [1] performed on cardiac fiber arrangement resulted in a translation of measurements into a mathematical formulation that can be applied to different cardiac geometries [17]. Thus, according to most detailed description for a whole heart available we generated the following *Streeter model* for the left ventricle (LV) extracted from our data for comparison with the structure tensor method. For each point $\mathbf{x}$ in the LV tissue, two distances are relevant: the smallest distance to the endocardium ($d_{endo}(\mathbf{x})$) and to the epicardium ($d_{epi}(\mathbf{x})$). From these two, a normalized distance map is constructed as:

$$e(\mathbf{x}) = \frac{d_{endo}(\mathbf{x})}{d_{endo}(\mathbf{x}) + d_{epi}(\mathbf{x})} \tag{3}$$

representing the normalized tissue thickness. The distance from the point to the base can be expressed with the polar angle $\Phi(\mathbf{x})$, with respect to the long axis, defined based on the mitral valve center $\mathbf{x}_{mc}$ and apex $\mathbf{x}_{ap}$ coordinates:

$$\Phi(\mathbf{x}) = \arccos\left(\frac{(\mathbf{x}_{ap} - \mathbf{x}_{mc}) \cdot (\mathbf{x} - \mathbf{x}_{mc})}{||\mathbf{x}_{ap} - \mathbf{x}_{mc}|| \cdot ||\mathbf{x} - \mathbf{x}_{mc}||}\right) \tag{4}$$

The relationship of the normalized distance and the base-distance with helix and transverse angle is now given with:

$$\alpha_F(\mathbf{x}) = a_1 e(\mathbf{x}) + a_0 \tag{5}$$
$$\alpha_T(\mathbf{x}) = b_2 \Phi((\mathbf{x}))^2 + b_1 \Phi(\mathbf{x}) + b_0 \tag{6}$$

The model parameters are $\mathbf{a} = [-1.90, 0.862]'$ and $\mathbf{b} = [-0.2149, 0.0089, -0.0930]'$ [18].

2.5   Analysis

Fiber orientations were assessed in a mid-LV equatorial slice, positioned at 70 % of the heart's height. Comparisons of the structure tensor method results with the rule-based fiber model in this slice were performed for their whole-LV-slice angle value distributions, and the transmural angle distributions in three 30° wide sectors of interest: lateral, anterior and posterior left ventricular free wall (LVFW) sector (with a 1° step between trajectories).

3   Results

The gradient structure tensor applied to the ESRF data and the mathematical model applied to geometry extracted from ESRF data resulted in LV fiber vector fields and angle maps for the analyzed fetal cardiac geometry. The spatial visualizations of the obtained results are given in Fig. 2. Equatorial angle maps for the mid-LV slice are given in Fig. 3(a). Distributions of fiber angles in the LV over the whole slice for both methods are shown in Fig. 3(b).

For quantifying the orientation, radial transmural profiles of fiber angles were extracted from the equatorial angle maps, from the center of the ventricle towards the epicardial surface. Figure 3(c) shows the average transmural transfer

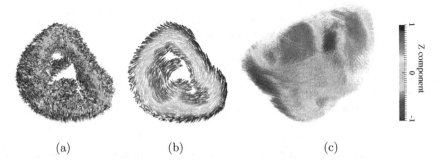

Fig. 2. Spatial vizualizations of the results for the fetal rabbit heart: (a) Vector fields obtained with structure tensors (LV only) in a mid-slice, (b) Mathematical model vector fields (LV only) in a mid-slice, (c) Myofiber tracking for all LV+RV slices in the heart. Images are color coded by z-component value of the unit fiber vector (Color figure online).

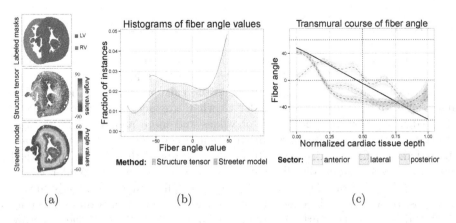

Fig. 3. Fiber orientation analysis: (a) columnwise: mask reflecting LV + RV equatorial geometry of the heart, LV angle maps obtained with structure tensor method and mathematical (Streeter) model, (b) whole-LV distribution of angle values, and (c) transmural LV sectorwise distributions of angle values (Color figure online)

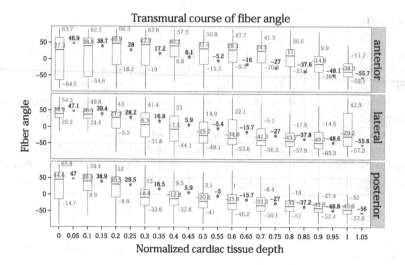

Fig. 4. Sectorwise boxplot of transmural fiber analysis obtained with structure tensors, compared with the Streeter model. The central numbers represent the median value per normalized tissue depth, and bottom/top numbers represent its first and third quartile. The Streeter model is represented with dots.

functions of fiber angles per sectors. Here, local regression smoothing was applied to trajectories within each sector in order to emphasize the principal direction in the defined region of interest. The width of the tissue was normalized to the local wall thickness.

Figure 4 provides the numerical sector-wise analysis, including median, its first and third quartile and 95 % confidence intervals.

4   Discussion

In this paper, we present the assessment of the spatial distribution of myocardial fibers in whole-heart high resolution phase-contrast CT images.

As can be observed in Fig. 3(b), while the calculated distribution of fiber angles over the whole tissue is similar to what is expected based on previous knowledge, it shows a wider range of values and more regional differences in comparison with the model-based distribution. We also observe that in our data circumferential fibers (with $\alpha_F \approx 0$) are less expressed than in the theoretical model, though the model distribution attains its minimum around similar values.

As can be seen from the angle maps in Fig. 3(a), fiber angles gradually change from positive values in the endocardium to negative in the epicardium, reaching $+60°$ and $-60°$ on the surfaces. The angle courses (Fig. 3(c)) for the structure tensor method and the mathematical model method are similar in the three sectors, with larger variability towards the endocardial surface in the anterior sector and towards epicardial surface in the lateral sector. Overall, the obtained angle courses indicate the helical trend of the fibers (longitudinal-circumferential-longitudinal) according to literature.

Most importantly, from Fig. 4 we observe that the substantial heterogeneity in local orientations is present within the cardiac tissue, while retaining the expected predominant orientation.

These differences are higher than most of the reported from DT-MRI image quantification. There are several reasons for this. Primarily, the biggest differences observed in the endocardial sites occur because the superior resolution of our data-set enables the visualisation and analysis of all trabecular structures within the cavity, especially at the endocardial side, where complex trabeculations are present and where there is little knowledge on how fibers are oriented locally. Next, the myocardium is interwoven with the complex coronary vessel tree, which locally might have a different orientation with regards to the fiber-structure. This information is also resolved in our data set. Last, the data set we have presented here is from an (end of gestation) fetal rabbit. Although it is expected that the myofiber structure has already matured, there might still be some changes as compared to adult hearts (where the Streeter-model was based on). Regardless of that, since the Streeter model is often used to create computational models, comparing this data to Streeter in a greater study can allow us to define where the description can be improved or should be complemented with novel and/or personalized data.

Some other aspects might have influenced our results: the signal-to-noise ratio of the image acquisition process (yet superior to most alternative approaches); the simplification of the geometrical model for angle calculations (assumed cylindrical shape of the LV); and ambiguity of the fiber definition, overlapping with the hypothesis of homogeneity captured in existing mathematical models.

5   Conclusion

The primary objective of our experiment was to examine and provide an insight into myocardial fiber orientation assessed with the new, high resolution synchrotron-based phase-contrast *ex vivo* acquisitions of a healthy fetal rabbit heart. We approached the problem using a gradient structure tensor method, and generated a theoretical model of fibers that incorporates knowledge of their expected distribution within cardiac geometry. Herewith, we showed that it is possible to effectively extract the information on micro-morphology from the presented dataset and that the superior details provide information on inter-individual differences that might be important for understanding or computational modelling of cardiac function.

Acknowledgements. The experiments were performed on the ID19 beamline at the European Synchrotron Radiation Facility (ESRF), Grenoble, France. We are grateful to Anne Bonnin at ESRF for providing assistance in using beamline ID19. This study was partly supported by Ministry of science, education and sports of the Republic of Croatia (036-0362214-1989); Ministerio de Economia y Competitividad (SAF2012-37196, TIN2012-35874); Instituto de Salud Carlos III (PI11/00051,

PI11/01709, PI12/00801,PI14/00226) integrados en el Plan Nacional de I+D+I y cofinanciados por el ISCIII-Subdirección General de Evaluación y el Fondo Europeo de Desarrollo Regional (FEDER) Otra manera de hacer Europa; the EU-FP7 for research, technological development and demonstration under grant agreement VP2HF (no611823); The Cerebra Foundation for the Brain Injured Child (Carmarthen, UK); Obra Social la Caixa (Barcelona, Spain); Fundació Mutua Madrileña and Fundació Agrupació Mutua (Spain).

References

1. Streeter, D.: Gross morphology and fiber geometry of the heart. In: Bethesda, B. (ed.) Handbook of Physiology, The Cardiovascular System, Chap. 4, vol. 1, pp. 61–112. American Physiological Society, US (1979)
2. Nielsen, P.M.F., Le Grice, I.J., Smaill, B.H., Hunter, P.J.: Mathematical model of geometry and fibrous structure of the heart. Am. J. Physiol. 260, H1365–H1378 (1991)
3. Lombaert, H., Peyrat, J.-M., Croisille, P., Rapacchi, S., Fanton, L., Cheriet, F., Clarysse, P., Magnin, I., Delingette, H., Ayache, N.: Human atlas of the cardiac fiber architecture: study on a healthy population. IEEE Trans. Med. Imaging 31(7), 1436–1447 (2012)
4. Geerts, L., Bovendeerd, P., Nicolay, K., Arts, T.: Characterization of the normal cardiac myofiber field in goat measured with MR-diffusion tensor imaging. Am. J. Physiol. Heart Circ. Physiol. 283, H13–H145 (2002)
5. Savadjiev, P., Strijkers, G., Bakermans, A., Piuze, E., Zucker, S., Siddiqi, K.: Heart wall myofibers are arranged in minimal surfaces to optimize organ function. PNAS 109(24), 9248–9253 (2012)
6. Scollan, D.F., Holmes, A., Zhang, J., Winslow, R.L.: Reconstruction of cardiac ventricular geometry and fiber orientation using magnetic resonance imaging. Ann. Biomed. Eng. 28(8), 934–944 (2000)
7. Healy, L.J., Jiang, Y., Hsu, E.W.: Quantitative comparison of myocardial fiber structure between mice, rabbit, and sheep using diffusion tensor cardiovascular magnetic resonance. J. Cardiovasc. Magn. Reson. 13(1), 74 (2011)
8. Piuze, E., Sporring, J., Siddiqi, K.: Maurer-Cartan forms for fields on surfaces: application to heart fiber geometry. IEEE TPAMI PP(99), 1 (2015)
9. Lee, W.N., Larrat, B., Pernot, M., Tanter, M.: Ultrasound elastic tensor imaging: comparison with MR diffusion tensor imaging in the myocardium. Phys. Med. Biol. 57(16), 5075–5095 (2012)
10. Papadacci, C., Pernot, M., Tanter, M., Fink, M.: Towards backscatter tensor imaging (BTI): analysis of the spatial coherence of ultrasonic speckle in anisotropic soft tissues. In: Conference Proceedings of the IEEE IUS, pp. 1208–1211 (2013)
11. Gilbert, S.H., Sands, G.B., LeGrice, I.J., Smaill, B.H., Bernus, O., Trew, M.L.: A framework for myoarchitecture analysis of high resolution cardiac MRI and comparison with diffusion tensor MRI. In: Conference Proceedings of the IEEE Engineering Medicine and Biology Society, pp. 4063–4066 (2012)
12. Jouk, P.S., Mourad, A., Milisic, V., Michalowicz, G., Raoult, A., Caillerie, D., Usson, Y.: Analysis of the fiber architecture of the heart by quantitative polarized light microscopy: accuracy, limitations and contribution to the study of the fiber architecture of the ventricles during fetal and neonatal life. Eur. J. Cardiothorac. Surg. 31(5), 91–921 (2007)

13. Ghafaryasl, B., Bijnens, B.H., van Vliet, E., Crispi, F., Cárdenes, R.: Cardiac microstructure estimation from multi-photon confocal microscopy images. In: Ourselin, S., Rueckert, D., Smith, N. (eds.) FIMH 2013. LNCS, vol. 7945, pp. 80–88. Springer, Heidelberg (2013)
14. Gan, Y., Fleming, C.P.: Extracting three-dimensional orientation and tractography of myofibers using optical coherence tomography. Biomed. Opt. Express 4(10), 2150–2165 (2013)
15. Varray, F., Wang, L., Fanton, L., Zhu, Y., Magnin, I.: High resolution extraction of local human cardiac fibre orientations. In: Ourselin, S., Rueckert, D., Smith, N. (eds.) FIMH 2013. LNCS, vol. 7945, pp. 150–157. Springer, Heidelberg (2013)
16. Ni, H., Castro, S.J., Stephenson, R.S., Jarvis, J.C., Tristan Lowe, T., Hart, G., Boyett, M.R., Zhang, H.: Extracting myofibre orientation from micro-CT images: an optimisation study. Comput. Cardiol. 40, 823–826 (2013)
17. Sebastian, R., Zimmerman, V., Sukno, F., Bijnens, B., Frangi, A.: Cardiac modelling for pathophysiology research and clinical applications. The need for an automated pipeline. In: Dössel, O., Schlegel, W.C. (eds.) World Congress on Medical Physics and Biomedical Engineering. IFMBE Proceedings, vol. 25/4, pp. 2207–2210. Springer, Heidelberg (2010)
18. Muñoz-Moreno, E., Cárdenes, R., Frangi, A.: Analysis of the helix and transverse angles of the muscle fibers in the myocardium based on diffusion tensor imaging. In: Conference Proceedings of the IEEE EMBS, pp. 5720–5723 (2010)

Sensitivity Analysis of Diffusion Tensor MRI in Simulated Rat Myocardium

Joanne Bates[1]([✉]), Irvin Teh[2], Peter Kohl[3,4], Jürgen E. Schneider[2], and Vicente Grau[1]

[1] Institute of Biomedical Engineering, University of Oxford, Oxford, UK
joanne.bates@eng.ox.ac.uk
[2] Division of Cardiovascular Medicine, Radcliffe Department of Medicine, University of Oxford, Oxford, UK
[3] Department of Computer Science, University of Oxford, Oxford, UK
[4] National Heart and Lung Institute, Imperial College London, London, UK

Abstract. A model of cardiac microstructure and diffusion MRI is presented. The results show a good correspondence between the simulated and experimental measurements. A sensitivity analysis shows that the diffusivity has the greatest effect on both the apparent diffusion coefficient and the fractional anisotropy. The cross-sectional area of the cells is the next most important factor; the aspect ratio of the cell cross-section also affects the fractional anisotropy. Neither the cell length nor the volume fraction of cells has a marked effect.

Keywords: Diffusion MRI · Model · Rat · Heart

1 Introduction

The normal microstructure of the heart is fundamental to its healthy mechanical and electrical function [1,2]. In disease there can be remodelling of this microstructure leading to myocardial disarray. In hypertrophic cardiomyopathy, there is evidence of myocardial disarray and it has been suggested that this may disrupt the normal propagation of the electrical signals within the heart, thus leading to the arrhythmias that cause sudden death [3]. It has also been shown in myocardial infarction that cell orientations change near the injured area [4].

Diffusion of water molecules in tissues is affected by the arrangement and structure of the cells, in particular diffusion occurs preferentially along the long axis of cells. Diffusion MRI uses a spatially-varying gradient to induce a phase change in moving water protons, thus sensitising the scan to diffusion along the gradient direction [5]. It can therefore be used to non-invasively characterise tissue structure, by measuring the direction and magnitude of diffusion [6].

Diffusion tensor imaging (DTI) fits the measured diffusion signal to a mono-exponential decay, and models water diffusion as a 3D diffusion tensor (DT) [6]. The tensor eigenvalues can be used to calculate the apparent diffusion coefficient (ADC) and fractional anisotropy (FA). These describe the magnitude

© Springer International Publishing Switzerland 2015
H. van Assen et al. (Eds.): FIMH 2015, LNCS 9126, pp. 120–128, 2015.
DOI: 10.1007/978-3-319-20309-6_14

and degree of directional preference of the water diffusion that occurs during a specified diffusion time interval. FA ranges from 0 (fully isotropic) to 1 (highly anisotropic).

Disease remodelling has an effect on the DTI outputs, but the relationship between microstructural changes and DTI is hard to measure experimentally [7]. Computational models are useful tools in the analysis of DTI, as they allow investigations that cannot be done experimentally. Inputs can be changed specifically and therefore the effect that each parameter change has on the output can be explored. Wang et al. [8] used a Monte Carlo model of cardiac tissue to investigate the effect of membrane permeability and water content on DTI measurements. Hall and Alexander [7] created a Monte Carlo model of brain white matter, comparing a simple version to analytical data and extending it to more realistic geometries for which there are no analytical solutions.

Sensitivity analysis measures the effect of changes in the input parameters on the output of a model. It enhances understanding of the model by revealing the parameters which have the greatest effect on the output. A simple and economic type of sensitivity analysis is the Morris method [9] which screens the input factors to determine which are the most important. It produces two sensitivity measures for each input factor: μ is the estimated mean of the elementary effects and estimates the overall effect of the factor on the output, and σ is the standard deviation of the elementary effects and estimates non-linear effects and interactions with other factors. Campolongo et al. [10] introduced a modified version, μ^*, which uses absolute values and avoids effects of opposite signs.

The aim of this study is, firstly, to demonstrate a modelling methodology in the heart and to qualitatively assess the similarity of experimental and simulated measurements. Secondly, we aim to determine which parameters related to the cell and tissue geometry of rat myocardium have the greatest influence on DTI measurements. This will inform future studies investigating the effect of cardiac microstructure on DTI results.

2   Methods

A model of a simplified volume of left ventricular myocardium was created using Smoldyn [11], as shown in Fig. 1. All anatomical and physiological parameters in the model are adopted from the literature, with their ranges reflecting the range of values reported in different publications. Cells are modelled as cuboids with a defined length (L), cross-sectional area (CSA) and aspect ratio of the cross-section (AR, thickness/width). The model consists of layers of cells where each layer is parallel to the epicardium and endocardium, that is, transverse angles (those between the circumferential axis and the projection of the cell's long axis onto the radial-circumferential plane [16]) are set to 0°. Cells within each layer are parallel to each other. The spacing of the cells and layers was chosen to give the desired volume fraction (VF) of cells, that is, the proportion of the tissue made up of myocytes. In the central layer, the helix angle (that between the cell's long axis projected onto a plane parallel to the epicardium, and the

Table 1. Sensitivity analysis parameters, based on [7,8,12–16]

Parameter	Abbreviation	Units	Minimum value	Maximum value
Cross-sectional area	CSA	μm^2	100	300
Length	L	μm	90	150
Aspect ratio (thickness/width)	AR		0.33	1.00
Volume fraction of myocytes	VF		0.75	0.90
Angle range	ANG	$°/mm$	50	100
Diffusivity	DIF	mm^2/s	0.5×10^{-3}	2.5×10^{-3}

radial-circumferential plane [16]) is set to 45°. The helix angle varies linearly across the voxel at a rate (ANG) based on both typical left ventricular (LV) wall thickness and the range of helix angles found across the myocardium. A single cubic voxel was analysed with sides of 200 µm. To avoid boundary effects, a cube with sides of 600 µm was modelled, with the 'voxel' at the centre of this. 4×10^6 molecules were randomly distributed across the simulation volume. At each time step, each molecule moves in a random direction with a step length normally distributed about a mean which is proportional to the diffusivity (DIF). If the molecule reaches a boundary, it reflects off that boundary. The chosen values for the sensitivity analysis parameters are in Table 1 and are based on values found in the literature [7,8,12–16].

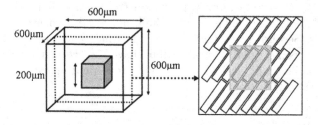

Fig. 1. Model geometry. 'Voxel' to be analysed is grey cube at centre of total model volume. Right image shows a typical arrangement of the cells (cuboids) in the central layer, with the grey area indicating the voxel. All layers are parallel to each other.

Custom-written Matlab code was used to calculate the total phase shift, Ψ, for each molecule from their locations. It is a function of the gradient strength that it experiences over every timestep j, and is calculated as [8]:

$$\Psi = \sum_{j=0}^{N} \Psi_j = \sum_{j=0}^{N} \gamma \langle G(t_j), r(t_j) \rangle \delta t \qquad (1)$$

where N is the total number of timesteps, $G(t_j)$ is the magnetic field gradient at timestep j, and $r(t_j)$ is the molecule position at timestep j. The MRI signal, S/S_0, for all M molecules is then:

$$\frac{S}{S_0} = \frac{1}{M} \sum_{i=1}^{M} \cos(\Psi) \tag{2}$$

The DT was calculated, the fibre angle was that between the primary eigenvector and the x-y plane, and the FA and ADC were calculated from [6]:

$$ADC = \frac{\lambda_1 + \lambda_2 + \lambda_3}{3} \tag{3}$$

$$FA = \sqrt{\frac{3}{2}} \sqrt{\frac{(\lambda_1 - \bar{\lambda})^2 + (\lambda_2 - \bar{\lambda})^2 + (\lambda_3 - \bar{\lambda})^2}{\lambda_1^2 + \lambda_2^2 + \lambda_3^2}} \tag{4}$$

where λ_i is the ith eigenvalue and $\bar{\lambda}$ is the mean of the eigenvalues.

Modified Morris method [10] parameters were generated and analysed using the Matlab (Mathworks, USA) code of [17]. A b-value of 1474 s/mm^2 was used (experiment 1 in Table 2), which is similar to the experimental data. The b-value combines the properties of the applied gradient pulse into a single factor. In the idealised free diffusion state for rectangular pulses, this is given by:

$$b = (G\gamma\delta)^2(\Delta - \frac{\delta}{3}) \tag{5}$$

where G is the magnetic field gradient strength, δ is the diffusion pulse duration, Δ is the time between pulse onsets, and γ is the gyromagnetic ratio (267.5 radμs^{-1}T^{-1} for water). For experimental data, cross-terms and ramp times also contribute to the b-value. To determine whether the b-value affects the sensitivity analysis, the Morris method was reassessed with 2 further b-values (382 and 748 s/mm^2) (experiments 2 and 3 in Table 2). Equation 5 shows that the b-value can be changed by changing the timings (δ or Δ) or the gradient strength (G). To analyse whether how the b-value is obtained affects the results, 2 further analyses were carried out at a b-value of 382 s/mm^2 (experiments 4 and 5 in Table 2). The simulation results are compared with experimental data from one fixed, embedded ex vivo rat heart. The data and results are in Table 2.

3   Results

Table 2 shows the ADC and FA results from the simulations and experimental data. The simulation values vary approximately 5-fold, but the mean value is similar to the experimental results.

Figure 2 shows the sensitivity analysis results for the b-value of 1474 s/mm^2. Parameters with a large μ^* (along the x-axis) have a large effect on the output of the model, those with a large σ have non-linear effects or interactions with other parameters. For the ADC the DIF is by far the most important factor, as its μ^* is much higher than the other factors. The σ is small so there is primarily a linear effect with little interaction with the other factors. Of the other factors, the CSA has the largest effect, followed by the ANG. L, AR and VF have the smallest effects.

Table 2. DTI sequence parameters and results

Simulation							Experimental		
No.	δ ms	Δ ms	G T/m	b-value s/mm²	ADC range (mean) × 10⁻³ s/mm²	FA range (mean)	mean b-value s/mm²	mean ADC × 10⁻³ mm²/s	FA mean
1	2.0	7.0	0.9	1474	0.4–1.8 (0.9)	0.12–0.53 (0.27)	1440	1.1	0.22
2	2.0	6.0	0.7	748	0.4–1.9 (1.0)	0.12–0.58 (0.27)	768	1.3	0.22
3	2.0	6.0	0.5	382	0.4–2.0 (1.0)	0.11–0.60 (0.27)	384	1.5	0.22
4	0.5	8.0	1.65	382	0.5–2.5 (1.3)	0.13–0.61 (0.29)			
5	1.0	2.0	1.79	382	0.5–2.4 (1.2)	0.07–0.41 (0.17)			

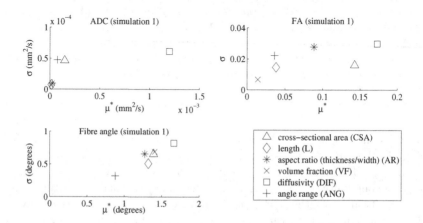

Fig. 2. Sensitivity analysis results (b-value of 1474 s/mm²). μ^* estimates the overall effect of the factor; σ estimates non-linear effects and interactions with other factors

For the FA, the DIF again has the largest effect, and has little interaction with other factors and a fairly linear effect. The CSA has the next largest effect followed by the AR. The σ for the AR is greater than for the CSA suggesting that there is more of a non-linear effect, or more interactions with other factors. The VF, L and ANG have the smallest effects.

The sensitivity analysis results for different b-values show that for the ADC, the DIF is always the most important factor by a large amount. The CSA and ANG are the next two most important factors, with their relative importance changing slightly between b-values but the μ^* and σ values are similar and always much larger than those of the L, AR and VF. For the FA, the order of DIF, CSA, AR is constant across all b-values, followed by ANG, with L and VF smallest. The values are fairly similar for all simulations, although simulation 5 (low b-value with small Δ) has a lower μ^* for all factors than the other b-values. For the fibre angle, DIF is again the most important factor, closely followed by VF, CSA, AR and L. However, the difference in angles between all the simulations is less than 5 degrees.

4   Discussion

The first aim of this study was to demonstrate a modelling methodology for the heart, and to assess its results compared with experimental data. The model geometry is simplified compared with biological tissue, with only helix angles modelled, uniformly sized cells, all layers parallel to each other, and the cell membranes as the only restriction to diffusion. The range of simulation results is wide, reflecting the broad range of input parameters found in the literature, however the mean value is similar to the experimental results suggesting that the model is a reasonable approximation. Tighter specification of biological input parameters would be required to narrow down the results. Further validation of the model using histology to give geometrical properties of the cells should be carried out.

The second aim of this study was to determine which of six factors concerning simplified cell shape and tissue structure have the greatest influence on cardiac DTI results. The DIF was by far the most important factor, especially for the ADC. As the diffusivity increases, the mean squared displacement of water in a given time increases thus increasing the ADC. In free space, this would be a linear relationship, but in the restricted case, as the step distance increases, the likelihood of interaction with a surface (cell membrane) increases, the measured distance in the direction of the applied magnetic field reduces, and therefore the ADC decreases. For the FA, this increase in molecule-surface interactions with increasing DIF leads to the structure having a greater effect on the direction of diffusion giving more anisotropic diffusion and therefore a higher FA. This increased directionality of diffusion is presumably why DIF also has the largest effect on the fibre angle. There are few values available for the diffusivity in the literature and previous simulation studies have not always justified the value that they have used [7,8]. Since the DIF has such a significant effect, this uncertainty is a significant limitation to the model. However, since the σ is low, there is little non-linearity or interaction with other factors and therefore the relative results of simulations shouldn't be affected significantly by the DIF.

The second most important factor is the CSA, although for the ADC its effect is much smaller than the DIF. A smaller CSA leads to an increase in molecule-surface interactions which will decrease the ADC and increase the FA. The AR is the third most important factor for the FA, but is one of the least important for the ADC. A smaller AR (greater difference between thickness and width) will cause a greater difference in the diffusion in the direction of the thickness compared with the width and hence a higher FA. For the mean ADC, the direction of the diffusion is not important and so the reduced diffusion in the thickness direction is compensated for by the increased diffusion in the width direction and thus the overall amount of diffusion is not much affected.

The L and the VF are amongst the least important factors for FA and ADC. The VF changes the size of the gaps between cells in the simulation. Since these are small compared with the size of the cells, the movement of the water in the extra-cellular space does not have much effect on the overall FA or ADC in this model. Since the cells are much longer than they are wide or thick, molecule-surface interactions with the ends of the cells are infrequent compared with those with the sides and hence the L does not have a large effect on the results.

The ANG is the third most important factor for the ADC, albeit much smaller than the DIF, while it is one of the least important for FA. In this model, an increased ANG means that a layer of cells is rotated more with respect to its adjacent layers. This could mean that there is more restriction in the diffusion perpendicular to the layers hence a lower ADC, and a higher FA, but would need further investigation.

There is very little difference in the ranking order of importance of the factors as the b-value changes, and the values of μ^* and σ remain similar. The exception is the FA in simulation 5 (b-value of $384 \, \text{s/mm}^2$ with small Δ) where, although the ranking order is maintained, the values are reduced. In this case, it may be that Δ is sufficiently small that molecules experience fewer collisions with cell surfaces, reducing the sensitivity of FA to the simulated tissue properties.

Diffusivity is assumed to be a single value, that is the same in all directions within the cell and outside it. Intracellular structures and the extracellular matrix will affect the rate of diffusion. This may not be the same intra and extra cellularly, nor the same in each direction in the cell. Disease may affect the intra and extra cellular environments and thus the diffusivity. Experimental investigation of these effects, especially in pathology, would enhance the model and the understanding of DTI in disease.

The model is a simplified geometry compared with biological tissue and does not include myocardial disarray, which can be found in disease. The introduction of myocardial disarray into the model would assist in the understanding of the effect of disease on DTI results. The analysis of fibre directions would be expected to become more informative with the introduction of disarray.

We have demonstrated a modelling method of DTI in the heart which gives similar ADC and FA measurements to experimental data, although further validation is needed. The sensitivity analysis suggests that the DIF is the most important factor affecting the FA and the ADC. CSA is the next most important, and AR affects the FA. While DTI can be a sensitive marker of pathological changes, it is not particularly specific. By better understanding the dominant tissue properties that affect the DTI results, we can prioritise development of models and acquisition strategies that enable estimation of such properties from DTI data. Improving the specificity of DTI measurements could augment the potential of DTI as an early biomarker of pathology in the heart, and our ability to distinguish between subtler forms of disease, such as different forms of disarray. Further work will introduce myocardial disarray into the model and investigate other output measures. The model could also be of use in the quantitative analysis of acquisition artefacts, including those caused by bulk motion.

Acknowledgements. This work is supported by a BBSRC grant (BB/I012117/1), an EPSRC grant (EP/J013250/1) and by funding from the British Heart Foundation (BHF) Centre for Research Excellence. JES and PK are BHF Senior Basic Science

Research Fellows (FS/11/50/29038; FS/12/17/29532). VG and PK are supported by BHF New Horizon Grant NH/13/30238. PK holds an ERC Advanced Grant (CardioNect). The authors acknowledge a Wellcome Trust Core Award (090532/Z/09/Z).

References

1. Waldman, L.K., Nosan, D., Villarreal, F., Covell, J.W.: Relation between transmural deformation and local myofiber direction in canine left ventricle. Circ. Res. **63**(3), 550–562 (1988)
2. Kanai, A., Salama, G.: Optical mapping reveals that repolarization spreads anisotropically and is guided by fiber orientation in guinea pig hearts. Circ. Res. **77**(4), 784–802 (1995)
3. Maron, B.: Contemporary insights and strategies for risk stratification and prevention of sudden death in hypertrophic cardiomyopathy. Circulation **121**(3), 445 (2010)
4. Wu, M.T., Tseng, W.Y.I., Su, M.Y.M., Liu, C.P., Chiou, K.R., Wedeen, V.J., Reese, T.G., Yang, C.F.: Diffusion tensor magnetic resonance imaging mapping the fiber architecture remodeling in human myocardium after infarction: Correlation with viability and wall motion. Circulation **114**(10), 1036–45 (2006)
5. Stejskal, E., Tanner, J.: Spin diffusion measurements: spin echoes in the presence of a time-dependent field gradient. J. Chem. Phys. **42**(1), 288–292 (1965)
6. Basser, P., Pierpaoli, C.: Microstructural and physiological features of tissues elucidated by quantitative diffusion tensor MRI. J. Mag. Res. B **111**(3), 209 (1996)
7. Hall, M., Alexander, D.: Convergence and parameter choice for Monte-Carlo simulations of diffusion MRI. IEEE Trans. Med. Imag. **28**(9), 1354 (2009)
8. Wang, L., Zhu, Y.-M., Li, H., Liu, W., Magnin, I.E.: Simulation of diffusion anisotropy in DTI for virtual cardiac fiber structure. In: Metaxas, D.N., Axel, L. (eds.) FIMH 2011. LNCS, vol. 6666, pp. 95–104. Springer, Heidelberg (2011)
9. Morris, M.: Factorial sampling plans for preliminary computational experiments. Technometrics **33**(2), 161–174 (1991)
10. Campolongo, F., Cariboni, J., Saltelli, A.: An effective screening design for sensitivity analysis of large models. Environ. Modell. Softw. **22**(10), 1509–1518 (2007)
11. Andrews, S.S., Addy, N.J., Brent, R., Arkin, A.P.: Detailed simulations of cell biology with Smoldyn 2.1. PLOS Comp. Biol. **6**(3), e1000705 (2010)
12. Satoh, H., Delbridge, L.M., Blatter, L.A., Bers, D.M.: Surface:volume relationship in cardiac myocytes studied with confocal microscopy and membrane capacitance measurements: species-dependence and developmental effects. Biophys. J. **70**(3), 1494–1504 (1996)
13. Chen, Y.-F., Redetzke, R.A., Sivertson, R.M., Coburn, T.S., Cypher, L.R., Gerdes, A.M.: Post-myocardial infarction left ventricular myocyte remodeling: are there gender differences in rats? Cardiovasc. Path. **20**(5), e189–e195 (2011)
14. Anversa, P., Ricci, R., Olivetti, G.: Quantitative structural analysis of the myocardium during physiologic growth and induced cardiac hypertrophy: a review. J. Am. Coll. Cardiol. **7**(5), 1140–1149 (1986)
15. McAdams, R.M., McPherson, R.J., Dabestani, N.M., Gleason, C.A., Juul, S.E.: Left ventricular hypertrophy is prevalent in sprague-dawley rats. Comp. Med. **60**(5), 357–363 (2010)

16. Hales, P.W., Schneider, J.E., Burton, R.A.B., Wright, B.J., Bollensdorff, C., Kohl, P.: Histo-anatomical structure of the living isolated rat heart in two contraction states assessed by diffusion tensor MRI. Prog. Biophys. Mol. Biol. **110**(2–3), 319 (2012)
17. Joint research centre: Institute for the protection and security of the citizen (2014). http://ipsc.jrc.ec.europa.eu/?id=756

3D Farnebäck Optic Flow for Extended Field of View of Echocardiography

A. Danudibroto[1,2(✉)], O. Gerard[1], M. Alessandrini[2], O. Mirea[2], J. D'hooge[2], and E. Samset[1]

[1] GE Vingmed Ultrasound, Oslo, Norway
adriyana.danudibroto@kuleuven.be
[2] Department of Cardiovascular Sciences, KU Leuven, Leuven, Belgium

Abstract. 3D echocardiography has enabled new clinical applications of ultrasound related to both interventional guidance and quantification of chamber characteristics (e.g. volumes, function). However, image quality may be hampered by dropouts in the image as well as limited field of view. By compounding data from several overlapping images, a volume with extended field of view can be formed. A 3D method based on Farnebäck optic flow is proposed to perform registration between ultrasound images taken from different orientations. It utilizes signal decomposition into polynomial basis functions and solves the transformation between the volumes analytically. Validation using synthetic data sets showed a registration error of 0.47 ± 0.05 mm. And testing on data sets of 50 real images showed promising results.

Keywords: Rigid registration of 3D echocardiography · Extension of field of view · Farnebäck optic flow

1 Introduction

3D echocardiography has brought added value to ultrasound imaging in clinical practice. In comparison to its 2D counterpart, 3D echo provides more anatomical information on the spatial relationship between structures. It assists in the diagnostic process and helps in training new echocardiographers. Its real time capability makes 3D echo a suitable imaging modality for intraoperative monitoring [11].

A. Danudibroto—Supported by Marie Curie Initial Training Network (USART-project, grant ID: PITN-GA-2012-317132).
M. Alessandrini—Supported by the Flemish Research Council (FWO), grant ID: 1263814N.

Electronic supplementary material The online version of this chapter (doi:10. 1007/978-3-319-20309-6_15) contains supplementary material, which is available to authorized users.

© Springer International Publishing Switzerland 2015
H. van Assen et al. (Eds.): FIMH 2015, LNCS 9126, pp. 129–136, 2015.
DOI: 10.1007/978-3-319-20309-6_15

One of the inherent limitations of 3D echo is its dependence on the incidence angle [8]. Signal dropouts often occurs in the presence of strongly reflective structures. Transesophageal Echocardiography (TEE) and Intracardiac Echocardiography (ICE) probes that have a clear access to the heart during minimally invasive procedures have limited field of view due to their small aperture size.

There have been attempts to overcome these limitations using post-processing techniques, aiming at both improvement of image quality as well as extension of field of view. Such approaches are mainly focused on compounding information from several acquisitions. When combining image data from several acquisitions, they must be both registered and fused. Registration is performed to optimally transform the images into one common coordinate system while fusion combines the redundant data for optimal visual. Some studies focused more on the accuracy of the registration [2,13,14], while others at the improvement of image quality through fusion [1,9,16]. This paper focuses on the registration part.

Some studies utilize tracker information for registration [2,16]. Either electromagnetic or optical trackers are commonly used. For electromagnetic tracking, there is a risk of disturbance in the electromagnetic field when ferro magnetic material is present. In the case of optical tracker, a clear line of sight from the optical marker to the sensor has to be kept.

Image based registration methods can be grouped into voxel based and feature based methods. Many available registration methods rely on optimization of a cost or similarity function that is computed directly from the voxel intensity values. Some of the known functions are: sum of absolute differences, sum of squared differences, normalized cross correlation (NCC) and mutual information [1,12,15]. Voxel based methods are to some extent dependent on pose initialization. Some works used tracker information for initial registration then refined it using a voxel based method [2,16]. Other approaches also perform optimization but on extracted features from the image instead of directly on the intensity values. For example, Ni et al. [13] utilized 3D SIFT which is a scale invariant and robust gradient based feature descriptor to perform registration on 3D ultrasound of the liver to build panoramic volumes. Similarly, Schneider et al. [14] used features based on Laplacian-of-Gaussian filtering to register images of a porcine heart in a water tank while Grau & Noble [9] used monogenic signal analysis.

In this work, a new approach is proposed. The proposed registration is based on signal decomposition into a polynomial basis using the optical flow method by Farnebäck [5]. Farnebäck optic flow was chosen for its ability to handle potential uncertainty in the images. In contrast to the voxel based methods that optimize global cost function by applying a series of transformations, the Farnebäck method attracts the floating image to the optimum position and orientation that matched the reference image based on the analytical solution of the optic flow problem. This approach was taken to allow for robust and fast registration in images with very small frame-to-frame transformations as well as larger transformations.

2 Methodology

The general overview of the proposed registration method can be seen in Fig. 1.

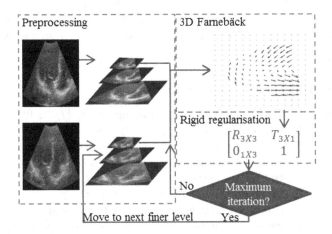

Fig. 1. Overview of registration method

2.1 Preprocessing

Before registration, three preprocessing steps were performed. First, to eliminate influence from the background and shadow regions, a high intensity mask was applied. In ultrasound images, the shape of the sector is one of the dominant image structures. However, this structure does not provide useful information for registration.

To ensure that the algorithm is not directionally biased, the images were resliced so that the voxel dimension is isometric. Lastly, to reduce computation time and increase the range of detectable transformations, Gaussian pyramids were built for both reference and floating images using an adapted method of *impyramid* function in MATLAB for 3D matrices.

2.2 Polynomial Decomposition

The registration approach is based on the work of Farnebäck [5,6]. Farnebäck utilized signal analysis using orientation tensors to compute disparity between a pair of images. It modeled the signal with polynomial basis functions weighted by an applicability function that determined the neighborhood's importance in signal analysis. The image was decomposed locally by means of normalized convolution as explained in [5] and approximated by a polynomial as in (1). In the case of 3D images, the position vector $\mathbf{x}$ consists of $[x, y, z]^T$. When expanded, it can be seen that the polynomial used 10 basis functions as seen in (2).

$$f(\mathbf{x}) \sim \mathbf{x}^T \mathbf{A} \mathbf{x} + \mathbf{b}^T \mathbf{x} + c \tag{1}$$

$$[1, x, y, z, x^2, y^2, z^2, xy, xz, yz] \tag{2}$$

The choice of applicability function is crucial to appropriately express the image description by a set of polynomial coefficients. In this case it should decay as it moves away from the center voxel of interest, and be isotropic to avoid biases towards any direction in the neighborhood, hence a Gaussian function was chosen. Ideally, the σ of the Gaussian applicability function would be comparable to the size of the feature of interest.

2.3   Rigid Transform Regularization

Based on the optic flow assumption, the floating image (f_2 in (3)) is a transformed version of the reference image (f_1 in (3)). The disparity between a pair of images was computed using this assumption. Once expanded with the expression on (1), the disparity can be solved by equating the coefficients and can be expressed as (4). Note that the disparity is independent from the constant term c which means the local brightness should not affect the result. In ultrasound image analysis, this could be an advantage because the appearance of structures is incident angle dependent. Moreover the inherent signal uncertainty consideration allowed for focusing the optic flow computation only on the overlap region excluding the signal dropouts area. Although the speckle appearance might differ due to varying probe positions, the second order polynomial representation of the image limited the optic flow consideration to more dominant structures. Furthermore the Gaussian filtering in the preprocessing smoothed it out to an extent.

$$f_2(\mathbf{x}) = f_1(\mathbf{x} - \mathbf{d}) \tag{3}$$

$$\mathbf{d} = -\frac{1}{2}\mathbf{A}_1^{-1}(\mathbf{b}_2 - \mathbf{b}_1) \tag{4}$$

The disparity was solved point by point over a neighborhood that is weighted by the same applicability function as in Sect. 2.2. This optic flow method provided a dense motion field between the image pairs. And this motion field indicated the correspondences between all points of interest in the two images. This is then regularized to rigid transformation using Procrustes analysis by finding the best rotations and translations that will fit the set of points in one image into their correspondences in another. It should be noted that cardiac deformation is not a concern in this case because only image pairs of corresponding point in the cardiac cycle were registered. Since the targeted application assumes that the images to be registered are taken in a small time window from nearby positions, rigid transformation consisting of translations and rotations should suffice.

To be able to capture a large range of transformations, the registration was implemented in a multiscale iterative scheme. Starting from the coarsest level of the Gaussian pyramid, the image pairs were registered iteratively for a fixed number of iterations and the registration was regularized to transform rigidly at each iteration. The number of iterations at each scale can be set arbitrarily as long as it captures the maximum range of motion. Usually the number of iterations decreases as the registration moves from the coarse to finer level.

2.4   Validation

The proposed method was tested on both synthetic and real data sets. The accuracy validation was done using a simulated dataset. Synthetic ultrasound volumes were obtained from a 3D cloud of point scatterers by applying the ultrafast ultrasound simulator COLE [7]. To have realistic visual appearance, as in [3], scattering amplitude was sampled from a real 3D recording. Hereto, a wide angle ($\Delta\theta \sim 90°$, $\Delta\phi \sim 90°$) 3D scan from a healthy volunteer was used. From the whole scatter map, a series of narrow angle views ($\Delta\theta \sim 45°$, $\Delta\phi \sim 90°$) were simulated by rotating the synthetic probe along θ. The overlap between the views was varied from 32.5° to 40.0° in a step of 2.5°.

The real 3D datasets were taken from 10 healthy volunteers (6 females, 4 males) using a Vivid E9 (GE Healthcare) ultrasound scanner with a 4 V 3D probe. On average, 6 images per subject were used for validation. The acquisition protocol was inspired by the work of Rajpoot et al. [1]. All images were acquired over 4 heart cycles and the sector width was set to 65°. Images were taken from several positions; first from the best position for imaging left ventricle (LV) in an apical view, second one aimed towards the lateral wall of the LV, third towards the interventricular septum, fourth aimed at the right ventricle, and the last two were taken from one intercostal space above and below the first position. Not all the volunteers had good acquisition window from all the aforementioned positions. In some cases, more than one image was taken at the same position but with slight tilt in the acquisition angle to differentiate the image content.

The registration of each subject's dataset was performed pairwise between one volume and the next volume taken from the closest physical position. For comparison, a voxel based rigid registration method using optimisation of NCC function was also applied on the validation dataset. A GPU implementation called AIRWC by Ansorge et al. [4] that has been adapted by Kiss et al. [10] was used for this purpose. Detailed explanation of the NCC method used by AIRWC can be seen in [4].

2.5   Implementation Details

The implementation was done using MATLAB. The temporal alignment was relied on ECG gating and the registration was applied on the end diastolic frame. For the preprocessing steps described in Sect. 2.1, the high intensity mask was obtained by thresholding. The threshold value was set to be 50 which is the typical intensity for myocardium in echocardiography. The reslicing of the volume was performed using trilinear interpolation. All dimensions of the voxels are resliced to its largest, which is 0.84 mm for synthetic data set and 1.0 mm for real data sets. For the Gaussian pyramid, the number of levels was adjusted empirically, depending on the image. For image pairs with clear contrast and at least 80 % content overlap, 3 levels would be sufficient.

In the signal decomposition as described in Sect. 2.2, the locality of the signal analysis was set to be in the 9^3 neighborhood. The σ of the Gaussian applicability was set to 1.5 pixel.

In the registration of the real data sets, the number of iterations were fixed to 20, 15, 10 and 0 from the coarsest to finest level. The fusion was done using maximum rule, where at each voxel overlap, the one with maximum intensity was used for the compounded volume.

3   Results

The computed registration error from the four pairs of synthetic images was found to be 0.54 ± 0.17 mm in the overlap region and $0.21 \pm 0.11°$, $0.05 \pm 0.02°$, $0.18 \pm 0.28°$ in the x, y, z rotation angle of the transformation matrix.

From 50 image pairs in the real dataset, the NCC was computed on the overlap area. The average result from the proposed method and NCC based registration method using AIRWC as well as the t-test evaluation between the two sets of results can be seen in Table 1. Figure 2 demonstrates the extension of field of view that can be obtained from data compounding when compared to one of the contributing volumes. On Fig. 2b, it can be seen that the registration still performed well in the presence of artifacts from the ribs. The mean computation time taken to process one pair of $162 \times 160 \times 150$ volumes is 48.7 s.

Table 1. Average $\pm$ standard deviation NCC comparison before and after registration between proposed method and AIRWC as well as the t-test of NCC results after both registrations

Before	After Farnebäck	After AIRWC	p-value
0.88 ± 0.04	0.93 ± 0.02	0.90 ± 0.04	<0.0001

4   Discussions

The registration was performed on the end diastolic volume obtained from ECG gating. So there were no expected cardiac deformation between volumes belonging to the same subject. Following this assumption, only rigid transformation was performed. In the presence of artifacts, the rigid transform helped to avoid erroneous non-anatomical deformation.

The results obtained from the synthetic images showed that the proposed registration method can achieve high accuracy. Table 1 shows that the registration improves the NCC value on the real dataset. It can also be seen that on average the proposed method managed to outperform NCC based registration in terms of the aforementioned metric of interest. The p-value from the paired t-test indicates that the results from the proposed method is significantly better than NCC based method. This is mainly caused by the reliance of global function based method on initialization. From the 50 images in the real dataset, by visual assessment, it was found that AIRWC failed to register four of them hence the average NCC result was brought down. However, even after the results from the four cases were removed, the p-value from the paired t-test between the two sets was still <0.0001.

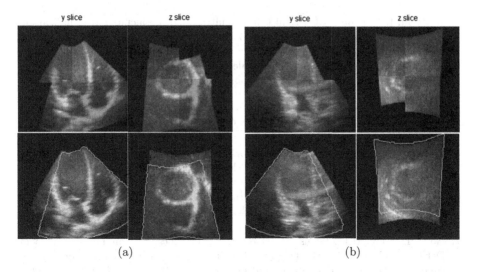

Fig. 2. Mid volume slices of compounded volume. Figure (a) shows the result from a data set of good quality image and figure (b) shows the result from a data set with poorer quality image and artifacts from the ribs. The compounded volume and one of the contributing volume are shown in alternating checkerboard pattern (top row) and in highlighting (bottom row).

It should be noted that this indication do not directly measure the registration accuracy. A more appropriate quantitative evaluation of image alignment which is still a topic of ongoing research would also need to be performed. This is linked to an automated stopping criteria for the registration iterations which is still a part of our ongoing research. It can also be noted that the viability of the proposed method to extend the field of view has been demonstrated. The lateral wall of LV that is not contained in the highlighted volume in Fig. 2 has been included in the compounded volume. The overall results have shown that Farnebäck optic flow can be used for rigid transformation of 3D echocardiography with high accuracy.

5 Conclusions

A new registration method for data compounding of 3D echocardiography has been proposed. The method utilized optical flow method by Farnebäck where the images were decomposed into polynomial basis functions and the transformation was computed as a function of the polynomial coefficients. The results have shown that the proposed registration method can align the volumes with high accuracy and would be viable for compounding the volumes to form wider field of view.

References

1. Rajpoot, K., Grau, V., Noble, J.A., Szmigielski, C., Becher, H.: Multiview fusion 3-D echocardiography: improving the information and quality of real-time 3-D echocardiography. UMB **37**(7), 1056–1072 (2011)

2. Brattain, L.J., Howe, R.D.: Real-time 4D ultrasound mosaicing and visualization. In: Fichtinger, G., Martel, A., Peters, T. (eds.) MICCAI 2011, Part I. LNCS, vol. 6891, pp. 105–112. Springer, Heidelberg (2011)

3. Alessandrini, M., Liebgott, H., Friboulet, D., Bernard, O.: Simulation of realistic echocardiographic sequences for ground-truth validation of motion estimation. ICIP 2012, 2329–2332 (2012)

4. Ansorge, R.E., Sawiak, S.J., Williams, G.B.: Exceptionally fast non-linear 3D image registration using GPUs. In: 2009 IEEE NSS/MIC, pp. 4088–4094. IEEE (2009)

5. Farnebäck, G.: Polynomial expansion for orientation and motion estimation. Ph.D. thesis, Linköping University, Sweden, SE-581 83 Linköping, Sweden (2002). Dissertation No 790, ISBN 91-7373-475-6

6. Farnebäck, G.: Two-frame motion estimation based on polynomial expansion. In: Bigun, J., Gustavsson, T. (eds.) SCIA 2003. LNCS, vol. 2749, pp. 363–370. Springer, Heidelberg (2003)

7. Gao, H., Choi, H.F., Claus, P., Boonen, S., Jaecques, S., van Lenthe, G., Van Der Perre, G., Lauriks, W., D'hooge, J.: A fast convolution-based methodology to simulate 2-D/3-D cardiac ultrasound images. IEEE Trans. UFFC 56(2), 404–409 (2009)

8. Grau, V., Becher, H., Noble, J.A.: Phase-based registration of multi-view real-time three-dimensional echocardiographic sequences. In: Larsen, R., Nielsen, M., Sporring, J. (eds.) MICCAI 2006. LNCS, vol. 4190, pp. 612–619. Springer, Heidelberg (2006)

9. Grau, V., Noble, J.A.: Adaptive multiscale ultrasound compounding using phase information. In: Duncan, J.S., Gerig, G. (eds.) MICCAI 2005. LNCS, vol. 3749, pp. 589–596. Springer, Heidelberg (2005)

10. Kiss, G., Asen, J., Bogaert, J., Amundsen, B., Claus, P., D'hooge, J., Torp, H.: Multi-modal cardiac image fusion and visualization on the GPU. In: 2011 IEEE International IUS, pp. 254–257, October 2011

11. Lang, R.M., Mor-Avi, V., Dent, J.M., Kramer, C.M.: Three-dimensional echocardiography: is it ready for everyday clinical use? JACC: Cardiovasc. Imaging 2(1), 114–117 (2009)

12. Mulder, H.W., van Stralen, M., van der Zwaan, H.B., Leung, K.Y.E., Bosch, J.G., Pluim, J.P.W.: Multiframe registration of real-time three-dimensional echocardiography time series. J. Med. Imaging 1(1), 014004 (2014)

13. Ni, D., Qu, Y., Yang, X., Chui, Y.P., Wong, T., Ho, S.S., Heng, P.A.: Volumetric ultrasound panorama based on 3D SIFT. In: Metaxas, D., Axel, L., Fichtinger, G., Székely, G. (eds.) MICCAI 2008, Part II. LNCS, vol. 5242, pp. 52–60. Springer, Heidelberg (2008)

14. Schneider, R.J., Perrin, D.P., Vasilyev, N.V., Marx, G.R., del Nido, P.J., Howe, R.D.: Real-time image-based rigid registration of three-dimensional ultrasound. Med. Image Anal. 16(2), 402–414 (2012)

15. Wachinger, C., Wein, W., Navab, N.: Three-dimensional ultrasound mosaicing. In: Ayache, N., Ourselin, S., Maeder, A. (eds.) MICCAI 2007, Part II. LNCS, vol. 4792, pp. 327–335. Springer, Heidelberg (2007)

16. Yao, C., Simpson, J.M., Jansen, C.H., King, A.P., Penney, G.P.: Spatial compounding of large sets of 3D echocardiography images. In: SPIE Medical Imaging, pp. 726515-1-8. International Society for Optics and Photonics (2009)

Towards Automatic Assessment of the Mitral Valve Coaptation Zone from 4D Ultrasound

Sandy Engelhardt[1][(✉)], Nils Lichtenberg[1,2][(✉)], Sameer Al-Maisary[3],
Raffaele De Simone[3], Helmut Rauch[4], Jens Roggenbach[4], Stefan Müller[2],
Matthias Karck[3], Hans-Peter Meinzer[1], and Ivo Wolf[1,5]

[1] Medical and Biological Informatics, DKFZ, Heidelberg, Germany
sandy.engelhardt@dkfz-heidelberg.de
[2] Institute for Computational Visualistics, University of Koblenz-Landau,
Koblenz, Germany
nlichtenberg@uni-koblenz.de
[3] Department of Cardiac Surgery, University Hospital Heidelberg,
Heidelberg, Germany
[4] Department of Anesthesiology, University Hospital Heidelberg,
Heidelberg, Germany
[5] University of Applied Science, Mannheim, Germany

Abstract. The coaptation zone is the part of the two mitral valve leaflets that collide during the cardiac cycle. It is an important parameter for the valve's function and closing capability, but difficult to assess. In this work, we present an automatic approach for leaflet segmentation from 4D ultrasound images, which incorporates steps for coaptation zone modelling and allows determining the coaptation zone from the resulting leaflet surface. The method segments the leaflets over the whole cardiac cycle given a previously segmented annulus model. To provide a meaningful analysis of the coaptation line assessment, the mean error between ground truth model and segmented model has been computed for each leaflet separately. For the anterior leaflet, we achieved a mean error of 1.16 ± 0.38 mm and 1.24 ± 0.37 mm for the posterior leaflet, respectively.

1 Introduction

The mitral valve (MV) prevents the backflow of blood into the left atrium during systole. MV insufficiency refers to valve leakage, which impairs the pumping function of the heart. 4D transesophageal echocardiography (TEE), a method based on ultrasound where the probe is inserted through the esophagus, is used for the diagnosis of the severity of MV insufficiency as well as for surgical treatment planning, e.g. it provides valuable information for deciding whether the

---

S. Engelhardt and N. Lichtenberg—These authors contributed equally to this work.

Electronic supplementary material The online version of this chapter (doi:10.1007/978-3-319-20309-6_16) contains supplementary material, which is available to authorized users. Videos can also be accessed at http://www.springerimages.com/videos/978-3-319-20308-9.

© Springer International Publishing Switzerland 2015
H. van Assen et al. (Eds.): FIMH 2015, LNCS 9126, pp. 137–145, 2015.
DOI: 10.1007/978-3-319-20309-6_16

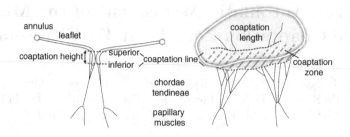

Fig. 1. Description of coaptation parameters.

MV needs to be replaced or whether it can be reconstructed. Reconstruction is preferable, because it mainly uses native tissue of the MV's components, which are the annulus, the anterior and posterior leaflets and the subvalvular apparatus. During such an intervention, the reestablishment of a sufficiently large coaptation zone is a major goal. During valve closure, the leaflet coaptation zone is formed by the colliding parts of the leaflets (cf. Fig. 1). The upper bound of the coaptation zone is the gently curved coaptation line evident from an atrial view. The height and length of the coaptation zone is viewed as an important assessment of MV function [4]. For instance, poor leaflet coaptation resulting from chordal retrains after MV reconstruction predicts disease recurrence [7].

So far, only a handful of works [3,4,10] have addressed the computation of the coaptation zone from TEE images. The presented approaches rely on more or less labor-intensive manual measurements. A method described by Gogoladze et al. [4] is based on the analysis of leaflet coaptation heights at three distinct locations along the coaptation line. However, 2D echocardiographic approaches are fundamentally limited due to the single planar view provided, and are prone to errors given the usually irregular 3D topography of the coaptation zone. Saito et al. measured the difference between mid- and end-systolic 3D MV leaflet surface areas in patients with dilated cardiomyopathy [10]. However, it has been demonstrated that the MV leaflet stretches during systole [2], possibly introducing errors in the calculation. The recent work of Cobey et al. [3] describes a manual approach for measuring the MV coaptation zone on an end-systolic frame. The method is based on time-consuming manual annotations of coaptation points (about 100 per time step) on rotational cross-sectional slices as well as subsequent summation of neighboring subareas. The main contribution of this work is to show that the coaptation zone, when indexed to the MV area is closely associated with severity of functional mitral regurgitation. Beside this, the authors state that they "expect that with continued technologic improvements in [...] quantitative software, future work on the MV coaptation zone will be more streamlined, if not largely automated" [3]. A general approach for automatic coaptation zone assessment is to create a model of the mitral valve from TEE. Existing (semi-)automatic segmentation methods, to name a few [6,9,11,13], are able to detect the coaptation line, but often fail to comprise the whole coaption zone, especially its height, in the final segmented model. In [13], the author states that the final model includes a detailed coaptation zone, but

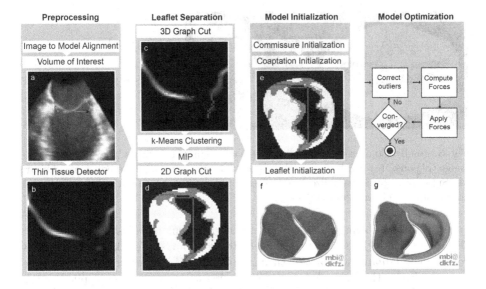

Fig. 2. Steps of our automatic segmentation pipeline (Color figure online).

a visual presentation of the coaptation and its quantitative parameters is not provided.

A segmentation should ideally capture patient-specific detail over the whole cardiac cycle and should be able to establish correspondences on valves of different subjects and over all time steps. A major challenge is to separate anterior from posterior leaflets upon coaptation, since there is no intensity-based demarcation and according to [11], "making this distinction [...] is extremely difficult even for a human observer." In this paper, we present a mitral valve segmentation method that integrates a leaflet separation step into the overall model initialization and specifically treats the free edges of the leaflets when they coapt in order to obtain a reliable coaptation height. The method is automatic in terms of leaflet delineation, given a semantically annotated annulus model.

2   Materials and Methods

The method requires the location of the mitral annulus, which we generate with minimal user input using our semi-automatic method [5]. The model contains semantic information about the location of the anterolateral and posteromedial annular commissure points and the saddle horn. On a volume of interest of the TEE (blue cylinder in Fig. 2a), the *Thin Tissue Detector* (TTD) [12] (cf. Fig. 2b) is applied using the same parameter configuration as in [12]. Higher TTD values describe the voxel's tendency to be located within a thin tissue region such as the MV leaflets. We construct a search space on the TTD that is comprised of an arc system centered on the annulus, as proposed in [11]. The Kolmogorov's min-cut/max-flow 3D graph cut method [1] is used to find a first estimation of the leaflet surfaces (cf. Fig. 2c), where the source and sink are connected to the

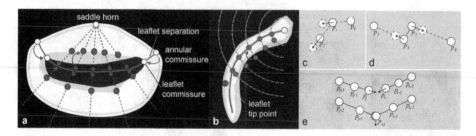

Fig. 3. (a+b) Commissure, leaflet initialization on C_l(white), C_u(grey), C_b(black) image regions. (c) Push forces shift points away from the preceding neighbour. (d) Smoothing forces equalize point distances. (e) Tips coapt, driven by push and gravity forces (Color figure online).

arcs [11]. The resulting surface S from the graph cut includes the leaflets, but also extends into the ventricle and needs to be trimmed. For this, the surface is voxelized and dilated with a structuring element with a diameter greater than the expected leaflet thickness. A *k-means* clustering with $k = 3$ is then applied on the grey values of the TTD masked with the dilated S, resulting in a classified image Ω that contains the classes *leaflet tissue* C_l, *uncertain* C_u and *background* C_b. A *Maximum Intensity Projection* (MIP) in the direction of the annulus plane normal is applied on Ω. Using the known annuluar commissure points from the annulus segmentation as separation points, a min-cut on the MIP result is calculated. The resulting cut line roughly delineates a **leaflet separation line** (cf. Fig. 2d). Using the MIP of the clustered image Ω as input can better ensure that the graph cut runs through the background, as the intensity difference between two labelled regions is higher than for the smooth grey value transitions in the original TTD image.

The goal of the subsequent initialization of the 3D model is to put the important landmarks into reasonable locations (cf. Fig. 2f) so that the following model optimization (cf. Fig. 2g) can work reliably. First, **leaflet commissure landmarks** are searched along the leaflet separation line on the MIP image of Ω (cf. Fig. 3a), beginning the traversal at the annular commissures. We are looking for transitions between two classes 1. $C_l \rightarrow C_b$, 2. $C_l \rightarrow C_u \rightarrow C_b$, 3. $C_u \rightarrow C_b$, 4. $C_l \rightarrow C_u$ with descending priority. The leaflet commissure points are set to the location of the transition with the highest priority (cf. Fig. 2e green and red point). For initialization of the **leaflet tip points**, the area between the two leaflet commissural points is uniformly sampled with 5 rotational slices in a fan-shaped manner, centered on the saddle horn (cf. Fig. 3a). Two leaflet tip points per slice are initialized on the anterior and posterior portion of the leaflet separation line. Then, the tip points move on their respective slices towards their leaflet root on the annulus (cf. Fig. 2e yellow points). Again, the location of the transition with the highest priority between the classes (1. $C_b \rightarrow C_l$, 2. $C_b \rightarrow C_u \rightarrow C_l$, 3. $C_b \rightarrow C_u$, 4. $C_u \rightarrow C_l$) decides where the movement stops. Finally, the points are backprojected to their original location in the 3D volume Ω (inverse MIP). After initializing the free leaflet edges, the remaining points of the leaflet body need to be defined on each fan slice. Starting with a uniformly subdivided straight connection from each 3D leaflet tip point to its annular root (cf. Fig. 3b), the position of each **leaflet body point** is then refined. Each voxel on the arc in the 3D clustered TTD image Ω is checked for its class value C as well as the presence of the actual graph cut surface S, and are sorted into the following categories: 1. C_l &S, 2. C_u &S, 3. C_l, 4. C_u. For each arc, the averaged location of the voxels in the highest priority group for the respective leaflet body point is taken. Outliers are replaced by interpolating neighboring leaflet body points.

The now initialized model is optimized regarding (1) finding the outermost tip of the free leaflet edge during diastole and (2) creating a coaptation of both leaflets during systole. The model optimization follows a modified active contour approach, where the external force F_{ext} tries to keep the model within the region of high TTD values (i.e., the leaflet) and the internal force F_{int} comprises forces that move the leaflet tip points further to the actual leaflet tip while maintaining reasonable smoothness. $F_{ext}(x) = \nabla_{x,norm} * \frac{(10-(1-|\nabla_x|)}{10} * (1 - \sqrt{TTD(x)})$ weights the TTD gradient with a term ensuring that the external force has less power in regions of high TTD intensity, i.e. gradients in tissue regions can be neglected, whereas gradients in darker regions, for example at the edge of the leaflets, have a strong influence and can keep the points inside the tissue region. The internal force is defined as $F_{int}(x) = P(x) + S(x) + G(x) + M(x)$, where $P(x)$ is a *Push Force* that leads to a stretching of the leaflet points. The connected points p in a single fan slice are labeled, where p_0 is the fixed annulus point, p_n the leaflet tip and the remainder are the leaflet body points. For $0 < i \leq n$ the push force is defined as $P(p_i) = (p_i - p_{i-1})_{norm} * \frac{\sum_{j=1}^{n} |p_j - p_{j-1}|}{|p_i - p_{i-1}|} * w_p$. The push direction is the vector to the point from its preceding neighbour (cf. Fig. 3c). The strength is determined by the current and average segment length, thus end points of shorter segments get pushed harder than end points of longer segments. $w_p = 0.2$ is a weight factor. To preserve an equal point distribution, the *Smoothing Force* $S(x)$ moves points towards their farther neighbour (cf. Fig. 3d). This force is only applied to the leaflet body points $0 < i < n$, let p_f be the farther neighbour of p_i and p_c the closer neighbor, then $S(p_i) = (p_f - p_i)_{norm} * \frac{|p_f - p_i|}{|p_c - p_i|}$. Since the free leaflet edges are usually directed towards the ventricle, a *Gravity Force* $G(x)$ is assigned to each leaflet tip p_n. Considering that the valve model and image are aligned in axial direction, the gravity has a fixed direction, $G(p_{n-1}) = [0, 0, -w_g]^T$ with weight $w_g = 2$. A last force is the *Momentum* $M(x) = F_{prev}(x) * w_m$ of a point, where $w_m = 0.1$ determines the momentum's strength. Here, a portion of the total force $F(x) = F_{ext}(x) + F_{int}(x)$ of the previous iteration is added to the current iteration to overcome small valleys in the TTD intensities.

In order to precisely model the coaptation zone, the forces are adjusted when coaptation of the leaflets is detected. Let p_i be a leaflet tip point and p_j be the opposite one on the same fan slice. Then the following adjustment is done

$$F_{coapt}(p_i) = \begin{cases} F(p_i), & \text{if } |p_i - p_j| > d \\ F(p_i) * (0.5 + a) + F(p_j) * (0.5 - a) + G_c, & \text{otherwise} \end{cases} \quad (1)$$

Table 1. Evaluation results: symmetric model-to-model distance D_s for the entire valve and separately for the anterior (A) and posterior (P) leaflets and their subsegments (A1-A3 and P1-P3, respectively).

		D_s	$D_{s,A}$	$D_{s,P}$	$D_{s,A1}$	$D_{s,A2}$	$D_{s,A3}$	$D_{s,P1}$	$D_{s,P2}$	$D_{s,P3}$
All timesteps	mean	0.94	1.16	1.24	1.22	1.09	1.18	1.59	1.03	1.09
	σ	0.3	0.38	0.37	0.56	0.47	0.36	0.48	0.5	0.48
	95 %	1.76	2.19	2.04	2.68	2.3	2.07	2.48	2.08	2.18
Diastolic timesteps	mean	0.96	1.03	1.12	1.14	0.9	1.04	1.4	0.95	1.02
	σ	0.26	0.29	0.37	0.5	0.32	0.3	0.46	0.62	0.45
	95 %	1.52	1.59	1.99	1.99	1.58	1.7	2.24	2.46	2.06
Systolic timesteps	mean	0.91	1.31	1.36	1.32	1.29	1.33	1.8	1.12	1.17
	σ	0.34	0.51	0.36	0.61	0.62	0.41	0.45	0.34	0.54
	95 %	1.83	2.41	2.01	2.75	2.54	2.23	2.58	1.8	2.34

where $a = \frac{|p_i - p_j|}{2*d}$ and $d = 5$ is a threshold that activates the coaptation treatment. The closer the coaptation points get to each other the more they share their forces and a coaptation of both leaflets is formed (cf. Fig. 3e). The gravitation force $G_c = [0, 0, w_{ga}]^T$ is adjusted to compensate for lower TTD values due to the slightly thicker tissue formed by the coapting leaflets. Finally, an outlier correction is applied to the leaflet tip points. Iteration stops when for all points the applied forces fall below a convergence threshold (cf. Fig. 2).

3   Experiments and Results

A iE33 xMatrix system (Philips Healthcare, Andover, MA, USA) with an X7-2t matrix array transducer in Live-3D mode was used to acquire full volume intraoperative TEEs, which show the left ventricle and the MV. The images were exported in DICOM file format with Cartesian coordinates to enable offline analysis in the open source software toolkit MITK [8]. For evaluation, data from five patients (with severity I-III of MV insufficiency) acquired in clinical routine with all in all 45 systolic and 45 diastolic time steps distributed over several cardiac cycles were used. The average dimensions of the exported files were $233.6 \times 217.6 \times 214.4$ voxels ($0.81 \times 0.8 \times 0.74 \, mm^3$ voxel size).

To obtain a ground truth, an expert without knowledge of the actual mitral valve pathology performed manual image segmentation by 1. definition of leaflet commissure and tip points; 2. definition of leaflet body on 2D slices sampled rotationally over the whole valve. The same annulus model was used during manual as well as automatic segmentation in order to evaluate the algorithm regarding performance of the leaflet segmentation. Most related work use a symmetric model-to-model distance D_s between ground truth and the segmented model computed globally for the entire model to measure the quality of the segmentation results. Following [8], we argue that the global computation of this

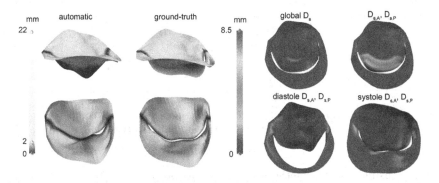

Fig. 4. Left: Coaptation area visualization (green region). Right, top row: Importance of the leaflet-based error measure: Example for an insufficient coaptation line identification that is only penalized using the leaflet-based distance $D_{s,A}, D_{s,P}$. Right, bottom row: Example results of an open and closed valve using $D_{s,A}, D_{s,P}$ (Color figure online).

measure usually gives too optimistic results and is inappropriate for evaluating the segmentation quality of important subparts of the valve, as the coaptation zone (for an example, see Fig. 4 right, top row). Therefore, we also computed the measure for subsurfaces of the valve, such as the leaflets and even the leaflet segments. The results are presented in Table 1. For the anterior leaflet, we achieved a mean error of 1.16 ± 0.38 mm and 1.24 ± 0.37 mm for the posterior leaflet. Leaflet segment P1 showed the highest mean error with 1.4 mm in diastole and 1.8 mm in systole, while the segmentation algorithm performed especially well in the middle segments A2 and P2. The automatic method turned out to be more than 10 times faster than an average manual segmentation. Several 4D quantifications can be obtained from the model, such as the leaflet or coaptation area (see Fig. 4 for a visual impression), which are exemplarily presented in Fig. 5 for two patients. The coaptation area is computed as the part of a leaflet with a symmetric distance smaller than 2 mm to the other leaflet. In Fig. 5, the size of the areas coincides with the division into systolic (S) and diastolic (D) time steps done by an expert.

4   Discussion

Our method is based on delineating the leaflet separation line, which is used during model initialization to obviate a malpositioning of the free leaflet ends. In the optimization step, a surface refinement is applied on different cut slices, which drives the leaflet tips towards the image information (push forces) and regulates the surface curvature (smoothing forces) at the same time. A specific treatment of the leaflet tips when they move closer to each other led to an improved delineation of the coaptation zone while preventing intersection of the two leaflet meshes. Our proposed method generates a surface mesh to represent the geometry with no additional user interaction. Different time steps can be analyzed individually without the need of geometric priors that are applied to other

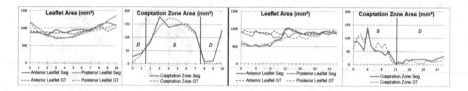

Fig. 5. Development of the leaflet and coaptation areas over time compared between segmentation (Seg) and ground truth (GT) for two patients (Color figure online).

timesteps using tracking methods (e.g., such as proposed in [13]) and might fail in datasets with a low temporal resolution. Furthermore, no atlas construction is necessary as in [9]. The mean symmetric distances D_s between ground truth and the segmented geometry evaluated on 90 time steps are comparable to other state-of-the-art methods. Our 95th percentile evaluation shows an improvement over the recent automatic approach proposed in [9], which may indicate that our algorithm is more robust against outliers using a given annulus model. We emphasize that evaluating parts of the output geometry $(D_{s,Ax}, D_{s,Px})$ against corresponding ground truth parts better penalizes suboptimal leaflet separation and gives therefore a better estimate on how good the coaptation was found, (see example Fig. 4 right). To conclude, our 4D MV model is able to quantify the clincial important coaptation zone for enhanced disease characterisation, diagnostics and therapy planning. In future work, we will evaluate and optimize the method on additional expert segmentations with a broader spectrum of valve pathologies, including, for example, prolapsed valves.

Acknowlegdements. This research received support from the German Research Foundation (DFG), grant ME 833/12-2, SI 1349/1-2 and by the Collaborative Research Centre SFB/TRR 125 *Cognition Guided Surgery* within project B01.

References

1. Boykov, Y., Kolmogorov, V.: An experimental comparison of min-cut/max-flow algorithms for energy minimization in vision. IEEE Trans. Pattern Anal. Mach. Intell. **26**(9), 1124–1137 (2004)
2. Chen, L., McCulloch, A.D., May-Newman, K.: Nonhomogeneous deformation in the anterior leaflet of the mitral valve. Ann. Biomed. Eng. **32**(12), 1599–1606 (2004)
3. Cobey, F.C., Swaminathan, M., Phillips-Bute, B., et al.: Quantitative assessment of mitral valve coaptation using three-dimensional transesophageal echocardiography. Ann. Thorac. Surg. **97**(6), 1998–2004 (2014)
4. Gogoladze, G., Dellis, S.L., Donnino, R., et al.: Analysis of the mitral coaptation zone in normal and functional regurgitant valves. Ann. Thorac. Surg. **89**(4), 1158–1161 (2010)
5. Graser, B., Wald, D., Al-Maisary, S., Grossgasteiger, M., de Simone, R., Meinzer, H.-P., Wolf, I.: Using a shape prior for robust modeling of the mitral annulus on 4D ultrasound data. Int. J. Comput. Assist. Radiol. Surg. **9**(4), 635–644 (2013)

6. Ionasec, R.I., Voigt, I., Georgescu, B., et al.: Patient-specific modeling and quantification of the aortic and mitral valves from 4-D cardiac CT and TEE. IEEE Trans. Med. Imaging **29**(9), 1636–1651 (2010)

7. Nielsen, S.L., Nygaard, H., Mandrup, L., et al.: Mechanism of incomplete mitral leaflet coaptation - interaction of chordal restraint and changes in mitral leaflet coaptation geometry. J. Biomed. Eng. **124**(5), 596–608 (2002)

8. Nolden, M., Zelzer, S., Seitel, A., Wald, D., et al.: The medical imaging interaction toolkit: challenges and advances: 10 years of open-source development. Int. J. Comput. Assist. Radiol. Surg. **8**(4), 607–620 (2013)

9. Pouch, A., Wang, H., Takabe, M., et al.: Fully automatic segmentation of the mitral leaflets in 3D transesophageal echocardiographic images using multi-atlas joint label fusion and deformable medial modeling. Med. Image Anal. **18**(1), 118–129 (2014)

10. Saito, K., Okura, H., Watanabe, N., Obase, K., et al.: Influence of chronic tethering of the mitral valve on mitral leaflet size and coaptation in functional mitral regurgitation. JACC: Cardiovasc. Imaging **5**(4), 337–345 (2012)

11. Schneider, R.J., Burke, W.C., Marx, G.R., del Nido, P.J., Howe, R.D.: Modeling mitral valve leaflets from three-dimensional ultrasound. In: Metaxas, D.N., Axel, L. (eds.) FIMH 2011. LNCS, vol. 6666, pp. 215–222. Springer, Heidelberg (2011)

12. Schneider, R.J., Perrin, D.P., Vasilyev, N.V., et al.: Mitral annulus segmentation from 3D ultrasound using graph cuts. IEEE Trans. Med. Imaging **29**(9), 1676–1687 (2010)

13. Schneider, R.J., Tenenholtz, N.A., Perrin, D.P., Marx, G.R., del Nido, P.J., Howe, R.D.: Patient-specific mitral leaflet segmentation from 4D ultrasound. In: Fichtinger, G., Martel, A., Peters, T. (eds.) MICCAI 2011, Part III. LNCS, vol. 6893, pp. 520–527. Springer, Heidelberg (2011)

Field-Based Parameterisation of Cardiac Muscle Structure from Diffusion Tensors

Bianca Freytag[1], Vicky Y. Wang[1]($^{(\boxtimes)}$), G. Richard Christie[1],
Alexander J. Wilson[1,2], Gregory B. Sands[1,2], Ian J. LeGrice[1,2],
Alistair A. Young[1,3], and Martyn P. Nash[1,4]

[1] Auckland Bioengineering Institute, University of Auckland, Auckland, New Zealand
{bfre608,awil220}@aucklanduni.ac.nz,
{vicky.wang,r.christie,g.sands,i.legrice,a.young,martyn.nash}
@auckland.ac.nz
[2] Department of Physiology, University of Auckland, Auckland, New Zealand
[3] Department of Anatomy with Radiology, University of Auckland,
Auckland, New Zealand
[4] Department of Engineering Science, University of Auckland,
Auckland, New Zealand

Abstract. This paper presents a robust method to directly construct
parametric representations of myocardial structure using a left ventricular (LV) finite element model customised to diffusion tensors derived
from cardiac diffusion tensor magnetic resonance images (DTMRI). This
method avoids the need to solve the eigenvector problem, and therefore avoids issues due to ambiguous eigenvector directions, and the non-
uniqueness of eigenvectors in regions of isotropic diffusion. Finite element
parameters describing the fibre orientations of a geometric model of the
LV are directly fitted to diffusion tensors using non-linear least squares
optimisation. The method was tested using *ex vivo* DTMRI data from
a Wistar-Kyoto rat and compared against the conventional eigenvector
analysis. Close agreement was found in most regions, except at some
boundary locations, and in regions with low fractional anisotropy.

Keywords: Model-based parameterisation · Myocardial fibre orientation · Diffusion tensor magnetic resonance imaging

1 Introduction

Finite element (FE) model-based descriptions of cardiac geometry and
microstructure have become key components in personalised heart modelling
frameworks designed for investigating the electrical [1–3], biomechanical [4–7],
and energetic function of the heart [8,9]. These models enable the integration
of structural and functional data acquired using various imaging modalities,
together with other measurements such as haemodynamic or electrophysiological recordings, to analyse the electro-mechanics of the heart on a subject-specific
basis.

© Springer International Publishing Switzerland 2015
H. van Assen et al. (Eds.): FIMH 2015, LNCS 9126, pp. 146–154, 2015.
DOI: 10.1007/978-3-319-20309-6_17

It is well established that cardiac shape and microstructural tissue organisation are important determinants of biomechanical function of the heart. While *in vivo* measurements of cardiac geometry are readily available via computed tomography (CT), magnetic resonance imaging (MRI) or ultrasound (US), *in vivo* microstructural measurements from the whole heart remain sparse and difficult to quantify. Diffusion tensor MRI (DTMRI) exploits the Brownian motion of the water molecules within myocardial tissue to determine local anisotropic diffusion in isolated heart preparations [10]. The direction of maximum water diffusion, represented by the primary eigenvector of the derived local diffusion tensor, has been found to correlate well with the local histologically-measured myofibre orientation [11,12], and is often represented as an elevation angle (or fibre angle) with respect to the short-axis plane of the heart. There are several approaches for estimating fibre orientations from the diffusion tensor data. For example, shape predictors [13] or analyses of shape-based transformations [14] have been used to personalise fibre orientations. Some studies have proposed probabilistic streamlines incorporating uncertainty [15] or geodesic tractography [16], others have proposed descriptions based on the data from Streeter et al. [17,18]. One of the most common techniques is the deterministic eigenanalysis, which estimates the fibre angle from the orientation of the primary eigenvector of the diffusion tensor. Spatial distributions of fibre angles are typically interpolated within FE models [3,5,6] although some studies have directly interpolated spatial distributions of the diffusion tensors in the log-Euclidean space to ensure that the interpolated tensors are positive-definite [4,19] (which is not guaranteed when independently interpolating the tensor components). In either case, eigenanalyses of diffusion tensors are required to construct local microstructural material axes [4].

The use of the primary eigenvector alone to derive myocardial fibre angles has two main disadvantages. Firstly, water can diffuse equally in opposite directions, but the primary eigenvector arbitrarily represents just one of these directions.

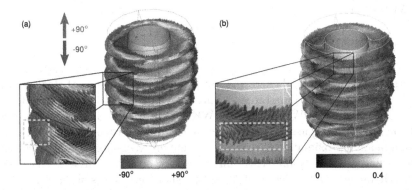

Fig. 1. Primary eigenvectors of the diffusion tensors in the left ventricle (LV) colour-coded by (a) their derived fibre angles showing discontinuities in spatial variation, and (b) fractional anisotropy (FA) showing regions of low FA with corresponding uncertain fibre orientation.

This representation can lead to large discontinuities in fibre angle distributions (see Fig. 1(a)), which can cause problems with FE interpolation of these data. In order to interpolate these spatially discontinuous data within FE models, the eigenvectors are usually phase-unwrapped according to a pre-defined smooth fibre field variation [6], however this method requires prior knowledge of the expected field, and does not eliminate all of the problems with these spatial discontinuities.

The second main issue with eigenanalysis of DTMRI is that in regions of apparent near-isotropic diffusion, the primary eigenvector may not reliably represent the local fibre orientation, as indicated in the inset of Fig. 1(b). Anisotropy in a diffusion process is generally quantified by fractional anisotropy (FA) which ranges from 0 (isotropic diffusion) to 1 (diffusion only along one axis) [20]. Low FA can arise from several factors, including low signal-to-noise ratio in the images [21], or that a single preferred direction does not adequately represent the local tissue microstructure. This can occur, for example, in some regions of myocardial infarct [22], or if there are multiple families of crossing fibres present in the tissue, such as near the inter-ventricular junction of the heart. In the case of crossing fibres, the assumption of a single direction to represent the fibres would need to be re-addressed. On the other hand, if regions of low FA are localised zones of poor image quality, then one could reduce the influence of these data during the fitting of fibre field parameters by incorporating a specialised weighting scheme, such as scaling the fitting error in each voxel by the FA value.

In this study, we present a workflow for parameterising myocardial fibre fields directly from the diffusion tensors without the need to compute eigenvectors or FA values for fibre orientation fitting. This framework not only circumvents issues associated with phase-unwrapping of eigenvectors prior to fibre fitting, but also helps to ensure that the interpolated fibre angles in regions with high FA are better representations of the diffusion tensors in those regions.

2   Methods

2.1   Experimental Data

The experimental study was approved by the Animal Ethics Committee of the University of Auckland and conforms to the National Institutes of Health Guide for the Care and Use of Laboratory Animals (NIH Publication No. 85–23).

The heart from a Wistar-Kyoto (WKY) rat was excised, perfused with St. Thomas' cardioplegic solution to relax the heart, and then fixed using Bouin's solution in an approximate end-diastolic state. One week later, DTMRI was performed using a 3D fast spin-echo pulse sequence on a Varian 4.7T MRI scanner. Each image set consisted of 11 or 12 short-axis slices with a thickness of 1.5 mm, and no gap between slices. Each slice contained one non-diffusion weighted anatomical image, and 30 diffusion weighted images. Other imaging parameters were as follows: TE = 15 ms; TR = 3 s; number of averages = 6; field of view (FOV) = 20 mm × 20 mm × 16 mm; in-plane resolution = 128 × 64 voxels (zero-pad interpolated to 128 × 128 voxels); in-plane voxel dimensions = 156 μm × 156 μm.

2.2   Fibre Field Parameterisation Workflow

The following workflow (Fig. 2) was developed to parameterise a spatially vary-ing myocardial fibre field directly from the diffusion tensors throughout the LV myocardium.

Step 1: Diffusion tensor calculation. From the non-diffusion weighted sig-nals (S_0) and the diffusion weighted signals (S_k) captured in 30 non-collinear directions (i.e. $k = 1...30$), a diffusion tensor, denoted $\mathbf{D}$ (a symmetric tensor with six independent components), was estimated for each voxel by solving the logarithm of the diffusion equation (Eq. 1) proposed by Basser et al. [10].

$$\log(S_k) = \log(S_0) - b\,\mathbf{Q}_{(k)}\mathbf{D} \text{ where } \mathbf{Q}_{(k)} = \mathbf{g}_{(k)} \otimes \mathbf{g}_{(k)} \tag{1}$$

where b is the diffusion weighting factor ($1462\,\text{s/mm}^2$ for this study). The outer-product $\mathbf{Q}_{(k)}$ of the diffusion gradient direction $(\mathbf{g}_{(k)})$ was pre-calculated before the system of equations was solved using least-squares to determine the six independent components of $\mathbf{D}$ at each voxel.

Step 2: Image segmentation. The analysis was limited to the LV for this study. The entire endocardial and epicardial surfaces of the LV were manually segmented from the non-diffusion images using MATLAB[1]. During the segmen-tation, papillary muscles and trabeculae were excluded. Three landmark points (the centroids of the LV base, LV apex, and RV base) were selected to construct an orthogonal cardiac coordinate system with the origin located one-third of the distance from base to apex along the long-axis of the LV, with which the x-axis was aligned. The y-axis was directed from the LV to RV centroids, and the z-axis was directed orthogonal to the x-axis and y-axis from the anterior to posterior of the LV.

Step 3: LV FE geometric model construction. To obtain a model repre-sentation of the LV geometry, a prolate ellipsoid-shaped 16-element (4 circum-ferential, 4-longitudinal and 1-transmural) hexahedral FE model was customised to the surface contours obtained from Step 2. The endocardial and epicardial surfaces of the model were simultaneously fitted using non-linear least squares to best match the corresponding surface data.

Step 4: Field-based parameterisation of LV fibre orientation. In order to parameterise the myofibre orientation field within the LV FE geometric model, we developed a new approach to estimate a smoothly continuous fibre field that best aligned with the diffusion tensors at all voxels within the LV. Firstly, we initialised the FE fibre field by setting the fibre angles $(\theta_{(n)})$ to be $+60°$ for the endocardial nodes, and $-70°$ for the epicardial nodes. Initial imbrication angles

---

[1] The MathWorks, Inc., Natick, Massachusetts, United States.

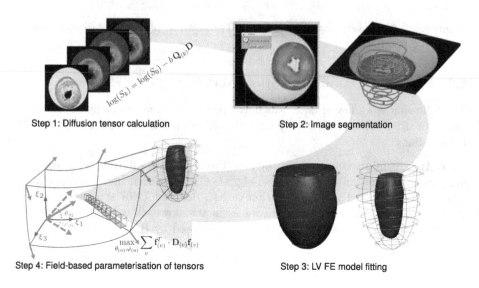

Step 1: Diffusion tensor calculation

Step 2: Image segmentation

Step 4: Field-based parameterisation of tensors

Step 3: LV FE model fitting

Fig. 2. Workflow for field-based parameterisation of diffusion tensors.

$(\varphi_{(n)})$ of all nodes were set to be $0°$. These angles were interpolated over the FE model using tricubic Hermite basis functions. Secondly, for each voxel (v), we determined its FE local coordinates within the LV geometric model, and, at that location, we computed the myofibre orientation $(\mathbf{f}_{(v)})$ defined by Euler angle rotation (using interpolated θ and φ) of vectors initially aligned with the local element axes [23]. Thirdly, a scalar objective function (Δ) was constructed using Eq. 2. At a given voxel, if $\mathbf{f}_{(v)}$ and the direction of maximal diffusion were perfectly aligned, then $\delta_{(v)}$, defined as the dot product of $\mathbf{f}_{(v)}$ with the projection of the diffusion tensor $(\mathbf{D}_{(v)})$ in the direction of $\mathbf{f}_{(v)}$, would be maximal. Conversely, $\delta_{(v)}$ would be minimal if $\mathbf{f}_{(v)}$ was aligned with the direction of minimal diffusion. Finally, this objective function was maximised using nonlinear optimisation (least-squares quasi-Newton) by modifying the nodal parameters $(\theta_{(n)}$ and $\varphi_{(n)})$. This procedure was implemented using the open-source Cmgui software package[2] [24,25].

$$\Delta = \sum_v \delta_{(v)} \text{ where } \delta_{(v)} = \mathbf{f}_{(v)}^T \cdot \mathbf{D}_{(v)} \mathbf{f}_{(v)} \tag{2}$$

3   Results

Based on the optimal fibre field fitted to the diffusion tensor data, we extracted the fibre orientations at all voxels from which the original tensors were calculated. Figure 3(a) shows the estimated fibre orientations colour-coded using the interpolated fibre angles. The resultant fibre field exhibits the expected smooth

---

[2] OpenCMISS-Cmgui application, www.opencmiss.org.

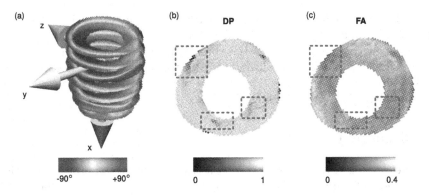

Fig. 3. (a) Fitted fibre orientations colour-coded by the interpolated fibre angles at all voxels within the LV myocardium. (b) A mid-ventricular slice of dot product (DP) values between interpolated fibre orientations and primary eigenvectors of the associated diffusion tensors (see text for details). (c) Map of the FA values for the same mid-ventricular slice. The dotted squares indicate areas of low correlation between the estimated fibre orientations and primary eigenvectors.

spatial variations throughout the LV, from negative angles (with respect to the short-axis plane) at the epicardium, to positive angles at the endocardium.

To quantitatively assess the ability of the new workflow to reconstruct local myofibre orientations, we performed eigenanalysis of the diffusion tensor for every voxel and examined the alignment between the estimated fibre orientation ($\mathbf{f}_{(v)}$) from our new workflow and the corresponding primary eigenvectors ($\mathbf{e}_{1(v)}$) on a voxel basis. When an estimated fibre vector is in perfect alignment with the primary eigenvector, their dot product is one. Conversely, if the two vectors are orthogonal, then the dot product is zero. We took the absolute value of the dot product as we were only interested in how well the vectors were aligned to each other regardless of their direction. To assess the overall degree of alignment throughout the model, we evaluated a normalised dot product (nDP; see Eq. 3), defined as the ratio of the sum of the dot products scaled by their associated FA values ($\text{FA}_{(v)}$) over all voxels (v), to the sum of FA values over all voxels. nDP is in the range from 0 to 1, with 1 representing best agreement of orientations. FA values were incorporated into the calculation of the nDP index to account for the differing degrees of confidence in the calculated eigenvectors (since an eigenvector in a voxel with low FA may not represent the underlying fibre orientation).

$$\text{nDP} = \frac{\sum_v (\text{FA}_{(v)} |\mathbf{f}_{(v)} \cdot \mathbf{e}_{1(v)}|)}{\sum_v \text{FA}_{(v)}} = 0.97 \tag{3}$$

The resulting nDP index was close to one, suggesting a high correlation between the estimated fibre orientations and primary eigenvectors of the diffusion tensors. Figure 3(b) shows a map of dot products for all voxels in a mid-ventricular slice. The alignment was generally good, except for some small localised regions in the interior, and near the boundaries of the LV. Boundary errors may be caused by the diffusion signals near the endocardial and epicardial surfaces being contaminated by partial-volume imaging artifacts, which may lead to unreliable diffusion tensor calculations near the ventricular surfaces. Internal regions with large disagreement between field and eigenvectors were generally associated with low FA, as highlighted in Fig. 3(c).

4   Discussion

We have developed a new method to construct a continuous fibre field within an FE model of the LV by directly fitting to cardiac diffusion tensor data. This approach was motivated by issues encountered with the eigenanalysis of diffusion tensors that can potentially lead to misrepresentation of the local myofibre orientation. Such misinterpretation can have a significant impact on electrophysiological and mechanical modelling studies since the orientations are used to construct local microstructural material axes from which electrophysiological, passive and contractile constitutive properties are derived.

We demonstrated that the fibre orientations estimated using this method agree well with the standard eigenvectors in regions with high FA, within which the primary eigenvector has been shown to reliably represent the local myofibre orientation. However, in regions with low FA, this method will still define continuously varying fibre orientations. If the main contributor to low FA is noise, then maintaining continuity in fibre field despite low FA is an important advantage of our method. On the other hand, if the image quality is high in regions with low FA, then the structural model should be carefully considered, and it may be appropriate to take the FA data into account when constructing biomechanical or electrophysiological constitutive models in these regions.

In previous studies that have explored direct interpolation of diffusion tensors [3,4,19], the tensors were first transformed into logarithmic space, and the six independent components of the logarithmic form were fitted as 3D scalar fields using FE interpolation. Fitted values were then transformed back to the Euclidean space to construct interpolated tensors throughout the models. To construct local fibre vectors, the primary eigenvectors were calculated from these interpolated tensors. Given that the tensors were interpolated using continuous functions, then this approach may improve the continuity of eigenvector directions, although this is not guaranteed. Furthermore, the above issues regarding variations in fractional anisotropy remain.

5   Conclusions

A field-based parameterisation scheme was developed to analyse *ex vivo* diffusion tensor MRI data. This scheme does not require the conventional calculation of primary eigenvectors of the diffusion tensors. Instead, myocardial fibre fields are fitted directly to spatial distributions of diffusion tensors. Results showed that the field-fitted fibre orientations correlated well with the primary eigenvectors in regions with high FA values. This workflow could be adapted to construct fibre fields using *in vivo* cardiac imaging data and associated geometric FE models for individualised analyses of heart mechanics.

References

1. Sermesant, M., Chabiniok, R., Chinchapatnam, P., Mansi, T., Billet, F., Moireau, P., Peyrat, J.M., Wong, K., Relan, J., Rhode, K., et al.: Patient-specific electromechanical models of the heart for the prediction of pacing acute effects in CRT: a preliminary clinical validation. Med. Image Anal. **16**(1), 201–215 (2012)
2. Sermesant, M., Delingette, H., Ayache, N.: An electromechanical model of the heart for image analysis and simulation. IEEE Trans. Med. Imaging **25**(5), 612–625 (2006)
3. Vadakkumpadan, F., Gurev, V., Constantino, J., Arevalo, H., Trayanova, N.: Modeling of whole-heart electrophysiology and mechanics: toward patient-specific simulations. In: Kerckhoffs, R.C. (ed.) Patient-Specific Modeling of the Cardiovascular System, pp. 145–165. Springer, New York (2010)
4. Krishnamurthy, A., Villongco, C.T., Chuang, J., Frank, L.R., Nigam, V., Belezzuoli, E., Stark, P., Krummen, D.E., Narayan, S., Omens, J.H., et al.: Patient-specific models of cardiac biomechanics. J. Comput. Phys. **244**, 4–21 (2013)
5. Walker, J.C., Ratcliffe, M.B., Zhang, P., Wallace, A.W., Hsu, E.W., Saloner, D.A., Guccione, J.M.: Magnetic resonance imaging-based finite element stress analysis after linear repair of left ventricular aneurysm. J. Thorac. Cardiovasc. Surg. **135**(5), 1094–1102 (2008)
6. Wang, V.Y., Lam, H., Ennis, D.B., Cowan, B.R., Young, A.A., Nash, M.P.: Modelling passive diastolic mechanics with quantitative MRI of cardiac structure and function. Med. Image Anal. **13**(5), 773–784 (2009)
7. Xi, J., Lamata, P., Niederer, S., Land, S., Shi, W., Zhuang, X., Ourselin, S., Duckett, S.G., Shetty, A.K., Rinaldi, C.A., et al.: The estimation of patient-specific cardiac diastolic functions from clinical measurements. Med. Image Anal. **17**(2), 133–146 (2013)
8. Niederer, S.A., Smith, N.P.: The role of the frank-starling law in the transduction of cellular work to whole organ pump function: a computational modeling analysis. PLoS Comput. Biol. **5**(4), e1000371 (2009)
9. Wang, V.Y., Ennis, D.B., Cowan, B.R., Young, A.A., Nash, M.P.: Myocardial contractility and regional work throughout the cardiac cycle using FEM and MRI. In: Camara, O., Konukoglu, E., Pop, M., Rhode, K., Sermesant, M., Young, A. (eds.) STACOM 2011. LNCS, vol. 7085, pp. 149–159. Springer, Heidelberg (2012)
10. Basser, P.J., Mattiello, J., LeBihan, D.: Estimation of the effective self-diffusion tensor from the NMR spin echo. J. Magn. Reson. Ser. B **103**(3), 247–254 (1994)
11. Hsu, E., Muzikant, A., Matulevicius, S., Penland, R., Henriquez, C.: Magnetic resonance myocardial fiber-orientation mapping with direct histological correlation. Am. J. Physiol. Heart Circ. Physiol. **274**(5), H1627–H1634 (1998)
12. Scollan, D.F., Holmes, A., Winslow, R., Forder, J.: Histological validation of myocardial microstructure obtained from diffusion tensor magnetic resonance imaging. Am. J. Physiol. Heart Circ. Physiol. **275**(6), H2308–H2318 (1998)
13. Lekadir, K., Hoogendoorn, C., Pereanez, M., Albà, X., Pashaei, A., Frangi, A.F.: Statistical personalization of ventricular fiber orientation using shape predictors. IEEE Trans. Med. Imaging **33**(4), 882–890 (2014)

14. Toussaint, N., Stoeck, C.T., Schaeffter, T., Kozerke, S., Sermesant, M., Batchelor, P.G.: In vivo human cardiac fibre architecture estimation using shape-based diffusion tensor processing. Med. Image Anal. **17**(8), 1243–1255 (2013)

15. Jones, D.K., Pierpaoli, C.: Confidence mapping in diffusion tensor magnetic resonance imaging tractography using a bootstrap approach. Magn. Reson. Med. **53**(5), 1143–1149 (2005)

16. Jbabdi, S., Bellec, P., Toro, R., Daunizeau, J., Pélégrini-Issac, M., Benali, H.: Accurate anisotropic fast marching for diffusion-based geodesic tractography. J. Biomed. Imaging **2008**, 2 (2008)

17. Bayer, J., Blake, R., Plank, G., Trayanova, N.: A novel rule-based algorithm for assigning myocardial fiber orientation to computational heart models. Ann. Biomed. Eng. **40**(10), 2243–2254 (2012)

18. Streeter, D.D., Spotnitz, H.M., Patel, D.P., Ross, J., Sonnenblick, E.H.: Fiber orientation in the canine left ventricle during diastole and systole. Circ. Res. **24**(3), 339–347 (1969)

19. Vadakkumpadan, F., Arevalo, H., Prassl, A.J., Chen, J., Kickinger, F., Kohl, P., Plank, G., Trayanova, N.: Image-based models of cardiac structure in health and disease. Wiley Interdisc. Rev. Syst. Biol. Med. **2**(4), 489–506 (2010)

20. Basser, P.J., Pierpaoli, C.: Microstructural and physiological features of tissues elucidated by quantitative-diffusion-tensor MRI. J. Magn. Reson. **111**, 209–219 (1996)

21. Farrell, J.A., Landman, B.A., Jones, C.K., Smith, S.A., Prince, J.L., van Zijl, P., Mori, S.: Effects of signal-to-noise ratio on the accuracy and reproducibility of diffusion tensor imaging-derived fractional anisotropy, mean diffusivity, and principal eigenvector measurements at 1.5 t. J. Magn. Reson. Imaging **26**(3), 756–767 (2007)

22. Fomovsky, G.M., Rouillard, A.D., Holmes, J.W.: Regional mechanics determine collagen fiber structure in healing myocardial infarcts. J. Mol. Cell. Cardiol. **52**(5), 1083–1090 (2012)

23. LeGrice, I.J., Hunter, P.J., Smaill, B.: Laminar structure of the heart: a mathematical model. Am. J. Physiol. Heart Circ. Physiol. **272**, H2466–H2476 (1997)

24. Christie, G., Bullivant, D., Blackett, S., Hunter, P.J.: Modelling and visualising the heart. Comput. Vis. Sci. **4**(4), 227–235 (2002)

25. Bradley, C., Bowery, A., Britten, R., Budelmann, V., Camara, O., Christie, R., Cookson, A., Frangi, A., Gamage, T., Heidlauf, T., Krittian, S., Ladd, D., Little, C., Mithraratne, K., Nash, M., Nickerson, D., Nielsen, P., Nordbø, T., Omholt, S., Pashaei, A., Paterson, D., Rajagopal, V., Reeve, A., Röhrle, O., Safaei, S., Sebastián, R., Steghfer, M., Wu, T., Yu, T., Zhang, H., Hunter, P.: OpenCMISS: a multi-physics & multi-scale computational infrastructure for the VPH/Physiome project. Prog. Biophys. Mol. Biol. **107**(1), 32–47 (2011)

Left Atrial Segmentation from 3D Respiratory- and ECG-gated Magnetic Resonance Angiography

Rashed Karim[1]([✉]), Henry Chubb[1], Wieland Staab[3], Shadman Aziz[1],
R. James Housden[1], Mark O'Neill[1,2], Reza Razavi[1,2], and Kawal Rhode[1]

[1] Department of Biomedical Engineering, King's College London, London, UK
rashed.karim@kcl.ac.uk
[2] Department of Cardiology, Guy's and St. Thomas' NHS Foundation Trust,
London, UK
[3] Department of Cardiology and Pneumology, University of Göttingen,
Göttingen, Germany

Abstract. Magnetic resonance angiography (MRA) scans provide excellent chamber and venous anatomy. However, they have traditionally been acquired in breath-hold and are not cardiac-gated. This has made it difficult to use them in conjunction with late gadolinium enhancement (LGE) scans for reconstructing fibrosis/scar on 3D left atrium (LA) anatomy. This work proposes an image processing algorithm for segmenting the LA from a novel MRA sequence which is both ECG-gated and respiratory-gated allowing reliable 3D reconstructions with LGE. The algorithm implements image partitioning using discrete Morse theory on digital images. It is evaluated in the context of creating 3D reconstructions of scar/fibrosis with LGE.

Keywords: Image segmentation · Delayed-enhancement MRI · Magnetic resonance angiography · Left atrium

1 Introduction

Cardiac magnetic resonance (CMR) has evolved in the last two decade as a technique for non-invasive assessment of cardiac anatomy and function. The focus has primarily been on the ventricles. Recent advancements in CMR techniques have made it possible to investigate the atria. There has been increasing interest to image the left atrium (LA) to derive biomarkers that are relevant to the management of atrial fibrillation. One such biomarker is fibrosis/scar pre- and post- ablation therapy. The gold standard method is to use late gadolinium enhancement (LGE) imaging. In this imaging sequence, the wash-out kinetics of the contrast agent within the fibrosis/scar regions allows it be imaged.

Accurate assessment of left atrial myocardial fibrosis/scar size and anatomy with LGE is challenging. The 3D reconstructions of fibrosis/scar on LA anatomy can be obtained for this purpose. A shape-based segmentation of

© Springer International Publishing Switzerland 2015
H. van Assen et al. (Eds.): FIMH 2015, LNCS 9126, pp. 155–163, 2015.
DOI: 10.1007/978-3-319-20309-6_18

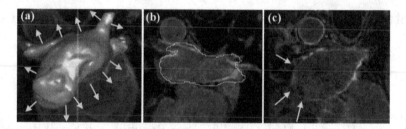

Fig. 1. (a) The shape-based MIP approach with surface normals (arrows) interrogating the thin atrial wall imaged with LGE. (b) Contour from high resolution 3D whole heart b-SSFP scan superimposed on LGE to demonstrate mis-fitting in less common variants of LA anatomy (c) Atrial wall imaged in LGE scan and arrows showing areas where a manual annotation of epi- and endo-cardium is challenging.

LA anatomy fused with LGE is the preferred technique at our institution [4]. This technique is based upon high resolution 3D whole-heart steady state free precessing (b-SSFP) MRI images and with shape-constrained deformable models (SmartHeart, Philips Research, Hamburg, Germany). The segmentation requires registration to the LGE sequence. An ensuing step interrogates the thin atrial wall LGE intensities using surface normals of the LA (see Fig. 1(a)). This method of 3D reconstructing fibrosis/scar on LA anatomy is also referred to as maximum intensity projection (MIP).

The shape-based MIP approach [4] provides a good visual assessment of fibrosis/scar. However, there are some disadvantages. Firstly, the b-SSFP image is a four-chamber whole heart scan. Separating the LA from the other chambers remains a challenging problem despite a decade of research. A recent challenge was organised by our institution to address limitations of current techniques [6]. Secondly, the shape-based approach can be inaccurate with less common variants of the LA anatomy.

An alternative to the shape-based MIP approach is the creation of 3D reconstructions of fibrosis/scar and LA anatomy solely from the LGE image [5]. This relies on accurate segmentation of LA. A common widely used manual approach is to use image processing software, such as Seg3D (NIH NCRR Centre for Integrative Biomedical, University of Utah, Utah, USA), ITK-Snap (University of Pennsylvania, Pennsylvania, USA) or Osirix (Pixmeo SARL, Bernex, Switzerland). The accuracy of segmentation is dependent upon operator experience and image quality. Annotating the thin epicardial and endocardial boundaries in LGE are time-consuming and prone to error (see Fig. 1(c)). Furthermore, obtaining smooth 3D reconstructions from manual annotations require further image processing steps. Often this involves smoothing filters, and finer details in the anatomy are lost.

In this work, we propose an image processing algorithm to segment the LA anatomy from a novel respiratory and cardiac ECG-gated magnetic resonance angiography (gated-MRA) sequence [1]. The gated-MRA scan provides for excellent atrial and venous anatomy. Unlike traditional MRA scans which

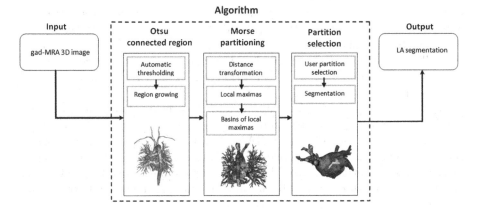

Fig. 2. A flow diagram showing the steps in the proposed algorithm.

are not cardiac-gated and acquired in breath-hold, they require registration to the LGE image sequence. With cardiac ECG-gating, both the gated-MRA and LGE images can be acquired in the same cardiac cycle, requiring no registration. With the proposed technique, the 3D reconstructions of fibrosis/scar and LA can be obtained with greater reliability than previous mentioned works.

2   Methods

2.1   Image Acquisition

All patients in this study ($n = 7$) were scanned on a commercial 1.5T Ingenia MRI scanner (Philips Healthcare, The Netherlands). Electrocardiographic electrodes were positioned for optimal gating before the study. A free breathing, contrast enhanced MRA using an ECG triggered, respiratory gated, inversion recovery prepared, 3D image sequence was acquired. The acquisition was started 90 s after initiation of an intravenous infusion of Gadolinium (Gadovist, Bayer, Germany) at a rate of 0.3 ml/s. The reconstructed image pixel resolution was: $1 \times 1 \times 2$ mm and typically 50 slices to cover the whole of the LA. A standard 3D LGE sequence was also performed using the same gating parameters. The reconstructed image pixel resolution was: $0.6 \times 0.6 \times 2$ mm and 45–50 slices.

2.2   Algorithm

The steps involved in the proposed algorithm are shown as a flow diagram in Fig. 2. The algorithm exploits *Morse* partitions of the image. In Morse theory, the topology of sub-level sets of real-valued functions is connected to its critical points. In binary images, real-valued functions can be derived from the distance transformation, and the sub-level set consists of all voxels x where $f(x) \leq c$. The proposed algorithm partitions and thus segments the LA by exploiting the *basins* surrounding the local maxima of $f(x)$. This is closely related to the watershed transform. In this work, the topology was defined by considering all voxels to be *connected* to its immediate neighbours (i.e. 26-neighbourhood or cubical complex). The algorithm was comprised of three steps: (1) Otsu region growing, (2) Morse partitioning, and (3) partition selection. Each step is herewith described in detail.

In the Otsu region growing step, a connected region was obtained from a user-selected seed point and region of interest (ROI). The threshold in the Otsu step was computed on the gated-MRA image; it is the optimal intensity which partitions the image into two tissue classes. This was ideal in gated-MRA where there is a high contrast between blood and surrounding structures. The Otsu step generated a well-connected binary mask containing the LA and surrounding chambers.

In the Morse partitioning step illustrated in Fig. 3, the distance transform $f(x)$ for the Otsu binary region was computed, where for every voxel x, f is the real-valued distance to the closest zero voxel. The local maxima of f was computed in an ensuing step. *Basins* surround each local maximum point in f. These basins are defined as voxels from which a *path* to the maxima exists. This path is the gradient flow $\phi(x, t)$ of f, where:

$$\phi(x, 0) = x$$
$$\frac{\partial}{\partial t}\phi(x, t) = \nabla f(\phi(x, t))$$

Each basin is a Morse partition B surrounding a critical point (i.e. local maximum) p:

$$B_p = \{x| \lim_{t \to \infty} \phi(x, t) = p\}$$

In the context of gated-MRA, the hollow LA chamber comprised of several local maxima on the distance transformation, yielding several such basins. The LA was generally connected to the aorta, coronaries and pulmonary artery due to partial voluming. Due to the nature of the distance transform and Morse partitions, rarely partitions spanned across chambers. For example, a partition spanning across the LA and aorta. It was this special feature of the algorithm that enabled the LA to be separated from its surrounding structures. However, a final partition selection step was necessary.

In the final partition selection step, the user selects the partitions (or basins) that should be included in the LA. The final segmentation is then computed from this selection.

2.3 Evaluation

The focus of this work was to obtain LA anatomy for quantifying fibrosis/scar in LGE. A time-consuming rigorous manual segmentation of the LA from LGE images generally gives the most reliable 3D reconstruction of LGE scar on LA

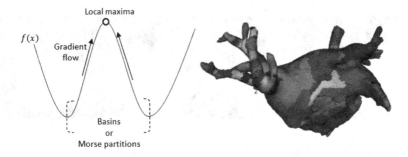

Fig. 3. (Left) A schematic showing how Morse partitions are computed from the distance transform $f(x)$ of the binary image. (Right) Morse paritions labelled on a 3D reconstruction of the LA.

anatomy. To evaluate the algorithm, segmentations were thus compared to LA anatomy obtained manually from LGE images. These manual segmentations were obtained from an observer with experience in LA anatomy. The observer analysed each corresponding LGE image, starting from the base of the LA, identifying the location of the mitral valve region, and working upwards to the roof. In each slice, the atrial wall boundary was annotated by selecting several points along the boundary and drawing a polygonal contour. The atrial wall contour was interpolated in regions where it was not visibly clear. Segmentations from the algorithm were also manually corrected, when necessary. However, the original segmentations obtained from the algorithm, without the manual correction step, were evaluated separately.

Segmentations were compared using two different techniques. The similarity between segmentations from the algorithm and manual method was quantified with a Dice similarity metric, separately computed on the LA boundary and inner chamber. A second technique evaluated smoothness of the segmentations. Smoothness was obtained by quantifying the disparity in the directions of the 3D LA surface normals. Measuring disparity of surface normals within a neighbourhood, as opposed to surface curvature, is more important. This is due to the method for creating 3D reconstructions of fibrosis/scar on LA anatomy, which relies on smooth surface normals interrogating the atrial wall in the LGE image. Furthermore, evaluating smoothness allows us to investigate the hypothesis that manual annotation of the LA on a slice-by-slice basis yields non-smooth surfaces, often difficult to recover using smoothing filters.

3   Results

Results from the algorithm in some selected cases can be seen in Fig. 4. The similarity between segmentations obtained from the algorithm (on gated-MRA) and manual annotation (on LGE) were compared with the Dice similarity metric. As the Dice metric on digital images is computed voxel-wise, it can be biased to large voxel volumes that differ mostly on the boundary. To eliminate this bias,

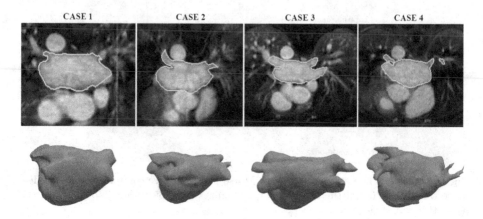

Fig. 4. Segmentations from the algorithm on gated-MRA showing contours on a selected slice, and the corresponding 3D reconstructions in four selected cases.

the Dice metric was computed for each image slice individually. The Dice metric is plotted in Fig. 5 for images before and after manual correction. A time limit of 2 min was set for manual correction of the algorithm's segmentations. This included the time taken for seed placement and Morse partitioning.

The aforementioned large voxel volume bias in the Dice metric can be further eliminated by investigating the segmentations along boundaries. This is the region where segmentations mostly differ. A boundary error metric in millimeters was computed on each slice individually. First, the boundary for each segmentation was computed. Next, the distance between each pixel on the two separate boundaries was computed, giving a mean distance. The distribution of

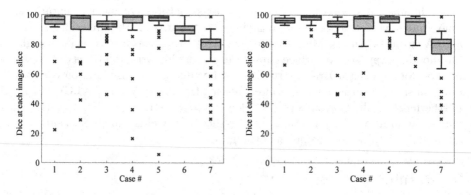

Fig. 5. Similarity between segmentations obtained from the algorithm on gated-MRA and manually annotated anatomy on LGE image on all seven cases, based on the Dice metric. Segmentations from the algorithm were manually corrected when necessary and the plots above show before (left plot) and after (right plot) correction.

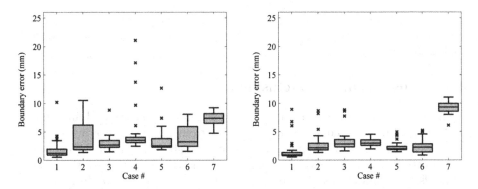

Fig. 6. The mean difference (in mm) along the boundary of the segmentations obtained from the algorithm on gated-MRA and manually annotated anatomy on LGE image on all seven cases. Segmentations from the algorithm were manually corrected when necessary and the plots above show before (left plot) and after (right plot) correction. (1 pixel $\sim$ 2 mm based on the image resolution)

this distance in each individual case is plotted in Fig. 6 for segmentations before and after manual correction.

The smoothness of the 3D reconstructions obtained both from algorithm and manual annotation was evaluated. The surface normal at every point was investigated by measuring the angular difference from other surface normals in its neighbourhood. The standard deviation of these angle-difference distributions were plotted and is shown for some example cases in Fig. 7. Finally, the total

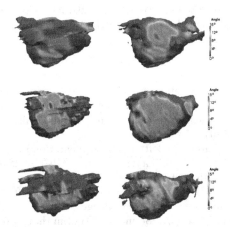

Fig. 7. Three cases where the smoothness of 3D reconstructed surfaces of manually annotated (left column) and algorithm segmentations (right column) are plotted. The smoothness is based on the distribution of angle differences between surface normal vectors in a neighbourhood.

time taken to run the algorithm, interact and manually correct the segmentations was measured as 3.2 ± 1.4 min. In comparison, manual annotations took 23.4 ± 2.4 min.

4   Discussion and Conclusion

Accurate assessment of left atrial myocardial fibrosis/scar size and anatomy is challenging. Shape-based segmentation of LA from high resolution whole-heart b-SSFP is a preferred technique at our institution [4]. Manual annotation of the LA directly from LGE images is also a well-known technique [5]. Both these methods have limitations. The proposed algorithm segments LA from a novel respiratory and ECG-gated MRA sequence. This enables 3D reconstructions of fibrosis/scar on LA anatomy with greater accuracy and reliability. Segmentation of LA from MRA images using distance transformations is not novel, and this has been investigated in earlier works [2,3]. The proposed work implements distance transform image partitioning with Morse theory and evaluates it in the context of LGE fibrosis/scar quantification.

Results demonstrated good similarity with manual annotations of the LA on LGE images. The analysis of smoothness is important in the context of 3D reconstructions of LGE-imaged fibrosis/scar. In the analysis, regions with highly variable surface normals (in terms of their direction) were identified. Results have demonstrated that these high curvature regions are more dense in annotated segmentations. This is since annotations are performed on the single axial view, which is typically the LGE acquisition orientation. The algorithm's segmentation achieved greater smoothness due to its 3D image processing pipeline.

The presented work has several limitations. The algorithm requires manual selection of partitions and an automation may be possible. Also, the algorithm was not validated with other existing MRA segmentation algorithms and thus it may be difficult to assess the quality of the results.

References

1. Groarke, J.D., Waller, A.H., et al.: Feasibility study of electrocardiographic and respiratory gated, gadolinium enhanced magnetic resonance angiography of pulmonary veins and the impact of heart rate and rhythm on study quality. J. Cardiovasc. Magn. Reson. **16**(1), 43 (2014)
2. John, M., Rahn, N.: Automatic left atrium segmentation by cutting the blood pool at narrowings. In: Duncan, J.S., Gerig, G. (eds.) MICCAI 2005. LNCS, vol. 3750, pp. 798–805. Springer, Heidelberg (2005)
3. Karim, R., Juli, C., et al.: Automatic segmentation of left atrial geometry from contrast-enhanced magnetic resonance images using a probabilistic atlas. In: Camara, O., Pop, M., Rhode, K., Sermesant, M., Smith, N., Young, A. (eds.) STACOM 2010. LNCS, vol. 6364, pp. 134–143. Springer, Heidelberg (2010)
4. Knowles, B., Caulfield, D., Cooklin, M., Rinaldi, C., Gill, J., Bostock, J., Razavi, R., Schaeffter, T., Rhode, K.: 3-D visualization of acute RF ablation lesions using MRI for the simultaneous determination of the patterns of necrosis and edema. IEEE Trans. Biomed. Eng. **57**(6), 1467–1475 (2010)

5. Oakes, R., Badger, T., et al.: Detection and quantification of left atrial structural remodeling with delayed-enhancement magnetic resonance imaging in patients with atrial fibrillation. Circulation **119**(13), 1758–1767 (2009)
6. Tobon-Gomez, C., Peters, J., et al.: Left atrial segmentation challenge: a unified benchmarking framework. In: Camara, O., Mansi, T., Pop, M., Rhode, K., Sermesant, M., Young, A. (eds.) STACOM 2013. LNCS, vol. 8330, pp. 1–13. Springer, Heidelberg (2014)

A Comprehensive Framework for the Characterization of the Complete Mitral Valve Geometry for the Development of a Population-Averaged Model

Amir H. Khalighi[1], Andrew Drach[1], Fleur M. ter Huurne[2], Chung-Hao Lee[1],
Charles Bloodworth[3], Eric L. Pierce[3], Morten O. Jensen[3], Ajit P. Yoganathan[3],
and Michael S. Sacks[1(✉)]

[1] The University of Texas at Austin, Austin, TX 78712, USA
{akhalighi,msacks}@ices.utexas.edu
[2] Eindhoven University of Technology, Eindhoven, The Netherlands
[3] Georgia Institute of Technology, Atlanta, GA 30332, USA

Abstract. Simulations of the biomechanical behavior of the Mitral Valve (MV) based on simplified geometric models are difficult to interpret due to significant intra-patient variations and pathologies in the MV geometry. Thus, it is critical to use a systematic approach to characterization and population-averaging of the patient-specific models. We introduce a multi-scale modeling framework for characterizing the entire MV apparatus geometry via a relatively small set of parameters. The leaflets and annulus are analyzed using a superquadric surface model superimposed with fine-scale filtered level-set field. Filtering of fine-scale features is performed in a spectral space to allow control of resolution, resampling and robust averaging. Chordae tendineae structure is modeled using a medial axis representation with superimposed filtered pointwise cross-sectional area field. The chordae topology is characterized using orientation and spatial distribution functions. The methodology is illustrated with the analysis of an ovine MV microtomography imaging data.

1 Introduction

The Mitral Valve (MV) is the bi-leaflet heart valve separating the left atrium and ventricle while regulating blood flow direction during cardiac cycles. The MV complex is comprised of anterior and posterior leaflets, annulus, chordae tendineae, and papillary muscles (PM). The geometry of MV leaflets is unique as it is the only heart valve with the natural dual-flap structure. The MV leaflets are attached to the left ventricle through chordae tendineae, which stem from PMs at the left ventricle wall and insert into the leaflets at multiple locations. The geometry of MV has a critical impact on the valve competence. Alterations in the native biological geometry of the valve can lead to non-homeostatic conditions such as Mitral Valve Prolapse (MVP) [1]. MV complications such as Mitral Valve Regurgitation (MVR) are likely to occur following the development of pathological geometries. MVR happens when the anterior and posterior leaflets cannot fully coapt to prevent backward blood flow into the atrium during ventricular contraction. More than two million people are affected by MVR in the United States [2].

© Springer International Publishing Switzerland 2015
H. van Assen et al. (Eds.): FIMH 2015, LNCS 9126, pp. 164–171, 2015.
DOI: 10.1007/978-3-319-20309-6_19

However, there is still no universal treatment for dysfunctional heart valves that is effective across the board, with valve replacement and valve repair being the most common treatments [3].

Computational simulations for studying heart valve behavior have been extensively pursued [4–7] since the pioneering work by Kunzelman et al. [8]. However, the complexity in their geometry hinders straightforward characterization and requires a rigorous shape analysis. In recent studies, the importance of anatomically-accurate geometry to achieve true predictive models has been underscored [9, 10]. The study by Swanson et al. [11] focused on analyzing the geometry of aortic valve based on a predefined set of geometric parameters. Recently, Haj-Ali et al. [12] developed a parametric representation of the aortic valve from 3D-TEE data that can be used for biomechanical simulations. Ryan et al. [13] extracted various geometric parameters of the MV from 3D Echocardiography images. Pouch et al. [14] performed Principal Component Analysis (PCA) to investigate the statistical variations in the average MV annulus geometry. Yeong et al. [15] evaluated the alterations in MV geometry among patients with pediatric rheumatic mitral regurgitation by investigating the annulus shape.

In this paper, we present a novel framework for analysis and characterization of MV geometry that facilitates population-averaging of the entire MV apparatus geometry. The proposed approach enables the quantification of the MV geometric features in a multi-scale framework suitable for investigating intra-patient variations. In the sections to follow, we review the data processing steps and discuss the techniques to perform high-fidelity geometry modeling of the MV.

2 Materials and Methods

2.1 Specimen Preparation and Data Acquisition

All experiments were performed on ovine MVs. Five fresh ovine hearts were obtained from a local slaughterhouse in Atlanta, GA. The MVs were excised and mounted in the extensively validated Georgia Tech Left Heart Simulator (GTLHS) [16, 17], preserving the annular and sub-valvular anatomy. Healthy hemodynamic conditions were simulated (70 beats/min, 5 L/min, 120 mmHg peak LVP), and PM locations were adjusted to achieve healthy MV leaflet geometric relationships as monitored by echocardiography: Anterior leaflet spanning 2/3 of the anteroposterior diameter, coaptation length of 4–5 mm, commissures in the 2 and 10 o'clock positions with minimal regurgitation. The ventricular chamber of the GTLHS was then dismounted from the pulse duplicator, and inserted into the bore of a Siemens Inveon micro-computed tomography (μCT) scanner (Siemens Medical Solutions USA, Inc., Malvern, PA). The MV was then scanned in the diastolic configuration (open state) using parameters optimized for imaging soft tissue (80 kV energy, 500 μA intensity, 650 ms integration time) at 43.3 μm isotropic voxel resolution. The acquired μCT images were stored in DICOM format for data processing and modeling.

2.2   Data Processing

The μCT image stacks were imported into Simpleware ScanIP (Simpleware Ltd., Exeter, UK). The images were then filtered to reduce noise and improve contrast, and subsequently used for segmentation and labeling of the entire MV apparatus. Thereafter, the tessellated surface representations of the MV leaflets for the Atrial and Ventricular sides were constructed. The chordal structure was modeled by the Centerline module in ScanIP, which builds the central axis representation following the approach described in [18]. The data processing pipeline was applied to the μCT datasets of five valves to extract the following major feature sets: (i) atrial surface mesh for the leaflets, (ii) ventricular surface mesh for the leaflets, and (iii) medial axis representation of the chordae structure. Due to the drastic difference in topology of leaflets (thin surfaces of revolution) and chordae (tubular structures with branching), we implement two different methods to model leaflets and chordal structure. We perform population-averaging based on the averaging of estimated model parameters.

2.3   Leaflet Geometry Decomposition and Modeling

We model the leaflet geometry using a two-scale model which distinguishes dimensional variations (large-scale features) from the detailed shape variations (fine-scale features), and thus enables us to accurately characterize geometric features of the leaflets and quantify them within an objective framework. The large-scale geometry is recovered using superquadrics surface fitting [19]. Superquadric shapes are governed by Eq. (1).

$$F(x,y,z) = \left[\left(\frac{x}{a_1} \right)^{2/\varepsilon_2} + \left(\frac{y}{a_2} \right)^{2/\varepsilon_2} \right]^{\varepsilon_2/\varepsilon_1} + \left(\frac{z}{a_3} \right)^{2/\varepsilon_1} = 1 \tag{1}$$

The parameters in Eq. (1) define the dimensions and regulate the shape of the superquadrics fit. For each valve, there are two surface meshes denoting the Atrial and Ventricular sides of the MV leaflets. The Levenberg–Marquardt non-linear programming algorithm [20] was employed to determine the superquadratic fits. The objective function used for optimization is denoted by Eq. (2).

$$R = \sqrt{a_1\, a_2\, a_3} \left(F^{\varepsilon_1} - 1 \right) \tag{2}$$

Fine-scale features are defined as normal residual field of the difference between the actual point cloud data and fitted surface. The residual scalar field is analyzed by discrete Fourier transform with low-pass filtering. Conformal Fourier Transform (CFT) method [21] using Non-Uniform Fourier Transform (NUFFT) technique [22] is used to accommodate irregular boundaries and non-uniform data structure. To reduce artifacts of spectral distortion due to the irregular sampling, iterative NUFFT approach is utilized [23]. The final leaflet surfaces are reconstructed by inverse Fourier transform of the filtered spectrum onto a resampled regularized mesh on the superquadric surface. In addition to volumetric representation of the leaflet tissue, this approach also allows to reconstruct the leaflet midsurface with superimposed thickness scalar field.

2.4 Modeling of Chordae Tendineae

The geometry of chordae tendineae is modeled using the medial axis approach. Skeletonization of the chordal structure provides a natural way to reconstruct the chordae tendineae geometry, and facilitates the characterization of their tubular topology. We model chordal structure by employing the concept of spatio-structural graphs used by Alhashim et al. [24]. The general topology of chordae tendineae is described by performing statistical analysis of the spatial distribution of origin and insertion points, development of distribution functions of lengthwise cross-sectional variations, and density of branching, and lengths of branches. By combining the statistical representation of the chordae structure with structural graph, we can perform averaging and consistent reconstruction of not only chordae structure, but also leaflet-chordae attachments. The latter is achieved by applying the Boolean operation on the reconstructed leaflet and chordae geometries with calibrated flaring parameter to avoid sharp corners in the transition zone. The distribution of origin points on the PM is analyzed in the context of PM classification introduced by [25]. This allows us to distinguish statistical variations in position vs. anatomical differences in the structure of PMs. Note that in contrast to the leaflet geometric model, the chordae are analyzed using the stochastic representation (vs. deterministic descriptors) due to the irregular topology of these MV components.

3 Results

In this section, the proposed modeling framework is used to analyze one of the considered ovine MV datasets. The 3D rendered representation of the segmented μCT images is depicted in Fig. 1a. Prior to segmentation, the image quality was enhanced with curvature flow smoothing filter. The major constituent parts of the MV apparatus (leaflets, chordae, PMs) are labeled with different colors. The markers placed on MV sample prior the imaging were used to capture the MV annulus shape in the reconstructed 3D image. The average of two fitted superquadric surfaces (leaflet midsurface) is illustrated in Fig. 1b, and corresponds to the large-scale model of the studied MV leaflets. The superimposed colormap (Fig. 2a) represents the absolute values of normal residuals prior to spectral filtering. Figure 2b indicates the projection of the filtered residual field (after Fourier analysis) onto the superquadric surface unfolded along the axis of revolution. The reconstructed MV geometry (prior to Boolean union operation) of both leaflets and chordae are illustrated in Fig. 3.

4 Discussions

Geometric models play a critical role in any computational simulation. In the context of cardiovascular biomechanics, the accuracy of geometric models directly affects the ability of the simulation to provide realistic predictions of MV behavior. In this paper, we propose a multi-scale approach for characterization of the MV geometry and decomposing the geometrical features is a systematic approach. The proposed framework

(a)   (b)

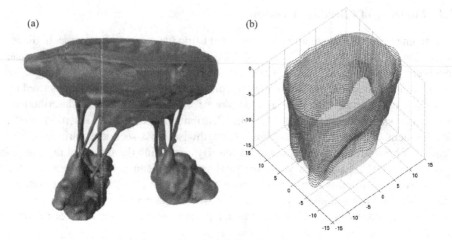

Fig. 1. (a) Labeled 3D representation of the segmented MV geometry. (b) Midsurface representation of the lealfets is shown as blue points on the superquadrics fit.

(a)   (b)

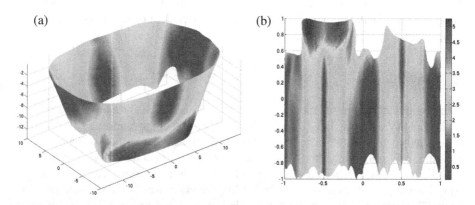

Fig. 2. (a) The residual field representing normal distance of midsurface from superquadrics fit. (b) Residual map unfolded onto the superquadrics natural parameterization space.

allows to not only quantify and accurately capture the intra-patient variations in geometry, but also to develop a representative population-averaged model of MV in a consistent manner. The methodology is independent of imaging modality and requires a limited set of parameters to quantify the geometry of full MV apparatus. In addition to deterministic population-averaged geometric model, this method allows to generate several realizations of the average model by modifying distributions of the stochastic parameters: spectral phases of leaflets and spatial characteristics of chordae. The reconstructed geometry can be easily imported in the mesh generation software packages to build Finite Element (FE) computational meshes with adjustable level of mesh discretization. Even though ex vivo and in vitro models provide high resolution and allow to incorporate fine details of the MVs, these models are not adequately precise for treatment planning simulations [26]. We are currently testing this methodology on human heart data to

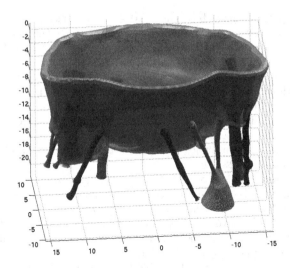

Fig. 3. The entire MV apparatus reconstructed by augmenting superquadrics surface with fine-scale features. Chordae tendineae are modeled using the medial axis approach.

develop a human population-averaged model. Our goal is to extend the proposed framework to allow multi-modal modeling, which uses high-resolution ex vivo data to augment in vivo imaging and enable high-fidelity computational models of patient-specific MVs in clinical applications.

Acknowledgments. Research reported in this publication was supported by National Heart, Lung, and Blood Institute of the National Institutes of Health under award number R01HL119297. The content is solely the responsibility of the authors and does not necessarily represent the official views of the National Institutes of Health.

References

1. Guy, T.S., Hill, A.C.: Mitral valve prolapse. Annu. Rev. Med. **63**, 277–292 (2012)
2. Enriquez-Sarano, M., Akins, C.W., Vahanian, A.: Mitral regurgitation. The Lancet. **373**(9672), 1382–1394 (2009)
3. Kheradvar, A., Groves, E.M., Dasi, L.P., Alavi, S.H., Tranquillo, R., Grande-Allen, K.J., Simmons, C.A., et al.: Emerging trends in Heart valve engineering: part I solutions for future. Ann. Biomed. Eng. **43**(4), 1–11 (2014)
4. Wang, Q., Sun, W.: Finite element modeling of mitral valve dynamic deformation using patient-specific multi-slices computed tomography scans. Ann. Biomed. Eng. **41**(1), 142–153 (2013)
5. Votta, E., Caiani, E., Veronesi, F., Soncini, M., Montevecchi, F.M., Redaelli, A.: Mitral valve finite-element modelling from ultrasound data: a pilot study for a new approach to understand mitral function and clinical scenarios. Philos. Trans. R. Soc. Lond., A: Math. Phys. Eng. Sci. **3669**(1879), 3411–3434 (2008)

6. Mansi, T., Voigt, I., Georgescu, B., Zheng, X., Mengue, E.A., Hackl, M., Ionasec, R.I., Noack, T., Seeburger, J., Comaniciu, D.: An integrated framework for finite-element modeling of mitral valve biomechanics from medical images: application to MitralClip intervention planning. Med. Image Anal. **16**(7), 1330–1346 (2012)

7. Choi, A., Rim, Y., Mun, J.S., Kim, H.: A novel finite element-based patient-specific mitral valve repair: virtual ring annuloplasty. Bio-Med. Mater. Eng. **24**(1), 341–347 (2012)

8. Kunzelman, K.S., Cochran, R.P., Chuong, C., Ring, W.S., Verrier, E.D., Eberhart, R.D.: Finite element analysis of the mitral valve. J. Heart Valve Dis. **2**(3), 326–340 (1993)

9. Votta, E., Le, T.B., Stevanella, M., Fusini, L., Caiani, E.G., Redaelli, A., Sotiropoulos, F.: Toward patient-specific simulations of cardiac valves: state-of-the-art and future directions. J. Biomech. **46**(2), 217–228 (2013)

10. Lee, C.-H., Oomen, P.J., Rabbah, J.P., Yoganathan, A., Gorman, R.C., Gorman III, J.H., Amini, R., Sacks, M.S.: A high-fidelity and micro-anatomically accurate 3D finite element model for simulations of functional mitral valve. In: Ourselin, S., Rueckert, D., Smith, N. (eds.) FIMH 2013. LNCS, vol. 7945, pp. 416–424. Springer, Heidelberg (2013)

11. Swanson, W.M., Clark, R.E.: Dimensions and geometric relationships of the human aortic value as a function of pressure. Circ. Res. **35**(6), 871–882 (1974)

12. Haj-Ali, R., Marom, G., Zekry, S.B., Rosenfeld, M., Raanani, E.: A general three-dimensional parametric geometry of the native aortic valve and root for biomechanical modeling. J. Biomech. **45**(14), 2392–2397 (2012)

13. Ryan, L.P., Jackson, B.M., Eperjesi, T.J., Plappert, T.J., St John-Sutton, M., Gorman, R.C., Gorman, J.H.: A methodology for assessing human mitral leaflet curvature using real-time 3-dimensional echocardiography. J. Thorac. Cardiovasc. Surg. **136**(3), 726–734 (2008)

14. Pouch, A.M., Vergnat, M., McGarvey, J.R., Ferrari, G., Jackson, B.M., Sehgal, C.M., Yushkevich, P.A., Gorman, R.C., Gorman, J.H.: Statistical assessment of normal mitral annular geometry using automated three-dimensional echocardiographic analysis. Ann. Thorac. Surg. **97**(1), 71–77 (2014)

15. Yeong, M., Silbery, M., Finucane, K., Wilson, N.J., Gentles, J.L.: Mitral valve geometry in paediatric rheumatic mitral regurgitation. Pediatr. Cardiol. **36**(4), 1–8 (2015)

16. Rabbah, J.-P., Saikrishnan, N., Yoganathan, A.P.: A novel left heart simulator for the multi-modality characterization of native mitral valve geometry and fluid mechanics. Ann. Biomed. Eng. **41**(2), 305–315 (2013)

17. Siefert, A.W., Rabbah, J.P.M., Koomalsingh, K.J., Touchton, S.A., Saikrishnan, N., McGarvey, J.R., Gorman, R.C., Gorman, J.H., Yoganathan, A.P.: In vitro mitral valve simulator mimics systolic valvular function of chronic ischemic mitral regurgitation ovine model. Ann. Thorac. Surg. **95**(3), 825–830 (2013)

18. Lee, T.-C., Kashyap, R.L., Chong-Nam, C.: Building skeleton models via 3-D medial surface axis thinning algorithms. CVGIP: Graph. Models Image Process. **56**(6), 462–478 (1994)

19. Jaklic, A., Leonardis, A., Solina, F.: Segmentation and Recovery of Superquadrics. Springer, The Netherlands (2000)

20. Moré, J.J.: The Levenberg-Marquardt algorithm: implementation and theory. In: Watson, G.A. (ed.) Numerical Analysis. LNCS, vol. 630, pp. 105–116. Springer, Heidelberg (1978)

21. Zhu, C.-H., Liu, Q.H., Shen, Y., Liu, L.: A high accuracy conformal method for evaluating the discontinuous Fourier transform. Prog. Electromagnet. Res. **109**, 425–440 (2013)

22. Greengard, L., June-Yub, L.: Accelerating the nonuniform fast Fourier transform. SIAM Rev. **46**(3), 443–454 (2004)

23. Gröchenig, K.: Reconstruction algorithms in irregular sampling. Math. Comput. **59**(199), 181–194 (1992)

24. Alhashim, I., Li, H., Xu, K., Cao, J., Ma, R., Zhang, H.: Topology-varying 3D shape creation via structural blending. ACM Trans. Graph. **33**(4), 158 (2014)
25. Berdajs, D., Lajos, P., Turina, M.I.: A new classification of the mitral papillary muscle. Med. Sci. Rev. **11**(1), 18–21 (2005)
26. Sun, W., Martin, C., Pham, T.: Computational modeling of cardiac valve function and intervention. Ann. Rev. Biomed. Eng. **16**, 53–76 (2014)

Very High-Resolution Imaging of Post-Mortem Human Cardiac Tissue Using X-Ray Phase Contrast Tomography

I. Mirea[1(✉)], F. Varray[1], Y.M. Zhu[1], L. Fanton[1], M. Langer[1,2], P.S. Jouk[3], G. Michalowicz[3], Y. Usson[3], and I.E. Magnin[1]

[1] Université de Lyon, CREATIS, CNRS #5220, Inserm U1044, INSA-Lyon, Université Lyon 1, Lyon, France
Iulia.Mirea@creatis.insa-lyon.fr
[2] European Synchrotron Radiation Facility, Grenoble, France
[3] Université Joseph Fourier, TIMC, CNRS #5220, Inserm U1044, Grenoble, France

Abstract. This paper investigates the 3D microscopic structure of *ex-vivo* human cardiac muscle. Usual 3D imaging techniques such as DMRI or CT do not achieve the required resolution to visualise cardio-myocytes, therefore we employ X-ray phase contrast micro-CT, developed at the European Synchrotron Radiation Facility (ESRF). Nine tissue samples from the left ventricle and septum were prepared and imaged at an isotropic resolution of 3.5 μm, which is sufficient to visualise cardio-myocytes. The obtained volumes are compared with 2D histological examinations, which serve as a basis for interpreting the 3D X-ray phase-contrast results. Our experiments show that 3D X-ray phase-contrast micro-CT is a viable technique for investigating the 3D arrangement of myocytes *ex-vivo* at a microscopic level, allowing a better understanding of the 3D cardiac tissue architecture.

Keywords: Phase-contrast imaging · Histology · Cardio-myocyte

1 Introduction

Cardiovascular diseases remain one of the most serious health problems in the world, motivating research that deepens our understanding of the myocardial function. This requires a good knowledge of the myocardial architecture, especially the myocyte architecture, to understand relations between mechanical function, hemodynamics and adaptive structural changes in cardiac diseases.

Cardiac muscle cells have diameters in the range of 10–20 μm and lengths around 100 μm. To image them, a resolution in the μm range is required. Current imaging techniques can deliver either very high resolution but only in a small field of view, or the entire heart can be imaged to see the overall organisation of the myocytes, but at a resolution that does not show how individual myocytes are arranged.

© Springer International Publishing Switzerland 2015
H. van Assen et al. (Eds.): FIMH 2015, LNCS 9126, pp. 172–179, 2015.
DOI: 10.1007/978-3-319-20309-6_20

Optical images of histological sections can deliver high resolution for *ex-vivo* 2D imaging, but they can suffer from distortions due to cutting the thin samples. Another optical technique, Polarized Light Imaging (PLI), can be used to obtain maps of the orientation of myocardiac cells. In [7,9], a fetal heart was divided into a set of parallel contiguous slices and each one was imaged with PLI, thus constructing a 3D map of the myocyte orientations in the entire heart at a resolution of 100 μm × 100 μm × 500 μm. This resolution remains insufficient for visualizing individual myocytes, and what is actually seen is the averaged orientation of a population of myocytes in each voxel.

Compared with the previous two techniques, Magnetic Resonance Imaging is non-invasive and can be used *in-vivo* to image the entire heart. Diffusion Tensor MRI (DT-MRI) can recover the dominant orientation of myocytes in each voxel because of the stronger diffusion of water molecules in the direction parallel to that of the myocytes. DT-MRI can thus map myocyte orientations throughout the entire heart *in-vivo* and can serve to detect abnormalities [8,15]. However, the typical spatial resolution is low (1 mm × 1 mm × 2 mm) therefore the 3D organization of myocytes at a microscopic level remains unknown. High-resolution MRI setups such as [1] can achieve 30 μm × 30 μm × 300 μm or 60 μm × 60 μm × 60 μm voxel sizes, however this remains above the diameter of cardio-myocytes.

A new method that has great potential for applications in biomedical imaging is X-Ray Phase Contrast Imaging (PCI). Just as conventional transmission-based X-ray imaging, PCI can be combined with Computed Tomography (CT) techniques to obtain 3D images, and a better contrast is obtained in soft tissue compared to classical CT. Several techniques are now available to exploit and visualize the phase-contrast, of which we mention propagation-based imaging (PBI), also known as in-line holography, which yields high spatial resolution without the need for complex instrumentation. PBI is useful for imaging cardiac tissue because it enhances contrast at myocyte boundaries [4].

Allowing high contrast 3D visualization of thick and complex samples at high spatial resolution, X-Ray PCI can image a wide range of tissues and organs not only *ex-vivo* but also *in-vivo*. In [10], angiographies in the mouse brain were performed at 6 μm resolution. However, most PBI studies use higher resolutions, such as [5] who have visualised *ex-vivo* 3D osteon morphology with a spatial resolution of 1.4 μm. Also interesting are the experiments of [10], who were able to visualize subtle details in the rat brain through a phase retrieval method that allows PBI-CT through a single distance PBI image, with a reduction in absorbed dose and acquisition time. *In-vivo* studies with X-Ray PCI on humans have also been conducted, the first clinical trials for mammography being described in [2]. They show that mammographies with synchrotron radiation can be used to clarify the diagnosis in patients with suspicious breast abnormalities identified by combined digital mammography and ultrasonography.

In our study, we investigate the 3D cardiac tissue structure using X-Ray micro-CT phase contrast imaging based on in-line propagation implemented at the European Synchrotron Radiation Facility (ESRF) [11]. The technique allows to detect the microscopic-scale boundaries of the myocytes in the heart tissue [4].

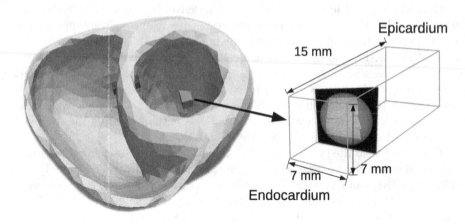

(a) heart sample section

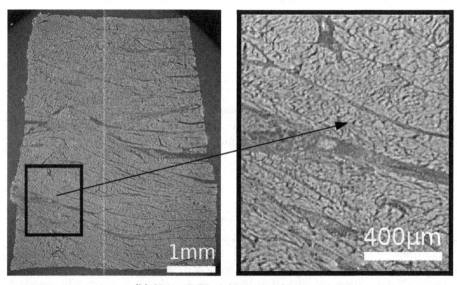

(b) X-ray PCI and a zoomed-in area

Fig. 1. (a): schematic representation of the heart sample section at the junction between the apical third and the equatorial third. The sample has a height of 15 mm, corresponding to the thickness of the left ventricle wall, and the sides of about 5–6 mm (7 mm including the outer edges of the image). It is imaged at a resolution of 3.5 μm. (b) left: X-ray PCI viewed in a plane orthogonal to the epicardium-endocardium direction. The bundle of myocardial cells grossly follows the curvature of the ventricular wall and corresponds mostly to a circular arrangement around the ventricle. (b) right: at a finer scale, the orientation of myocardial cells can be refined, they are obliquely cut. Myocardial cells appear in white, separated by short and thin dark gaps. The larger, roughly linear dark gaps separate bundles of myocardial cells. The lumen of a blood vessel is also visible as a thick dark gap in the middle of the image.

2   Materials and Methods

Human heart samples were supplied by the Medico-Legal Institute of Lyon IML HCL ($n°DC - 2012 - 1588$) and were collected during a medical-legal autopsy of a subject who suffered a violent death. From this heart, 9 samples were collected from the septal, anterior, lateral, and posterior left ventricle, with heights of 11–20 mm and 6 mm sides for imaging.

The samples were first immersed in a formalin solution for sterilization. However, previous experience showed that an immersion in alcohol results in better image quality in soft tissue. Therefore, the formaldehyde was rinsed with water, then the samples were immersed in pure water for rehydration. Finally, they were immersed in successive baths of increasing alcoholic concentrations, from 10 % to 70 %, before being imaged. Image acquisition was performed at ESRF on beam-line ID19 using X-ray micro-CT PCI based on in-line propagation [3]. Unfiltered undulator radiation was used, with an average energy of ~ 19 keV and an energy bandwith of $< 10\%$. The detector was placed 1 m from the sample to achieve phase contrast. Image reconstruction was performed using Paganin's method for phase retrieval and filtered backprojection for tomographic reconstruction [13]. Compared with [3] who studied the silicon single crystal and metal matrix composites, we now apply the method to human cardiac tissue and use more recent materials with higher spatial resolution. Our acquisitions are made at a resolution of 3.5 μm × 3.5 μm × 3.5 μm.

Due to the maximum height allowed by the detector, 3 or 4 successive acquisitions with an overlap of 97 planes were made by shifting the sample vertically. These were merged to recreate the entire volume. Scan time for each successive acquisition is approximately 25 min. For the 15 mm × 6 mm × 6 mm sample in Fig. 1, the total acquisition time is 1 h 17 min. The imaging setup used in this experiment is non-destructive for the sample.

3   Image Interpretation

For illustration purposes we choose a sample from the inferior part of the left ventricle, at the junction between the apical third and the equatorial third, as illustrated in Fig. 1a.

Figure 2a shows a 3D view of a small part of the chosen sample. This part contains $0.44 \cdot 10^9$ voxels for a file size of 1.75 GB. The 3D visualisation was made with CreaTools [6], a software tool created in the CREATIS lab. Three orthogonal planes are shown in order to illustrate different types of details that can be seen. The plane orthogonal to the X axis illustrates projections of myocyte orientations appearing as long, thin filaments. The lumen of a blood vessel can be seen with round shaped erythrocytes.

Figure 3a illustrates a zoomed-in area of another plane orthogonal to the X axis. Again, we see the lumen of a blood vessel (arrow 1), a capillary with red blood cells (arrow 2), projections of myocyte orientations in high detail, as thin white filaments (arrows 3), and intercellular space, as black filaments (arrows 4).

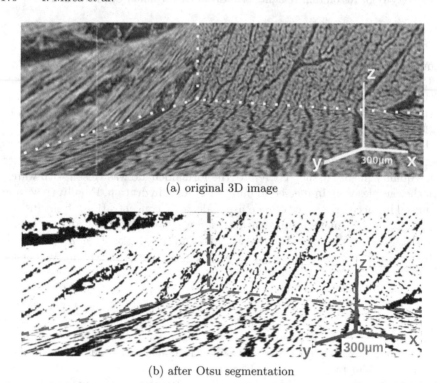

(a) original 3D image

(b) after Otsu segmentation

Fig. 2. (a): X-ray phase contrast 3D image of a heart sample from the inferior part of the lateral wall of the left ventricle. (b): segmentation of myocardial cells using the Otsu thresholding method with two thresholds [12].

Figure 3c is also from the same volume as Fig. 2a, but in a plane orthogonal to the Y axis. In this figure myocytes are separated by short, thin, curved dark areas, and bundles of myocytes separated by thicker, long and straight dark zones. A blood vessel is also visible as a thick dark band in the left part of the image.

The obtained phase contrast images are suitable for automatic analysis. For illustration purposes, the Otsu thresholding method [12] was applied to segment myocytes in Fig. 2b. This result is already good considering the 3D aspect of the problem, the large number of voxels and the simplicity of the method. More complex segmentation methods could highlight the separation between myocytes even better. Thresholding is easier in 2D, as shown in Fig. 4. Other types of information can also be extracted from phase contrast images, such as the 3D orientation of myocytes throughout the volume [14].

It is interesting to compare the X-ray phase contrast images with optical images of histological sections. To this end, Fig. 3b depicts such a histological image taken from approximately the same area of the heart as Fig. 3a and with a similar slice orientation. Figure 3d is another histological section, but with approximately the same localisation and orientation as Fig. 3c. The two image

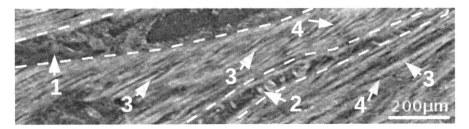

(a) X-ray phase contrast image, X plane

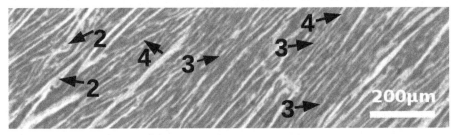

(b) histological section, X plane

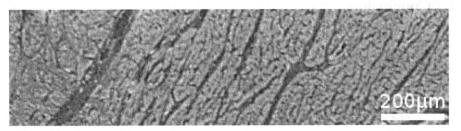

(c) X-ray phase contrast image, Y plane

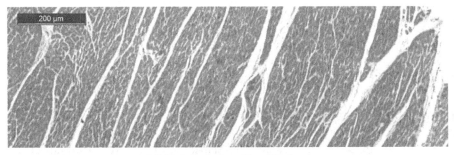

(d) histological section, Y plane

Fig. 3. (a): X-ray phase contrast image section perpendicular to the X axis (see Fig. 2a); *1* - blood vessel; *2* - capillary with red blood cells; *3* - projections of myocyte orientations (white); *4* - intercellular space (black). (b): histological section similar to (a). Both (a) and (b) show myocytes as thin long filament-like structures. (c): X-ray phase contrast image section perpendicular to the Y axis. (d): histological section similar to (c). Both (c) and (d) show myocytes and bundles of myocytes (Color figure online).

Fig. 4. Binary thresholding of Fig. 3c with a manually-chosen threshold.

pairs in Fig. 3 are structurally similar. In the above pair, myocytes appear as long, filament-like structures (white in the X-ray phase-contrast images, red in the optical images). The separation between myocytes appears as thin, black lines in Fig. 3a and as thin, white lines in Fig. 3b. For the other image pair, myocytes are separated by thin, short, curved areas, and bundles of myocytes are separated by long, straight areas.

4   Conclusion

We have shown that X-ray phase contrast images are a viable tool for examining the microstructure of the heart. They can show the same structures as those seen in histological sections, without having to carefully prepare a set of slices. The greatest advantage of X-ray phase contrast images is the ability to capture 3D structures, from which much more information can be extracted, either through visual inspection by a human expert or through analysis with automatic algorithms.

Acknowledgements. We would like to thank F. Peyrin, C. Olivier, L. Wang and M. Ozon for technical support at ESRF Grenoble. This study was funded by the French National Research Agency (ANR) through the MOSIFAH project (Multimodal and multiscale modeling and simulation of the fibre architecture of the human heart, ANR-13-MONU-0009).

References

1. Baltes, C., Radzwill, N., Bosshard, S., Marek, D., Rudin, M.: Micro mri of the mouse brain using a novel 400 mhz cryogenic quadrature rf probe. NMR Biomed. **22**(8), 834–842 (2009). http://dx.doi.org/10.1002/nbm.1396
2. Castelli, E., Tonutti, M., Arfelli, F., Longo, R., Quaia, E., Rigon, L., Sanabor, D., Zanconati, F., Dreossi, D., Abrami, A., Quai, E., Bregant, P., Casarin, K., Chenda, V., Menk, R.H., Rokvic, T., Vascotto, A., Tromba, G., Cova, M.A.: Mammography with synchrotron radiation: first clinical experience with phase-detection technique. Radiology **259**(3), 684–694 (2011)

3. Cloetens, P., Pateyron, M., Buffière, J.Y., Peix, G., Baruchel, J., Peyrin, F., Schlenker, M.: Observation of microstructure and damage in materials by phase sensitive radiography and tomography. J. Appl. Phys. **81**(9), 5878–5886 (1997)
4. Cloetens, P., Ludwig, W., Baruchel, J., Guigay, J.P., Pernot-Rejmnkov, P., Salom-Pateyron, M., Schlenker, M., Buffire, J.Y., Maire, E., Peix, G.: Hard x-ray phase imaging using simple propagation of a coherent synchrotron radiation beam. J. Phys. D Appl. Phys. **32**(10A), A145–A151 (1999)
5. Cooper, D.M.L., Erickson, B., Peele, A., Hannah, K., Thomas, C.D.L., Clement, J.G.: Visualization of 3D osteon morphology by synchrotron radiation micro-CT. J. Anat. **219**(4), 481–489 (2011)
6. Dávila Serrano, E.E., Guigues, L., Roux, J.-P., Cervenansky, F., Camarasu-Pop, S., Riveros Reyes, J.G., Flórez-Valencia, L., Hernández Hoyos, M., Orkisz, M.: CreaTools: a framework to develop medical image processing software: application to simulate pipeline stent deployment in intracranial vessels with aneurysms. In: Bolc, L., Tadeusiewicz, R., Chmielewski, L.J., Wojciechowski, K. (eds.) ICCVG 2012. LNCS, vol. 7594, pp. 55–62. Springer, Heidelberg (2012)
7. Desrosiers, P.A., Michalowicz, G., Jouk, P.-S., Usson, Y., Zhu, Y.: Modeling of the optical behavior of myocardial fibers in polarized light imaging. In: Camara, O., Mansi, T., Pop, M., Rhode, K., Sermesant, M., Young, A. (eds.) STACOM 2012. LNCS, vol. 7746, pp. 235–244. Springer, Heidelberg (2013)
8. Ferreira, P., Kilner, P., McGill, L.A., Nielles-Vallespin, S., Scott, A., Ho, S., McCarthy, K., Haba, M., Ismail, T., Gatehouse, P., de Silva, R., Lyon, A., Prasad, S., Firmin, D., Pennell, D.: In vivo cardiovascular magnetic resonance diffusion tensor imaging shows evidence of abnormal myocardial laminar orientations and mobility in hypertrophic cardiomyopathy. J. Cardiovasc. Magn. Reson. **16**(1), 87 (2014)
9. Jouk, P.S., Mourad, A., Milisic, V., Michalowicz, G., Raoult, A., Caillerie, D., Usson, Y.: Analysis of the fiber architecture of the heart by quantitative polarized light microscopy: accuracy, limitations and contribution to the study of the fibre architecture of the ventricles during fetal and neonatal life. Eur. J. Cardio-Thorac. Surg. **31**(5), 915–921 (2007)
10. Kidoguchi, K., Tamaki, M., Mizobe, T., Koyama, J., Kondoh, T., Kohmura, E., Sakurai, T., Yokono, K., Umetani, K.: In vivo x-ray angiography in the mouse brain using synchrotron radiation. Stroke **347**, 1856–1861 (2006)
11. Langer, M., Cloetens, P., Pacureanu, A., Peyrin, F.: X-ray in-line phase tomography of multimaterial objects. Opt. Lett. **37**(11), 2151–2153 (2012)
12. Otsu, N.: A threshold selection method from gray-level histograms. IEEE Trans. Syst. Man Cybern. **9**(1), 62–66 (1979)
13. Paganin, D., Mayo, S.C., Gureyev, T.E., Miller, P.R., Wilkins, S.W.: Simultaneous phase and amplitude extraction from a single defocused image of a homogeneous object. J. Microsc. **206**(1), 33–40 (2002)
14. Varray, F., Wang, L., Fanton, L., Zhu, Y.-M., Magnin, I.E.: High resolution extraction of local human cardiac fibre orientations. In: Ourselin, S., Rueckert, D., Smith, N. (eds.) FIMH 2013. LNCS, vol. 7945, pp. 150–157. Springer, Heidelberg (2013)
15. Wang, L., Zhu, Y., Li, H., Liu, Y., Magnin, I.E.: Multiscale modeling and simulation of the cardiac fiber architecture for dmri. IEEE Trans. Biomed. Eng. **59**(1), 16–19 (2012)

Viewpoint Recognition in Cardiac CT Images

Mehdi Moradi[1]([✉]), Noel C. Codella[2], and Tanveer Syeda-Mahmood[1]

[1] IBM Almaden Research Center, San Jose, CA, USA
mmoradi@us.ibm.com
[2] IBM Thomas J. Watson Research Center, Yorktown Heights, NY, USA

Abstract. Position and orientation information is often lacking in
DICOM datasets. This creates a need for human involvement or computationally expensive 3D processing for any analytical tool, such as
a software-based cognitive assistant, to determine the viewpoint of an
input 2D image. We report a solution for cardiac CT viewpoint recognition to identify the desired images for a specific view and subsequent
processing and anatomy recognition. We propose a new set of features to
describe the global binary pattern of cardiac CT images characterized by
the highly attenuating components of the anatomy in the image. We also
use five classic image texture and edge feature sets and devise a classification approach based on SVM classification, class likelihood estimation,
and majority voting, to classify 2D cardiac CT images into one of six
viewpoint categories that include axial, sagittal, coronal, two chamber,
four chamber, and short axis views. We show that our approach results
in an accuracy of 99.4 % in correct labeling of the viewpoints.

1 Introduction

Coronary heart disease is the most common cause of mortality in the United
States and contributes to one in every five deaths, according to the American Heart Association. Acute coronary symptoms result in hospitalization of
nearly 900,000 Americans every year. Cardiac catheterization under CT or X-ray
angiography provides definitive evidence for plaque build-up in coronary arteries.
However, the invasive nature of such procedures prohibits their use for screening purposes in low to intermediate risk individuals. This has created a growing
interest in cardiac computed tomography (CT) as an imaging technology to
study the heart vessels and chambers for screening purposes. Several studies
have shown that cardiac CT, without the use of a contrast agent, has a very
high specificity and provides a negative predictive value of nearly 100 % and can
be used to rule out a large number of low and intermediate risk patients without
the need for invasive methods [1].

The effective and wide-spread use of CT as a screening methodology for
cardiovascular disease could be facilitated by the introduction of an end-to-
end cardiology/radiology "cognitive assistant". A cognitive assistant is a software system with the ability to automatically complete the pre-processing steps,
recognize or generate the appropriate views within a complete scan, extract relevant features and concepts from an image and the text associated with the

© Springer International Publishing Switzerland 2015
H. van Assen et al. (Eds.): FIMH 2015, LNCS 9126, pp. 180–188, 2015.
DOI: 10.1007/978-3-319-20309-6_21

image, run image analysis methods to extract relevant features, and generate a clinically relevant outcome, such as the calcium score or likelihood of disease. These kinds of systems have the ability to reduce the workload, prevent errors, and enable population screening. As an example, previous work has reported a decision support system for cardiology that derives the consensus opinions of other physicians who have looked at similar cases [2]. The system generates a report that summarizes possible diagnoses based on statistics from similar cases. In deploying a system of this type, one needs to retrieve the relevant or similar images and activate the image analytics processes that are often dependent on the modality and viewpoint of the image. In cardiac imaging, the viewpoint of the image is an essential input for any algorithm designed to measure clinical features of the heart, such as detection of left ventricle, valves, thickness of the pericardium. Since viewpoint recognition is often the first step in the analytic pipeline within a cognitive assistant system, a nearly perfect classification accuracy is needed. Even though DICOM headers provide optional tags to store modality information, viewpoint is often not recorded. Also, as several investigators have reported, one can not rely on the accuracy and completeness of DICOM headers for image categorization [3] particularly on optional and manually entered tags [4]. The introduction of a machine learning approach to slice/viewpoint recognition could also facilitate the use of 2D technics in segmentation and anatomy recognition within the cognitive assistant system, providing savings in terms of computational resources compared to 3D.

Much of the previous work in cardiac viewpoint detection focuses on echocardiography images [5,6]. Due to the small field of view, the free-hand nature of ultrasound images, and the fundamentally different nature of ultrasound image texture, the methods can not be directly applied to CT imaging. In this paper, we present a method for solving the problem of viewpoint recognition in cardiac CT images. We explore the utility of a number of different types of texture and edge characterizing features for image classification. We also propose a new set of features that rely on the anatomic context of the CT images, particularly the pattern of the appearance of bone structures of the rib cage and the vertebral column. This new set of features provides a solution for global binary characterization of cardiac images. We combine the conventional image classification features and the global binary pattern features in an innovative machine learning framework based on support vector machine classification and voting to determine the correct image viewpoint from six different viewpoints. We report very accurate performance in cardiac CT viewpoint recognition.

2 Materials and Methods

2.1 The Data and the Classes

The most common protocol for cardiac CT is multi-slice imaging with a minimum of 64 axial slices. Volumetric re-sampling is performed to obtain any arbitrary plane. Most commonly, clinicians use the three standard orthogonal planes that are parallel to the cardinal planes. As such, a viewpoint recognition system for

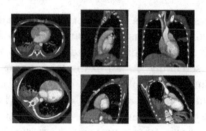

Fig. 1. Top row: standard cardinal views (from left: axial, sagittal, coronal). Bottom row: non-cardinal (from left: four chamber (4C), short axis (SHA), two chamber (2C)).

CT imaging should correctly label sagittal, axial and coronal planes. There are also oblique planes that are obtained to assess cardiac chamber morphology, size and function. Short axis view (SHA) through the entire left ventricle is useful in calculation of left ventricle volume and ejection fraction, whereas the function of the left ventricle should be reviewed in long axis views which include the two chamber (2C) and four chamber (4C) views. There are also three chamber and five chamber views that are useful to study the aortic valve and left ventricle outflow. However, these two views are generated with the maximum intensity projection technic as opposed to re-sampling [7]. In this work, we study the effectiveness of our proposed methods on three different viewpoint recognition problems: classification of standard orthogonal planes (axial/sagittal/coronal), classification of the most commonly used re-sampled non-cardinal views (SHA, 2C, 4C), and a six-class classification problem that includes both the cardinal and non-cardinal planes. Figure 1 illustrates these two groups of views. It is also important to note that each of these three non-cardinal planes are fairly close to one of the cardinal planes (4C to axial, 2C to coronal and SHA to sagittal).

The data in this study consists of cardiac CT data obtained in standard axial planes, with 2 mm slice thickness. We used the open source software package called TurtleSeg reported in [8] to re-sample the sagittal and coronal images, and also the 2C, 4C, and SHA oblique views. These were examined and confirmed by an experienced radiologist. A total of 168 images, equally distributed between the six viewpoint types are available. These are from 28 contrast-enhanced 3D scans, each from a different patient.

2.2   Features

Group 1 - Statistical Image Texture Features: These include features calculated directly from the image (minimum value, maximum value, mean, range, standard deviation, median, central moment, square sum, average top quartile, average bottom quartile), and also those extracted from the co-occurrence matrices (entropy, energy, homogeneity). These features are calculated at different levels of granularity. Namely, we have included both the global feature calculated over the entire image, and also over image partitions that divide the image into 2×2, 3×3, 5×5, 7×7 grids. The resulting features are concatenated to create a feature vector, per image.

Group 2 - Curvelet Features: Curvelet features are proposed as a solution to overcome the limitations of wavelet as a multi-scale transform. We used the implementation reported in [9]. This curvelet methodology builds a multiscale pyramid in a Fourier-Mellin polar transform. Pyramid levels in the radial dimension consist of 1, 2, 4, and 8, and in the angular dimension the levels are 1, 2, 4, 8, 16, and 32 sections. Due to the large size of the feature vector, we only used the global granularity for this group of features.

Group 3 - Wavelet Features: This group consists of 120 texture features obtained by discrete wavelet transform at each granularity.

Group 4 - Edge Histogram Features: These are from a histogram of edge directions in the image in 64 bins resulting in 64 features, calculated per global, 2×2 and 3×3 granularity levels.

Group 5 - Local Binary Pattern (LBP) Features: LBPs are calculated by dividing the image into cells, and comparing the center pixel with neighboring pixels in the window [10]. A histogram is built that characterizes how many times, over the cell, the center pixel is larger or smaller than the neighbors. In the implementation used here, the histogram is built on different scales $(1, 1/2, 1/4$ and $1/8$ of the image), and a combined 59 dimensional histogram is produced. In this implementation, the LBP features are weighted by the inverse of the scale [9].

Group 6 (Proposed in this Work) - A Global Binary Pattern (GBP) of the Image: This proposed set of features relies on the pattern of high intensity components of the anatomy of the chest, including the ribs and vertebrae. The images are first pre-processed with histogram equalization to a range routinely utilized by radiologists to maximize the contrast in cardiac chambers and vessels. Then multi-level Otsu thresholding, with four levels, is applied [11]. Otsu thresholding calculates the optimal thresholds to minimize intra-level variance. The highest intensity level is then subjected to connected component clustering. The resulting connected components are then filtered based on the size of the area. An area size of 30 pixels is used in images of size 512×512. Samples of the resulting binary images are presented in Fig. 2. The binary image is then re-sized and down-sampled to obtain an $m \times m$ matrix, where m is chosen by experimentation from values of $m = 2, 4, 8, 16, 20$. In Fig. 2, examples of this matrix for $m = 4$ are shown. The feature vector used for this method is the $m^2 \times 1$ vector generated by concatenating the columns of this matrix.

2.3 Classification

We used a support vector machine (SVM) classifier *for each feature category*. Given the large size of the feature vectors and the relatively small size of the dataset, we do not combine the six different categories of features into a single vector. Instead, we use individual SVMs and combine with voting. SVM training optimizes w and b to obtain the hyperplane $\mathbf{w}^\top \phi(\mathbf{x}) + b$ where $\mathbf{x}$ is the feature

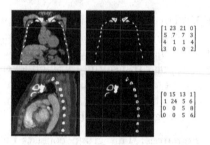

Fig. 2. Examples of converting a coronal (top row) and sagittal (bottom row) cardiac CT image to a set of 16 GBP features.

vector and ϕ is the kernel function, to maximize the distance between the hyperplane and the closest samples (support vectors) in the two classes. SVM is by definition a binary classifier. In the current work, we need to solve three and six-class classification problems. We used a one-versus-all approach to decide the label of each image for each feature group. In this approach, for an n class classification, we train n classifiers each separating one of the viewpoint types from the rest of images. Each test sample is classified by the n classifiers, and "class likelihood" is calculated as described below for each of the n classifiers. The label with the largest class likelihood obtained from its corresponding one-versus-all classifier is chosen as the viewpoint suggested by the feature group for the test sample. In order to calculate the class likelihood, we use the method described in [12]. Given the SVM hyperplane obtained in training, the class likelihood (L_c) for class c for test sample $\mathbf{x}_i$ is computed using a sigmoid function of form:

$$L_c = \frac{1}{1 + exp(\alpha(\mathbf{w}^\top \phi(\mathbf{x}_i) + b) + \beta)} \tag{1}$$

where α and β are calculated using maximum likelihood estimation on the training data. We experimented with three different kernel functions. These were the linear kernel, radial basis function (RBF), and the polynomial kernel. There was no advantage in terms of accuracy when RBF or polynomial kernel were used. We report the results obtained using the linear kernel where $\phi(\mathbf{x}) = \mathbf{x}$.

Voting: The resulting six classifiers return six potentially different viewpoints for each image. In order to determine the final image label and maximize the accuracy, we used a majority voting scheme where the class label is determined as the most frequently returned label among the six classifiers. The result is reported as "correct" only if at least three classifiers return the correct label. There was no tie (3 to 3) vote in our experiments. However, a potential tie can be decided based on the average class likelihoods.

Training and Testing Strategy: We used a leave-one-sample out strategy for training and testing. Each sample serves as the hold-out once. For each of the classification problems, all feature specific classifiers are trained using all but the hold-out sample and the hold-out is used as the test sample. This is repeated until all samples are tested. In the six-class problem, six one-versus-all classifiers are trained per feature type which adds up to 36 SVMs to be trained for each leave-one-out step. In the three-class problem, there are 18 SVMs to be trained. Note that in all of these training and testing experiments, a linear SVM with similar parameters is employed. The SVM slack variable is set to $C = 1$.

3   Results

Three-class Classification of the Cardinal Planes: Table 1 shows the results of the three-class classification of the axial, sagittal, and coronal images. For each feature group, we have listed the accuracy per class and also the combined accuracy over the three classes. All the feature groups return over-all accuracies over 90 %. Curvelet, edge histograms, and GBP features approach or surpass 99 % accuracy. The GBP matrix size is set to 4×4. Using the majority voting scheme, a consistently perfect classification is obtained.

Table 1. Accuracy of three-class classification - cardinal planes.

	Statistics	Curvelet	Wavelet	Edge	LBP	GBP
Axial	100 %	100 %	100 %	100 %	100 %	100 %
Coronal	85.7 %	100 %	100 %	100 %	92.9 %	100 %
Sagittal	92.9 %	100 %	92.9 %	100 %	92.9 %	96.4 %
Combined	92.9 %	100 %	97.6 %	100 %	95.2 %	98.8 %
Voting	100 %					

Three-Class Classification of the Non-cardinal Planes: Table 2 reports the results of classification of 2C, 4C and SHA viewpoints. For this classification problem, the SVM trained on statistical texture features results in the most accurate classification with an overall accuracy of 100 %, with the proposed simpler GBP being a close second at 96.4 %. The GBP matrix size is set to 4×4. When the six classifiers are combined using voting, 83 out of 84 images are correctly classified resulting in an accuracy of 98.8 %.

Six-Class Classification of all Images: The accuracies in the six-class classification of the viewpoints are reported in Table 3. In this problem, the edge

Table 2. Accuracy of three-class classification - non-cardinal planes.

	Statistics	Curvelet	Wavelet	Edge	LBP	GBP
2C	100 %	92.9 %	92.9 %	92.9 %	85.7 %	96.4 %
4C	100 %	100 %	100 %	100 %	100 %	96.4 %
SHA	96.4 %	92.9 %	85.7 %	100 %	92.9 %	96.4 %
Combined	99.4 %	95.2 %	92.9 %	97.6 %	92.8 %	96.4 %
Voting	98.8 %					

Table 3. Accuracy of six-class classification.

	Statistics	Curvelet	Wavelet	Edge	LBP	GBP
Axial	100 %	100 %	100 %	100 %	100 %	100 %
Coronal	78.6 %	100 %	100 %	100 %	85.7 %	100 %
Sagittal	78.6 %	92.9 %	64.3 %	100 %	92.9 %	92.9 %
2C	85.7 %	85.7 %	92.9 %	100 %	85.7 %	85.7 %
4C	100 %	100 %	100 %	100 %	100 %	85.7 %
Short axis	78.6 %	92.9 %	85.7 %	92.9 %	92.9 %	85.7 %
Combined	85.7 %	95.2 %	90.5 %	98.8 %	91.6 %	91.6 %
Voting (three out of 6)	99.4 %					

histogram and curvelet features return the highest accuracy and LBP and GBP are next, all with accuracies above 91 %. The GBP matrix size is set to 16×16. The voting results in an accuracy of 99.4 % where only one image is misclassified. This single case of mis-classification is a sagittal image classified as a short axis image by five of the six feature group. Only the GBP SVM correctly classified this image. The most common mis-classification in the feature specific SVM classifiers is the classification of two chamber images as coronal images. In all cases, however, this is rectified by voting. It is notable that all classifiers return 100 % accuracy on the axial viewpoint which is the most commonly used view in cardiac CT imaging in clinical practice.

4   Discussion and Conclusion

In the area of medical imaging, big data is still elusive. A host of legal and ethical issues bar the free sharing of data. Gold standard labeling is also expensive. As a result, methods based on deep learning [13] are difficult to tune for many medical imaging applications. In this work, we show that a curated set of features including context-sensitive anatomic features, can provide a very accurate classification of cardiac CT viewpoints. In a recent work, IBM scientists have also developed a generalized framework for medical image classification and recognition which uses a large set of visual texture features extracted from image patches at various levels of granularity [14]. These are used along with the ensembling method described in [15]. Given the very large set of features and classifiers used in [14], a fair evaluation of this generalized solution for the specific problem of CT viewpoint recognition requires a larger dataset to avoid over-fitting.

An important feature of our current work is the use of GBP that constitutes a set of context-sensitive features as opposed to the general purpose features used in [14] and elsewhere. It should be noted that in contrast-enhanced CT, depending on the time of imaging, high intensities could appear within heart chambers as well. Our dataset was from contrast-enhanced CT and this issue is likely to have contributed to some of the errors in GBP classification. More sophisticated methods of segmenting the bones, including atlas-based approaches, could provide a more accurate binary image. Nevertheless, despite the simplicity of the binarization approach, we obtained a highly effective set of features through this method which ranks among the top three groups of features in terms of accuracy in all experiments. The GBP features calculated with this approach are also the least computationally expensive group of features.

The specific machine learning framework developed here is based on linear SVMs. While SVM is inherently binary, we have effectively built a multiclass solution with class likelihood estimation and voting. An alternative solution is using random forests. At this stage, we have not been able to obtain the same level of accuracy from random forests, probably due to the fact that unlike linear SVMs, random forests require the user to set a large number of parameters for optimized performance. The need for nearly perfect classification of the viewpoints, as the first step in the deployment of a cognitive assistant system for cardiologists, requires a very robust solution. Therefore, we opted for a system based on several classifiers and voting. The top contributors to the classification accuracy in the six way problem were edge, curvelet, GBP and LBP features.

In order to build a fully automatic cognitive assistant, the classification problem needs to be expanded to include not only other viewpoints such as three or five chamber views, but also the slice level. Our future work will address these more complicated problems with GBPs and classical features.

References

1. Budoff, M.J., Achenbach, S., Blumenthal, R.S., Carr, J.J., Goldin, J.G., Greenland, P., Guerci, A.D., Lima, J.A., Rader, D.J., Rubin, G.D., Shaw, L.J., Wiegers, S.E.: Assessment of coronary artery disease by cardiac computed tomography. Circulation **114**, 1761–1791 (2006)
2. Syeda-Mahmood, T., Wang, F., Beymer, D., Amir, A., Richmond, M., Hashmi, S.: Aalim: Multimodal mining for cardiac decision support. Comput. Cardiol. **34**, 209–212 (2007)
3. Gueld, M.O., Kohnen, M., Keysers, D., Schubert, H., Wein, B.B., Bredno, J., Lehmann, T.M.: Quality of DICOM header information for image categorization. In: Proceedings of SPIE Medical Imaging, vol. 4685, pp. 280–287 (2002)
4. Yoshimura, H., Inoue, Y., Tanaka, H., Fujita, N., Hirabuki, N., Narumi, Y., Nakamura, H.: Operating data and unsolved problems of the dicom modality worklist: an indispensable tool in an electronic archiving environment. Radiat. Med. **21**(2), 68–73 (2003)
5. Park, J., Georgescu, S.Z.B., Simopoulos, J., Otsuki, J., Comaniciu, D.: Automatic cardiac view classification of echocardiogram. In: ICCV, pp. 1–8 (2007)
6. Kumar, R., Wang, F., Beymer, D., Syeda-Mahmood, T.: Echocardiogram view classification using edge filtered scale-invariant motion features. In: IEEE CVPR, pp. 723–730 (2009)
7. Halpern, E.: Clinical Cardiac CT, 2nd edn. Thieme Medical Publishers Inc., USA (2011)

8. Top, A., Hamarneh, G., Abugharbieh, R.: Active learning for interactive 3D image segmentation. In: Fichtinger, G., Martel, A., Peters, T. (eds.) MICCAI 2011, Part III. LNCS, vol. 6893, pp. 603–610. Springer, Heidelberg (2011)

9. Cao, L., Chang, S., Codella, N., Cotton, C., Ellis, D., Gong, L., Hill, M., Hua, G., Kender, J., Merler, M., Mu, Y., Smith, J.: IBM Research and Columbia University TRECVID-2012 Multimedia Event Detection (MED), Multimedia Event Recounting (MER), and Semantic Indexing (SIN) Systems. In: NIST TRECVID Workshop, pp. 1–18 (2012)

10. Ojala, T., Pietikinen, M., Harwood, D.: A comparative study of texture measures with classification based on featured distributions. Pattern Recogn. **29**(1), 51–59 (1996)

11. Liao, P.S., Chen, T.S., Chung, P.C.: A fast algorithm for multilevel thresholding. J. Inf. Sci. Eng. **17**, 51–59 (2001)

12. Platt, J.: Probabilistic outputs for support vector machines and comparisons to regularized likelihood methods. Adv. Large Margin Classifiers **10**(3), 61–74 (1999)

13. Krizhevsky, A., Sutskever, I., Hinton, G.E.: Imagenet classification with deep convolutional neural networks. In: Proceedings of the Neural Information Processing Systems (NIPS), pp. 1–9 (2012)

14. Codella, N., Connell, J., Pankanti, S., Merler, M., Smith, J.R.: Automated medical image modality recognition by fusion of visual and text information. In: Golland, P., Hata, N., Barillot, C., Hornegger, J., Howe, R. (eds.) MICCAI 2014, Part II. LNCS, vol. 8674, pp. 487–495. Springer, Heidelberg (2014)

15. Caruana, R., Niculescu-Mizil, A., Crew, G., Ksikes, A.: Ensemble selection from libraries of models. In: Proceedings of the Twenty-First International Conference on Machine Learning, ICML 2004, p. 18. ACM, New York (2004)

Data-Driven Feature Learning for Myocardial Segmentation of CP-BOLD MRI

Anirban Mukhopadhyay[1], Ilkay Oksuz[1(✉)], Marco Bevilacqua[1],
Rohan Dharmakumar[2], and Sotirios A. Tsaftaris[1,3]

[1] IMT Institute for Advanced Studies Lucca, Lucca, Italy
ilkay.oksuz@imtlucca.it
[2] Biomedical Imaging Research Institute, Cedars-Sinai Medical,
Los Angeles, CA, USA
[3] Department of Electrical Engineering and Computer Science,
Northwestern University, Evanston, IL, USA

Abstract. Cardiac Phase-resolved Blood Oxygen-Level-Dependent
(CP-BOLD) MR is capable of diagnosing an ongoing ischemia by detecting changes in myocardial intensity patterns at rest without any contrast
and stress agents. Visualizing and detecting these changes require significant post-processing, including myocardial segmentation for isolating the
myocardium. But, changes in myocardial intensity pattern and myocardial shape due to the heart's motion challenge automated standard CINE
MR myocardial segmentation techniques resulting in a significant drop
of segmentation accuracy. We hypothesize that the main reason behind
this phenomenon is the lack of discernible features. In this paper, a multi
scale discriminative dictionary learning approach is proposed for supervised learning and sparse representation of the myocardium, to improve
the myocardial feature selection. The technique is validated on a challenging dataset of CP-BOLD MR and standard CINE MR acquired in
baseline and ischemic condition across 10 canine subjects. The proposed
method significantly outperforms standard cardiac segmentation techniques, including segmentation via registration, level sets and supervised
methods for myocardial segmentation.

Keywords: Dictionary learning · CP-BOLD MR · CINE MR · Segmentation

1 Introduction

CP-BOLD MR is a truly noninvasive (without contrast or stress agents and
ionizing radiation) method for early diagnosis of ongoing ischemia. CP-BOLD
identifies the ischemic myocardium by examining changes in myocardial signal intensity patterns as a function of cardiac phase [14]. However, visualizing and quantifying such changes requires significant post-processing, including

---

The first two authors contributed equally to this work.

© Springer International Publishing Switzerland 2015
H. van Assen et al. (Eds.): FIMH 2015, LNCS 9126, pp. 189–197, 2015.
DOI: 10.1007/978-3-319-20309-6_22

myocardial segmentation to isolate the myocardium from the rest of the anatomy. In particular, although CP-BOLD is a cine type acquisition, automated myocardial segmentation and registration algorithms developed for standard CINE under-perform, due to the spatio-temporal intensity variations of the myocardial BOLD effect [9], an example of which is shown in Fig. 1. Thus, in CP-BOLD in addition to violations of shape invariance (as with standard CINE MRI) the principal assumption of appearance invariance (consistent intensity) is violated as well.

As a result, no automated CP-BOLD MR myocardial segmentation algorithms exist, and semi-automated methods based on tracking are currently employed [13]. We hypothesize that it is due to the lack of appropriate features, which are invariant yet unique and descriptive under the particular type of appearance and shape deformation observed in CP-BOLD images. Rather than relying on low-level features used often for myocardial segmentation of standard CINE MR which are inconsistent for CP-BOLD MR, a more generalized feature learning method should be developed to accommodate the myocardial BOLD effect while still being reliable in the CINE MR case.

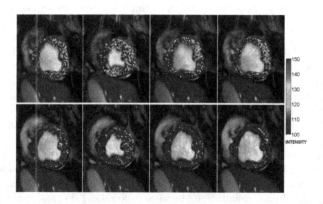

Fig. 1. Exemplary cardiac phases of CP-BOLD MR (top row) and standard CINE MR (bottom row) obtained from the same subject under baseline conditions (absence of ischemia) where the myocardium is color coded to underline the challenge of appearance variation in CP-BOLD MR which is minimal in the case of standard CINE MR (Color figure online).

We adopt a patch-based discriminative dictionary learning technique (which has been used also in echocardiography [6]) to learn features from previously segmented data in a fully supervised manner. The motivation behind the choice of a sparse dictionary is to employ a compact and high-fidelity low-dimensional subspace representation which is able to extract semantic information of the myocardium as well [16]. The key observation behind this strategy is that, though the patch intensity level varies significantly across the cardiac cycle, sparse representations based on learnt dictionaries are invariant across the cardiac cycle, as well as unique and robust. Briefly described, during training two separate dictionaries are learnt at multiple scales for the myocardium and background. In this regard, we also introduce a discriminative initialization step (discarding patches with high values in intra-class Gram matrix) to promote diversity in initialization, and a discriminative pruning step (discarding training patches with high values in inter-class Gram matrix) to further boost the discriminative abilities of the dictionaries. During testing, multiscale sparse features are used.

The main contributions of the paper are twofold. First, we experimentally demonstrate that BOLD contrast significantly affects the accuracy of segmentation algorithms (including segmentation via registration of an atlas, level sets, supervised classifier-based and other dictionary-based methods) which instead perform well in standard CINE MR. Second, to address our hypothesis we design a set of compact features using Multi-Scale Discriminative Dictionary Learning, which can effectively represent the myocardium in CP-BOLD MR. The method has been evaluated on canine subjects, which makes the problem even more challenging (lower accuracy is expected) due to the smaller size of myocardium. The remainder of the paper is organized as follows: Sect. 2 discusses related work, Sect. 3 presents the proposed method, whereas results are described in Sect. 4. Finally, Sect. 5 offers discussions and conclusion.

2 Related Work

Automated myocardial segmentation for standard CINE MR is a well studied problem [10]. Most of these algorithms can be broadly classified into three categories based on whether the methodology is segmentation-only, level set or Atlas-based segmentation with inherent registration. Recently, Atlas-based segmentation techniques have received significant attention. The myocardial segmentation masks available from other subject(s) are generally propagated to unseen data in Atlas-based techniques [2] using non-rigid registration algorithms, e.g., diffeomorphic demons (dDemons) [15], FFD-MI [5] or probabilistic label fusion [2]. Level set class of techniques uses a non-parametric way for segmenting myocardium with weak prior knowledge [3, 7].

Segmentation-only class of techniques mainly focuses on feature-based representation of the myocardium. Texture information is generally considered as an effective feature representation of the myocardium for standard CINE MR images [17]. The patch-based static discriminative dictionary learning technique (SJTAD) [11] and Multi-scale Appearance Dictionary Learning technique [6] have achieved high accuracy and are considered as state-of-the-art mechanisms for supervised learning of discernible myocardial features from previously segmented data. In this paper, we follow the segmentation-only approach with the major feature of considering multi-scale appearance and texture information as the input of a discriminative dictionary learning procedure.

3 Method

General image segmentation strategies are developed on the assumption that both appearance and shape do not vary considerably across the images of a given sequence. Cardiac motion affects the shape invariance assumption, and varying CP-BOLD signal intensities violate the appearance invariance assumption as well. To overcome this issue, dictionary learning techniques can be leveraged to learn better representative features. To this end, we propose a Multi-Scale Discriminative Dictionary Learning (MSDDL) method (detailed in Algorithm 1). The features learnt via dictionary learning are tested in a rudimentary classification scheme solely for the purpose of comparing to other methods.

Feature generation with Multi-scale Discriminative Dictionary Learning (MSDDL): Given some sequences of training images and corresponding ground truth labels (i.e. masks), we can obtain two sets of matrices, $\{Y_k^B\}_{k=1}^K$ and $\{Y_k^M\}_{k=1}^K$, where the matrix Y_k^B contains the background information at a particular scale k (each scale is characterized by a different patch size), and Y_k^M is the corresponding matrix referring to the myocardium. Information is collected from image patches: squared patches are sampled around each pixel of the training images. More precisely, the i-th column of the matrix Y_k^B (and similarly for the matrix Y_k^M) is obtained by concatenating the normalized patch vector of pixel intensities at scale k, taken around the i-th pixel in the background, along with the Gabor and HOG features of the same patch. Our MSDDL method takes as input these two sets of training matrices, to learn, at each scale k, two dictionaries, D_k^B and D_k^M, and two sparse feature matrices, X_k^B and X_k^M. E.g. , the i-th column of the matrix X_k^B, $x_{k,i}^B$, is considered as the discriminative feature vector for the particular pixel corresponding to the i-th column in Y_j^B. Dictionaries and sparse features are trained via the well known K-SVD algorithm [1]. One main modification to K-SVD is the use of the "intra-class Gram matrix" to promote diversity in the initialization step. The idea is to have a subset of patches as much diverse as possible to train dictionaries and sparse features. For a given class considered (let us say background) and a given scale k, we can define the intra-class Gram matrix as $G_k^B = (Y_k^B)^T Y_k^B$. To ensure a proper discriminative initialization, patches that correspond to high values in the Gram matrix are discarded from the training before performing K-SVD. Notably, we sort the training patches w.r.t. the sum of their related coefficients in the Gram Matrix, and we prune them by choosing a certain percentage.

A second proposed modification relates to a pruning step, which is performed after K-SVD. In this case, at each scale k, an "inter-class Gram matrix" is computed ($G_k^{BM} = (D_k^B)^T D_k^M$): the atoms of each dictionary are sorted according to their cumulative coefficients in G^{BM}, and a chosen percentage of them is discarded to ensure mutual exclusiveness between the dictionaries of the two different classes. The philosophy behind this operation is similar to the one of the discriminative dictionary learning algorithm proposed in [8], where the norm of the inter-class Gram matrix appears in the optimization formulation as a constraint to be minimized. By pruning the undesired dictionary atoms all at one time, we actually adopt a greedier and low-complexity approach to the same problem. Moreover, we believe that, instead of globally minimizing the Gram matrix norm, directly removing the most "problematic" patches, which create ambiguity between background and myocardium, is more effective in our case.

Algorithm 1. Multi-scale Discriminative Dictionary Learning (MSDDL)

Input: Multi-scale training patches for background and the myocardium:
$\{Y_k^B\}_{k=1}^K$ and $\{Y_k^M\}_{k=1}^K$
Output: Multi-scale dictionaries for background and the myocardium:
$\{D_k^B\}_{k=1}^K$ and $\{D_k^M\}_{k=1}^K$
 1: **for** $k = 1...K$ **do**
 2:    **for** C={B,M} **do**
 3:       Evaluate Y_k^C
 4:       Compute the intra-class Gram matrix G_k^C
 5:       Discard atoms with high values in G_k^C
 6:       Learn dictionary and sparse feature matrix with the K-SVD algorithm
 7:

$$\operatorname*{minimize}_{D_k^C, X_k^C} \| Y_k^C - D_k^C X_k^C \|_2^2 \quad \text{s. t.} \quad \| x_{k,i}^C \|_0 \leq L$$

 8:    Compute the inter-class Gram matrix G_k^{BM}
 9:    Discard from D_k^B and D_k^M atoms with high values in G_k^{BM}

Building a Rudimentary Classifier for Segmentation: When considering the same patch-based approach in a segmentation problem, we have a set of test matrices $\{\hat{Y}_k\}_{k=1}^K$, obtained by sampling patches at multiple scales from the test image, and concatenating intensity values of these patches, along with Gabor and HOG features. The goal is to assign to each pixel of the test image a label, i.e. establish if the pixel is included in the background or the myocardial region. To perform this classification, we use the multi-scale dictionaries, $\{D_k^B\}_{k=1}^K$ and $\{D_k^M\}_{k=1}^K$, previously learnt with MSDDL. The Orthogonal Matching Pursuit (OMP) algorithm [12] is used to compute, at each scale k, the two sparse feature matrices $\hat{X}_k^B$ and $\hat{X}_k^M$. A certain patch at scale k, $\hat{y}_{k,i}$ will be assigned to the class that gives the smallest dictionary approximation error. More precisely, if $\| \hat{y}_{k,i} - D_k^B \hat{y}_{k,i}^B \|_2$ is larger than $\| \hat{y}_{k,i} - D_k^M \hat{y}_{k,i}^M \|_2$, at scale k the patch is assigned to the background; otherwise, it is considered belonging to the myocardial region. In this study, we employed a simple majority voting across all scales to obtain the final classification for each pixel of the test image.

4   Results

This section offers a qualitative and quantitative assessment of our proposed method w.r.t. state-of-the-art methods, to demonstrate its effectiveness for myocardial segmentation. It is particularly important to note that our method significantly outperforms all methods from current literature in both baseline and ischemia cases of CP-BOLD MR, whereas yields state-of-the-art results for both baseline and ischemia cases of standard CINE MR.

4.1   Data Preparation and Parameter Settings

2D short-axis images of the whole cardiac cycle were acquired at baseline and severe ischemia (inflicted as stenosis of the left-anterior descending coronary artery (LAD)) on a 1.5T Espree (Siemens Healthcare) in the same 10 canines along mid ventricle using both standard CINE and a flow and motion compensated CP-BOLD acquisition within few minutes of each other. All quantitative experiments are performed in a strict leave-one-subject-out cross-validation setting.

As for the parameters of MSDDL, in this paper we have empirically chosen a dictionary of 1000 atoms for foreground and background respectively, a sparsity of 4, a number of scales $K = 3$, and $9 \times 9, 11 \times 11$ and 13×13 as patch sizes. We tested the parameter sensitivity within a reasonable range, but a detailed performance chart is beyond the scope of this paper.

4.2 Visual Comparison of the Discriminativeness of the Learnt Dictionaries and Features

The feature patches learnt by MSDDL are discriminative enough for representing the myocardium separately from the background. In particular a set of feature patches of size 11×11 (without HOG and Gabor) learnt for the myocardium and background are shown in Fig. 2 to illustrate the discriminativeness of the learnt feature patches.

The motivations behind choosing each step of the proposed MSDDL strategy and the effectiveness of the features learnt by this technique are highlighted in Fig. 3, where the Cosine Similarity metric [4] is used to determine the most similar patches to a given patch in the MSDDL feature space. When selecting a patch inside the myocardium, without texture and Gram filtering, similar patches are found outside the myocardium too. Adding texture improves somewhat localization, but when considering also Gram filtering, the discriminative strengths of the approach are more evident, since few similar patches are found only within the myocardium. Similar observations hold also for the case of images from standard CINE as well (not shown for brevity).

4.3 Quantitative Comparison

As segmentation quality metric, the Dice coefficient, which measures the overlap between ground truth segmentation masks and those obtained by the

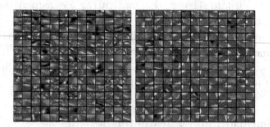

Fig. 2. Exemplar set of dictionary atoms (without HOG and Gabor) for Background (left) and Myocardium (right) learnt from patches of size 11×11 on CP-BOLD MR.

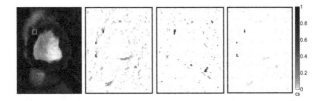

Fig. 3. Cosine Similarity (CS) between the learnt features showing the advantage of adding texture and Gram filtering. Test patch denoted by a green square in the raw image (first column), MSDDL only on appearance (second column), with texture (third column), and with proposed Gram filtering (final column) (Color figure online).

algorithm(s), is employed. For our implementation of Atlas-based segmentation methods, the registration algorithms dDemons [15] and FFD-MI [5] were used to propagate the segmentation mask of the end-diastole image from all other subjects to the end-diastole image of the test subject, followed by a majority voting to obtain the final myocardial segmentation. For level-set class of methods, a hybrid approach of [3] for endocardium and [7] for epicardium is used. For supervised classifier-based methods, namely Appearance Classification using Random Forest (ACRF) and Texture-Appearance Classification using Random Forest (TACRF) we used random forests as classifiers to get segmentation labels from different features. To provide more context we compare our approach with dictionary-based methods, SJTAD and RDDL. SJTAD is an implementation of the method in [11], whereas for RDDL we used the discriminative dictionary learning of [8] within the same classification framework that we described in Sect. 3. Finally to showcase the strengths of our design choices we considered two additional variants of MSDDL, one without Gram filtering (MSDDL No GF) and one without texture information as well (MSDDL No GF No Texture). Note that the former is similar to [6] without level-set refinement.

As Table 1 shows, overall, when standard CINE acquisition is used, most algorithms perform adequately and the presence of ischemia slightly reduces performance. However, when BOLD contrast is present, other approaches fail to accommodate changes in appearance due to contrast, but MSDDL obtains consistent performance. Specifically, Atlas-based methods are shown to perform well in standard CINE cases but poorly in CP-BOLD. ACRF and TACRF, instead, show very low performance in both standard CINE MR and CP-BOLD MR. Among dictionary-based techniques, SJTAD performs well in standard CINE MR, but underperforms in CP-BOLD MR. Our MSDDL method outperforms all approaches. When comparing it with its variants, it shows that both texture and appearance are important and that the pruning steps based on the Gram matrix are extremely beneficial. Even when we replaced our dictionary learning algorithm with RDDL, an algorithm that forces discrimination by explicitly penalizing the inter-class Gram matrix norm, the results are unimpressive. These findings are also statistically significant using a paired t-test between the results of MSDDL and the second-best performing one, i.e. SJTAD [11]. For both baseline

Table 1. Dice coefficient (mean(std)) for segmentation accuracy in %.

Methods	Baseline		Ischemia	
	Standard CINE	CP-BOLD	Standard CINE	CP-BOLD
Atlas-based methods				
dDemons [15]	60(8)	55(8)	56(6)	49(7)
FFD-MI [5]	60(3)	54(8)	54(8)	45(6)
Level set-based methods				
CVL [3,7]	50(8)	43(11)	45(9)	37(10)
Supervised classifier-based methods				
ACRF	57(3)	25(2)	52(3)	21(2)
TACRF	65(2)	29(3)	59(1)	24(2)
Dictionary-based methods				
SJTAD [11]	71(2)	32(3)	66(3)	23(4)
RDDL [8]	42(15)	50(20)	48(13)	61(12)
MSDDL No GF No Texture	52(8)	51(7)	45(4)	51(6)
MSDDL No GF	62(5)	52(4)	53(5)	57(7)
MSDDL	75(3)†	75(2)⋆	72(2)‡	71(2)⋆

and ischemia cases of CP-BOLD MR, MSDDL shows improved performance compared to SJTAD ($\star, p < 0.001$). In the case of standard CINE MR although differences appear small they are still statistically significant, i.e. ($\dagger, p < 0.05$) and ($\ddagger, p < 0.01$) for baseline and ischemia respectively.

5   Discussions and Conclusion

Rethinking the assumptions underlying the design of analysis algorithms for standard CINE MR is critical for successfully developing the appropriate analytical tools necessary to meet the new challenges posed by myocardial CP-BOLD MR. In particular, this study pin-pointed the challenges the BOLD effect poses on these assumptions made when segmenting the myocardium and quantitatively analyzed the adverse effect on algorithmic performance. In addition, in this study we showed that by learning appropriate features to best represent texture and appearance in CP-BOLD, it is possible to improve the performance of automated algorithms for myocardial segmentation. This study also showed overall low performance of state-of-the-algorithms even for standard CINE MR in canine subjects, which can be attributed to the small size of the myocardium. The proposed algorithm does not exploit the temporal information across cardiac phases and doing so should increase performance in future extensions. Finally, such post-processing tools are expected to be instrumental in advancing the utility of cardiac CP-BOLD MR towards effective clinical translation.

Acknowledgments. This work was supported by the National Institutes of Health under Grant 2R01HL091989-05.

References

1. Aharon, M., et al.: K-SVD: an algorithm for designing overcomplete dictionaries for sparse representation. IEEE TSP **54**(11), 4311–4322 (2006)
2. Bai, W., et al.: A probabilistic patch-based label fusion model for multi-atlas segmentation with registration refinement: application to cardiac MR images. IEEE TMI **32**(7), 1302–1315 (2013)
3. Chan, T.F., et al.: Active contours without edges. IEEE TIP **10**(2), 266–277 (2001)
4. Chang, L.-H., et al.: Achievable angles between two compressed sparse vectors under norm/distance constraints imposed by the restricted isometry property: a plane geometry approach. IEEE T Inf. Theory **59**(4), 2059–2081 (2013)
5. Glocker, B., et al.: Dense image registration through MRFs and efficient linear programming. MIA **12**(6), 731–741 (2008)
6. Huang, X., et al.: Contour tracking in echocardiographic sequences via sparse representation and dictionary learning. MIA **18**, 253–271 (2014)
7. Li, C., et al.: Distance regularized level set evolution and its application to image segmentation. IEEE TIP **19**(12), 3243–3254 (2010)
8. Ramirez, I., et al.: Classification and clustering via dictionary learning with structured incoherence and shared features. In: IEEE CVPR, pp. 3501–3508 (2010)
9. Rusu, C., et al.: Synthetic generation of myocardial bloodoxygen-level-dependent MRI time series via structural sparse decomposition modeling. IEEE TMI **7**(33), 1422–1433 (2014)
10. Tavakoli, V., et al.: A survey of shape-based registration and segmentation techniques for cardiac images. CVIU **117**, 966–989 (2013)
11. Tong, T., et al.: Segmentation of MR images via discriminative dictionary learning and sparse coding: application to hippocampus labeling. NeuroImage **76**, 11–23 (2013)
12. Tropp, J.A., et al.: Signal recovery from random measurements via orthogonal matching pursuit. IEEE T Inf. Theory **53**(12), 4655–4666 (2007)
13. Tsaftaris, S.A., et al.: A dynamic programming solution to tracking and elastically matching left ventricular walls in cardiac CINE MRI. In: IEEE ICIP, pp. 2980–2983 (2008)
14. Tsaftaris, S.A., et al.: Detecting myocardial ischemia at rest with cardiac phaseresolved blood oxygen leveldependent cardiovascular magnetic resonance. Circ.: Cardiovasc. Imaging **6**(2), 311–319 (2013)
15. Vercauteren, T., Pennec, X., Perchant, A., Ayache, N.: Non-parametric diffeomorphic image registration with the demons algorithm. In: Ayache, N., Ourselin, S., Maeder, A. (eds.) MICCAI 2007, Part II. LNCS, vol. 4792, pp. 319–326. Springer, Heidelberg (2007)
16. Wright, J., et al.: Sparse representation for computer vision and pattern recognition. Proc. IEEE **98**(6), 1031–1044 (2010)
17. Zhen, X., Wang, Z., Islam, A., Bhaduri, M., Chan, I., Li, S.: Direct estimation of cardiac bi-ventricular volumes with regression forests. In: Golland, P., Hata, N., Barillot, C., Hornegger, J., Howe, R. (eds.) MICCAI 2014, Part II. LNCS, vol. 8674, pp. 586–593. Springer, Heidelberg (2014)

Cardiac Fibers Estimation from Arbitrarily Spaced Diffusion Weighted MRI

Andreas Nagler[1], Cristóbal Bertoglio[1,2(✉)], Christian T. Stoeck[3,4],
Sebastian Kozerke[3,4], and Wolfgang A. Wall[1]

[1] Institute for Computational Mechanics, Technische Universität München,
Munich, Germany
[2] Center for Mathematical Modeling, Universidad de Chile, Santiago, Chile
cbertoglio@dim.uchile.cl
[3] Institute for Biomedical Engineering, University and ETH Zurich,
Zurich, Switzerland
[4] Division of Imaging Sciences and Biomedical Engineering,
King's College of London, London, UK

Abstract. We propose a framework for estimating fiber fields in the
heart from arbitrarily spaced diffusion weighted MRI. The approach is
based on a parametric and space-dependent mathematical representation
of the helix angles across the heart, leading to a semi-analytical formula of
the diffusion tensor, without any particular assumption on the ventricular
shape. Then, by solving an nonlinear inverse problem, the degrees of
freedom of the model can be estimated from measured diffusion weighted
data. We illustrate the methodology using synthetic data and compare
it with previously reported fiber reconstruction techniques.

1 Introduction

Diffusion Tensor Magnetic Resonance Imaging (DTMRI) allows observing tissue
fiber structure non-invasively [1,9]. Unfortunately, a full three-dimensional acquisition of the fiber data in the in-vivo heart through DTMRI requires extremely
long scan times, and therefore has mostly been applied to *ex-vivo* heart samples
e.g. [5] and references therein.

Recently, *in-vivo* DTMRI acquisitions in a limited number of slices along the
heart were reported [2,7,8]. Hence, a three-dimensional reconstruction of the
fiber architecture across the heart from the DTMRI data is required, which has
to deal not only with low signal-to-noise ratios (SNR) encountered in DTMRI,
but also with the spatial missmatch of the diffusion weighted images (DWIs), to
compensate for breath hold inconsistencies during acquisition and eddy current
effects.

However, to the authors best knowledge, all tensor reconstruction schemes
from a set of DWIs rely on the co-existance of the images in the same spatial
location, see e.g. [4] and references therein. For the specific case of the heart,
recently in [11] a tensor interpolation method based on curvelinear coordinates
was proposed for a left ventricular geometry for reconstructing the cardiac fiber

H. van Assen et al. (Eds.): FIMH 2015, LNCS 9126, pp. 198–206, 2015.
DOI: 10.1007/978-3-319-20309-6_23

architecture from a set of DWI-slices. Since its starting point is a voxelwise estimation of the tensor, it also relies in the spatial matching of the DWIs, what until now has been achieved by in-plane registration of the DWIs onto "mean slices" not accounting for though plane motion.

In this work, we propose a tensor estimation scheme which can handle arbitrarily located DWIs. The method is based on a parametric representation of the diffusion tensor across the left ventricular volume – inspired from [6] – which is based on previously reported data of the cardiac fiber architecture. This allows to relate sparsely located diffusion weighted information by estimating the cardiac fiber orientations using a nonlinear least squares approach.

The rest of the paper is organized as follows. Section 2 describes the proposed diffusion tensor model and introduces the estimation formulation. Then in Sect. 3 we exemplfiy the method using a synthetic fiber data set, and hence we compare the performance of the methods against the tensor-interpolation of [11]. Finally, we give some conclusions and perspectives in Sect. 4.

2   The Rule-Based DT-model and Estimation Algorithm

Let us denote the computational domain of the left-ventricle $\Omega \subset \mathbb{R}^3$, where we define a transverse isotropic diffusion tensor model of the form:

$$\boldsymbol{D}(\theta(\boldsymbol{x}), \lambda_1, \lambda_2) = (\lambda_1 - \lambda_2)\boldsymbol{f}(\theta(\boldsymbol{x}), \boldsymbol{x}) \otimes \boldsymbol{f}(\theta(\boldsymbol{x}), \boldsymbol{x}) + \lambda_2 \boldsymbol{I} \qquad (1)$$

$$\boldsymbol{f}(\theta(\boldsymbol{x}), \boldsymbol{x}) = \cos(\theta(\boldsymbol{x}))\boldsymbol{c}(\boldsymbol{x}) + \sin(\theta(\boldsymbol{x}))\boldsymbol{\ell}(\boldsymbol{x}) \qquad (2)$$

for all $\boldsymbol{x} \in \Omega$, with $\theta(\boldsymbol{x})$ the helix angle of the fiber and $\boldsymbol{c}, \boldsymbol{\ell}$ the local circumferential and long-axial directions, respectively (see Fig. 1(a)–(b)). The values $\lambda_1 > \lambda_2 > 0$ are the diffusivities in fiber and cross-fiber direction, respectively.

We assume the helix angle distribution is given by a set of degrees-of-freedom $\boldsymbol{\Theta} \in \mathbb{R}^\kappa$ (DOF), such that the following linear relation holds: $\theta(\boldsymbol{x}_j) = \boldsymbol{H}_j\boldsymbol{\Theta}$, $\boldsymbol{x}_1, \ldots \boldsymbol{x}_p \in \Omega$, with $\boldsymbol{x}_1, \ldots \boldsymbol{x}_p$ the set of discrete mesh points. Each element of $\boldsymbol{\Theta}$ is associated to one vertex of the surface "patchs" depicted in Fig. 2(a)–(b). Hence, the operator $\boldsymbol{H} \in \mathbb{R}^{p \times \kappa}$ summarizes the linear interpolations from

(a) Circumferential
direction $\boldsymbol{c}$

(b) Long-axis direction $\boldsymbol{\ell}$

Fig. 1. Circumferential and long-axis coordinate system.

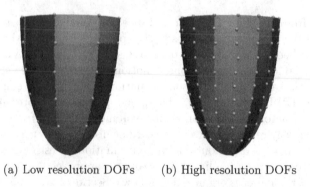

(a) Low resolution DOFs   (b) High resolution DOFs

Fig. 2. Patches definition on the prolate spheroid geometry (white spheres at the corners represent the DOFs used for the diffusion tensor model).

the patchs-corners to the rest of the surface, and then from the surfaces to the heart's volume using Poisson interpolation. Two sets of degree of freedom were used in the numerical examples, a low resolution set with 82-DOFs (Fig. 2(a)) and a high-resolution set with 322 DOFs (Fig. 2(b)). We assume that we have a set of measured diffusion weighted signal values $A_{i,j}$, for $j = 0, \ldots, N_i$ voxels for each diffusion encoding direction $b_0 = 0$, $\|b_i\| = 1$, $i = 1, \ldots, N_g$. The fiber estimation method consists then of the following steps:

Step1: Being A_0 the values of the zero-weighted images at the measured voxels, we first reconstruct these values in the whole heart domain as

$$\alpha_0 = H(H_0^\mathsf{T} H_0)^{-1} H_0^\mathsf{T} A_0.$$

The linear operator H_0 consists of the rows of H, which maps the DOFs to the zero-weighted slices.

Step2: We then minimize the functional

$$\mathrm{argmin}_{\Theta,\lambda_1,\lambda_2} J(\Theta, \lambda_1, \lambda_2) = \sum_{i=1}^{N_g} \sum_{j=1}^{N_i} \left(exp\left(-b_i^\mathsf{T} D(\theta_j(\Theta), \lambda_1, \lambda_2) b_i\right) - \frac{A_{i,j}}{\alpha_{0,j}} \right)^2 \tag{3}$$

Note that since we directly compare the mismatch in the graduation itself, each voxel and diffusion gradient are equally weighted in the functional.

3   Numerical Example with Synthetic Data

3.1   Reference Fiber Orientation

To show that the proposed estimation scheme can properly rebuild a given fiber orientation and detect systematic changes from noisy and arbitrarily spaced DWIs, we constructed two reference fiber families (and hence two tensor families)

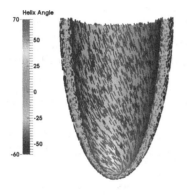
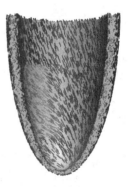

Fig. 3. Synthetic fibers organization for healthy (left) and infarcted (right) cases. The infarction is located in the area with more horizontal (yellow) fibers (Color figure online).

on a prolate spheroid geometry named *healthy* and *infarcted*, see Fig. 3. This synthetic fiber orientation was defined as $\hat{f} = \cos(\hat{\theta})g_1 + \sin(\hat{\theta})g_2$, with the unit vectors of the prolate spheroid coordinates g_1 and g_2 tangential to the surface, and the helix angle $\hat{\theta}$ defined via the following rules:

– Quadratic variation of the helix-angle in apex-to-base direction from $40°$ to $60°$ and $-40°$ to $-60°$ on the endocardium and epicardium, respectively.
– Slightly non-linear in-wall deviation of the helix angle using a tanh function, what the diffusion model presented above cannot exactly reproduce.
– A constant helix angle of $35°$ was imposed in a part of the endocardium ranging to the ventricular midwall in order to represent an infarcted region (see more yellow fibers in Fig. 3-right).

(a) Perfectly matching DWIs at each slice (blue wireframe)

(b) Desired position of the DWIs for a given slice (red) at missalinged slices (blue)

Fig. 4. Positions of the measured DWIs (Color figure online).

3.2 Synthetic DWIs

We construct from the healthy and infarcted fiber families diffusion weighted images (DWI) for 16 diffusion encoding directions at five equally distributed slice positions, see Fig. 4(a), by

1. Averaging the helix angles of the computational domain to each pixel of size $2 \times 2 \times 8\,\text{mm}^3$.
2. The diffusivities are chosen constant throughout the domain Ω as $\lambda_1 = 0.8$ and $\lambda_2 = 0.5$ (diffusivities normalized with b-value) for both healthy and infarcted cases.
3. At each voxel j of DWI i, from the mean angle $\theta_{\text{mean};i,j}$ and λ_1, λ_2 we compute the signal intensity $A_{i,j}$ for each gradient direction $\boldsymbol{b}_i$ via

$$A_{i,j} = A_0 \exp(-\boldsymbol{b}_i^\mathsf{T} \boldsymbol{D}(\theta_{\text{mean};i,j}, \lambda_1, \lambda_2)\boldsymbol{b}_i)$$

with $A_0 = 100$, as it usually assumed in all tensor reconstruction methods as the relation between the diffusion tensor and the measured values through DWMRI, see e.g. [10].

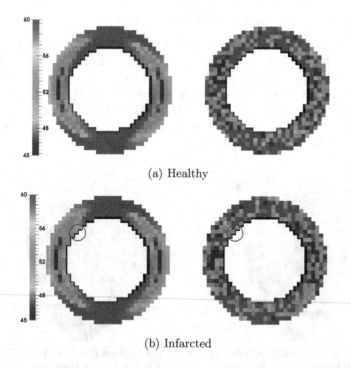

(a) Healthy

(b) Infarcted

Fig. 5. Sample DWI slice for the healthy (a) and infarcted (b) case without (left) and with (right) noise. Infarcted region in (b) is indicated with a red circle. The voxels are coloured by the signal intensity ($A_{i,j}$-value) (Color figure online).

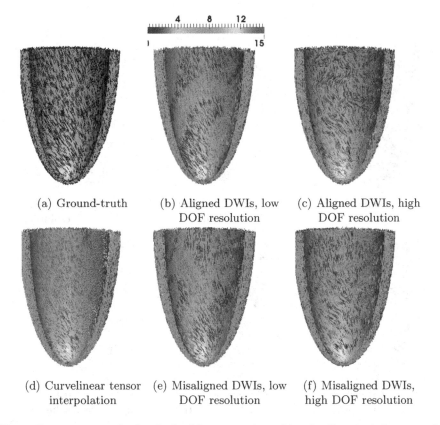

(a) Ground-truth (b) Aligned DWIs, low DOF resolution (c) Aligned DWIs, high DOF resolution

(d) Curvelinear tensor interpolation (e) Misaligned DWIs, low DOF resolution (f) Misaligned DWIs, high DOF resolution

Fig. 6. Estimation results for the healthy case, coloured by the fiber angle between the reconstruction and the ground truth (Color figure online).

Since in real DWMRI each DWI is located in different positions with respect to the heart, we also perturb the DWIs at each slice to emulate this misalignement, see Fig. 4(b). We also consider a rician noise [3] with complex gaussian deviation $\sigma_{\text{Gauss}} = 7$ (see Fig. 5) in order to emulate a more realistic testing scenario.

3.3 Estimation Results and Discussion

Figures 6 and 7 present the estimation results, in terms of the estimated fibers coloured by the angle error with respect to the ground-truth, for the two sets of data (*aligned* and *missaligned*), our method with two different DOFs resolution (*low* and *high*) and the results of the tensor interpolation method from [11] for the perfectly aligned DWIs at each slice only. The results can be interpreted as follows:

– In the healthy case and aligned DWIs, our estimation method with a low DOF resolution seems to returns smaller error compared to the curvelinear tensor interpolation method.

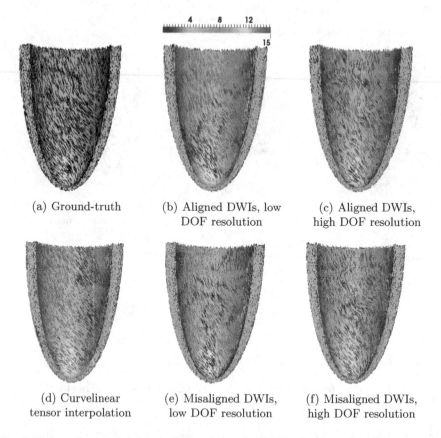

Fig. 7. Estimation results for the infarcted, coloured by the fiber angle between the reconstruction and the ground truth (Color figure online).

- In the infarcted case, this still holds, but in some extent of the infarcted area curvelinear tensor interpolation returns more accurate values. This may result from the constant helix angle within the infarction, which can be more easily reproduced by such a method. However, the healthy tissue near the infarct seems to be more accurately reproduced by our estimation framework.
- The resulting fiber organization using a high DOF resolution show a more irregular behaviour due to the overfitting of the noisy DWIs.
- For the misaligned DWIs, the curvelinear tensor interpolation method is not applicable and therefore the results can not be compared against the estimation. However, the error remains of the same order compared with the results with aligned DWIs, but with a different spatial distribution. The exception are the regions close to the base, were in fact almost no data is available and the fibers result from a simple extrapolation.

4   Conclusions and Perspectives

We have presented a three-dimensional fiber reconstruction scheme, which can handle arbitrarly spaced diffusion weighted MR images. The benefits of the method, compared with previously reported ones are also that we can fully deal with every heart shapes deviating from the ideal spheroid, and we do not require to morph the left-ventricular shape onto, e.g., a spheroid. This would allow to easily deal with biventricular geometries (as it was done in [6] for the estimation of the helix angle from 3D-DTMRI) or pathological heart shapes. Moreover, images that are acquired in different scanning sessions can be integrated into the estimation, and of different expected signal intensities (e.g., acquired using 1.5 T and 3T machines) by choosing different weighting of the measurements in the functional. Moreover, as it can be appreciated from the results above, the framework would enable to adjust the DOFs resolution and amount of data required (i.e. the number of slices and diffusion encoding directions) in order to estimate the fiber architecture up to a certain precision. Another relevant aspect is that the flexibility of the choice of the diffusion encoding directions and with respect to slices orientation may allow increasing spatial resolution of fiber fields without significantly increase inquisition duration or to speed-up the acquisition of diffusion MR sequences.

Ongoing work consists in the extension to the estimation of the sheet direction and further comparisons with the curvelinear tensor interpolation method in real geometries, with both synthetic and real data.

Acknowledgements. The results presented in this article are part of the *Advanced Cardiac Mechanics Emulator*, an initiative supported by the Institute for Advanced Study (TU München). This support is gratefully acknowledged. We also thank Radomír Chabiniok and Jack Harmer (King's College London) for the valuable discussions.

References

1. Basser, P., Mattiello, J., Lebihan, D.: Estimation of the effective self-diffusion tensor from the NMR spin echo. J. Magn. Reson. Ser. B **103**(3), 247–254 (1994)
2. Gamper, U., Boesiger, P., Kozerke, S.: Diffusion imaging of the in vivo heart using spin echoesconsiderations on bulk motion sensitivity. Magn. Reson. Med. **57**(2), 331–337 (2007)
3. Gudbjartsson, H., Patz, S.: The Rician distribution of noisy MRI data. Magn. Reson. Med. **34**(6), 910–914 (1995)
4. Koay, C.G., Chang, L.C., Carew, J.D., Pierpaoli, C., Basser, P.J.: A unifying theoretical and algorithmic framework for least squares methods of estimation in diffusion tensor imaging. J. Magn. Reson. **182**(1), 115–125 (2006)
5. Lombaert, H., Peyrat, J., Croisille, P., Rapacchi, S., Fanton, L., Cheriet, F., Clarysse, P., Magnin, I., Delingette, H., Ayache, N.: Human atlas of the cardiac fiber architecture: study on a healthy population. IEEE Trans. Med. Imaging **31**(7), 1436–1447 (2012)

6. Nagler, A., Bertoglio, C., Gee, M., Wall, W.: Personalization of cardiac fiber orientations from image data using the unscented kalman filter. In: Ourselin, S., Rueckert, D., Smith, N. (eds.) FIMH 2013. LNCS, vol. 7945, pp. 132–140. Springer, Heidelberg (2013)

7. Nguyen, C., Fan, Z., Sharif, B., He, Y., Dharmakumar, R., Berman, D.S., Li, D.: In vivo three-dimensional high resolution cardiac diffusion-weighted MRI: a motion compensated diffusion-prepared balanced steady-state free precession approach. Magn. Res. Med. **72**(5), 1257–1267 (2013)

8. Nielles-Vallespin, S., Mekkaoui, C., Gatehouse, P., Reese, T.G., Keegan, J., Ferreira, P.F., Collins, S., Speier, P., Feiweier, T., Silva, R., Jackowski, M.P., Pennell, D.J., Sosnovik, D.E., Firmin, D.: In vivo diffusion tensor MRI of the human heart: reproducibility of breath-hold and navigator-based approaches. Magn. Reson. Med. **70**(2), 454–465 (2013)

9. Scollan, D.F., Holmes, A., Winslow, R., Forder, J.: Histological validation of myocardial microstructure obtained from diffusion tensor magnetic resonance imaging. Am. J. Physiol. Hear. Circ. Physiol. **275**(6), H2308–H2318 (1998)

10. Stejskal, E., Tanner, J.: Spin diffusion measurements: spin echoes in the presence of a time-dependent field gradient. J. Chem. Phys. **42**(1), 288–292 (1965)

11. Toussaint, N., Stoeck, C.T., Schaeffter, T., Kozerke, S., Sermesant, M., Batchelor, P.G.: In vivo human cardiac fibre architecture estimation using shape-based diffusion tensor processing. Med. Image Anal. **17**(8), 1243–1255 (2013)

Cardiac Motion Estimation Using Ultrafast Ultrasound Imaging Tested in a Finite Element Model of Cardiac Mechanics

Maartje M. Nillesen[1]([✉]), Anne E.C.M. Saris[1], Hendrik H.G. Hansen[1], Stein Fekkes[1], Frebus J. van Slochteren[2], Peter H.M. Bovendeerd[3], and Chris L. De Korte[1]

[1] Medical UltraSound Imaging Center, Department of Radiology and Nuclear Medicine, Radboud University Medical Center, Nijmegen, The Netherlands
maartje.nillesen@radboudumc.nl
[2] University Medical Center Utrecht, Utrecht, The Netherlands
[3] Department of Biomedical Engineering, Eindhoven University of Technology, Eindhoven, The Netherlands

Abstract. Recent developments in ultrafast ultrasound imaging allow accurate assessment of 3D cardiac deformation in cardiac phases with high deformation rates. This paper investigates the performance of a multiple spherical wave (SW) ultrasound transmission scheme in combination with a motion estimation algorithm for cardiac deformation assessment at high frame rates. Ultrasound element data of a realistically deforming 3D cardiac finite element model were simulated for a phased array transducer, transmitting five SWs (PRF 2500 Hz). After delay-and-sum beamforming, coherent compounding of multiple SW transmissions was performed to generate radiofrequency data (frame rate 500 Hz). Axial and lateral displacements were determined using a normalized cross-correlation-based technique. Good agreement was obtained between estimated and ground truth displacements derived from the model over the cardiac cycle. This study indicates that high frame rate displacement estimation using multiple SWs is feasible and serves as an important step towards high frame rate 3D cardiac deformation imaging.

Keywords: Ultrafast ultrasound imaging · Cardiac deformation imaging · Cardiac modeling

1 Introduction

Echocardiography is a widely used imaging modality for monitoring the heart function. Although echocardiography provides rich information on the geometry of the heart and can be used to assess function of the valves using Doppler based techniques, it does not provide quantitative information on the local contractile function of the myocardium. During the past decades, strain imaging techniques have been developed to estimate the local deformation of the heart muscle [1–3]. Deformation of cardiac tissue can be estimated either using speckle tracking techniques [2, 4] or radio-frequency-based correlation techniques [1, 3]. The first technique only uses the amplitude of the signal and has proven to be robust when inter-frame strains are high, however less accurate

© Springer International Publishing Switzerland 2015
H. van Assen et al. (Eds.): FIMH 2015, LNCS 9126, pp. 207–214, 2015.
DOI: 10.1007/978-3-319-20309-6_24

displacements are obtained for small strains. The latter technique uses the phase information that is present in the radio-frequency (RF) signal and outperforms speckle tracking techniques if frame rate is high enough with respect to deformation rate [5]. Due to recent developments in ultrafast imaging, cardiac imaging at high frame rates is feasible, making RF-based strain imaging the designated technique to accurately assess cardiac deformation.

For accurate assessment of 3D deformation of the heart over the entire cardiac cycle, high frame rate 3D ultrasound imaging is required. Conventional focused ultrasound imaging techniques use line-by-line image acquisition and therefore have limited temporal resolution. Frame rates vary from 40 to 50 3D volumes per second, depending on imaging depth and field of view. A recent development that is expected to allow volumetric cardiac imaging at frame rates of up to 5000 frames per second is ultrafast imaging. In ultrafast imaging the entire image view is insonified using unfocused ultrasound transmissions. Ultrafast imaging can be of great advantage for accurate assessment of 3D cardiac deformation, in particular in cardiac phases with high deformation rates. Although frame rate increases markedly using ultrafast imaging, image quality is lowered due to the use of unfocused transmit waves. The lower contrast and signal-to-noise ratio complicate deformation estimation in these images. A number of studies [6–9] have investigated the potential of coherent compounding of multiple unfocused transmit waves to improve the image quality while at the same time preserving high frame rates.

In this study, the benefit of coherent compounding of spherical waves for tissue motion estimation was investigated. Imposed by the large imaging depth and the large field of view that is needed in (transthoracic) cardiac imaging, spherical (diverging) waves are preferred over plane waves which are often used in non-cardiac applications. This paper reports the results of cardiac motion estimation using ultrafast spherical wave imaging in a simulation study using a realistic cardiac model. For more robust motion estimation, a stepwise varying beam forming grid was constructed and different time steps were used for axial and lateral displacement estimation respectively. The feasibility of the method was evaluated using ground truth deformation fields as provided by the model.

2   Methods

For simulation of cardiac ultrasound data, realistic 3D deformation fields of the cardiac muscle were generated using a 3D finite element model describing the mechanics of a healthy left ventricle [9, 10]. The model correctly predicts circumferential, longitudinal and radial strain. To obtain more realistic circumferential-radial shear strain values, active stress development perpendicular to the myofibers was included in this model by setting active cross-fiber stress to 25 % of the active fiber stress (ACT simulation model as described in [10]).

The deformation fields of the cardiac model (dt = 2 ms) served as an input for simulating ultrasound data for a 2D apical view over the entire cardiac cycle. Scatterers were initially placed randomly inside the cardiac muscle and were then displaced according to the 3D deformation fields obtained from the model. To ensure fully developed speckle patterns, a density of 250 scatterers per mm^3 was used. Temporal interpolation of the scatterer positions was performed between each deformation state of the model (dt = 2 ms) to match the 5 times higher pulse repetition time (dt = 0.4 ms) required for the ultrafast ultrasound acquisitions.

A phased array transducer (f_c = 2.5 MHz, 64 elements, pitch = 320 μm) was defined to generate ultrasound radiofrequency (RF) element data using the ultrasound simulation package FIELD II [11, 12]. The element data are unfocused radio-frequency (RF) data per element and were sampled at 18 MHz. The speed of sound was set to 1540 m/s. The virtual transmit point of the spherical wave (Fig. 1, upper right panel) was positioned behind the transducer [13]. This increases the amount of emitted energy as a group of elements now participates in the spherical wave transmission. To optimize image quality, element data obtained from multiple transmits, originating from different positions behind the transducer, can be combined.

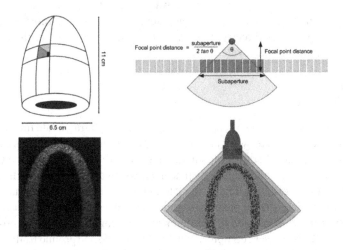

Fig. 1. Schematic overview of the ultrasound simulation set-up: geometry of the cardiac model (*upper left*), spherical wave transmit using a virtual source (*upper right*), example of multiple spherical wave imaging of the left ventricle (apical view) filled with scatterers (*lower right)* and resulting ultrasound image after coherent compounding *(lower left)*.

The amount of spherical waves to be combined (s = 5), the opening angle θ (90°) and the size of the transmit subaperture (number of elements (n = 21) that are active for each spherical wave transmission) were chosen in accordance with the findings in [13, 14]. These settings ensure a good trade-off between lateral resolution, transmit pressure and frame rate. The individual spherical waves were transmitted at a pulse repetition frequency (PRF) of 2500 Hz. Element data of each spherical transmit were reconstructed into RF data using delay-and-sum beamforming. The grid for constructing the RF data was adjusted stepwise over imaging depth because of the depth dependent lateral resolution. Coherent summation of RF data from 5 spherical waves was then performed to obtain compounded RF data at an effective frame rate of 500 Hz. A schematic overview of the simulation set-up is shown in Fig. 1.

A 2D coarse-to-fine cross-correlation based displacement estimation algorithm [15] was used to compute axial and lateral inter-frame displacements. In the first 'coarse' iteration, large RF data kernels of 4×1.3 mm^2 [axial x lateral] were used for cross-correlation calculation. The second iteration used a kernel size of 3×1.3 mm^2 to obtain 'finer' displacement estimates. The axial kernel size corresponds to 78 (iter. 1) and 58 (iter. 2) RF data points respectively. As the grid for constructing the RF data was adjusted stepwise over imaging depth because of the depth dependent lateral resolution, the number of lines corresponding to 1.3 mm varies (stepwise) over imaging depth.

As no phase information is available in the lateral direction, lateral displacement estimation is by definition less accurate than axial displacement estimation. Lateral displacements were estimated at a 10 times lower frame rate (i.e., 50 Hz) to obtain higher displacements between frames and thus more robust estimates. Cumulative displacements were derived from the inter-frame displacements using bilinear interpolation to track the cardiac tissue over time. Tracking was started at end-diastole. A comparison was made between the cumulative axial and lateral displacement estimates and ground truth displacement values directly derived from scatterer positions in the myocardial tissue over the entire cardiac cycle.

3   Results

To evaluate the deformation estimation technique using high frame rate imaging, cumulative displacement estimates were compared to ground truth displacements as obtained from the model. Figure 2 shows the cumulative axial displacement estimates (upper panel) and ground truth cumulative displacements (lower panel) for 6 characteristic phases of the cardiac cycle. In Fig. 3, the lateral displacement estimates (upper panel) and ground truth values (lower panel) are depicted for the same cardiac phases. To gain more insight in the performance of the deformation estimation algorithm, root-mean-squared-errors (RMSEs) were computed between estimated and ground truth cumulative displacement values for each frame in the cardiac cycle. Resulting RMSE curves (solid lines) are given in Fig. 4. This figure also shows the maximum cumulative displacement values over the cardiac cycle to allow better interpretation of the RMSE values. It can be derived from this figure (left panel) that RMSEs for the cumulative axial displacement estimates are small (<1.2 mm). RMSEs for cumulative lateral displacement are below 1.3 mm, however the relative lateral RMSE values (compared to maximal cumulative displacement value) are larger than the relative axial RMSEs.

A more detailed analysis of the accuracy of the cumulative displacement estimates is shown in Fig. 5, where the apical view of the left ventricle was divided into 7 regions (according to the segment model of the American Heart Association (AHA)). Mean absolute motion errors and standard deviation were computed. It should be noted that the cardiac model is rotational-symmetric and thus anterior and inferior curves are comparable.

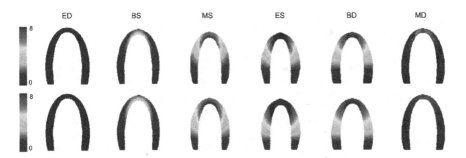

Fig. 2. Cumulative axial displacement estimations [mm] *(upper panel)* for selected frames over the cardiac cycle. *Lower panel:* cumulative axial displacements [mm] as derived directly from the finite element model (ground truth).

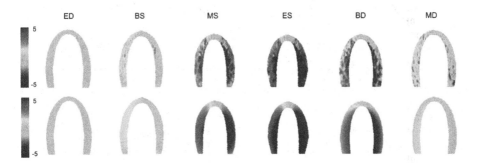

Fig. 3. Cumulative lateral displacement estimations [mm] *(upper panel)* for selected frames over the cardiac cycle. *Lower panel:* cumulative lateral displacements [mm] as derived directly from the finite element model (ground truth).

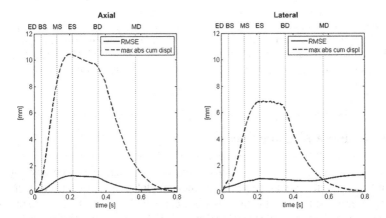

Fig. 4. Comparison between ground truth cumulative displacement values and cumulative displacement values computed by the high frame rate deformation estimation method. Root-mean-squared-errors [mm] are shown (solid) and related to the maximum absolute cumulative displacement value [mm] (dashed) for each frame in the cardiac cycle.

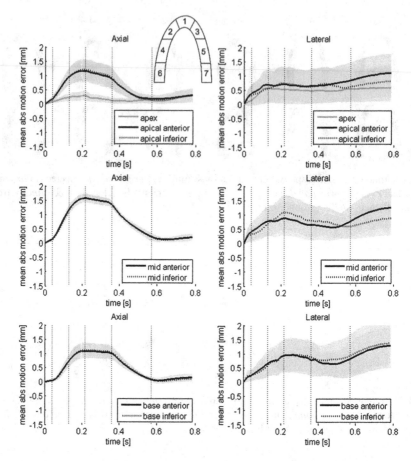

Fig. 5. Mean absolute motion errors (in [mm]) of the cumulative axial and lateral displacement estimates for 7 cardiac segments of the apical view (apex (1), apical-anterior (2), apical-inferior (3), mid-anterior (4), mid-inferior (5), base-anterior (6), base-inferior (7)). Mean values (line plots) and standard deviations (gray area) are shown over the cardiac cycle for each cardiac segment (Color figure online).

4   Discussion and Conclusion

An ultrafast imaging method for cardiac motion assessment has been developed and tested in a simulation study using a finite element model of the left ventricle. Cumulative axial and lateral displacement estimates were compared to the reference displacement values directly derived from the cardiac model. Good agreement was observed between the displacement estimates and reference values (i.e., low RMSEs). Motion of the cardiac muscle could thus be tracked over the entire cardiac cycle using the presented method.

Lateral displacement estimates were less accurate than axial displacement estimates. This can be explained by the fact that the resolution in this direction is much lower and no phase information is available. Another challenge the low amount of energy at larger imaging depths using spherical waves. This can be partly overcome by combining even more spherical waves transmits, although this will affect frame rate. Furthermore, the time delay between the transmits of the individual spherical waves might cause compounding errors due to motion of the cardiac muscle in between the subsequent transmits. As frame rates up to 6000 Hz are feasible for imaging depths normally used in cardiac applications, this compounding error could be reduced by using a higher PRF, while keeping the effective frame rate (i.e., delay between each group of 5 transmits) unaltered. At this high PRF, up to 12 spherical waves could be transmitted in order to increase the energy at larger imaging depths.

In this study, the performance of the technique was judged by analyzing cumulative displacement estimates as the estimates could be compared directly to the ground truth displacements derived from the deformation fields of the model. Cumulative cardiac strain will be computed and analyzed in near future to investigate the performance of the method from a more clinical point of view.

A limitation of the current study is that no noise was incorporated in the simulation set-up and surrounding structures were not taken into account. For a more thorough evaluation of cardiac motion estimation technique using high frame rate imaging, the simulated cardiac images could be made more realistic by including techniques as described in [16]. Further research is required to extend the method to 3D. It should be noted that using ultrafast imaging techniques the entire 3D field of view can also be insonified and obtained 3D volume rates can in principle thus be as high as obtained 2D frame rates. We therefore expect that 3D motion estimation will also be feasible based on the proposed technique.

In conclusion, 2D cardiac displacement estimation using multiple spherical waves is feasible and produces accurate motion estimates. This technique might serve as an important predecessor for high frame rate 3D cardiac deformation imaging.

Acknowledgements. This research is supported by the Dutch Technology Foundation STW (NKG 12122), which is part of the Netherlands Organization for Scientific Research (NWO), and which is partly funded by the Ministry of Economic Affairs.

References

1. Konofagou, E.E., D'hooge, J., Ophir, J.: Myocardial elastography-a feasibility study *in vivo*. Ultrasound Med. Biol. **28**, 475–482 (2002)
2. Leitman, M., Lysyansky, P., Sidenko, S., Shir, V., Peleg, E., Binenbaum, M., Kaluski, E., Krakover, R., Vered, Z.: Two-dimensional strain-a novel software for real-time quantitative echocardiographic assessment of myocardial function. J. Am. Soc. Echocardiogr. **17**, 1021–1029 (2004)
3. Lopata, R.G., Nillesen, M.M., Thijssen, J.M., Kapusta, L., de Korte, C.L.: Three-dimensional cardiac strain imaging in healthy children using RF-data. Ultrasound Med. Biol. **37**(9), 1399–1408 (2011)

4. Bohs, L.N., Trahey, G.E.: A novel method for angle independent ultrasonic imaging of blood flow and tissue motion. IEEE Trans. Biomed. Eng. **38**(3), 280–286 (1991)
5. Céspedes, E.I., de Korte, C.L., van der Steen, A.W.: Echo decorrelation from displacement gradients in elasticity and velocity estimation. IEEE Trans. Ultrason. Ferroelectr. Freq. Control **46**, 791–801 (1999)
6. Montaldo, G., Tanter, M., Bercoff, J., Benech, N., Fink, M.: Coherent plane-wave compounding for very high frame rate ultrasonography and transient elastography. IEEE Trans. Ultrason. Ferroelectr. Freq. Control **56**(3), 489–506 (2009)
7. Hasegawa, H., Kanai, H.: High-frame-rate echocardiography using diverging transmit beams and parallel receive beamforming. J. Med. Ultrasound **38**(33), 129–140 (2011)
8. Tong, L., Gao, H., Choi, H.F., D'hooge, J.: Comparison of conventional parallel beamforming with plane wave and diverging wave imaging for cardiac applications: a simulation study. IEEE Trans. Ultrason. Ferroelectr. Freq. Control **59**(8), 1654–1663 (2012)
9. Kerckhofs, R.C.P., Bovendeerd, P.H.M., Kotte, J.C.S., Prinzen, F.W., Smits, K., Arts, T.: Homogeneity of cardiac contraction despite physiological asynchrony of depolarization: a model study. Ann. Biomed. Eng. **31**, 536–547 (2003)
10. Bovendeerd, P.H.M., Kroon, W., Delhaas, T.: Determinants of left ventricular shear strain. Am. J. Physiol. Heart Circ. Physiol. **297**, 1058–1068 (2009)
11. Jensen, J.A., Svendsen, N.B.: Calculation of pressure fields from arbitrarily shaped, apodized, and excited ultrasound transducers. IEEE Trans. Ultrason. Ferroelectr. Freq. Control **39**, 262–267 (1992)
12. Jensen, J.A.: FIELD: a program for simulating ultrasound systems. Med. Biol. Eng. Comput. **34**(1), 351–353 (1996)
13. Lockwood, G.R., Talman, J.R., Brunke, S.S.: Real-time 3-D ultrasound imaging using sparse synthetic aperture beamforming. IEEE Trans. Ultrason. Ferroelectr. Freq. Control **45**(4), 980–988 (1998)
14. Papadacci, C., Pernot, M., Couade, M., Fink, M., Tanter, M.: High-contrast ultrasound imaging of the heart. IEEE Trans. Ultrason. Ferroelectr. Freq. Control **61**(2), 288–301 (2014)
15. Lopata, R.G.P., Nillesen, M.M., Hansen, H.H.G., Gerrits, I.H., Thijssen, J.M., de Korte, C.L.: Performance of two dimensional displacement and strain estimation techniques using a phased array transducer. Ultrasound Med. Biol. **35**(12), 2031–2041 (2009)
16. De Craene, M., Alessandrini, M., Allain, P., Marchesseau, S., Waechter-Stehle, I., Weese, J., Saloux, E., Morales, H.G., Cuingnet, R., Delingette, H., Sermesant, M., Bernard, O., D'hooge, J.: Generation of ultra-realistic synthetic echocardiographic sequences. In: Proceedings of IEEE International Symposium on Biomedical Imaging, pp. 73–76 (2014)

Quantification of Gaps in Ablation Lesions Around the Pulmonary Veins in Delayed Enhancement MRI

Marta Nuñez Garcia[1](✉), Catalina Tobon-Gomez[2], Kawal Rhode[2],
Bart Bijnens[1,3], Oscar Camara[1], and Constantine Butakoff[1]

[1] PhySense, DTIC, Universitat Pompeu Fabra, Barcelona, Spain
{marta.nunez,oscar.camara,Constantine.butakoff}@upf.edu
[2] Division of Imaging Sciences and Biomedical Engineering,
Kings College London, London, UK
{catalina.tobon-gomez,kawal.rhode}@upf.edu
[3] ICREA, Barcelona, Spain
bart.bijnens@upf.edu

Abstract. We propose a method for measuring the quality of pulmonary vein isolation in delayed enhancement MRI images for the patients that underwent atrial radiofrequency ablation. To that end we construct a graph from an anatomy independent representation of the atrium, where every node represents a scar lesion and the edges are the distances between the lesions. Subsequently we search for the shortest path in this graph. The total amount of gap between the scar lesions is measured as the fraction of the path's length that passes through the healthy tissue. We illustrate the proposed technique using pre-segmented atria from a freely available database and show that the proposed approach is able to measure the amount of gaps in the scar isolating the PV's as well as provide a meaningful definition of the gap in cases where the scar lesions are patchy and not continuous.

Keywords: Radiofrequency ablation · Left atrium · Gaps · Repeated ablation · Graphs · Shortest path

1 Introduction

Radiofrequency (RF) ablation is a common technique in clinical routine to treat atrial fibrillation (AF) where the goal is to electrically isolate one region of the myocardium from another, in particular the pulmonary veins (PV) in the case of the left atrium (LA) [1]. However, the success rate for some ablation procedures ranges from only 40 to 80 % [2]. One of the reasons for the low success of the therapy is incomplete isolation of PV due to punctual ablation, leading to the presence of gaps in ablation lesion. In the current ablation procedures, there is no easy way to immediately visualize changes in atrial tissue properties and the ablation success is judged based on local electrogram attenuation, power

© Springer International Publishing Switzerland 2015
H. van Assen et al. (Eds.): FIMH 2015, LNCS 9126, pp. 215–222, 2015.
DOI: 10.1007/978-3-319-20309-6_25

delivered and temperature at the catheter tip. Moreover, some gaps may not be significant enough to be detected right after the therapy but only after the atrium remodels (that can take several months).

One way to analyze ablation lesions and plan the repeated ablation procedure, that gained popularity recently, is the delayed enhancement magnetic resonance imaging (de-MRI). The ability of de-MRI to visualize gaps in ablation lesions is actively being investigated [3–6] and even though there are some controversies in the relation between de-MRI and electroanatomical data, complete PV isolation seems to be correlated to the success of the ablation outcome, suggesting the importance of introducing a quantitative measure of isolation for the analysis of the ablation success.

Most of the current techniques for gap characterization [3,4] are based on a visual inspection of 3D renderings of the atrial anatomy and subsequent manual identification of gaps looking for discontinuities in de-MRI bright intensities corresponding to previous ablation lesions. This process is quite time-consuming and observer-dependent, suffering as a consequence from poor reproducibility. In addition, it is not straightforward to relate the ablation information from de-MRI at follow-up with data from the first ablation or with complementary data such as electro-anatomical maps since they are measured using different sensors and/or different images and therefore reside in different reference spaces. Very recently [7] we developed a method to automatically generate a 2D map of the left atrium, called Standardized Unfold Map (SUM), which provides a simplified and easy to process representation of the information, spatially distributed on the atrial myocardium.

In this paper we propose a method for gap quantification in de-MRI images, based on the SUM representation, that provides an anatomy independent measure of the amount of healthy tissue inbetween the scar lesions. In our approach we transform the regions around every PV in the SUM representation into a rectangular domain using polar coordinates (see Fig. 1). We hypothesize that if the vein is completely isolated by the ablation, there must be a continuous scar running from one side of the rectangular domain to another side. On the other hand if the scar is not continuous, the proposed method automatically detects and quantifies the existing gaps based on the minimum path connecting the nodes in a graph constructed from the scar lesions around each vein. The proposed methodology is applied to 45 atrial fibrillation patients from the Comprehensive Arrhythmia Research and Management (CARMA) Left Atrial MRI database [8].

2   Methods

2.1   Database

The Comprehensive Arrhythmia Research and Management Left Atrial MRI database is a publicly available database composed of MRI data from patients with AF who have undergone RF catheter ablation. For this study we have made use of late-gadolinium enhanced MRI of 45 patients available exactly 3 months

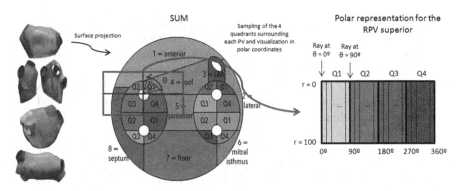

Fig. 1. From left to right: LA region definition; surface projection to SUM; Radial sampling of the four quadrants around each PV; Representation in polar coordinates (only one vein in this case).

after the ablation procedure. These de-MR scans were obtained on either a 1.5 Tesla Avanto or a 3.0 Tesla Verio clinical scanner (Siemens Medical Solutions, Erlangen, Germany). Typical acquisition parameters were: free-breathing using navigator gating, a transverse imaging volume with voxel size of 1.25 x 1.25 x 2.5 mm (reconstructed to 0.625 x 0.625 x 1.25 mm). For more details on the MR acquisition parameters the reader is referred to the CARMA Left Atrial MRI database website [8]. Manual segmentations of the endocardial surface of the left atrium and the left atrial wall are also available for each patient.

2.2 Scar Segmentation

The classification of voxel intensities of the LA wall into scar and healthy tissue was done using the unsupervised clustering method based on normalized intensity [9]. This algorithm is implemented in the module "Automatic Left Atrial Scar" of the Cardiac MRI Toolkit [10] for 3D Slicer [11,12]. The output of the algorithm is a binary mask including LA wall points where scar tissue has been detected. The segmented scar tissue was then projected onto the LA endocardial surface mesh (built using Marching Cubes [13]) by sampling the mask labels along the normals to the wall.

2.3 Standardized Unfold Map (SUM) of the Left Atrium

The scar information obtained in the previous step (Sect. 2.2) is represented as a 2-manifold embedded in $\mathbb{R}^3$ (scar on LA endocardial surface mesh), where it is not straightforward to analyze and compare data. In order to facilitate the analysis of scar information for gap quantification we represent the atrium in a flat domain using the Standardized Unfold Map (SUM) of the LA [7]. The SUM representation is a disk divided into 24 left atrial regions and with a number of holes corresponding to the pulmonary veins, as it is illustrated in Fig. 1.

The SUM is calculated by a quasi-conformal flattening of the LA surface mesh under the constraints that the mitral annulus is mapped to the exterior boundary of a disk, while the PV and the LA appendage are mapped to predefined holes within that disk. The position and size of these holes are defined as to facilitate the analysis in accordance with clinical requirements. The consistency of mappings from different patients is handled through the currents-based surface registration technique recently proposed by Durrleman et al. [14], which is implemented in the freely available Open-Source software Deformetrica [15]. The whole pipeline to generate a SUM for the LA can be summarized by the following steps, given an already defined template atrium (e.g. an average atrium) and its flattening (calculated here using Graphite [16]): (a) The PV of the LA surface are automatically clipped at 10 mm from the ostia to ensure the same length of the PV trunks in all LA anatomies. The method requires manual selection of four seed points corresponding to each PV; (b) currents-based surface registration of every atrium to the template atrium; (c) Closest point mapping of scar information (scar binary mask) from the analyzed atrium to the template atrium and to its flattened representation.

2.4   Gap Quantification

In the SUM, the anatomical regions around each PV are represented in four quadrants (see Fig. 1). An ideal ablation lesion isolating the PV from the rest of the atrium is expected to be a connected scar surrounding the PV. However the scar is not always connected and some gaps of healthy tissue may appear. We propose to transform the surroundings of every PV into a rectangular domain using a polar coordinate transformation centered at each PV hole in the SUM, as shown in Fig. 1. An eventual continuous ablation lesion running from left to right in this rectangular map would correspond to a successful ablation completely isolating PV from the rest of the atria.

To construct the rectangular map every quadrant around each PV is radially sampled using a set of rays connecting the centre of the PV with the boundaries of the quadrant at radial angle intervals of $\Delta\Theta = 1°$ producing 90 rays per quadrant. We represent the radial angles along the horizontal axis, and the distance from the PV origin along the vertical axis. Since PV regions are not circular, ray segments will have different lengths. In order to have a convenient representation (rectangle) we use the same number of samples (100) along every ray irrespectively of its length. Once the scar information around each PV is mapped onto the rectangular domain, these maps are post-processed using morphological opening and closing to eliminate the possible small noisy structures. Several examples of these rectangular maps are shown in Fig. 2.

The identification and quantification of gaps is carried out in the rectangular map by estimating the shortest path that isolates the PV going through as little healthy tissue as possible starting from the left ($\Theta = 0°$) and finishing at the right side ($\Theta = 360°$). To this end a graph is constructed where each node represents every connected patch of scar in the rectangular map. The pairwise distances between all patches, computed using the distance transform, are associated to

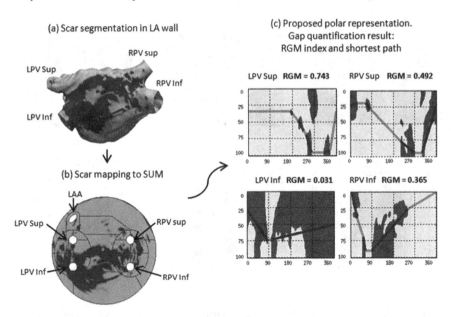

Fig. 2. Pipeline of the method. (a) Scar segmentation in the LA wall mapped to the LA surface. Dark color denotes tissue classified as scar. (b) Scar mapped to the SUM. (c) Visualization of the scar in the polar representation. The neighborhood of each PV is shown separately. The lines in (c) show the shortest isolating path and the RGM for each PV are shown on top of every subplot.

graph edges. Two auxiliary nodes on the horizontal extremes of the rectangular map (left and right sides) are added. Finally the Dijkstra algorithm [17] is used to estimate the shortest isolating path connecting these auxiliary nodes.

As a measure of PV isolation we propose to use the total length of healthy tissue crossed along the shortest isolating path, normalized by the total length of the path. Therefore, we define the Relative Gap Measure (RGM) as: $RGM = \frac{Gap\ length}{Total\ length}$. Values of RGM equal to 0 mean that there are no gaps along the shortest isolating path and the PV is then electrically isolated. Any non-zero value of RGM would represent the presence of a gap around the studied PV, larger values of RGM corresponding to more gaps between scar patches. Figure 2 shows the rectangular maps corresponding to four pulmonary veins of a left atria together with the estimated shortest isolating paths and their associated RGM values. One can see two distinct situations in these atria: while the left inferior PV (LPV Inf) is almost completely surrounded by scar tissue, which is represented with a RGM value of 0.031, the left superior PV (LPV Sup) is surrounded by healthy tissue, which is characterized by a RGM value equal to 0.743. Therefore we can say that the LPV Inf is quite well electrically isolated while the LPV Sup has a big amount of gap.

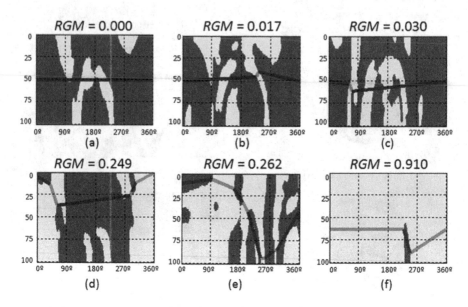

Fig. 3. Results of the proposed gap quantification method. Dark color represents scar tissue. On top of every subplot the correspoding RGM is shown.

3 Results

The proposed gap quantification method was used to measure the isolation of PV's in 45 de-MRI images of patients that underwent ablation therapy. Several relevant examples can be seen in Fig. 3, where image (a) corresponds to a completely isolated PV ($RGM = 0$); (b) and (c) correspond to PV with a single small gap; (d) and (e) are more general cases with several gaps of different shape and relative position; and (f) corresponds to a PV almost completely surrounded by healthy tissue, where the RGM is close to its possible maximum, 1. We can see that the more overall scar there is, the smaller is the RGM.

Table 1. Gap comparison between patients with and without AF recurrence.

	LPV superior		LPV inferior		RPV superior		RPV inferior		TOTAL	
	Mean	STD	Mean	STD	Mean	STD	Mean	STD	Mean	STD
Recurrence	0.336	0.231	0.145	0.182	0.424	0.185	0.381	0.206	0.322	0.023
No recurrence	0.410	0.231	0.145	0.132	0.384	0.198	0.301	0.190	0.310	0.041

Subsequently we tried using the results of the gap quantification to see if the proposed measure could be used to differentiate between the patients who had reported recurrence (18 out of 45) of AF after the LA ablation and who had not. Table 1 shows the mean and standard deviation of the RGM index for these two categories. The *RGM* is calculated for each PV separately and the last column, labeled TOTAL, shows the average of all the *RGM*s. Unfortunately no significant differences among the groups was observed. For the explanation please refer to the Discussion.

4   Discussion and Conclusions

In this paper we presented a novel method for gap quantification. With the proposed method (SUM + polar representation) we achieve a subject and anatomy independent visualization and characterization of the tissue around the PVs. This representation allows automatically finding the presence of gaps and provide a measure of their extent in a convenient polar representation. Furthermore in the cases where the scar is not continuous and is patchy, where it is hard to define what is a gap, our method provides a way of defining the gaps using the minimum path of the graph description and makes the method independent of the spatial distribution of the scar patches. In particular our algorithm does not require the scar patches to be separated by the vertical gaps in order to identify the gaps.

There are two major limitations to this work however. First of all the segmentation of the scar has been performed without a proper validation using an automatic algorithm and therefore the amount of gaps presented in Table 1 may be incorrect. Just by visual analysis of the images, in many of them we did not see a clear scar, the result of the ablation, and in many cases the scar did not make much sense from the clinical point of view.

The second limitation is that we do not have information about the source of the recurrence of the AF for this dataset. We clearly cannot average the patients for which the AF originated in different PV's, and where different PV's were isolated. The average would be completely misleading and therefore a better way of correlating the gaps with the recurrence is required.

Nevertheless we would like to emphasize that the intention of this work was to propose a gap quantification algorithm, which we illustrated using examples of different scar distribution and we are currently working on a proper validation of the technique using a clinically relevant database.

Acknowledgments. This study was partially supported by the EU FP7 for research, technological development and demonstration under grant agreement VP2HF (no 611823) and by the Spanish Ministry of Science and Innovation (TIN2011-28067).

References

1. Calkins, H., Kuck, K.H., Cappato, R., Brugada, J., John Camm, A., et al.: HRS/EHRA/ECAS expert consensus statement on catheter and surgical ablation of atrial fibrillation: recommendations for patient selection, procedural techniques, patient management and follow-up, definitions, endpoints, and research trial design. J. Interv. Card. Electrophysiol. **33**(2012), 171–257 (2012)

2. Weerasooriya, R., Khairy, P., Litalien, J., Macle, L., Hocini, M., Sacher, F., Lellouche, N., Knecht, S., Wright, M., Nault, I., et al.: Catheter ablation for atrial fibrillation: are results maintained at 5 years of follow-up? J. Am. Coll. Cardiol. **57**(2), 160–166 (2011)
3. Peters, D.C., Wylie, J.V., Hauser, T.H., Nezafat, R., Han, Y., Woo, J.J., Taclas, J., Kissinger, K.V., Goddu, B., Josephson, M.E., et al.: Recurrence of atrial fibrillation correlates with the extent of post-procedural late gadolinium enhancement: a pilot study. JACC: Cardiovasc. Imaging **2**(3), 308–316 (2009)
4. Bisbal, F., Guiu, E., Cabanas-Grandío, P., Berruezo, A., Prat-Gonzalez, S., Vidal, B., Garrido, C., Andreu, D., Fernandez-Armenta, J., Tolosana, J.M., et al.: CMR-guided approach to localize and ablate gaps in repeat af ablation procedure. JACC: Cardiovasc. Imaging **7**(7), 653–663 (2014)
5. Harrison, J.L., Sohns, C., Linton, N.W., Karim, R., Williams, S.E., Rhode, K.S., Gill, J., Cooklin, M., Rinaldi, C.A., Wright, M., et al.: Repeat left atrial catheter ablation: cardiac magnetic resonance prediction of endocardial voltage and gaps in ablation lesion sets. Circ.: Arrhythm. Electrophysiol. **8**(2), 270–278 (2015). CIRCEP-114
6. Nazarian, S., Beinart, R.: CMR-guided targeting of gaps after initial pulmonary vein isolation. JACC: Cardiovasc. Imaging **7**(7), 664–666 (2014)
7. Tobon-Gomez, C., Zuluaga, M.A., Chubb, H., Williams, S.E., Butakoff, C., Karim, R., Camara, O., Ourselin, S., Rhode, K.: Standardised unfold map of the left atrium: regional definition for multimodal image analysis. J. Cardiovasc. Magn. Reson. **17**(1), 1–3 (2015)
8. Comprehensive arrhythmia research management center. University of Utah Health Sciences, January 2015. http://hdl.handle.net/1926/1755
9. Perry, D., Morris, A., Burgon, N., McGann, C., MacLeod, R., Cates, J.: Automatic classification of scar tissue in late gadolinium enhancement cardiac MRI for the assessment of left-atrial wall injury after radiofrequency ablation. In: SPIE Medical Imaging, pp. 83151D–83151D. International Society for Optics and Photonics (2012)
10. Slicer cardiac MRI extension for atrial fibrillation research and management comprehensive arrhythmia research and management center, January 2015.http://capulet.med.utah.edu/namic/cmrslicer/
11. 3DSlicer, January 2015. http://www.slicer.org
12. Fedorov, A., Beichel, R., Kalpathy-Cramer, J., Finet, J., Fillion-Robin, J.C., Pujol, S., Bauer, C., Jennings, D., Fennessy, F., Sonka, M., Buatti, J., Aylward, S., Miller, J.V., Pieper, S., Kikinis, R.: 3D slicer as an image computing platform for the quantitative imaging network. Magn. Reson. Imaging **30**, 1323–1341 (2012)
13. Lorensen, W.E., Cline, H.E.: Marching cubes: a high resolution 3D surface construction algorithm (1987)
14. Durrleman, S., Prastawa, M., Charon, N., Korenberg, J.R., Joshi, S., Gerig, G., Trouvé, A.: Morphometry of anatomical shape complexes with dense deformations and sparse parameters. NeuroImage **101**, 35–49 (2014)
15. Deformetrica, January 2015. http://www.deformetrica.org
16. Graphite, January 2015. http://alice.loria.fr/
17. Dijkstra, E.W.: A note on two problems in connexion with graphs. Numerische Mathematik **1**, 269–271 (1959)

Probabilistic Edge Map (PEM) for 3D Ultrasound Image Registration and Multi-atlas Left Ventricle Segmentation

Ozan Oktay[1]([✉]), Alberto Gomez[2], Kevin Keraudren[1], Andreas Schuh[1], Wenjia Bai[1], Wenzhe Shi[1], Graeme Penney[2], and Daniel Rueckert[1]

[1] Biomedical Image Analysis Group, Imperial College London, London, UK
o.oktay13@imperial.ac.uk
[2] Division of Imaging Sciences and Biomedical Engineering, King's College London, London, UK

Abstract. Automated left ventricle (LV) segmentation in 3D ultrasound (3D-US) remains a challenging research problem due to variable image quality and limited field-of-view. Modern segmentation approaches (shape, appearance and contour model based surface fitting) require an accurate initialization and good image boundary features to obtain reliable and consistent results. They are therefore not well suited for this problem. The proposed method overcomes those limitations with a novel and generic 3D-US image boundary representation technique: Probabilistic Edge Map (PEM). This new representation captures regularized and complete edge responses from standard 3D-US images. PEM is utilized in a multi-atlas LV segmentation framework to spatially align target and atlas images. Experiments on data from the MICCAI CETUS challenge show that the proposed approach is better suited for LV segmentation than the active contour, appearance and voxel classification approaches, achieving lower surface distance errors and better LV volume estimates.

Keywords: Structured decision forest · Probabilistic edge map · Multi-atlas label fusion · Left ventricle segmentation · Ultrasound image analysis

1 Introduction

Cardiac ultrasound remains the primary imaging modality in the assessment of left ventricular systolic function, mass and volume to assess the morphology and function of the heart. Automated tools to analyse three-dimensional ultrasound (3D-US) images are important to ensure reproducibility as well as consistency of segmentations and to reduce the workload of clinicians. The development of such tools is still an ongoing research problem due to limitations posed by low image quality, restricted field-of-view and anatomical variations. For these reasons, accurate and generic image analysis techniques are crucial.

Related Work: Automated left ventricle (LV) segmentation techniques can be broadly categorized into two groups: (1) image-driven and (2) model-driven

© Springer International Publishing Switzerland 2015
H. van Assen et al. (Eds.): FIMH 2015, LNCS 9126, pp. 223–230, 2015.
DOI: 10.1007/978-3-319-20309-6_26

approaches. Level-set approaches such as phase asymmetry [13] are part of the first category. They calculate 3D LV surfaces with weak or no shape constraints and do not require the fitting of a model to a large number of images. Also the B-spline active surface approach proposed in [4] does not require model fitting. Instead, the surface is initialized with an ellipsoid and B-splines are used to regularize the deformation of the surface model. Approaches in the second group use additional a-priori information by analyzing intensity patterns in training samples and manually traced contours. This includes approaches such as appearance models (AAM) [15] and semantic labelling of voxels using a classifier such as a decision forest [9]. Another method proposed in [10] uses labeled atlases and image registration to segment the LV volume. It does not require the training of a shape model, but makes an implicit use of such model through the atlases.

Research Motivation and Method Proposal: Active contour and level-set approaches require an accurate estimate of LV shape and position for initialization. This is because final segmentation results are sensitive to initializations obtained either manually [7,10] or through ad-hoc solutions such as Hough transform of edges [4] or through selection of image center points [15]. Such approaches depend on the acquisition field-of-view and cannot be generalized to acquisitions from different acoustic windows such as apical and parasternal views together.

Similarly, these approaches [4,13,15] make use of intensity and phase based features to delineate ventricle borders. Since phase features rely on the agreement of phases between different Fourier components (and are therefore insensitive to contrast), less importance is given to local energy information. This causes these features to be sensitive to noise. Likewise, intensity based approaches are sensitive to low image quality, shadowing, speckle and clutter.

This paper proposes a fully automatic multi-atlas LV segmentation framework for US images. Additionally, a novel robust 3D boundary representation method, Probabilistic Edge Map (PEM), is presented and utilized within this framework to address the challenges outlined above. PEMs delineate object boundaries in the input images by using a trained structured decision forest (SDF) classifier [6]. With this method, we are extending the structural representation proposed in [7], applied on 2D cardiac short-axis slices, to a 3D structural analysis together with the use of US related image features. In this way, discontinous and spurious edge responses in through plane direction can be eliminated, while achieving smooth and regularized tissue boundaries, as shown in Fig. 1.

In the proposed multi-atlas LV segmentation framework (PEM-MA), the PEMs are used in robust affine registration [11] and non-rigid registration [14] to spatially align multiple atlas images to the target. PEM based US image registration provides more reliable initialization between target and atlas images, and achieves better atlas selection [1] and LV segmentation performance. The proposed segmentation framework is evaluated on a benchmark dataset used in the MICCAI 2014 CETUS segmentation challenge. The results collected from the online evaluation platform show that PEM-MA achieves state-of-the-art LV segmentation accuracy in both surface distance and volumetric measure metrics, while outperforming all other challenge participants [3,7,15] in terms of the used evaluation criteria.

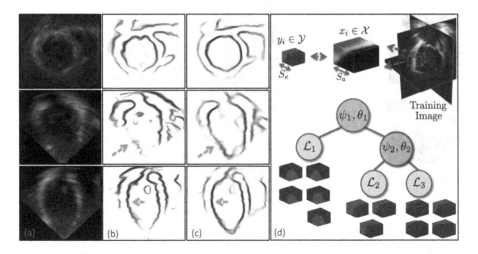

Fig. 1. (a) 3D cardiac US image, (b) phase congruency [13], and (c) PEM which captures missing structures (orange arrows) and provides smoother edge response (green arrows). In (d) SDF training is illustrated, where the label patches (y_i) are clustered at each node split, and the weak learners (ψ_i) search for the optimal threshold value (θ_i) and feature (x_i) to separate the two clusters (Color figure online).

2 Methodology

2.1 Probabilistic Edge Map (PEM) Representation

In cardiac imaging, 3D-US images outline an anatomical representation of the heart chambers. Further image analysis typically requires an accurate and smooth object boundary delineation. Data driven approaches may fail due to severe intensity artefacts and missing boundaries. A machine learning approach such as a structured decision forest (SDF) [6] can cope with these difficulties as the training data guides the boundary extraction. This is shown in Fig. 1, where the proposed PEM captures the missing boundaries and delineates them accurately.

The US images are initially resampled to isotropic voxel size. Furthermore, speckle noise is reduced using a sparse coding approach: The K-SVD algorithm [8] is used to learn an over-complete dictionary from US image patches. After the learning stage, the image is reconstructed from a sparse combination of the learned dictionary atoms to remove speckle patterns. Finally, a SDF classifier for the PEM is trained from the preprocessed images. While SDFs are similar to decision forests, they possess several unique properties and advantages.

In the tree structure of SDF, the output space ($\mathcal{Y}$) is assumed to be structured. In our case, this means that the output labels ($y_i \in \mathcal{Y}$) of size (S_e)^3 represent the edge labelling for image patches. In general, any type of multi-dimensional output can be stored at each tree leaf node, as long as labels can be clustered into two or more subsets by determining the optimal splitting function (ψ) at each tree branch, as shown in Fig. 1(d). In the PEM classifier training,

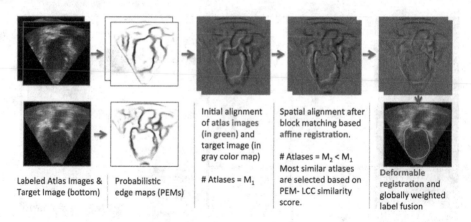

| Labeled Atlas Images & Target Image (bottom) | Probabilistic edge maps (PEMs) | Initial alignment of atlas images (in green) and target image (in gray color map)

Atlases = M_1 | Spatial alignment after block matching based affine registration.

Atlases = $M_2 < M_1$
Most similar atlases are selected based on PEM- LCC similarity score. | Deformable registration and globally weighted label fusion |

Fig. 2. A block diagram of the proposed multi-atlas segmentation framework.

this is achieved by mapping each image patch label to an intermediate space ($\Theta : \mathcal{Y} \to \mathcal{Z}$) where label clusters can be generated based on the Euclidean distance in $\mathcal{Z}$ (cf. [6]). Similar to decision forests, SDFs operate on standard input feature space which is defined by the high dimensional appearance features ($x_i \in \mathcal{X}$) extracted from image patches of fixed size $(S_a)^3$. These features are computed in a multi-scale fashion and correspond to image intensities, gradient magnitudes, soft-binning based histogram of oriented gradients, and local phase features. Weak classifiers $\psi(x_i, \theta)$, e.g., 1D and 2D decision stumps, are trained by maximizing the entropy based information gain criterion at each tree node with one of the selected image features. The parameter vector θ contains the stump threshold value and selected feature identifier. At testing time, each target image voxel is voted for $(S_e)^3 \times N_t$ times by N_t number of trees and these votes are aggregated by averaging all the predictions. Multiple and overlapping patch label predictions are the main advantage of PEMs, as these result in smooth, regularized and complete delineations of the cardiac chambers.

2.2   Multi-atlas Left Ventricle Segmentation

Next, we detail our proposed multi-atlas LV segmentation framework as outlined in Fig. 2, employing the generated edge maps. Initial affine alignment, atlas selection and deformable registration between target (I) and atlas images (J_i) are performed based on the PEMs (P^I, P_i^J) generated from the US images. A dataset consisting of a number of manually annotated US images is used in the atlas formation. The annotations for these atlases contain only the LV endocardial labels. The composite spatial transformations transfer the atlas labels to the target, followed by a globally weighted label fusion based on PEM similarity.

Global Alignment: The PEMs from both target image and atlases are first aligned using a block matching technique [11] which maximizes the normalized correlation coefficient between image blocks. The set of vectors defined by the displacement of each block is regularized before finding the global affine transformation A_i. A least trimmed squared regression based regularization (cf. [11]) removes the influence of displacements for the atlas blocks which have no target block correspondence due to missing features in the images. For this reason, this approach is robust to shadowing and anatomical variations and can provide an accurate spatial alignment for atlas selection and good initial segmentation.

Atlas Selection: It was shown in multi-atlas brain segmentation [1], that a selection of most similar atlases is beneficial. Therefore, after affine registration, all M_1 atlases are ranked according to their average local correlation coefficient [5] score, $LCC(P^I, P_i^J \circ A_i)$, and the $M_2 < M_1$ top scoring atlases in the upper quartile are selected. The LCC similarity metric is defined in (1), where Ω denotes the target voxels within a region defined by the dilated LV mask.

$$LCC(P^I, P^J) = \frac{1}{|\Omega|} \sum_{x \in \Omega} \frac{|\langle P^I, P^J \rangle_x|}{\sqrt{\langle P^I, P^I \rangle_x \langle P^J, P^J \rangle_x}} \tag{1}$$

A Gaussian window G_σ with variance σ^2 locally weights the PEMs and $\langle P^I, P^J \rangle_x = G_\sigma * (P^I.P^J)[x] - (G_\sigma * P^I)[x](G_\sigma * P^J)[x]$, where . denotes the Hadamard product, and $*$ the convolution. As the SDF classifier makes use of image intensities in node splits ψ, local intensity changes in the input images can influence the edge probabilities in PEMs. For this reason, LCC is a more suitable similarity measure for PEMs than global metrics such as sum of squared differences.

Local Alignment: To correct for residual misalignment, a registration based on free-form deformations (FFDs) [14] follows the atlas selection. The total energy $E(T_i) = -LCC(P^I, P_i^J \circ T_i \circ A_i) + \lambda BE(T_i)$ is minimised in a multi-resolution scheme, where BE is the bending energy of the cubic B-spline FFD T_i and λ defines the trade-off between local PEM alignment and deformation smoothness.

Label Fusion: Finally, the transferred atlas labels are fused using a globally weighted voting[1] [2] based on the dissimilarity $m_i = 1 - LCC(P^I, P_i^J \circ T_i \circ A_i)$. The LV segmentation of the target image is then given by the labelling function $S^I(x) = \arg\max_{l \in \{0,1\}} \sum_{i=1}^{M_2} w_i \cdot \delta(S_i^J(x) - l)$, where δ is the Dirac delta function and global weights $w_i = \exp(-m_i / \frac{1}{M_2} \sum_{j=1}^{M_2} m_i)$. In this fusion strategy, atlases more similar (higher LCC score) to the target image have a stronger influence on the final segmentation and those with a relatively lower score are downgraded.

3   Algorithm Evaluation

The proposed segmentation framework is evaluated on a benchmark dataset used in the MICCAI 2014 CETUS challenge [12]. It consists of 4D echo sequences acquired from an apical window in healthy volunteers and patients with myocardial infarction and dilative cardiomyopathy. The dataset is divided into 15 training and 30 testing image sequences. Contours of the heart chambers were outlined

---

[1] Locally weighted and majority voting fusion methods were also evaluated in the experiments, and the best results were obtained with the global fusion method.

Table 1. LV segmentation results on 30 subjects (CETUS challenge testing dataset [12]). Mean distance (MD [mm]), Hausdorff distance (HD [mm]) and Dice coefficient (DC [%]) results are listed separately for ED and ES frames.

	MD_{ED}	MD_{ES}	HD_{ED}	HD_{ES}	DC_{ED}	DC_{ES}
Manual [12]	$1.01 \pm .30$	$1.01 \pm .38$	$3.37 \pm .87$	$3.30 \pm .94$	$0.949 \pm .15$	0.938 ± 0.21
AAM [15]	$2.44 \pm .xx$	$2.79 \pm .xx$	$8.45 \pm x.xx$	$8.65 \pm x.xx$	$0.879 \pm .xx$	$0.835 \pm .xx$
BEAS [3,4]	$2.26 \pm .xx$	$2.43 \pm .xx$	$8.10 \pm x.xx$	$8.13 \pm x.xx$	$0.894 \pm .xx$	$0.856 \pm .xx$
SE-MA [10]	$2.18 \pm .70$	$2.47 \pm .74$	7.55 ± 1.76	8.57 ± 2.96	$0.894 \pm .03$	$0.849 \pm .04$
SDF-LS [7]	$2.09 \pm .xx$	$2.20 \pm .xx$	$9.31 \pm x.xx$	$8.35 \pm x.xx$	$0.894 \pm .xx$	$0.871 \pm .xx$
PEM-MA	$1.94 \pm .55$	$2.23 \pm .60$	7.00 ± 1.99	7.53 ± 2.23	$0.904 \pm .02$	$0.874 \pm .04$

by three experts, but only those of the training set are publicly available. Therefore, the CETUS web site[2] is used for evaluation. Submissions are automatically evaluated based on surface distance errors and clinical LV volumetric indices.

In all experiments, segmentations are computed only for end-diastolic (ED) and end-systolic (ES) phases. Table 1 lists the surface distance errors obtained in the first experiment. The proposed PEM-MA framework achieves better results than the challenge top performing algorithms: AAM [15] (active appearance model), BEAS [3,4] (B-spline active contours), SDF-LS (structured decision forest followed by level-set segmentation), and SE-MA [10] (spectral embedding multi-atlas method). The inter-observer manual segmentation [12] variations are reported for comparison. We can conclude that PEMs provide a better boundary representation than spectral features [10] based on mean ($p < 0.01$) and Hausdorff distance ($p < 0.01$). Moreover, the proposed approach does not require landmark selection [10] or manual affine alignment of LV surface template to initialize the segmentation [7].

The difference in segmentation accuracy between PEM-MA and model based surface fitting methods (AAM, BEAS) can be explained as follows. The proposed approach employs affinely aligned atlas labels as shape priors which are selected based on LCC similarity of PEMs, whereas the other methods use less data specific priors such as mean LV shape [15] and ellipsoid [4] shape assumption. Similarly, in PEM-MA, the LV segmentation is initialized with position priors obtained through a robust affine block matching of PEMs. This delineates the left ventricle position in the image more accurately than Hough transform [4] and the mean LV position of the training images [15].

In the second experiment, clinical indices, such as ejection fraction (EF), ED and ES volume values, are computed from the proposed segmentation approach. The obtained results are compared against their reference values using the aforementioned web site. The results in Table 2 show that PEM-MA achieves a better agreement with the ground truth compared to the other methods. As PEM-MA delineates LV boundaries more accurately, better volume estimates are obtained. Additionally, we observe that PEM-MA displays a consistent performance in

---

[2] https://miccai.creatis.insa-lyon.fr/miccai/community/1.

Table 2. Segmentation results on 30 images (CETUS testing data [12]). Pearson's correlation coefficient (corr) and Bland-Altman ($\mu \pm 1.96\sigma$) limit of agreement (LOA) between ground-truth and estimated LV volume values are reported.

	ED_{corr}	ED_{LOA}	ES_{corr}	ES_{LOA}	EF_{corr}	EF_{LOA}
Manual [12]	0.981	-0.636 ± 18.2	0.987	-0.50 ± 14.4	0.959	0.13 ± 6.07
AAM [15]	0.966	-15.42 ± 32.1	0.964	-13.2 ± 28.9	0.611	3.69 ± 17.58
SE-MA [10]	0.945	-6.02 ± 41.6	0.924	-0.42 ± 41.2	0.780	-1.55 ± 13.88
BEAS [3,4]	0.965	-4.99 ± 35.3	0.967	-6.78 ± 27.7	0.889	2.88 ± 10.48
SDF-LS [7]	0.917	8.73 ± 49.9	0.956	-5.16 ± 31.7	0.819	8.33 ± 14.46
PEM-MA	0.961	-4.14 ± 34.0	0.973	-3.47 ± 26.7	0.892	0.48 ± 10.78

both LV surface fitting and volume estimation in contrast to SDF-LS. The performance difference between the two can be linked to the improved structural representation and the choice of different surface fitting algorithm.

All experiments were carried out on a 3.00 GHz quad-core machine. The average computation time per image pair was 74 s for non-rigid registration, 16 s for affine alignment and 20 s to compute each PEM. The training of the SDF (70m per tree) and atlas PEM computation were performed offline prior to target segmentations. The segmentation of the LV takes in total 16 m per image. The proposed approach is computationally more complex than the methods in [4,7] due to the multitude of registrations. However, a parallel implementation of these registrations significantly reduces the total runtime.

Implementation Details: In total $N_t = 8$ PEM decision trees are trained using 20 US sequences plus rotated versions of these images. PEM quality was not improved further by including more trees. Patch sizes for training features and ground-truth edges are chosen as $S_a = 20$ and $S_e = 10$ per dimension. For global alignment, blocks of size 5^3 voxels were used with search radius equal to the block size as in [11]. A multi-scale optimization strategy was employed to capture large displacements and to improve convergence. A total of $M_1 = 30$ ED and ES atlases were aligned to each subject. Of these, on average $M_2 = 6.3$ were selected based on their LCC score, with a standard deviation of the Gaussian $\sigma = 7$ voxels in each dimension.

4   Conclusion

We presented a novel US image representation (PEM) which achieves state-of-the-art cardiac US image registration and LV segmentation results within a multi-atlas framework. The proposed framework outperforms all other methods participating in the MICCAI CETUS challenge based on the obtained surface mesh evaluation criteria. The main contributions of the paper are: (1) highly accurate 3D edge map representation for cardiac US images, and (2) block-matching based robust and accurate initialization technique for automatic LV segmentation. The proposed PEM representation is generic and modular. It has the potential of being applied to echo images acquired from other organs and does not make assumptions on the acquisition window and image orientation. Additionally, the multi-atlas segmentation framework is shown to be applicable for clinical routine as it can estimate functional indices very accurately.

References

1. Aljabar, P., Heckemann, R.A., Hammers, A., Hajnal, J.V., Rueckert, D.: Multi-atlas based segmentation of brain images: atlas selection and its effect on accuracy. NeuroImage **46**(3), 726–738 (2009)
2. Artaechevarria, X., Munoz-Barrutia, A., Ortiz-de Solórzano, C.: Combination strategies in multi-atlas image segmentation: application to brain MR data. IEEE Trans. Med. Imag. **28**, 1266–1277 (2009)
3. Barbosa, D., Friboulet, D., D'hooge, J., Bernard, O.: Fast tracking of the left ventricle using global anatomical affine optical flow and local recursive block matching. In: Proceedings of MICCAI CETUS Challenge (2014)
4. Barbosa, D., et al.: Fast and fully automatic 3-D echocardiographic segmentation using B-spline explicit active surfaces: feasibility study and validation in a clinical setting. Ultrasound Med. Biol. **39**(1), 89–101 (2013)
5. Cachier, P., Pennec, X.: 3D non-rigid registration by gradient descent on a Gaussian windowed similarity measure using convolutions. In: IEEE Workshop on Mathematical Methods in Biomedical Image Analysis, pp. 182–189 (2000)
6. Dollár, P., Zitnick, C.L.: Structured forests for fast edge detection. In: ICCV, pp. 1841–1848. IEEE (2013)
7. Domingos, J.S., Stebbing, R.V., Leeson, P., Noble, J.A.: Structured random forests for myocardium delineation in 3D echocardiography. In: Wu, G., Zhang, D., Zhou, L. (eds.) MLMI 2014. LNCS, vol. 8679, pp. 215–222. Springer, Heidelberg (2014)
8. Elad, M., Aharon, M.: Image denoising via sparse and redundant representations over learned dictionaries. IEEE Trans. Image Process. **15**(12), 3736–3745 (2006)
9. Lempitsky, V., Verhoek, M., Noble, J.A., Blake, A.: Random forest classification for automatic delineation of myocardium in real-time 3D echocardiography. In: Ayache, N., Delingette, H., Sermesant, M. (eds.) FIMH 2009. LNCS, vol. 5528, pp. 447–456. Springer, Heidelberg (2009)
10. Oktay, O., Shi, W., Caballero, J., Keraudren, K., Rueckert, D.: Sparsity based spectral embedding: application to multi-atlas echocardiography segmentation. In: Proceedings of MICCAI STMI Workshop (2014)
11. Ourselin, S., Roche, A., Pennec, X., Ayache, N.: Reconstructing a 3D structure from serial histological sections. Image Vis. Comput. **19**(1), 25–31 (2001)
12. Papachristidis, A., et al.: Clinical expert delineation of 3D left ventricular echocardiograms for the CETUS segmentation challenge. In: Proceedings of MICCAI CETUS Challenge, pp. 9–16 (2014)
13. Rajpoot, K., Grau, V., Alison Noble, J., Becher, H., Szmigielski, C.: The evaluation of single-view and multi-view fusion 3D echocardiography using image-driven segmentation and tracking. MedIA **15**(4), 514–528 (2011)
14. Rueckert, D., Sonoda, L., Hayes, C., Hill, D.L., Leach, M., Hawkes, D.J.: Nonrigid registration using free-form deformations: application to breast MR images. IEEE Trans. Med. Imag. **18**(8), 712–721 (1999)
15. Stralen, M.V., Haak, A., Leung, K., Burken, G.V., Bosch, J.: Segmentation of multi-center 3D left ventricular echocardiograms by active appearance models. In: Proceedings of MICCAI CETUS Challenge, pp. 73–80 (2014)

Fuzzy Segmentation of the Left Ventricle in Cardiac MRI Using Physiological Constraints

Tasos Papastylianou[1]([⊠]), Christopher Kelly[1], Benjamin Villard[1], Erica Dall' Armellina[2], and Vicente Grau[1]

[1] Institute of Biomedical Engineering, University of Oxford, Oxford, UK
tasos.papastylianou@kellogg.ox.ac.uk
http://tpapastylianou.com
[2] Acute Vascular Imaging Centre, John Radcliffe Hospital, Oxford, UK

Abstract. We describe a general framework for adapting existing segmentation algorithms, such that the need for optimisation of intrinsic, potentially unintuitive parameters is minimized, focusing instead on applying intuitive physiological constraints. This allows clinicians to easily influence existing tools of their choice towards outcomes with physiological properties that are more relevant to their particular clinical contexts, without having to deal with the optimisation specifics of a particular algorithm's intrinsic parameters. This is achieved by a structured exploration of the parameter space resulting in a subspace of relevant segmentations, and by subsequent fusion biased towards segmentations that best adhere to the imposed constraints. We demonstrate this technique on an algorithm used by a validated, and freely available cardiac segmentation suite (**Segment** – http://segment.heiberg.se).

Keywords: cineMRI · Heart · Probabilistic · Segmentation

Abbreviations: *MRI* – Magnetic Resonance Imaging, *CT* – Computed Tomography, *SSFP* – Steady-State Free Precession, *LGE* – Late Gadolinium Enhancement, *EF* – Ejection Fraction, *SV* – Stroke Volume, *SA* – Short Axis, *LV* – Left Ventricle, *PCA* – Percutaneous Coronary Angioplasty, *MI* – Myocardial Infract

1 Introduction

Cardiac cineMRI is rapidly becoming one of the leading investigations in the assessment of cardiac disease, due to its ability to capture good quality images of the whole heart throughout the cardiac cycle, allowing for a more dynamic assessment of the heart. Any measurements relating to the ventricles that can be made from these images, such as stroke volume and ejection fraction, require segmenting the blood pool from the myocardium and background, for use in further calculations or processing. Segmentation of the ventricles by manual delineation from an expert clinician is prohibitively time-consuming, therefore much research has focused on automated techniques for segmentation.

© Springer International Publishing Switzerland 2015
H. van Assen et al. (Eds.): FIMH 2015, LNCS 9126, pp. 231–239, 2015.
DOI: 10.1007/978-3-319-20309-6_27

A large number of approaches to cardiac segmentation from cardiac MRI (and other modalities) have been developed over the past few decades, such as approaches based on Deformable models or Shape models, Registration-based techniques, Image-/Voxel-based classification methods, or combinations of the above [1]. A common limitation of all these methods is that they tend to be reliant on a careful selection of parameters for optimal performance. However, such parameters tend to reflect more the algorithm's inner workings, rather than the clinical task at hand, and the choice of an optimal set of parameters can therefore be fairly unintuitive to the clinician.

Segmentation algorithms traditionally produce deterministic results (i.e. a clearly defined label per voxel); however, algorithms that produce fuzzy or probabilistic results are becoming increasingly common, as they have several advantages, such as being able to represent more complex situations relating to partial volume voxels, and therefore lead to more accurate estimation of clinical parameters. An example of this fuzzy approach is "ensemble" methods, such as Adaboost [2], where a stronger classifier is built from a weighted collection of weaker classifiers, or such as STAPLE [3] which attempts to produce a better result from an existing pool of segmentation results (where these are already considered to be of reasonable quality, i.e. derived from strong classifiers) via some form of weighted consensus.

1.1   Our Approach

We propose a framework which aims to enhance existing segmentation algorithms, with the following goals in mind: (1) in the case of deterministic algorithms, propose a generalisable method to produce a probabilistic equivalent, by exploring the parameter space in a structured manner so as to produce a segmentation space, and then fusing the results appropriately using an ensemble approach; (2) reduce the need for predefined, optimal, problem-specific parameter sets, by weighing each segmentation in the segmentation space according to its compatibility with respect to intuitively defined physiological constraints, before fusing them together into a fuzzy segmentation result. (3) Use the above to improve the output and usability of *existing* algorithms from a *clinical* viewpoint, by allowing clinicians to guide segmentation algorithms towards results that are more relevant to their particular clinical context by simply defining such a context in intuitive clinical terms, rather than attempt algorithmic optimization via trial and error over an unintuitive set of parameters.

We demonstrate the framework on an existing segmentation algorithm proposed by Heiberg et al. [4]. This algorithm was chosen both because of its simplicity, making demonstration of the concept straightforward, and also because an implementation is made freely available online by the authors (Segment — http://segment.heiberg.se), which has been shown to be robust and produce accurate results (an extensive list of publications citing the algorithm is made available on the project website).

Paper Organisation: The remainder of the paper is organised as follows: Sect. 2 provides a brief background for the Heiberg algorithm, which is used to demonstrate the proposed framework; Sect. 3 expands on the motivation behind our approach; Sect. 4 details the method used to obtain the fused segmentation; Sect. 5 presents the results with respect to clinician-provided manual segmentations used as a gold standard.

2   Background

The 2005 algorithm proposed by Heiberg et al. [4] (henceforth called the Heiberg algorithm) is essentially a *deformable model*-based segmentation approach. The model, consisting of a set of 2D active contours (one per slice), seeks to achieve an equilibrium between two competing sets of forces acting on its surface, while taking into account within-slice and temporal information; at each iteration, external forces guide the evolution of the model towards image-dependent features, whereas internal forces constrain the evolution, such that model smoothness and shape are relatively preserved.

The model has two external forces, an inflating *Balloon force*, and an *Edge force*. The Balloon force is dependent on local intensity, favouring expansion of the contour in areas closer to the estimated object's average intensity (as initialised by the user by selecting a single voxel lying within the left ventricle from the image). The Edge force is defined in terms of edge images derived from the image. Four edge images are produced, corresponding to estimating image edges in 4 different directions. At the point of calculation of the Edge force, the most appropriate edges to evolve towards are chosen given the direction of evolution of the model. Temporal information is introduced to the model by smoothing the edge force at each node-point of the model over several timeframes.

There are four internal forces with the purpose of ensuring spatial and temporal smoothness: a *Curvature force* which promotes smoothness in the overall contour shape, a *Damping force* and an *Acceleration force*, which ensure spatial continuity of the model's nodes within timeframes, and a *Slice force* which relatively discourages node movement between the slices (i.e. in the z-plane).

The above six forces are then combined in a "modality dependent" manner to control model evolution; here, "modality dependent" means choosing a set of modifiers for each force (i.e. the algorithm's parameters), which are most effective at leading the model towards a successful segmentation, given a particular investigation or image type.

Details of the mathematical implementation of these forces are beyond the scope of the present paper — particularly in the context of proposing a generalised framework aiming to minimize the role played by individual parameters, and by extension their particular role in the underlying mechanics of the algorithm in question; we refer the interested reader to the original paper for implementation details.

3   Motivation

As with most segmentation algorithms, the Heiberg algorithm relies on a careful selection of parameters. To a large extent, the choice of parameters represents partial knowledge about the nature of the problem, or about the environment in which segmentation is to take place. For example, for algorithms that are generalisable such that they can be used in more than one modality, object, or clinical problem, a common approach is to find a generally optimal set of parameters for each scenario, suitably defined on a test database through trial and error or by machine learning. In the case of the Heiberg algorithm and their implementation, provided freely online, a selection of pre-defined parameter sets is provided, each optimised for a particular *general* scenario: different types of MRI, segmentation of Left Ventricle vs Right Ventricle, segmentation from CT, etc. There are drawbacks to such 'scenario-based' approaches: Firstly, while a parameter set optimised on a training set adhering to a particular scenario serves as a good starting point, as we will demonstrate further on, it does not guarantee optimal results on particular images (even within the limits of the particular algorithm), or for particular setups and clinical contexts. Secondly, selecting an optimal set of parameters is normally a process which is largely intrinsic to the inner workings of an algorithm, offering little to no intuition on how they should be adjusted to accommodate changes in clinical context to ensure a more relevant outcome. Therefore, if a particular clinical environment has a slightly different setup to the one used for the algorithm training phase, and therefore has slightly different parameter requirements for optimal results (within the limits of the algorithm) than the ones provided by the manufacturer, tweaking that default parameter set to adjust it for their own setup is usually beyond the abilities of the clinician, because it does not translate to relatable clinical information. Therefore the clinician is more likely to simply accept the suboptimal parameter set (and by extension, a suboptimal segmentation result) *as is*, and simply try to take this into account clinically when weighing up the information. The main motivation behind our approach, therefore, is to enable the clinician to steer a segmentation algorithm towards results which are more relevant to their particular clinical context, by allowing them to introduce intuitive and clinically relatable information to the process; this could be performed once to adjust the default parameter set to one more suitable to a particular clinical setup, or it could be performed on a per-case basis as required.

The intuition for our approach lies in the following key observation: Segmentation results which are 'better' — better, here, defined as results that are closer, in a mathematical sense, to the gold standard — will also produce estimates of physiological parameters — such as Ejection Fraction (EF) and Stroke Volume (SV) — which are 'better'. Our first premise, therefore, is derived by following this logic in reverse:

Premise 1: A segmentation result producing a large number of physiological parameters, which both individually and as a group are all 'better', is more likely to correspond to a 'better' segmentation. — Fig. 1 demonstrates this graphically.

If we had the theoretical ability to explore all the possible values and combinations for each of the algorithm's parameters, we would obtain a set of segmentation results, covering all possible segmentation outcomes that are possible

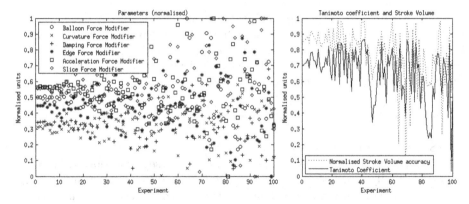

Fig. 1. Generating 100 segmentations using a default parameter set and increasing noise. Left: Distribution of algorithm parameters (normalised for comparison) for all experiments. Right: Comparison of physiological parameter accuracy and segmentation accuracy for all experiments; here stroke volume (SV) accuracy (A) is defined with respect to the Gold Standard (SV_G) as follows: $A = 1 - \frac{f(SV)}{f(SV)_{max}}$, where $f(SV) = |SV - SV_G|$ and $f(SV)_{max}$ is the maximum f value obtained within this set. There are three pertinent things to note: (1) As the parameters become increasingly noisy, accuracy tends to drop, but there are *still* occasions which produce good results. (2) The 'default' parameter set is not necessarily the best one; but it's difficult to predict a priori which one is. (3) Sets with even just *one* estimated physiological parameter being closer to the true value are visibly more likely to have higher accuracy; this is regardless of whether they originated from the less or more 'noisy' part of the experiment.

for a particular algorithm on a given image. We refer to this finite set, as the algorithm's *Segmentation Space*. Equivalently, the complete *Parameter Space* is the set consisting of all possible parameter sets, each mapping to a segmentation in the segmentation space.

While exploration of the full parameter space is generally infeasible, we can select samples from a focused region, which is most likely to correspond to more accurate segmentations. If we treat a set of N parameters as an N-dimensional vector, then a simple way of doing this is by selecting samples with an N-dimensional gaussian probability function centered at a point of interest. A reasonable choice for this would be the default parameter set/vector suggested by the algorithm itself. This hopefully should restrict the segmentation space to a subset of generally more accurate segmentations, which we could then fuse to obtain a fuzzy segmentation.

It follows from Premise 1, that if we introduce a bias in the fusion process, to favour segmentations that are 'better', the fused result should logically be biased towards a 'better' fused result.

Premise 2: In the presence of a segmentation subspace, biasing segmentation fusion towards results associated with better physiological parameters, should result in a better fused result overall, compared to an unbiased fusion

The practical implication of applying the above insights to any algorithm, is that we shift the focus from having to optimise highly unintuitive parameters intrinsic to the segmentation algorithm, to something that is more intuitive within the context of the task at hand – i.e. the physiological parameters – and which is therefore easier, and more relevant to non image-analysis specialists.

4   Methods

The Segment cardiac segmentation suite by Heiberg et al. (http://segment. heiberg.se) [4] was used to obtain left ventricle (LV) segmentations from a set of images, kindly provided by the University of Oxford Centre for Clinical Magnetic Resonance Research at the John Radcliffe Hospital, Oxford. This set was produced on a 3.0T Siemens Tim Trio whole-body MRI scanner using a 4D (i.e. 3D+time) TrueFISP cineMRI protocol, from a patient undergoing a post-PCA investigation, following a diagnosis of an Inferior MI; the set was anonymised appropriately and no other clinical or radiological details were available. The image set consisted of 25 timeframes of 8 Short-Axis (SA) slices at a resolution of 256×176 voxels, of size $1.5625 \times 1.5625 \times 8$ mm. Manual segmentations of the left ventricle were provided by an expert clinician, which were used as a gold standard; this was obtained as per-slice 2D contours, drawn at 4×4 subresolution accuracy per in-slice image voxel, using the CMR42 cardiac imaging suite [5]. Full diastole was identified in timeframe 1, and full systole at timeframe 10. Data was processed using Matlab [6]; images were extracted from the DICOM files using a modified version of Laszlo Balkay's DICOM reader [7]; all other processing (including extraction of contours from CMR42 files) is the work of the authors.

A set of 100 segmentations was obtained by applying normally distributed random noise of linearly increasing standard deviation, on each of the default parameters (i.e. force modifiers; see Fig. 1) provided by Segment for the case of SSFP MRI; the noise was generated with mean $\mu = initial\ modifier\ value$ for each parameter, and standard deviation σ taking values linearly from 0 to μ over the 100 experiments. Experiment 70, which was the best outcome in this set was retained as a reference to the best segmentation obtainable with this algorithm for this particular image. For each of the resulting segmentations, the following physiological parameters were derived:

– Volumes in systole (V_s) and diastole (V_d), defined as the number of voxels in the set of LV-labeled voxels in systole (LV_s) and diastole (LV_d) respectively
– Stroke Volume (SV) = $V_d - V_s$
– Ejection Fraction (EF) = SV/V_d
– Centre of mass (systole): A 3D-coordinate vector $C_s = [\bar{x}_i, \bar{y}_i, \bar{z}_i]^T, \forall\ i \in LV_s$
– Centre of mass (diastole): $C_d = [\bar{x}_i, \bar{y}_i, \bar{z}_i]^T, \forall\ i \in LV_d$
– Combined centre of mass: $C = (C_d + C_s)/2$.

The weight each segmentation carries within the fusion process is determined by a measure of how close each of their physiological parameters is to a reference value; in particular, a suitable range around this reference value acts as a fuzzy constraint, that prevents bad segmentations, from a physiological-estimates point of view, from exerting much influence on the fused end-result. In practice, such values might be already available clinically (i.e. from a previous echocardiogram), or from known values. However, for the purposes of this paper, three different types of constraints were generated:

- Reference range derived from the 'default' segmentation (i.e. the segmentation resulting from the 'default' parameter set), using the median and inter-quartile range to define lower (l) central (c) and upper (u) reference values: This should produce the fuzzy analogue closest to the default case.
- Reference values derived from a very quick and crude initialisation process, where the user draws rough squares outside and inside the blood pool; thereby defining lower (l) and upper (u) constraint values for the reference range, with their average representing the central reference value (c).
- Reference values derived from the known Gold Standard. This should produce the best outcome which is possible from the algorithm, with respect to the known gold standard. Lower (l) and upper (u) constraint values for this case were set as $\pm\,10\,\%$ of the central value (c) for all physiological parameters, except for the distance from the centroid, which was set at the range of 0–10 voxels apart.

Weights were then calculated for each of n segmentations (S_n) from these reference values, by evaluating a fuzzy membership function on each of the estimated physiological parameters. We found that a good membership function was a gaussian membership function, with mean $\mu = c$ and standard deviation $\sigma = (u - l)/2$. A total weight (w) was then obtained by fuzzy conjunction of all the weights; this was evaluated separately for two Triangular-Norms [8]: Product (involving multiplication of all terms), and Gödel (involving selecting the minimum of the set as the weight, i.e. the "weakest link"). Segmentations were then fused by a simple weighted averaging process: $S_{fuzzy} = \sum_1^n w_n S_n$, and the fused result thresholded at 0.5; the resulting binary segmentation mask S was validated against the Ground Truth G using the Tanimoto Coefficient: $\frac{G \cap S}{G \cup S}$.

5 Results and Discussion

The accuracy of the different segmentations is shown in Table 1. For comparison, the Tanimoto coefficient of the original algorithm with default parameters was 0.7016; the best possible outcome for the algorithm yielded a Tanimoto coefficient of 0.8671. While the Fuzzy equivalent of the default parameter set seems to be a bit lower for both Product and Gödel cases, it is very close (and indeed this is also the case visually; see Fig. 2), and it is in fact a better fuzzy equivalent than the simple averaging of all 100 segmentations without weighting, which resulted in a Tanimoto coefficient of 0.6507. However, with respect to the best outcome, both the default case and its fuzzy analogues are poor by comparison.

There is clear improvement when more appropriate physiological parameters are provided as constraints. In the case of parameters derived from the Gold

Fig. 2. Segmentation results on representative slice (5th): Continuous lines: inside – original default segmentation; outside – manual gold standard. Dotted lines, from innermost to outermost: (a) result from physiological constraints from default case; (b) result from crude manual initialisation (c) result from initialisation from gold standard (d) best segmentation from initial set.

Table 1. Tanimoto coefficient of resulting fuzzy segmentations at diastole, after thresholding at 0.5 to obtain a binary result, and as compared against the gold standard (i.e. manual segmentations), for Product and Gödel fuzzy logic.

	Default	Manual	Gold Standard
Product fuzzy logic	0.6745	0.7953	0.8183
Gödel fuzzy logic	0.6768	0.7536	0.8102

Standard (which would be equivalent, for instance to having those parameters provided by the clinician, e.g. via a different investigation or from prior knowledge), this comes very close to the 'best' result of the set. Furthermore, the rough manual initialisation is not far behind in terms of accuracy. In other words, even in the abscence of perfect physiological parameters, a rough estimate can still lead to a markedly better result. It is worth pointing out that it was only possible to identify the 'best' result via validation against the gold standard; therefore, in the absence of a gold standard, it would be very difficult to confidently identify the optimal parameter set. Our results demonstrate that by using the more intuitively generated physiological constraints in this fashion, we can achieve similarly good results as the best possible segmentation obtained through an optimal parameter set.

Figure 2 demonstrates the resulting contours, and shows the effect of our approach visually. Rather unsurprisingly, we note that the algorithm seems to retain its shape properties; in other words, if all segmentations in the set share common shape characteristics, the fused result is unlikely to produce a result which is structurally very different than the best result in the set. However, since the resulting surface is biased towards having similar physiological parameters as the gold standard, the final outcome should favour surfaces that are generally closer to it.

Conclusion

We have demonstrated a framework for producing a fuzzy equivalent segmentation from an existing algorithm, by exploring its parameter space to produce a segmentation space. This can then be fused in a weighted scheme, constrained by physiological parameters which can either be introduced by a clinician, much more intuitively than intrinsic algorithm parameters, or can be approximated by rough initialisation. The concept and framework can be generalised to any algorithm, such that instead of focusing on optimising intrinsic parameter sets for general cases, one would only need to explore the parameter space appropriately, and provide appropriate physiological constraints, which can be more intuitively defined, to produce better segmentations. The framework is particularly suited for medical images where the object in question has particular physiological properties that can then be represented via a fuzzy membership function and incorporated as a constraint; heart segmentation lends itself naturally to this problem, as it offers both physiological and anatomical constraints. Further work could focus on automating initialisation further, such as by using Haar features; introducing further types of physiological constraints, such as correctness of anatomical position based on other landmarks (e.g. defined as fuzzy spatial relationships of being "below the lung", "above the diaphragm" etc.); and improving efficiency through parallelisation or a convergent approach to the acquisition of segmentation weights.

Acknowledgements. TP and BV acknowledge the support of the RCUK Digital Economy Programme grant number EP/G036861/1 (Oxford Centre for Doctoral Training in Healthcare Innovation). VG is supported by a BBSRC grant (BB/I012117/1), an EPSRC grant (EP/J013250/1) and by BHF New Horizon Grant NH/13/30238.

References

1. Petitjean, C., Dacher, J.-N.: A review of segmentation methods in short axis cardiac MR images. Med. Im. Anal. **15**, 169–184 (2011)
2. Schapire, R.E.: The boosting approach to machine learning: an overview. In: Denison, D.D., Hansen, M.H., Holmes, C.C., Mallick, B., Yu, B. (eds.) Nonlinear Estimation and Classification. LNS, pp. 149–172. Springer, New York (2003)
3. Warfield, S.K., Zou, K.H., Wells, W.M.: Simultaneous truth and performance level estimation (STAPLE): an algorithm for the validation of image segmentation. IEEE Trans. Med. Imaging **23**(7), 903–921 (2004)
4. Heiberg, E., Wigstrom, L., Carlsson, M., Bolger, A. F., Karlsson, M.: Time resolved three-dimensional automated segmentation of the left ventricle. In: Computers in Cardiology, pp. 599–602. IEEE, September 2005
5. cmr[42]. [software]. Circle Cardiovascular Imaging Inc., Calgary, Canada
6. MATLAB, v8.2 (R2013b). Natick, Massachusetts: The MathWorks Inc., 2012
7. Balkay, L.: DICOMDIR reader. University of Debrecen (2011). http://www.mathworks.co.uk/matlabcentral/fileexchange/7926-dicomdir-reader
8. Klement, E.P., Mesiar, R., Pap, E.: Triangular Norms. Kluwer, Dordrecht (2000)

Subject Independent Reference Frame for the Left Ventricular Detailed Cardiac Anatomy

Bruno Paun[1]([✉]), Bart Bijnens[1,2], and Constantine Butakoff[1]

[1] Universitat Pompeu Fabra, Barcelona, Spain
bruno.paun@upf.edu
[2] ICREA, Barcelona, Spain

Abstract. Mapping of surfaces to a parametric domain is a widely used tool in medical imaging for analysis and localization of injured tissue. By assigning the same coordinate values to specific anatomical landmarks, parametrization allows us putting into correspondence surfaces of anatomical shapes with inherently different geometry and facilitates integration of data acquired by different imaging modalities. In this paper we propose a method for subject independent anatomical parametrization of the left ventricular (LV) wall that includes trabeculations, papillary muscles and false tendons. The method relies on a disk parametrization of the LV smooth epicardium and mapping the interior of the ventricular cavity using ray casting. In this way we define a common reference frame whereupon any LV is mapped in a consistent way thus allowing for statistical analysis and comparisons between different patients.

Keywords: Left ventricle · Parametrization · Coordinate system · Surface flattening · Cardiac trabeculations

1 Introduction

The role of trabeculations and their normal morphological expression in a heart is still unclear. The amount of trabeculae increases in both cardiac ventricles from base to apex everywhere except for the outflow tracts which are smooth [1]. The interest in the trabeculations started to arise mostly due to the phenomenon of excessive trabeculations known as LVNC (Left Ventricular Non Compaction), which is currently considered a heart disorder and can lead to heart failure, arrhythmias, including sudden cardiac death [2]. Nevertheless every heart has a trabeculated inner layer whose function is not completely clear. Additionally to the lack of understanding of the role of the trabeculations, there is also a lack of *in vivo* imaging data where the trabeculations are visible and could be either monitored through imaging or extracted to be included in cardiac model studies. Modeling those structures could provide an alternative way of analyzing their function and their relation to cardiomyopathies. Although advances in imaging techniques made extraction of detailed cardiac structures feasible, their accurate

© Springer International Publishing Switzerland 2015
H. van Assen et al. (Eds.): FIMH 2015, LNCS 9126, pp. 240–247, 2015.
DOI: 10.1007/978-3-319-20309-6_28

segmentation is still a challenging task. For instance, recently, M. Gao et al. proposed a method for a topologically accurate segmentation by restoring missing topological structures of a initially given segmentations [3] which allowed them to preform morphological analysis of such structures in LV [4].

Due to the above difficulties, when it comes to the computational models of the heart, most of them use a smooth endocardial surface for the cardiac chambers. An exception is the work of Bishop et al. [5] where the authors attempted to investigate the effect of trabeculations in the rabbit heart. The analysis is carried out on the data extracted *ex vivo* as the only way of obtaining the highly detailed dataset. However if one wants to include detailed cardiac anatomy in the modeling studies of human hearts, most of the time one is limited to the *in vivo* data acquired using a 1.5T/3T MRI scanner for a reasonably short period of time. In the latter case, the resolution of the images is not sufficient to visualize the trabeculations and a different approach is necessary.

One of the possibilities to overcome this limitation is the extraction of the trabeculations from the *ex vivo* studies and using them in the computational models. This approach however requires establishing a mapping between the domains represented by the meshes. There is a wide spectrum of mesh parametrization methods applicable to different shape topologies: Least Squares Conformal Maps [6], Discrete Surface Ricci Flow [7], Intrinsic Parameterizations [8] and Teichmüller Maps [9] to name a few. Surface parametrization is an active research field with applications to a wide range of problems. In the medical field, they have been applied to the problems of analysis and visualization of brain surfaces [10–12], creation of inter-patient liver coordinate systems [13], visualization and mapping of functional information from ventricular cavities onto a discretized or continuous bulls-eye plot [14], creation of normalized parametric domain for comparison of LV function across subjects [15].

In this paper we propose a method for mapping of detailed cardiac anatomy extracted from human hearts acquired *ex vivo* to any smooth surface either representing different cardiac geometry or a patient independent geometry. The proposed approach produces a *reference frame* where the detailed cardiac anatomy of different subjects can be represented in a manner independent of the overall shape of the heart. We show how this representation can be used for statistical analysis of the distribution and differences in trabeculations.

2   Methods

The proposed method operates on two triangle surface meshes: M and S where M represents the LV endocardial structures (trabeculations) and S represents its bounding surface (tight surface on which the trabeculations reside). The preprocessing steps to obtain these meshes are explained in Subsect. 2.4. Depending on the application, mesh M can represent just endocardial structures of interest or all structures within the ventricular cavity, and can include a part or the whole myocardial wall which implies that a different base mesh S has to be used. The whole process can be represented by 3 steps:

1. Mapping S to the planar domain with the LV apex at the center
2. Due to symmetry of the LV, orientation ambiguity is eliminated
3. Mapping M into the cylinder defined by the flattened S.

Prior to applying a mapping function ρ to different subject meshes M, we are locating two anatomical landmark points on their corresponding base meshes S. As cardiac landmarks we use the apex and the mid-septal point of the *mitral annulus* δS. The mid-septal point of δS is manually defined by the user, while the apex is defined as the furthest point from the centroid $c_{\delta S}$ of δS.

2.1   Mapping of the Anatomical Base Mesh to a Planar Domain

As the planar domain for S we used a unit disk following the approach of De Craene et al. [16]. As implied by the Riemann Mapping Theorem, any surface homeomorphic to a disk can be conformally mapped into any simply-connected region of the plane. The only requirement is that our mapping is harmonic, meaning that every surface coordinate has to have a vanishing Laplacian. Therefore we are computing a bijective mapping $\varphi : S \subset \mathbb{R}^3 \longrightarrow D \subset \mathbb{R}^2$, where S is the anatomical base surface mesh of the LV and D is a unit disk. The boundary ∂D of D is defined by uniformly sampling a unit circle with the number of samples equal to the number of points on the boundary ∂S and the following system of linear equations is solved for the coordinates of the points inside the disk:

$$\begin{cases} L_{S \setminus \partial S} \cdot \mathbf{x}_{D \setminus \partial D} = 0 \\ \mathbf{x}_{\partial D} = \mathbf{x} \end{cases} \tag{1}$$

The boundary x, y coordinates are given by the columns of matrix $\mathbf{x}$. $L_{S \setminus \partial S}$ represents the Laplacian matrix of the mesh S with the rows corresponding to its boundary ∂S removed. $\mathbf{x}_{\partial D}$ and $\mathbf{x}_{D \setminus \partial D}$ are the matrices of coordinates of the points on the disk (the one we are calculating and that define our mapping) corresponding to the boundary and the interior, respectively. The desired mapping is then given by (1), while the connectivity information is retained from the mesh S. An example of the mapping can be seen in Fig. 1.

As a result of this mapping, the position of the apex in the planar domain will still be variable and depend on the specific anatomy. However in order to define a subject independent frame we must make sure that the apex is consistently mapped to the same point which is achieved using thin plate splines, mapping the boundary to itself and the apex to the center of the disk. This additional step will enforce a more consistent localization of the cardiac regions among different subjects, but reduce the conformal map to quasi-conformal.

2.2   Elimination of Orientation Ambiguity

To assure correspondence of parametrized meshes N and D between subjects we are removing orientation ambiguity by assigning the same coordinate values to their mid-septal landmarks. Thus during the mapping to a unit disk we are placing these landmarks at the same location $\mathbf{x}_l = (1, 0)$ (l is the index of the landmark in the array of all the points).

2.3 Mapping the Detailed LV Anatomy

Once we have the outer surface S of the LV mapped to the disk D, we have provided a base of a subject independent reference frame on top of which we want to map a detailed mesh M. To do so, we first project the vertices of M onto S. The common way of achieving this would be to locate the closest point v'_S for every vertex v_M. As the LV has structures traversing thorough the whole cavity, the closest point projection will not project the neighbourhoods of v_M's to the neighbourhoods of v'_S's. Such projection will cause the mapping function ρ to be non-bijective and result in a highly distorted mesh N.

To alleviate the above problem, we project the points by casting rays through all vertices v_M from a fixed point $c_{\delta S}$ and locate the corresponding intersection points v'_S. As the fixed point $c_{\delta S}$, we take a centroid of the surface boundary δS. The choice of the centroid of the *mitral annulus* as the origin of the rays is motivated by the fact that its location is free of any detailed structures we want to parametrize.

Subsequently, v'_S points are mapped to v'_D points using the map calculated in Subsect. 2.1. Finally, for every v_M, we calculate its distance $d_{v_M v'_S}$ along the ray to the corresponding v'_S normalized by the length $d_{c_{\delta S} v'_s}$ of the ray segment between v'_S and $c_{\delta S}$. Then the vertices v_M are mapped to our reference frame by placing them along the normal direction of v'_D at the distance $d_{v_N v'_D} = d_{v_M v'_S}/d_{c_{\delta S} v'_s}$. The transformed vertices v_N of N have the same connectivity as v_M of M.

2.4 Datasets and Preprocessing

The input meshes were obtained from 3T MRI datasets (Siemens Tim Trio, Erlangen, Germany) with 0.44×0.44 mm in-plane resolution and slice thickness

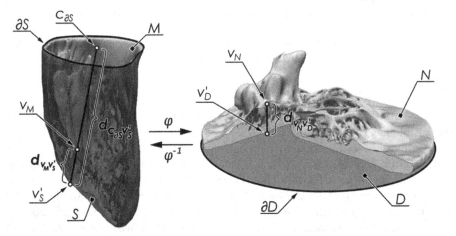

Fig. 1. Illustration of the mapping of a trabeculated mesh vertex v_M to the reference frame

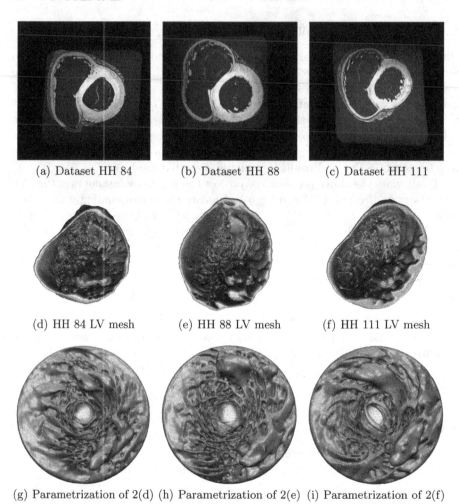

(a) Dataset HH 84   (b) Dataset HH 88   (c) Dataset HH 111

(d) HH 84 LV mesh   (e) HH 88 LV mesh   (f) HH 111 LV mesh

(g) Parametrization of 2(d) (h) Parametrization of 2(e) (i) Parametrization of 2(f)

Fig. 2. (a), (b), (c) - middle stack short axis view slices of MRI human heart datasets; (d), (e), (f) - corresponding LV meshes; (g), (h), (i) - corresponding parametrized LV meshes.

of 1–1.17 mm. The Dicom datasets (as in Fig. 2(a), (b), (c)) were provided by the Visible Heart® Laboratory [17].

The meshes of the detailed LV anatomy including the part of the myocardial wall, are shown in Fig. 2(d), (e), (f). The meshes were generated from the segmentations of the MRI data (Seg3D [18]). The segmentations were cut at the basal part of the heart at the level of the valves, meshed using marching cubes of ParaView [19] and smoothed and uniformly remeshed using ReMesh [20].

Base surface meshes S were extracted from the above meshes of detailed anatomy using VTK [21] introducing a cut in the basal part along the boundary rim, separating the mesh into inner and outer open surfaces.

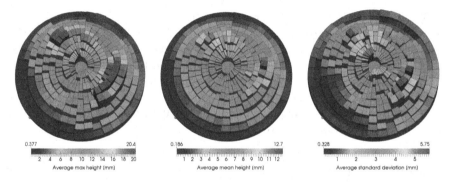

Fig. 3. Max, mean and standard deviation of parametrized meshes heights per segment averaged over all the cases.

3   Results

The method proposed in the previous chapter was applied to six human LV datasets. Parametrized meshes N of Fig. 2(d), (e), (f) are shown in Fig. 2(g), (h), (i). For every subject, max, mean and standard deviation of a height of N (namely distance $d_{v_N v_D'}$ of their corresponding vertices v_N) was calculated for the defined segments and averaged over all subjects. The corresponding plots are shown in Fig. 3. Calculations were obtained from non-normalized distances $(d_{v_N v_D'} = d_{v_M v_S'})$. The vertices v_N residing on the outer surface of N were excluded. The segments were arbitrarily defined to best show variability of measurements in plots. As the meshes N include part of myocardial wall, we corrected the results by subtracting the value of the segment with min value which represent the cardiac wall.

From Fig. 2(g), (h), (i) one can observe unique trabecular morphology inherent to each subject. From the plots in Fig. 3 we observe an increase in the amount of trabeculations as we move from the base towards the apex of the LV. The basal part of the septal wall is free of trabeculations and they start to emerge in its middle part while, on the lateral wall, we have them present along the whole wall. The trabeculations present in the basal part of the lateral wall are attached to the wall along their whole length and they form big prominent ridges. The segments with maximal values correspond to locations where papillary muscles reside. Having this in mind, one can observe from the standard deviation plot that coarseness of trabeculations or sponginess of the heart increases toward the apex. That coincides with observations of a highly trabeculated apical region where trabeculations form a complex interwoven network.

4   Discussion

We proposed an bijective mapping of the detailed LV anatomy to a common reference frame represented by a cylinder with unit disk base. Every point inside the LV cardiac chamber is characterized by 3 coordinates: two coordinates characterizing the point's projection onto the anatomical bounding surface of the chamber and the normalized distance to that surface along a ray. The surface is then flattened by a quasi-conformal transformation to a disk whereupon any point is characterized by two coordinates. The choice of the disk as a parametric domain was motivated by the fact that such domain is well suited, and is traditional, for visualizations or mapping of any information from LV cavities onto a discretized or continuous bulls-eye plot [22].

5   Conclusion

The proposed method represents a major step forward in the analysis and quantification of different detailed cardiac morphologies among patients in a common framework with the potential application in computational cardiac models and construction of the atlas of trabeculations.

Acknowledgments. The Dicom datasets were provided by the Visible Heart® Laboratory [17], they were obtained by MRI scanning perfusion fixed hearts that were graciously donated by the organ donors and their families through LifeSource. B. Paun is supported by the grant FI-DGR 2014 (2014 FI B01238) from the Generalitat de Catalunya. The research leading to these results has received funding from the EU FP7 for research, technological development and demonstration under grant agreement VP2HF (no. 611823). This study was partially funded by the Spanish Ministry of Science and Innovation (TIN2011-28067).

References

1. Wilcox, B., Cook, A., Anderson, R.H.: Anatomy of the cardiac chambers. In: Surgical Anatomy of the Heart, pp. 11–44 (2004)
2. Oechslin, E., Jenni, R.: Left ventricular non-compaction revisited: a distinct phenotype with genetic heterogeneity? Eur. Heart J. **32**(12), 1446–1456 (2011)
3. Gao, M., Chen, C., Zhang, S., et al.: Segmenting the papillary muscles and the trabeculae from high resolution cardiac CT through restoration of topological handles. In: Gee, J.C., Joshi, S., Pohl, K.M., Wells, W.M., Zöllei, L. (eds.) IPMI 2013. LNCS, vol. 7917, pp. 184–195. Springer, Heidelberg (2013)
4. Gao, M., Chen, C., Zhang, S., et al.: Morphological analysis of the papillary muscles and the trabeculae. In: Proceedings of IEEE 11th International Symposium on Biomedical Imaging (ISBI), pp. 373–376 (2014)
5. Bishop, M., Plank, G., Burton, R., et al.: Development of an anatomically detailed mri-derived rabbit ventricular model and assessment of its impact on simulations of electrophysiological function. Am. J. Physiology-Heart Circulatory Physiol. **298**(2), 699–718 (2010)
6. Lévy, B., Petitjean, S., Ray, N., et al.: Least squares conformal maps for automatic texture atlas generation. ACM Trans. Graph. **21**(3), 362–371 (2002)
7. Jin, M., Kim, J., Luo, F., Gu, X.: Discrete surface ricci flow. IEEE Trans. Vis. Comput. Graph. **14**(5), 1030–1043 (2008)

8. Desbrun, M., Meyer, M., Alliez, P.: Intrinsic parameterizations of surface meshes. Comput. Graph. Forum **21**(3), 209–218 (2002)
9. Ng, T.C., Gu, X., Lui, L.M.: Computing extremal teichmüller map of multiply-connected domains via beltrami holomorphic flow. J. Sci. Comput. **60**, 249–275 (2014)
10. Gu, X., Wang, Y., Chan, T.F., et al.: Genus zero surface conformal mapping and its application to brain surface mapping. IEEE Trans. Med. Imaging **23**(8), 949–958 (2004)
11. Joshi, A.A., Shattuck, D.W., Thompson, P.M., et al.: Surface-constrained volumetric brain registration using harmonic mappings. IEEE Trans. Med. Imaging **26**(12), 1657–1669 (2007)
12. Wang, Y., Shi, J., Yin, X., et al.: Brain surface conformal parameterization with the ricci flow. IEEE Trans. Med. Imaging **31**(2), 251–264 (2012)
13. Vera, S., Ballester, M.A.G., Gil, D.: Anatomical parameterization for volumetric meshing of the liver. In: Proceedings of SPIE, vol. 9036, pp. 903605–903605 (2014)
14. Karim, R., Ma, Y., Jang, M., et al.: Surface flattening of the human left atrium and proof-of-concept clinical applications. Comput. Med. Imaging Graph. **38**(4), 251–266 (2014)
15. Garcia-Barnes, J., Gil, D., Badiella, L., et al.: A normalized framework for the design of feature spaces assessing the left ventricular function. IEEE Trans. Med. Imaging **29**(3), 733–745 (2010)
16. De Craene, M., Tobon-Gomez, C., Butakoff, C., et al.: Temporal diffeomorphic free form deformation (TDFFD) applied to motion and deformation quantification of tagged MRI sequences. In: Camara, O., Konukoglu, E., Pop, M., Rhode, K., Sermesant, M., Young, A. (eds.) STACOM 2011. LNCS, vol. 7085, pp. 68–77. Springer, Heidelberg (2012)
17. T.V.H. Laboratory, January 2015. http://www.vhlab.umn.edu
18. CIBC, "Seg3d: Volumetric image segmentation and visualization," Scientific Computing and Imaging Institute (SCI) (2015)
19. Squillacote, A.H.: The Paraview Guide: A Parallel Visualization Application. Kitware Inc., New York (2007)
20. Attene, M., Falcidieno, B.: Remesh: an interactive environment to edit and repair triangle meshes. In: Proceedings of IEEE International Conference on Shape Modeling and Applications, pp. 41–41 (2006)
21. Schroeder, W., Martin, K., Lorensen, B.: The Visualization Toolkit, 4th edn. Kitware Inc., Clifton Park (2006)
22. Cerqueira, M.D., Weissman, N.J., Dilsizian, V., et al.: Standardized myocardial segmentation and nomenclature for tomographic imaging of the heart a statement for healthcare professionals from the cardiac imaging committee of the council on clinical cardiology of the american heart association. Circulation **105**(4), 539–542 (2002)

Application of Diffuse Optical Reflectance to Measure Myocardial Wall Thickness and Presence of Infarct Scar: A Monte Carlo Simulation Study

Yee Chia Tang and Martin J. Bishop[✉]

Department of Biomedical Engineering, Division of Imaging Sciences,
King's College London, London, England
martin.bishop@kcl.ac.uk

Abstract. Catheter ablation in patients suffering from chronic arrhythmias often requires detailed knowledge of the specific myocardial anatomy underlying the catheter tip to guide the delivery of the ablating RF energy. Such information is often lacking in a clinical procedure. In this study, we present a proof-of-concept computational investigation into the potential for using an optical strategy, based-on diffuse optical reflectance, to provide quantitative anatomical measures of underlying myocardial wall thickness and presence of scar. In detailed Monte Carlo simulations of light scattering, significant changes in peak height and FWHM of radial profiles of diffusely reflected light were seen with both changes in tissue thickness and presence of underlying scar. Such changes were seen to occur to different degrees for different wavelengths of light. In conclusion, our findings suggest that examining the ratio of these changes between different wavelengths may provide the potential basis for an optical-catheter to map underlying cardiac anatomy and guide ablation in the clinic.

1 Introduction

Catheter ablation is a common clinical therapy to treat patients suffering from chronic arrhythmias of both the atria and ventricles. Ablation involves the application of RF-energy through a catheter tip to destroy the electrical functioning of the underlying myocardial tissue, thus removing important arrhythmic conduction pathways.

During atrial ablations, it is important to know the thickness of myocardial tissue at the exact point at which the RF-energy is delivered; too little energy and the ablation legion may not be sufficient to prevent the arrhythmia; too much energy and potentially lethal perforation of the wall may occur [1]. Due to its thin nature, accurate measures of atrial wall thickness from imaging modalities such as CT or MR are highly problematic; even if present, such data would be hard to correlate or register the anatomy from a pre-operative scan to the real-time catheter location during a procedure.

H. van Assen et al. (Eds.): FIMH 2015, LNCS 9126, pp. 248–255, 2015.
DOI: 10.1007/978-3-319-20309-6_29

In the ventricles, ablation procedures attempt to interrupt essential slow conduction isthmuses through infarct scar tissue that provide the arrhythmic substrate [2]. In this scenario, detailed knowledge of the location, depth and thickness of scar tissue beneath the catheter is vital to identify ablation targets with optimal lesion creation. Identifying scar location is currently done with pre-operative MR scans, but is only suitable for patients without implanted electrical devices. During the procedure in the cath-lab, electro-anatomical maps may also be created by roaming catheters recording extracellular potentials from the endocardial surface [3], with regions of low-potential being identified as scar. However, the resolution of such recordings is limited, and it is also not fully understood how subsurface scar tissue (with healthy tissue on the endocardial surface above it) is represented in such maps.

Novel catheters using ultrasound probes have recently showed promising results for assessing wall thickness and monitoring lesion formation during ablation [4]. However, despite its widespread use in basic science anatomical imaging, optical methods have, to our knowledge, so far not been investigated for use in this context. In this study, we investigate the potential of diffuse optical reflectance as a method of reliably detecting myocardial tissue wall thickness and the presence of underlying scar.

2   Methods

Theory. Diffuse optical reflectance refers to light which exits from the initial illuminated surface after having scattered (and diffused) through the tissue. Although every path taken by the photons through the tissue is different (due to stochastic scattering events), light which exits the tissue at a further distance from the initial source location will have, on average, passed into deeper tissue depths on its journey before returning to exit the illuminated surface. Such an effect, shown in Fig. 1A, is often termed the 'banana effect' due to the shape of the most common photon paths. The specific shape of these 'banana' profiles and the intensity of light exiting the surface at different radial distances from the source are governed by the optical absorption and scattering properties of the tissue through which the light passes. Due to the different path lengths and depths of light exiting the surface at different locations, information regarding localised differences in anatomical structure and optical properties of the tissue in the vicinity of the photon source may be transduced as changes in the profile of detected light at different radial distances from the initial source. In addition, different wavelengths of light often have correspondingly very different optical absorption and scattering properties within different biological tissue [5]. For example, light with longer wavelengths typically penetrates more deeply, undergoing fewer absorption and scattering events, thus altering the shape of the 'banana' and the profiles of the recorded intensity with distance from the source. These properties of diffuse reflectance have seen it used in other areas of biomedical imaging such as brain and breast imaging [6].

Photon Propagation. A Monte Carlo algorithm of the step-by-step propagation and interaction of photons through cardiac tissue was used to simulate diffuse optical reflectance. The fundamental Monte Carlo algorithm used was based-on that of Wang et al. (1995) [7] for use in cardiac tissue [8]. The algorithm describes the transport of photons through multi-layered biological tissue within a structured domain discretized into equal cubic optical elements into which physical quantities are stored. Photons are moved as 'packets' containing large numbers of individual photons represented by the packet weight, W. During propagation, individual scattering, absorption, transmission and reflection events are simulated, governed by the optical properties of the tissue.

Tissue Optical Properties. Optical properties of the tissue are characterised by the absorption coefficient (μ_a), the scattering coefficient (μ_s), the anisotropy scattering coefficient (g) and the refractive index (n). Three different wavelengths of light were considered based-on their known optical properties: green (532 nm), red (660 nm) and near infra-red (NIR, 745 nm). Specific optical properties of healthy myocardium at these wavelengths were taken from [9], being: green $\mu_a = 0.89\,\text{mm}^{-1}$, $\mu_s = 14.8\,\text{mm}^{-1}$; red $\mu_a = 0.07\,\text{mm}^{-1}$, $\mu_s = 11.8\,\text{mm}^{-1}$; NIR $\mu_a = 0.05\,\text{mm}^{-1}$, $\mu_s = 10.8\,\text{mm}^{-1}$. At all wavelengths, $g = 0.95$ and $n = 1.4$ for myocardium was used, with $n = 1.0$ for the surrounding medium. Optical properties of scar tissue were approximated by reducing μ_a and μ_s by 50 % [10].

Model Setups and Protocols. A 3D wedge of myocardium was used measuring $30 \times 30 \times 5\,\text{mm}$. Photon packets were incident perpendicular to the $z = 0$ surface in the centre of the xy-plane (at $(15, 15, 0)\,\text{mm}$). A total of 10^6 photon packets (each with initial weight of $W = 1.0$) were incident for each simulation. The specific nature of the device responsible for inputting photons (laser, LED, etc.) is not relevant to the model, so long as the photons are of the same wavelength. Those packets exiting from the initial illuminated surface (i.e. the $z = 0$ plane) were detected, with total packet weight binned according to the radial distance at the point of exit (r) from the initial source of illumination, shown in Fig. 1A. Later, scar tissue was represented within the tissue as a layer parallel to the illuminating surface of varying thickness and depth beneath the surface, as shown in Fig. 1B.

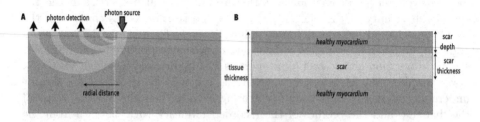

Fig. 1. (A) Demonstration of the 'banana' effect. (B) Tissue setup including scar.

3   Results

3.1   Ability to Discern Tissue Thickness of Healthy Myocardium

Initially, we examined how light of different wavelength, with significantly different associated optical properties, may produce different radial profiles of diffusely reflected light. Figure 2 shows radial profiles of detected photons for the three different wavelengths for healthy tissue thickness of 3 mm. After initially peaking, all profiles have a characteristic decrease in detected weight as radial distance increases. Such profiles are expected as photons exiting the tissue surface from increasingly further distances from the source have been attenuated to a greater extent. However, as the wavelength increases (from green to NIR), both the peak height is seen to noticeably increase (more total photons are detected) and the width of the profiles broadens. Specifically, peak height increases from approximately 1.5×10^3 at green to 2.4×10^4 at NIR, with FWHM increasing from 1.5 mm at green to 2.4 mm at NIR. The increased peak height at longer wavelengths and increased profile spread is due to the decreased absorption coefficient, meaning photon packets travelling similar distances are attenuated less compared to shorter wavelengths.

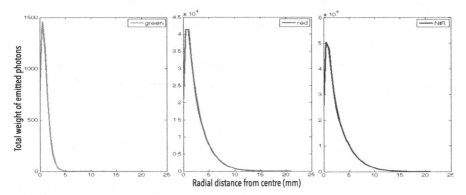

Fig. 2. Profiles of total emitted photon weight versus radial distance from source for green (*left*), red (*centre*) and NIR (*right*) light for a tissue thickness of 3 mm (Color figure online).

We now consider how these respective radial profiles for different wavelengths may change upon changing the thickness of the tissue. Figure 3 shows that the peak heights of the radial profiles (panel A) decrease and the FWHMs (panel C) increase for all three wavelengths as the thickness of the tissue sample increases from 1–7 mm. For thin geometries, backscattering of photon packets from the lower surface increase the intensity of light that successfully exits the illuminated surface due to diffuse reflectance. At longer wavelengths, this continues to have an effect for thicker geometries, due to the relatively lower absorption coefficient compared to green light. Consequently, this decrease in peak height with tissue thickness is by far most significant in the case of the green light, but is still pronounced for red and NIR light.

Given these significant changes, we now examine how the ratios of peak heights (panel B) and FWHMs (panel D) between pairs of wavelengths changes with tissue thickness. Although little differences are seen in the ratio of NIR/red, the peak height ratio of both red/green and NIR/green show changes of over 3 orders of magnitude over this tissue thickness range (panel B). An increase in FWHM ratio was similarly seen (panel D), although this was less significant than the peak height.

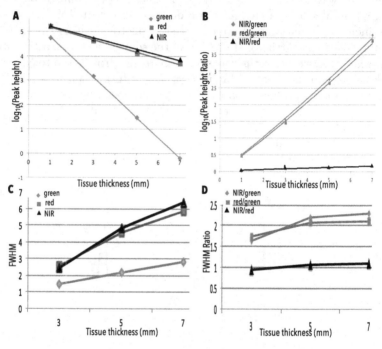

Fig. 3. (A) Changes in peak heights of radial profiles for the three wavelengths as a function of tissue thickness. (B) Ratio of peak heights between wavelength pairs as a function of tissue thickness. Note: log_{10} on y-axes. (C) and (D) show similar plots for FWHM.

3.2   Ability to Discern Presence, Location and Thickness of Scar Tissue

Figure 4A shows the radial profiles of detected photons for the three wavelengths in the cases of two different scar depths (1 mm and 2 mm beneath the surface) and a constant scar thickness of 3 mm. As in the case of healthy myocardium (no scar) above (Fig. 2), the red and NIR profiles exhibit much higher peaks and broader decay-profiles compared to the green light. For different scar depths, little discernible difference is seen in the green light profiles; however, for red and NIR light, the profiles reach a slightly higher peak and decay slightly faster in the case of scar depth 2 mm compared to 1 mm depth.

Upon varying the thickness of the scar, all wavelengths showed very little difference in the peak heights as the thickness of the scar was varied (at either scar depths of 1 mm or 2 mm), as shown in Fig. 4B. Similarly, little change in FWHM with scar thickness was also seen.

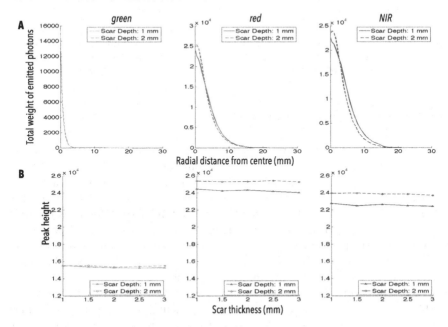

Fig. 4. (A) Profiles of total emitted photon weight versus radial distance from source for green (*left*), red (*centre*) and NIR (*right*) light for scar depths of 1 mm (solid lines) and 2 mm (dashed-lines). All plots are for scar thickness of 3 mm. (B) Peak heights from similar radial profiles as scar thickness is varied between 1–3 mm (Color figure online).

However, the actual presence of scar tissue itself did affect both peak height and FWHM relative to the case of healthy tissue. In the presence of scar, peak height of the radial profiles were consistently higher than the case of healthy tissue for all wavelengths and all scar thicknesses and depths. Figure 5 compares the relative peak height (panel A) and FWHM (panel B) for the case of a scar being present compared to healthy myocardium (no scar). Here we notice that this change in peak height is particularly apparent for green light, where the peak height was > 500 times larger in the presence of scar compared to healthy tissue. For red and NIR light, the peak height in the presence of scar was only approximately 1.9 and 1.4 times higher in the presence of scar, respectively. Interesting, in the case of green light, the FWHM was approximately 40 % smaller in the presence of scar compared to healthy tissue, whereas for red and NIR the FWHM was 50–60 % larger. In all cases, as in Fig. 4B, there was little variation with different scar thicknesses and depths.

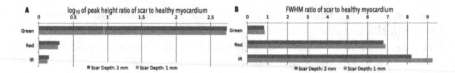

Fig. 5. Comparison of ratio of peak heights (panel A) and FWHM (panel B) of radial profiles in the presence of scar tissue compared to healthy tissue for the three different wavelengths. Note: log_{10} scale in panel A. Scar depths of 1 mm (red) and 2 mm (blue) are considered. Scar thickness is 3 mm in all cases (Color figure online).

4   Discussion and Conclusions

In this proof-of-concept study, we have used computational simulations of photon transport within tissue to demonstrate the strong potential of using optical techniques based-on diffuse reflectance to derive important quantitative measures of myocardial wall anatomy. We have shown that the parameters defining the profiles of diffusely-reflected light show significant relative differences between different wavelengths of light due to their inherently different optical properties and tissue penetration depths. As these relative differences were seen to depend on underlying tissue thickness and presence/absence of scar, they could be used as reliable indicators of these anatomical properties.

As calibration of exact values of peak height or FWHM at a specific wavelength may be problematic, here we focused on calculating ratios between different combinations of pairs of wavelengths of these parameters and examined how these relative ratios changed with different underlying myocardial wall anatomy. For example, although it may not be possible to use such a technique to find an exact quantitative value of myocardial wall thickness, the noted significant change in relative peak height and FWHM between green light compared to red or NIR with changing wall thickness may be able to provide a relative normalised representation of wall thickness. This would therefore provide guidance as to regions of the myocardium which have relatively thicker or thinner walls than elsewhere.

Unfortunately, we did not see any significant difference in relative peak height or FWHM for different scar depths or scar thicknesses; thus, using such a method to accurately discern these parameters may not be possible. However, relative changes in both peak height and FWHM between healthy myocardium and where scar was present were significant, particularly at the green wavelength. This suggests that looking for these changes may be a reliable method to identify the presence of subsurface scar. Such a technique may be of particular benefit in identifying intramural scar in cases where standard surface-based electro-anatomical mapping strategies may not perform well. We note, however, that values used for the optical properties of the scar were estimated in this present study [10]; future work will involve their more detailed experimental measurement at specific wavelengths which could alter our findings related to scar.

The simulations here were conducted in highly simplified models of myocardium and scar. Application of similar methodology to more anatomically-complex MR-based geometries, utilising newly-developed Monte Carlo photon scattering simulation methods on unstructured finite element geometries [11], will be essential to assess the validity of our findings in more structurally-realistic scenarios. Such an approach will also allow careful validation of these results with experimental optical imaging using optrodes to deliver and record light in wedge preparations.

Ultimately, the application of these findings will depend upon the technological advancement in a number of areas. However, such advances are currently taking place. Therefore, this work, and its future developments and validation, will play an important role in the potential development of an optical catheter for quantifying relative myocardial wall thickness and presence or absence of underlying scar for use in clinical ablation procedures.

References

1. Bunch, T., Asirvatham, S., Friedman, P., Monahan, K., Munger, T., Rea, R., Sinak, L., Packer, D.: Outcomes after cardiac perforation during radiofrequency ablation of the atrium. J. Cardiovasc. Electrophysiol. 16, 1172–1179 (2005)
2. Stevenson, W.: Ventricular tachycardia after myocardial infarction: from arrhythmia surgery to catheter ablation. J. Cardiovasc. Electrophysiol. 6, 942–950 (1995)
3. Stevenson, W.G., Soejima, K.: Catheter ablation for ventricular tachycardia. Circulation 115, 2750–2760 (2007)
4. Wright, M., Harks, E., Deladi, S., Suijver, F., Barley, M., van Dusschoten, A., Fokkenrood, S., Zuo, F., Sacher, F., Hocini, M., Haissaguerre, M., Jais, P.: Real-time lesion assessment using a novel combined ultrasound and radiofrequency ablation catheter. Heart Rhythm 8, 304–312 (2011)
5. Cheong, W.F., Prahl, S.A., Welch, A.J.: A review of the optical properties of biological tissues. IEEE Quant. Electron. 26, 2166–2185 (1990)
6. Gibson, A., Hebden, J., Arridge, S.: Recent advances in diffuse optical imaging. Phys. Med. Biol. 50, R1–R43 (2005)
7. Wang, L., Jacques, S.L., Zheng, L.: MCML–Monte carlo modeling of light transport in multi-layered tissues. Comput. Methods Programs Biomed. 47, 131–146 (1995)
8. Bishop, M.J., Bub, G., Garny, A., Gavaghan, D.J., Rodriguez, B.: An investigation into the role of the optical detection set-up in the recording of cardiac optical mapping signals: a Monte Carlo simulation study. Physica D 238, 1008–1018 (2009)
9. Walton, R.D., Benoist, D., Hyatt, C.J., Gilbert, S.H., White, E., Bernus, O.: Dual excitation wavelength epifluorescence imaging of transmural electrophysiological properties in intact hearts. Heart Rhythm 7, 1843–1849 (2010)
10. Splinter, R., Svenson, R.H., Littmann, L., Tuntelder, J.R., Chuang, C.H., Tatsis, G.P., Thompson, M.: Optical properties of normal, diseased, and laser photocoagulated myocardium at the Nd: YAG wavelength. Lasers Surg. Med. 11, 117–124 (1991)
11. Bishop, M.J., Plank, G.: Simulating photon scattering effects in structurally detailed ventricular models using a Monte Carlo approach. Front. Physiol. 5, 1–14 (2014)

Automated Quantification of Myocardial Infarction Using a Hidden Markov Random Field Model and the EM Algorithm

M. Viallon[1,2], Joel Spaltenstein[3], C. de Bourguignon[2], C. Vandroux[1],
A. Ammor[1], W. Romero[1], O. Bernard[1], P. Croisille[1,2], and P. Clarysse[1](✉)

[1] Université de Lyon, CREATIS, CNRS UMR5220, Inserm U1044,
INSA-Lyon, Université Lyon 1, Lyon, France
patrick.clarysse@creatis.insa-lyon.fr
[2] CHU Saint Etienne, Université Jean Monnet, Saint-Étienne, France
[3] Spaltenstein Natural Image, Geneva, Switzerland

Abstract. Infarct size has been recognized as a good indicator of the functional status of the ischemic heart and to evaluate the impact of myocardial infarction therapies. Its assessment can be performed from late gadolinium enhancement magnetic resonance images. A number of methods have been proposed for the semi-automatic and automatic quantification of necrosis. We developed an automatic method based on a Markov random field framework and a region growing approach within an EM optimization, which enables segmentation of both necrosis and microvascular obstructions. The method has been evaluated on both synthetic data and 10 clinical cases in 3D and lead to the best results as compared to other conventional approaches and expertise.

1 Introduction

Characterization of myocardial tissues is essential to establish an accurate diagnosis and to predict the functional outcome of ischemic patients. Cardiac magnetic resonance imaging (CMRI) is particularly efficient in this respect. Late Gadolinium Enhancement (LGE) CMRI is recognized as a reference for myocardial viability assessment [1]. However, infarct size quantification can be a tedious and subjective task. Reperfusion can lead to deleterious additional lesions such as microvascular obstruction (MVO or no reflow phenomenon) which has been shown to be associated with adverse Left Ventricular (LV) remodeling [2]. Figure 1 shows a typical LGE image where necrosis appears as a hyperenhanced area surrounding a hypointense core corresponding to a lesion.

Several methods have been proposed to estimate the relative part of normal and altered tissues. They can be categorized in 3 types. *Threshold based methods* evaluate a unique threshold to partition pixels in two classes: pixels belonging to normal myocardium (intensity below the threshold) and pixels associated to necrosis (intensity above the threshold). The most common method evaluates a mean μ and standard deviation σ intensity in a remote myocardial region.

© Springer International Publishing Switzerland 2015
H. van Assen et al. (Eds.): FIMH 2015, LNCS 9126, pp. 256–264, 2015.
DOI: 10.1007/978-3-319-20309-6_30

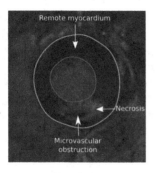

Fig. 1. (left) Typical short axis LGE-MR image where the different types of tissues can be seen within the endocardial (inner) and epicardial (outer) contours. (right) Polar deployment of the LV in the middle image with angle along the vertical axis (myocardium is included between the two contours with LV cavity on the left part)

The threshold is then defined as $T = \mu + c\sigma$ with c a constant between 2 and 10. The *Full Width at Half Maximum* evaluates the threshold as half of the highest intensity in the myocardium: $T = I_{max}/2$. The gaussian mixture model (GMM) defines the threshold as the intersection of two gaussian distributions fitted from the pixel histogram within the myocardium. *Classification methods* partition the myocardium into several components, basically healthy myocardium, infarct and microvascular obstruction. Fuzzy C-means [3] and support Vector Machine [4] methods have been proposed in this context. *Combined methods* associate several technics within a workflow. Hsu et al. [5,6] and Valindria et al. [7] combine threshold based methods and a characteristic analysis to limit false positive. Surprisingly, none of these methods makes use of the local neighborhood and *a priori* regarding the location of the respective tissue categories. The proposed approach combines a region growing approach within a *maximum a posteriori* (MAP) optimisation through an Expectation-Maximization (EM) algorithm.

2 Method

Our approach is inspired from the HMRF-EM method by Zhang et al. for the segmentation of brain images [8]. Zhang's method estimates the parameters of a statistical model from a maximum likelihood estimation with an expectation maximization (EM) algorithm. Let a 2D or 3D image $\mathbf{y} = \{y_1, ..., y_N\}$ and a hidden variable $\mathbf{x} = \{x_1, ..., x_N\}$, called the label map with N the number of pixels. The endocardial and epicardial contours are supposed to be known as a result from a preliminary manual tracing or (semi)automatic segmentation. In the context of cardiac tissue segmentation, a domain transform is first applied. An initial label map and the corresponding parameters are computed.

2.1 Cartesian to Polar Transformation

We chose to work in polar space for the following reasons (Fig. 1(right)). First, a polar development is well adapted to the LV morphology in short axis views (see Fig. 1(left)). Second, the same weight is attributed to inner and outer pixels/voxels in the polar system while inner layers tend to be underweighted in the Cartesian system. Also, axes for neighboring operations (e.g. morphological) are aligned with the radial and circumferential anatomical directions in the polar system. Last, the 2D image size in Cartesian space is generally 256×256 while in the transformed polar space, it is around 50×360 making the processing faster in 2D and 3D. In 3D, this transformation is applied to all the image slices to produce a 3D stack of the unfolded LV myocardium.

2.2   Initial Classification

A first guess for the label map is obtained by extracting the pixel with the most likelihood to belong to the necrosis class. This is obtained with a drastic threshold to keep hyper intense pixels. In practice, we took $T = 4I_{max}/5$ with I_{max} the maximum intensity over the whole myocardium. Initial mean and standard deviation for the two zones are computed leading to the initial label map $\mathbf{x}^0$ and models' parameters $\mathbf{\Theta}^{(0)}$.

2.3   EM Algorithm

The classification problem consists in determining the optimal label map $\mathbf{x}^*$ and the statistical model parameters $\mathbf{\Theta}^* = \{\theta_l = (\mu_l, \sigma_l), l \in \mathcal{L}\}$, with $\mathcal{L}$ being the set of labels given the observed data $\mathbf{y}$. The EM algorithm seeks for a maximum likelihood estimate in two steps. The *expectation* step computes the expectation of the log likelihood function relatively to the conditional distribution $P(\mathbf{x}|\mathbf{y})$ with the current estimates of the statistical model parameters $\mathbf{\Theta}^{(t)}$. The *maximization* step updates the models' parameters that maximize the expectation. The whole algorithm is summed up in Fig. 2.

Step E. The E step determines the conditional expectation $Q(\mathbf{\Theta}|\mathbf{\Theta}^{(t)})$ given by [8]:

$$Q(\mathbf{\Theta}|\mathbf{\Theta}^{(t)}) = E(\; ln[\; P(\mathbf{x}, \mathbf{y}|\mathbf{\Theta})\;]\;|\;\mathbf{y}, \mathbf{\Theta}^{(t)}\;) \tag{1}$$

$$= \sum_x P(\mathbf{x}|\mathbf{y}, \mathbf{\Theta}^{(t)}) ln P(\mathbf{x}, \mathbf{y}|\mathbf{\Theta}) \tag{2}$$

According to the Bayes rule, the *a posteriori* probability of $\mathbf{x}$ given $\mathbf{y}$ and $\mathbf{\Theta}^{(t)}$ reads:

$$P(\mathbf{x}|\mathbf{y}, \mathbf{\Theta}^{(t)}) = \frac{P(\mathbf{y}|\mathbf{x}, \mathbf{\Theta}^{(t)}) P(\mathbf{x}, \mathbf{\Theta}^{(t)})}{P(\mathbf{y}, \mathbf{\Theta}^{(t)})} \tag{3}$$

Within a Markov Random Field (MRF) framework, both $\mathbf{x}$ and $\mathbf{y}$ are considered as realization of random fields $X = \{X_i, i \in \mathcal{S}\}$ and $Y = \{Y_i, i \in \mathcal{S}\}$, respectively ($\mathcal{S}$ is the set of sites of the MRF, and can be assimilated to pixels here). Using the conventional Gaussian distribution for the probabilities and assuming the conditional independence of sites, the maximization of (3) comes to the minimization of a sum of energies at each site i:

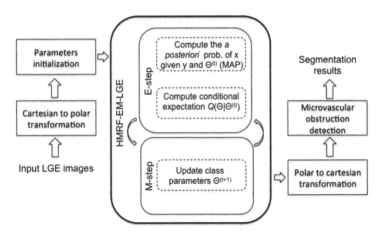

Fig. 2. Flowchart of the proposed method for the identification of the infarct size and the MVO

$$x_i^{k+1} = \underset{l \in \mathcal{L}}{argmin}\{U(y_i|x_i^k, l) + U(x_{N_i})\} \tag{4}$$

where i, k, N_i represent the site index, the MAP iteration and the list of neighboring sites of site i, respectively. With the gaussian assumption, the conditional likelihood energy can be written as:

$$U(y_i|x_i^k, l) = \frac{(y_i - \mu_l)^2}{2\sigma_l^2} + ln(\sigma_l) \tag{5}$$

Within MRF framework, the *a priori* energy of the label map $U(x_{N_i})$ is computed from the clique potential $V_c(l, x_j)$ which insures the local coherence of the labels:

$$U(x_{N_i}) = \sum_{\substack{j \in N_i \\ c \in \mathcal{C}}} V_c(l, x_j) \tag{6}$$

With $\mathcal{C}$ the set of cliques. In 3 dimensions, the clique potential is taken as all the pairs constituted by the current site i and each of its neighbors $j \in N_i$ with N_i the neighborhood of site i (26 resp. 8 neighbors in 3D resp. 2D).

$$V_c(l, x_j) = \frac{1}{2}(1 - I_{l,x_j}) \tag{7}$$

$$I_{l,x_j} = \begin{cases} 0 \ si \ l = x_j \\ 1 \ si \ l \neq x_j \end{cases} \tag{8}$$

Equation (4) is minimized iteratively over all the sites. Bordering pixels of the current necrosis regions in the label map are screened to update the total energy and label for those pixels is attributed (only pixels connected to a necrosis region are considered). Iterations are repeated until the stagnation of the energy or a maximum of K iterations is attained.

From the new label map, the conditional expectation (2) is estimated.

Step M. The parameters of the classes are updated to maximize the conditional expectation $Q(\Theta|\Theta^{(t)})$:

$$\mu_l^{(t+1)} = \frac{\sum_{i\in\mathcal{S}} P^{(t)}(l|y_i)y_i}{\sum_{i\in\mathcal{S}} P^{(t)}(l|y_i)} \tag{9}$$

$$(\sigma_l^{(t+1)})^2 = \frac{\sum_{i\in\mathcal{S}} P^{(t)}(l|y_i)(y_i - \mu_l^{(t+1)})^2}{\sum_{i\in\mathcal{S}} P^{(t)}(l|y_i)} \tag{10}$$

A *posteriori* probability for label l given y_i at each site i is computed from the Bayes rule:

$$P^{(t)}(l|y_i) = \frac{G^{(t)}(y_i, \theta_l)P^{(t)}(l|x_{N_i})}{P(y_i)} \tag{11}$$

With

$$G^{(t)}(y_i, \theta_l) = \frac{1}{\sqrt{2\pi\sigma_l^2}}exp(-\frac{(y_i - \mu_l)^2}{2\sigma_l^2}) \tag{12}$$

and

$$P^{(t)}(l|x_{N_i}) = \frac{1}{Z}exp(-\sum_{j\in N_i} V_c(l, x_j)) \tag{13}$$

The clique potential is computed the same way as before. Also

$$P(y_i) = \sum_{l\in L} G^{(t)}(y_i, \theta_l)P^{(t)}(l|x_{N_i}) \tag{14}$$

Both E and M steps are iterated until convergence or a maximum iteration number T is attained.

2.4   Post-processing and Identification of Microvascular Obstructions

Cluster Selection. Polar to cartesian transformation is applied to get back to the image domain. In order to reduce false positives, a connected component analysis is applied to remove small clusters (i.e. less than 0.1g) or isolated clusters too distant from the endocardium (distance above 1.5mm).

Microvascular Obstructions Detection. MVO detection proceeds on a slice by slice basis. A MVO region can be completely included within a necrosed region or adjoin to myocardial borders. Connected components of the myocardial region less the necrosis are labeled. Those that are bigger than 50 % the myocardium size are removed. Components such that the percentage of external bordering pixels that are not labeled necrosis is above a certain threshold (set to 50 %) are removed as well.

3   Results

The algorithm, named HMRF-EM-CMR, has been implemented within an OsiriX[1] plugin dedicated to the segmentation of CMR images: CMRSegTools[2]. Prior extraction of the endocardial and epicardial contours is achieved using the BEAS algorithm [9] followed eventually with some manual adjustments at each slice level. The respective percentage of healthy myocardium, necrosis and MVO is delivered for the whole stack of slices as can be seen in Fig. 3. HMRF-EM-CMR has been confronted to manual evaluation and compared to most of the above cited classical methods: mean $\pm$ x-standard deviations, Full Width at Half Maximum, GMM and an implementation of the Hsu's method. Evaluation has been conducted on both synthetic and clinical data. In the experiments, initial threshold is 80 % of the maximum intensity, maximum MAP and EM iteration numbers are set to 7 (K) and 25 (T) respectively.

Evaluation on Synthetic Data. 3D synthetic data has been designed from the contouring of a clinical image stack. Myocardium is approximated by an annulus. Several data sets were generated with varying infarct size, transmurality, percent signal enhancement (%SE between 1.5 and 9) and contrast-to-noise ratio (CNR between 1 and 31 defined as $(I_{necrosis} - I_{remote})/\sigma_{noise}$). The results

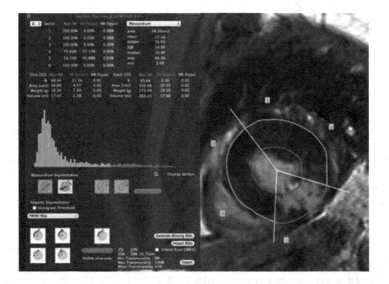

Fig. 3. View of the result window at one slice level. Myocardial histogram on the left and identified necrosis within the angular sector delimited by the acute angle on the right. Amount of respective tissues within 6 anatomical sectors is given in the table on the top left.

---

[1] www.osirix-viewer.com.

[2] https://www.creatis.insa-lyon.fr/CMRSegTools/.

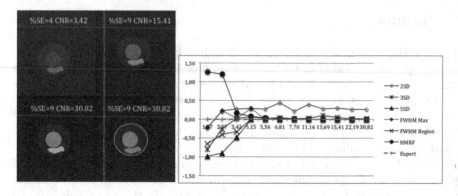

Fig. 4. (left) Three example images of the digital phantom with various signal enhancement and CNR and the reference contours, notably the necrosis in pink. (right) Results obtained with HMRF-EM-CMR as compared to classical methods (threshold based with 2, 3 and 5 standard deviations, maximum and region based FWHM). Accuracy is expressed as the relative error in necrosis size as a function of CNR

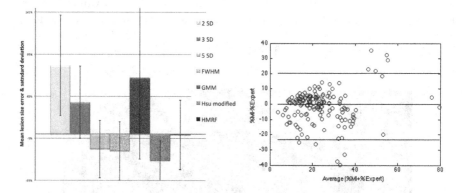

Fig. 5. (left) Mean lesion size error (in % of the total infact size including necrosis (hyperenhanced) and MVO (hypoenhanced core)) over 10 cases of the HMRF-EM-CMR algorithm as compared to classical methods in the field: threshold based with 2, 3 and 5 standard deviations, FWHM, Gaussian mixture model (GMM), and an implementation of the Hsu's method. (right) Bland Altman plot for HMRF-EM-CMR method for the 10 cases (each dot corresponds to one slice level)

obtained with HMRF-EM-CMR are directly compared to the exact regions definition used to build the phantom (Fig. 4). As shown on Fig. 4(right), HMRF-EM-CMR obtained almost perfect results for CNR>4 (CNR for standard clinical CMR-LGE images can be estimated around 10).

Evaluation on Clinical Data. Ten multi-slice (about 20 slices) CMR-LGE sequences (from datasets of late contrast enhanced images obtained using a gradient echo sequence) have been processed with HMRF-EM-CMR. The results are compared to those obtained by two experts who delineated the endocardial and epicardial contours, and the contours of the necrosed and MVO regions. Figure 5(left) shows the performance of our method over the 10 cases compared to several 'conventional' methods. The quality index is computed as the mean total lesion size (including MVO) error over the 10 cases relatively to the expertise. This shows the very good behavior of HMRF-EM-CMR method relatively to others with a tendency to slightly underestimate the necrosis size. The mean difference in infarct size, segmented by two independent experts without redrawing the endocardial and epicardial contours was $4.32\% \pm 5.6\%$.

4 Conclusion

The proposed algorithm combines several specific features that make it especially suitable for the myocardial tissue categorization for CMR-LGE images. First, the transformation from cartesian to polar coordinate system eases and speeds up the neighboring operations. The classification by itself relies on an elegant MRF framework which has demonstrated its power in other applications, notably brain tissues segmentation. Also, priors regarding the size and location of lesions has been taken into account to reduce false positives. The results obtained on synthetic data with various configurations of intensity contrast and noise are very good. On a 10 case database, the results obtained by our method compares quite favorably with the most known ones with the advantage of being fully automatic. The validation has to be pursued on a larger database. One important prerequisite is the availability of the endocardial and epicardial contours with a good accuracy. This is the reason why we conducted the comparison between methods from the same contours. MVO detection has to be improved as well since some parts are still missed. Computing time to process a 3D stack of about 20 levels is less than a minute and could be reduced with heuristics and limited calls to costly operations such as region growing. The OsiriX plugin integrating HMRF-EM-CMR method as well as other methods is an open source software. It allows for the first time common improvement and validation of new methods to better answer the urgent need for advanced automatic software within large clinical trials.

References

1. Arai, A.: The cardiac magnetic resonance approach to assessing myocardial viability. J. Nucl. Cardiol. **18**(6), 1095–1102 (2011)
2. Wu, K.C.: CMR of microvascular obstruction and hemorrhage in myocardial infarction. J. Cardiovasc. Magn. Reson. **14**, 68 (2012)
3. Kachenoura, N., Redheuil, A., Herment, A., Mousseaux, E., Frouin, F.: Robust assessment of the transmural extent of myocardial infarction in late gadolinium-enhanced MRI studies using appropriate angular and circumferential subdivision of the myocardium. Eur. Radiol. **18**(10), 2140–2147 (2008)

4. Positano, V., Pingitore, A., Giorgetti, A., Favilli, B., Santarelli, M.F., Landini, L., Marzullo, P., Lombardi, M.: A fast and effective method to assess myocardial necrosis by means of contrast magnetic resonance imaging. J. Cardiovasc. Magn. Reson. 7(2), 487–494 (2005)
5. Hsu, L.-Y., Natanzon, A., Kellman, P., Hirsch, G.A., Aletras, A.H., Arai, A.E.: Quantitative myocardial infarction on delayed enhancement MRI. Part I: animal validation of an automated feature analysis and combined thresholding infarct sizing algorithm. J. Magn. Reson. Imaging 23(3), 298–308 (2006)
6. Hsu, L.-Y., Ingkanisorn, W.P., Kellman, P., Aletras, A.H., Arai, A.E.: Quantitative myocardial infarction on delayed enhancement MRI. Part II: clinical application of an automated feature analysis and combined thresholding infarct sizing algorithm. J. Magn. Reson. Imaging 23(3), 309–314 (2006)
7. Valindria, V. V., Angue, M., Vignon, N., Walker, P. M., Cochet, A., Lalande, A.: Automatic quantification of myocardial infarction from delayed enhancement MRI. In: 2012 Eighth International Conference on Signal Image Technology and Internet Based Systems, pp. 277–283 (2011)
8. Zhang, Y., Brady, M., Smith, S.: Segmentation of brain MR images through a hidden markov random field model and the expectation-maximization algorithm. IEEE Trans. Med. Imaging 20(1), 45–57 (2001)
9. Barbosa, D., Dietenbeck, T., Schaerer, J., D'Hooge, J., Friboulet, D., Bernard, O.: B-spline explicit active surfaces: an efficient framework for real-time 3-D region-based segmentation. IEEE Trans. Image Process. 21(1), 241–251 (2012)

Cross-Frame Ultrasonic Color Doppler Flow Heart Image Unwrapping

Artem Yatchenko[✉] and Andrey Krylov

Laboratory of Mathematical Methods of Image Processing, Faculty of Computational Mathematics and Cybernetics, Lomonosov Moscow State University, Moscow, Russia
{yatchenko,kryl}@cs.msu.ru
http://imaging.cmc.msu.ru

Abstract. Ultrasonic color Doppler flow image unwrapping algorithm that uses cross-frame connection is proposed and compared with other unwrapping methods. An original complex phase preliminary filtration is used to suppress a false-aliasing artifact and to improve the results. Flow variances are used as weight coefficients in the minimization energy function. For the comparison a test data series is constructed. It uses an anatomic 3D left ventricle region model for the simulation of the blood flow. Experiments show that cross-frame weights significantly improve the quality of unwrapping.

Keywords: Ultrasound color doppler flow mapping · Cross-frame · Phase unwrapping · Heart · Blood flow model · Algorithms comparison

1 Introduction

Ultrasonic Color Doppler Flow Mapping (CDFM) is an imaging technique that uses ultrasonic pulse-echo signal processing to generate color-coded maps of tissue and blood velocity overlaid on grey-scale images of tissue anatomy. It assigns a given color to the direction of flow: red is flow toward and blue is flow away from the probe. CDFM has been found to be useful in many areas of cardiovascular investigation.

When the pulse frequency is low frame frequency is small and low-flow states might not be displayed. Conversely, when the pulse frequency is high a phase of the returned signal may wrap and flow aliasing occurs. In some regions wrapped and unwrapped areas join into one-color area. For example, it occurs in the case when an area with high positive velocity is wrapped and falsely detected as a part of the close negative velocity region. In these cases sharp border between regions can disappear and a smooth gradient occurs (we call this situation as a "hidden aliasing"). Contrariwise, due to the device noises a false aliasing may appear in the boundary between positive and negative velocities (we call this as a "false aliasing"). These cases will be later illustrated in the Sect. 2.

Solving the velocity aliasing problem is a key to the quantification purpose of the color Doppler blood flow images. The goal of the solution is to restore

H. van Assen et al. (Eds.): FIMH 2015, LNCS 9126, pp. 265–272, 2015.
DOI: 10.1007/978-3-319-20309-6_31

the original velocity values from the aliased ones and to estimate true blood velocities.

There are many different approaches for phase unwrapping problem solution: using graph-cut algorithm [1,2]; the minimization of the L_2 norm [3] and L_p norm [4] of partial field derivatives; using complex-valued random Markov field [5] and using regularizing conditions [6].

The aim of this work is to compare graph-cut and regularization based unwrapping methods and to analyze the influence of the cross-frame correlation on the unwrapping result. The comparison uses the generated test data set. The article starts with the Sect. 2 on the suggested CDFM test data generation; it is followed by a short Sect. 3 the filtering procedure for the CDFM data [7]; the cross-frame unwrapping methods are described in the Sect. 4; the article is finished with the experimental results and conclusions.

2   CDFM Test Data Generation

It is impossible to obtain real CDFM data with the precise knowledge of the blood flow dynamics to verify the unwrapping results. Thus, to compare the quality of phase unwrapping methods we generated artificial ultrasound slices. The goal of the modeling was to simulate the color Doppler flow artifacts and the model does not give the flow model possible to be used for other tasks. The used parameters give a realistic model of CDFM data wrapping.

We used a real patient series of 250 CT slices with 0.6 mm resolution at XY and 0.45 mm at Z to build a 3D heart model (see Fig. 1(a)). A surface of the inner border of the left ventricle, left atrium and beginning of the aorta were extracted as a housfield isosurface using marching-cubes algorithm [8]. Than the polygonal mesh was transformed to a set of parametric surfaces. 25 vertical volumetric stripes were cut from the ventricle to imitate papillary muscles. The heart geometry was static for the flow calculations, therefore additional open areas were added in the apex of the ventricle to let the redundant blood exit and enter the chamber during the systole and diastole (see Fig. 1(b)).

ANSYS CFX 15.0 was used for blood flow modeling. The blood was set as a Newton liquid with density $1080 \, kg/m^3$ and viscosity $0.00388 \, Pa \cdot s$. These parameters correspond to normal blood parameters [9]. The Reynolds number is about 25000, therefore laminar and Reynolds-Averaged Navier-Stokes (RANS) models are not suitable for this case [10] (pp. 292–303). Thus, Large Eddy Simulation turbulent model (LES) was used in our approach to take into account small vortices [10] (pp. 277–291). The finite element mesh was regular with the element size 0.5 mm, in total 2400881 elements. 7 total heart cycles (1 s each cycle) were simulated. The 0.0005 s time step was selected (in total 14000 steps) to satisfy the Courant number less than 1 [11]. Inflow and outflow with the sine-form shape 0.5 s duration and 1 m/s peak velocity were used.

For the calculated flow streams (see Fig. 1(c)) a single 2D-plane was fixed and virtual US scans of a convex probe were simulated. The scanning plane imitated the 3-chamber probe position (see Fig. 2(a)). The 3-chamber view position is the

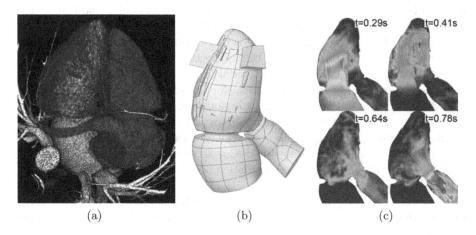

Fig. 1. Model of heart: (a) anatomical volume rendering of CT slices, (b) heart model, (c) different time moments of cardiac cycle flow.

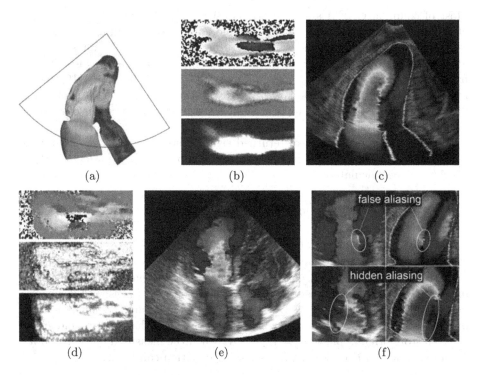

Fig. 2. (a) selected plane and area of virtual US scan for $t = 0.33$ s, (b) generated fields, top to bottom: real phase F, amplitude A, power P, (c) artificial color Doppler flow mapping image, (d) real image fields, (e) real ultrasound Doppler image, (f) false aliasing and hidden aliasing artifacts for real (left) and artificial (right) series.

most challenging for the unwrapping task because it shows peak blood velocities for both inflow and outflow. Frame rate was limited to 14.29 fps (total 100 frames per 7 cycles) It is a typical frame rate for real Doppler investigation. 5 different virtual pulses were modeled for each beam direction. Real phase F, amplitude A and power P (see Sect. 3) were constructed with the resolution 80 beams and 200 pixels per beam (see Fig. 2(b)). Artificial color Doppler flow image was generated (Fig. 2(c)) with the following added distortions corresponding to the real probe conditions: random shift of a measured point with maximum displacement 4 mm, region averaging, 7 dB white noise and velocity wrapped to $[-0.5, 0.5]$ m/s.

Undistorted images were also stored as a set of phase F fields with the same resolution as noisy fields. This set is a reference data to validate the results of unwrapping methods.

Generated artificial images have all drawbacks inherent to real ultrasonic Doppler scans: phase wrapping, noise, false aliasing and hidden aliasing (see Fig. 2(f)).

3   Complex Phase Restoration

Most of common CDFM systems rely on estimating the rate of phase change of the signal returning from a given sample volume [12]. Scanning transducer estimates phases and amplitudes of the returned echoes. Several impulses (8–16) are formed for better stability. For the blood flow the unwanted clutter signal from solid tissues can exceed the signal from blood by 40–60 dB [13–15] due to rejection of echoes from stationary or nearly stationary tissue. This problem is difficult to resolve in CDFM applications because of the low number of samples available for analysis.

In each image pixel a set of returned signal echoes are given as n complex values $R_k = I(k) + i \cdot Q(k)$. The power P and complex phase $R = R_x + i \cdot R_y = A \cdot e^{i \cdot F}$ can be calculated:

$$P = \frac{1}{n} \sum_{k=1}^{n} \overline{R_k} \cdot R_k \ , \quad R = \frac{1}{n-1} \sum_{k=2}^{n} \overline{R_{k-1}} \cdot R_k,$$

where A is the amplitude of R and the colorized real phase F is the basic information for the device output blood flow image.

The use of clutter rejection filters will color the system noise leading to a bias in the mean frequency estimate [16, 17]. The phase for pixels with low power becomes close to a bias value $M = M_x + i \cdot M_y$. This value can be estimated as the mean of complex phases of pixels with the power P less than a threshold P_{low}.

To restore original phase $R' = R'_x + i \cdot R'_y$ when the measured phase R, power P and bias point M are known the following equation can be used [7]:

$$R'_x = R_x \cdot P + (1 - M_x) \cdot (1 - P) \ , \quad R'_y = R_y.$$

This preprocessing procedure suppresses the noise so the low velocities become visible and the amplitude becomes more reliable. It results in the better quality of unwrapping methods.

4   Graph-Based Unwrapping Algorithms

As a graph cut based method we used the method based on energy minimization via graph cut described in [2]. In this method the preprocessed amplitude field representing the accuracy of the detected phase is used as a weight function for each edge in the graph. Weights of all edges that are used to realize the energy between two nodes p and q in frames i and j respectively are multiplied by $w(i, p, j, q) = (A_i(p) + A_j(q))/2$.

Blood flow is smooth and continuous in the time domain, therefore the velocity in a specific pixel can not change significantly from frame to frame. To limit the variance of neighbor-frame pixels additional edges can be added to graph. These edges connect a pixel in a frame with the same position pixels in neighbor frames. The following energy function is minimized:

$$M = \underbrace{\sum_i \sum_{p,q} |F_i(p) - F_i(q)| w(i, p, i, q)}_{\text{in-frame edges}} + \underbrace{\sum_{i,j} \sum_p |F_i(p) - F_j(p)| w(i, p, j, p)}_{\text{cross-frame edges}},$$

where p and q are neighbor pixels, and i and j are the connected frames.

Different i and j connections present the following methods (see Fig. 3):

- PF (Per-Frame unwrapping) – no cross-frame connections,
- CFN (Cross-Frame unwrapping for Neighbor frames) – all neighbor frames are connected,
- CFP (Cross-Frame unwrapping for Parts of series) – only neighbor frames within a single cardiac cycle are connected,
- CFC (Cross-Frame unwrapping with cross-Cycles connections) – each frame is connected to both neighbor frames and to all frames closest to this frame cycle time position (synchronized by ECG) from other cardiac cycles.

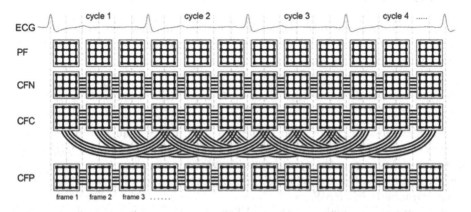

Fig. 3. Scheme of cross-frame connections. Each box with grid denotes graph edges within a single frame. Lines between block denotes cross-frame edges.

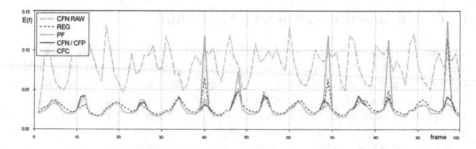

Fig. 4. $E(\mathrm{f})$ comparison.

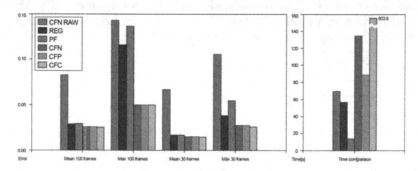

Fig. 5. Error and computation time comparison.

Graphs in PF, CFN and CFC have a simple regular edge structure and can be represented as a regular 3-dimensional 6-connected grids. Therefore, faster algorithms for grid cut [18] can be used for these cases.

5 Experiments

For different unwrapping methods the restored velocities were compared. To estimate the quality of restored velocities $F(p)$ knowing an exact values $F_0(p)$ mean and max error values were computed for whole series:

$$
E_{\mathrm{mean}} = \underset{\text{all frames}}{\mathrm{mean}}\, E(\mathrm{f})\,, \quad E_{\mathrm{max}} = \underset{\text{all frames}}{\mathrm{max}}\, E(\mathrm{f})\,, \quad E(\mathrm{f}) = \underset{p\in\mathrm{f}}{\mathrm{mean}}\frac{|F(p) - F_0(p)|}{\max F_0}.
$$

Comparison of $E(\mathrm{f})$ for each frame of a 100 frames sequence for different graph cut methods (PF, CFN, CFP, CFC) and regularizing method (REG) [6] is in Fig. 4. Also a processing of raw series (without complex phase restoration) was carried out to test the influence of complex phase restoration on the result of unwrapping (CFN RAW).

To compare the quality of the real flow detection a merged cycle was also used. The cycle merging is a typical method for increasing resolution in other cardiovascular imaging modalities such as 3D-echo, computed tomography, or

Table 1. Error and time costs. CPhR – complex phase restoration without unwrapping; unwrapping methods: PF – per-frame, RAW1 – PF without complex phase preprocessing, RAW2 – CFN without complex phase preprocessing; BK – Boykov and Kolmogorov graph cut realization [20], JS – Jamriska and Sykora grid cut realization [18].

Method	CDFM	CPhR	RAW1	RAW2	REG	PF	CFN	CFP	CFC
E_{mean} 100	0.1430	0.0789	0.0865	0.0827	0.028	0.029	0.025	0.025	0.0251
E_{max} 100	0.4558	0.4177	0.1872	0.1430	0.115	0.136	0.049	0.049	0.0498
E_{mean} 30	0.1199	0.0663	0.0690	0.0667	0.069	0.016	0.014	0.014	0.0145
E_{max} 30	0.3482	0.3176	0.1129	0.1056	0.113	0.054	0.027	0.027	0.0258
Time [s], BK			8.2	122.5		15.2	220.9	140.2	803.8
Time [s], JS			6.7	69.9		13.6	134.8	88.5	
Time [s]					57.0				

MR [19]. A series of 100 frames of 7 cardiac cycles was merged into one cycle with 30 frames. For each of the compared unwrapping methods an initial 100 frames series was processed first and then unwrapped frames were used for a merged cycle construction. The merged circle was compared with the merged reference data. E_{mean} and E_{max} for series of 100 frames and averaged cardiac cycle of 30 frames are listed in Table 1 and are graphically illustrated in Fig. 5.

6 Conclusion

The constructed CDFM test data enabled us to perform the quality comparison of different unwrapping methods. All previous works in this area had not a valid comparison. It has been found that the data preprocessing method [7] increases the quality of all considered unwrapping methods. The regularization approach gives better results than graph-cut method, but it can not use cross-frame dependencies due to the long processing time. If the cross-frame correlation is not used than hidden aliasing artifact is undetected for some frames like peak-systole and diastole. CFN, CFP and CFC methods remove aliasing even in very complex frames with false and hidden aliasing but due to the time reasons CFP looks the most efficient for the practical use. Cross-cycle connection does not give a significant improvement of the results but makes computation very slow and cannot be used in the case of arrhythmia. A real CDFM series processing also confirmed these conclusions. Nevertheless, a combined method using also regularization ideas can be a good candidate for the best practical method. Also, frame rate can be used for more precise selection of cross-frame weights in the minimizing energy function.

Additional materials for the test and real practical data are available at http://imaging.cs.msu.ru/files/medical/CDFM_Unwrapping.zip.

Acknowledgments. The work was supported by Russian Science Foundation grant #14-11-00308.

References

1. Bioucas-Dias, J.M., Valadao, G.: Phase unwrapping via graph cuts. IEEE Trans. Image Process. **16**(3), 698–709 (2007)
2. Yatchenko, A., Krylov, A., Gavrilov, A., Arkhipov, I.: Graph-cut based antialiasing for doppler ultrasound color flow medical imaging. In: IEEE VCIP, pp. 1–4 (2011)
3. Pritt, M.D., Shipman, J.S.: Least-squares two-dimensional phase unwrapping using FFTs. IEEE Trans. Geosci. Remote Sens. **32**(3), 706–708 (1994)
4. Ghiglia, D.C., Romero, L.A.: Minimum Lp-norm two-dimensional phase unwrapping. J. Opt. Soc. Am. A **13**(10), 1999–2013 (1996)
5. Yamaki, R., Hirose, A.: Singular unit restoration in interferograms based on complex-valued Markov random field model for phase unwrapping. Geosci. Remote Sens. Lett. **6**(1), 18–22 (2009)
6. Yatchenko, A., Krylov, A., Sandrikov, V., Kulagina, T.: Regularizing method for phase antialiasing in color doppler flow mapping. Neurocomputing **139**, 77–83 (2014)
7. Yatchenko, A., Krylov, A., Gavrilov, A., Sandrikov, V., Kulagina, T.: Image preprocessing for color Doppler flow antialiasing using power and complex phase data. In: IEEE International Conference on Signal Processing, pp. 1072–1076 (2014)
8. Lorensen, W.E., Cline, H.E.: Marching cubes: a high resolution 3D surface construction algorithm. Comput. Graph. **21**, 163–169 (1987)
9. Brown, S., Wang, J., Ho, H., Tullis, S.: Numeric simulation of fluid-structure interaction in the aortic arch. In: Wittek, A., Miller, K., Nielsen, P.M.F. (eds.) Computational Biomechanics for Medicine, pp. 13–23. Springer, New York (2013)
10. Ferziger, J.H., Peric, M.: Computational Methods for Fluid Dynamics, 3rd edn, pp. 277–303. Springer, Heidelberg (2002)
11. Davidson, L.: How to estimate the resolution of an LES of recirculating flow. Qual. Reliab. Large-Eddy Simul. **16**, 269–286 (2011)
12. Evans, D.H., Jensen, J.A., Nielsen, M.B.: Ultrasonic colour Doppler imaging. Interface Focus **1**(4), 490–502 (2011)
13. Jensen, J.A.: Stationary echo canceling in velocity estimation by time-domain cross-correlation. IEEE Trans. Med. Imaging **12**, 471–477 (1993)
14. Brands, P.J., Hoeks, A.P., Hofstra, L., Reneman, R.S.: A noninvasive method to estimate wall shear rate using ultrasound. Ultrasound Med. Biol. **21**, 171–185 (1995)
15. Heimdal, A., Torp, H.: Ultrasound Doppler measurements of low velocity blood flow: limitations due to clutter signals from vibrating muscles. IEEE Trans. Ultrason. Ferroelectr. Freq. Control **44**(4), 873–881 (1997)
16. Kadi, A.P., Loupas, T.: On the performance of regression and step-initialized IIR clutter filters for color doppler systems in diagnostic medical ultrasound. IEEE Trans. Ultrason. Ferroelectr. Freq. Control **42**(5), 927–937 (1995)
17. Tysoe, C., Evans, D.H.: Bias in mean frequency estimation of Doppler signals due to wall clutter filters. Ultrasound Med. Biol. **21**, 671–677 (1995)
18. Jamriska, O., Sykora, D., Hornung, A.: Cache-efficient graph cuts on structured grids. In: IEEE Transactions on Computer Vision and Pattern Recognition, pp. 3673–3680 (2012)
19. Garcia, D., del Alamo, J.C., et al.: Two-dimensional intraventricular flow mapping by digital processing conventional color-Doppler echocardiography images. IEEE Trans. Med. Imaging **29**(10), 1701–1713 (2010)
20. Boykov, Y., Kolmogorov, V.: An experimental comparison of min-cut/max-flow algorithms for energy minimization in vision. IEEE Trans. Pattern Anal. Mach. Intell. **26**(9), 1124–1137 (2004)

Orthogonal Shape Modes Describing Clinical Indices of Remodeling

Xingyu Zhang[1](✉), Bharath Ambale-Venkatesh[2], David A. Bluemke[3], Brett R. Cowan[1],
J. Paul Finn[4], William G. Hundley[6], Alan H. Kadish[5], Daniel C. Lee[5], Joao A.C. Lima[2],
Avan Suinesiaputra[1], Alistair A. Young[1], and Pau Medrano-Gracia[1]

[1] Department of Anatomy with Radiology, University of Auckland, Auckland, New Zealand
zha238@aucklanduni.ac.nz
[2] The Donald W. Reynolds Cardiovascular Clinical Research Center, The Johns Hopkins
University, Baltimore, USA
[3] National Institute of Biomedical Imaging and Bioengineering, Bethesda, MD, USA
[4] Department of Radiology, UCLA, Los Angeles, USA
[5] Feinberg Cardiovascular Research Institute, Northwestern University Feinberg School
of Medicine, Chicago, USA
[6] Section of Cardiology Medical Center Blvd, Winston Salem, NC, USA

Abstract. Quantification of the left ventricle (LV) shape changes (remodeling) is of great importance for therapeutic management of myocardial infarction. Orthogonal shape modes derived from principal component analysis (PCA) often do not describe clinical remodeling indices. We developed a method for deriving orthogonal shape modes directly from any set of clinical indices. Cardiac magnetic resonance images of 1,991 asymptomatic volunteers from the MESA study (age 44−84, mean age 62, 52 % women) and 300 patients with myocardial infarction from the DETERMINE study (age 31−86, mean age 63, 20 % women) were obtained from the Cardiac Atlas Project. Clinical indices of LV size, sphericity, wall thickness and apical conicity were calculated. For each index, cases outside two standard deviations of the mean, but within one standard deviation for all other indices, were chosen as a representative subgroup. Orthogonal modes were defined sequentially, using the first principal component of each subgroup. At each step, the contribution of the previous mode was removed mathematically from the shape description, similar to Gram–Schmidt orthogonalization. Correlation analysis and logistic regression were performed to show the effectiveness of these features to characterize remodeling due to myocardial infarction.

Keywords: Cardiac remodeling · Magnetic resonance imaging · Principal component analysis

1 Background

Left ventricular (LV) remodeling refers to the process by which ventricular size, shape and function are regulated by mechanical, neurohormonal and genetic responses to insult [1]. Myocardial infarction leads to LV remodeling of the heart, which provides important diagnostic information for the therapeutic management of ischemic heart disease [2–5].

© Springer International Publishing Switzerland 2015
H. van Assen et al. (Eds.): FIMH 2015, LNCS 9126, pp. 273–281, 2015.
DOI: 10.1007/978-3-319-20309-6_32

Remodeling associated with LV size is an important predictor of mortality after myocardial infarction [6], and changes of sphericalization of heart shape are linked with decreased survival [5]. LV wall thickness [1] and apical conicity [7] changes are also important indicators for LV infarction remodeling due to myocardial infarction.

Standard clinical indices used to describe remodeling are typically simple measures of mass and volume, such as end-diastolic (ED) volume (largest volume), end-systolic (ES) volume (smallest volume) or left ventricular mass. However, these ignore much of the available shape information. Principal component analysis (PCA) [8] is currently one of the most widely used feature extraction techniques. PCA projects the data onto a linear space of maximum-variation directions (known as modes) with orthogonal transformations. After the projection the first mode accounts for the maximum variance, and each succeeding mode in turn has the highest residual variance possible under the linear orthogonality constraint. PCA analysis of cardiac remodeling has been explored previously [9] showing that combined PCA modes can be more powerful descriptions of remodeling than traditional indices. Typically, the first PCA mode corresponds with LV size and the second typically with LV sphericity. Mode orthogonality is important for maintaining simplicity in many mathematical applications (e.g., flow computations [10]). However, PCA modes do not generally correspond well with clinically established indices of remodeling (e.g., wall thickness). We developed a novel methodology which captures the shape characteristics of a given clinical indicator while maintaining mode orthogonality. Clinically-defined indices of size, sphericity, wall thickness and conicity, known from the literature to be important in the management of myocardial infarction, were used to create a corresponding orthogonal linear space from the shape parameters.

Cardiac magnetic resonance images of 1,991 asymptomatic volunteers from the Multi-Ethnic Study of Atherosclerosis (MESA) [11] and 300 patients with myocardial infarction contributed from the Defibrillators to Reduce Risk by Magnetic Resonance Imaging Evaluation (DETERMINE) [12] study were obtained through the Cardiac Atlas Project [13] to create these novel clinically-defined modes. A logistic regression model on the MESA and DETERMINE data was established to explore the effectiveness of these modes to predict LV remodeling due to myocardial infarction.

2   Methods

Finite element models were customized to model the shape and function of each case using a standardized procedure [9]. For the MESA cohort, short-axis hand-drawn contours on the inner and outer surfaces of the left ventricle were available from the MESA MRI core laboratory. These contours were fitted by the finite element model by linear least squares as described previously [14, 15]. For the DETERMINE cohort, expert observers performed the analysis using guide-point modeling [16] to interactively customize a time-varying 3D cardiac finite element model of the LV to MR images using custom software (CIM version 6.0, University of Auckland, New Zealand). The shape models were evenly sampled at sufficient resolution to capture all visible features, which resulted in 2,738 Cartesian (x_i, y_i, z_i) points.

LV mass and volume at ED were subsequently calculated from the cardiac LV shape models. The sphericity index was calculated as the EDV divided by the volume of a sphere with a diameter corresponding to the major axis at end-diastole in LV long view [17]. The apical conicity index was calculated as the ratio of the apical axis (defined as the diameter of the endocardium one third above the apex) over the basal diameter [7]. Wall thickness was calculated as the mean distance between the corresponding endo- and epi-cardium surfaces.

For each clinical index, cases outside two standard deviations of the mean (95 % of variance). But within one standard deviation for all other indices, were chosen to form a patient subgroup linked with each clinical index. Orthogonal modes were defined sequentially, using the first principal component of each subgroup. At each step, the contribution of the previous mode was removed mathematically from the shape description, similar to the Gram–Schmidt orthogonalization algorithm [18], prior to the calculation of the principal component. Modes were defined in the following order: (1) LV size, (2) sphericity, (3) conicity, and (4) wall thickness.

Mathematically, let X^1 represent the shape space as a matrix where each column contains the coordinates of 3D points describing the shape of one case. Selecting cases with high and low EDV as previously described, a principal decomposition of this subgroup yielded a matrix of modes (M_k^1), so that

$$X_m^1 = \overline{X^1} + \sum_{k=1}^{K} \alpha_{mk}^1 M_k^1 \tag{1}$$

for each case m in the subgroup, where $\overline{X^1}$ represents the Euclidean mean, K is the total number of modes M_k^1, and α_{mk}^1 their corresponding projections (these are also referred to as weights or scores). The first mode (M_1^1) was used to describe the LV size variation. The projections of the first mode were then removed from the initial space for all cases, creating a new space X^2. A PCA of the resulting shapes for the subgroup consisting of high and low sphericity was then performed. The mode relating to sphericity was then defined to be the first principal component of this subgroup, i.e. M_1^2:

$$X_m^2 \stackrel{\text{def}}{=} X_m^1 - \alpha_{m1}^1 M_1^1 = \overline{X^2} + \sum_{k=1}^{K} \alpha_{mk}^2 M_k^2. \tag{2}$$

New mode M_1^2 was then subtracted from the shape space X^2, and the procedure repeated for the conicity (M_1^3) and wall thickness modes (M_1^4). Note that by construction M_1^{i+1} is orthogonal to M_1^i [18] and $\langle M_1^{1,2,3,4} \rangle$ generate an orthogonal linear sub-space of X^1.

3   Results

The orthogonal modes corresponding to size, sphericity, conicity and wall thickness are shown in Fig. 1. Linear correlation coefficients were calculated between the clinical indices and the shape-mode scores ($\alpha_1^{1,2,3,4}$) in the combined MESA and DETERMINE population. The linear correlation coefficient between the size score and EDV, sphericity score and LV sphericity index, wall thickness score and LV wall thickness, conicity sore and LV conicity index were 0.93, −0.87, 0.75 and −0.63 respectively (Table 1). The linear correlation coefficients among the clinical indices (Table 2), correlation coefficients among the mode scores (Table 3) and correlation coefficients between clinical indices and scores of the first four PCA modes of the original dataset (standard PCA in Table 4) were also calculated as a reference. It can be seen that, although the shape modes are orthogonal (their dot products are zero), some remodeling indices and corresponding mode scores were significantly correlated. In particular, the wall thickness score was significantly correlated with LV size, with larger hearts having thinner walls, consistent with eccentric remodelling. Also, although the first two standard principal components correspond with size and sphericity respectively, other components do not correspond with any particular remodeling index.

The clinically-derived shape modes described 39 %, 9 %, 6 % and 4 % of the total shape variation respectively, compared with 50 %, 10 %, 8 % and 7 % for the first four standard PCA modes. A logistic regression model was performed to evaluate the discriminatory power of the clinically-derived modes to characterize LV remodeling due to myocardial infarction. Age, sex, height, weight, systolic blood pressure (SBP), smoking status and diabetes status were used to establish a baseline model. The baseline model was established to give a control model consisting of common clinical factors known to be associated with myocardial infarction.

These variables were included in all the models since they may be confounding factors between the disease and shape features. The scores from the clinical modes show significant odds ratios in the myocardial infarction model (Table 5). The odds ratio of size, sphericity, wall thickness and conicity indicate that myocardial infarction patients tend to have larger and more spherical LV shapes with thicker walls, and a less conical shape. The area of receiver operating characteristic (ROC) curve (Fig. 2) for the clinical mode logistic classification model is 90.87 %, similar to the model using scores of the first four standard PCA modes (92.05 %).

Table 1. Correlation coefficients between the clinical indices and the clinical modes scores

	Size score	Sphericity score	Conicity score	Wall thickness score
EDV	0.93*	0.03	−0.03	−0.66*
Sphericity index	−0.06*	−0.87*	−0.05*	−0.04
Conicity index	−0.07*	0.29*	0.75*	0.08*
Wall thickness	0.37*	0.11*	−0.08*	−0.63*

Note: *indicates p value < 0.05 in all the tables of this paper

Table 2. Correlation coefficients between the clinical indices

	EDV	Sphericity index	Conicity index	Wall thickness
EDV	1.00	0.22*	−0.15*	0.14*
Sphericity index		1.00	−0.22*	−0.15*
Conicity index			1.00	−0.03
Wall thickness				1.00

Table 3. Correlation coefficients between the clinical mode scores

	Size score	Sphericity score	Conicity score	Wall thickness
Size score	1.00*	0.28*	0.001	−0.76*
Sphericity score		1.00	0.23*	−0.12*
Conicity score			1.00	0.09*
Wall thick score				1.00

Table 4. Correlation coefficients between the clinical indices and the first four modes of variation of X^1 (standard PCA).

	EDV	Sphericity index	Conicity index	Wall thickness
PC1	−0.93*	0.06*	0.07*	−0.36*
PC2	−0.25*	−0.84*	0.39*	0.02
PC3	−0.05*	−0.18*	−0.25*	0.36*
PC4	0.06*	0.11*	−0.21*	−0.07*

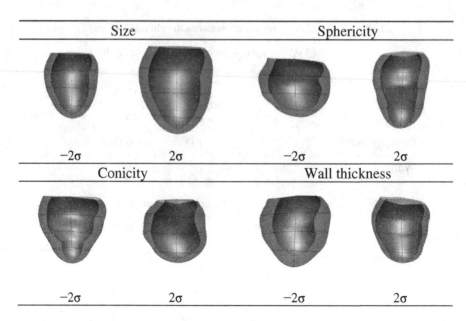

Fig. 1. Plot of the clinical modes

Table 5. Logistic regression analysis of the clinical modes to myocardial infarction

Parameter	Coefficient	Standardized coefficient	Odds ratio(OR)	OR 95 % Confidence interval	
Age*	0.041	0.231	1.042	1.023	1.060
Gender	−0.430	−0.118	0.651	0.394	1.075
Height*	0.031	0.178	1.032	1.007	1.057
Weight*	−0.041	−0.415	0.960	0.950	0.970
SBP*	−0.015	−0.173	0.985	0.977	0.994
Diabetes*	1.231	0.249	3.425	2.315	5.070
Smoke	−0.363	−0.066	0.696	0.420	1.153
Size*	0.020	1.408	1.020	1.018	1.023
Sphericity*	−0.009	−0.297	0.991	0.989	0.994
Conicity*	−0.008	−0.177	0.992	0.987	0.996
Wall thick-ness*	0.010	0.263	1.010	1.005	1.016

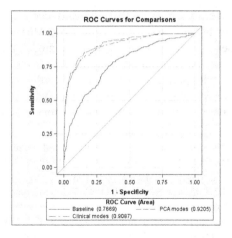

Fig. 2. ROC curve for the three logistic regression models

4   Discussion

Patients with myocardial infarction have significant shape differences with respect to the normal population, due to cardiac remodeling. An atlas-based analysis of cardiac remodeling has previously shown superior performance to traditional mass and volume analysis in large data sets [9]. The framework consists of three steps: (1) fitting a finite element model to the LV MR images, (2) feature extraction of the aligned shape parameters, and (3) quantification of the association between the features and disease using logistic regression. Although PCA provides orthogonal shape features, which describe the maximum amount of variation for the fewest number of modes, these modes typically do not correspond with clinical indices of cardiac remodeling. To avoid this problem, and maintain the advantage of orthogonal shape features, we developed a method to generate orthogonal shape modes from any set of clinical indices.

In this paper, we generated a linear shape sub-space from the finite-element parameters encoding the clinical indices of size, sphericity, wall thickness and conicity. These orthogonal modes derived from traditional remodeling indices may be used to partition shape variation in a similar way to PCA, but with a clinical rationale. Unlike PCA, correlation analysis shows that these clinically derived modes have high correspondence with traditional remodeling indices (Table 1). However, the absolute correlation decreases with increasing orthogonal modes, as the shape space is reduced by subtracting previous modes (Table 2). The clinical indices were moderately correlated (Table 2). Although the shape modes are orthogonal (zero dot product between different mode shape vectors) the mode scores are not guaranteed to be orthogonal in the statistical sense (uncorrelated), since size can be correlated with wall thickness in the human heart (Table 3). Furthermore, the proposed approach is dependent on the selection order of the clinical indices. Future work could look into the decorrelation of these scores by means of signal whitening, and also the difference of the shape modes obtained in different order.

In summary, we have demonstrated that clinically-derived modes quantitatively characterize remodeling features associated with myocardial infarction with similar accuracy to PCA modes. Compared with the baseline model, there was a significant incremental discriminatory power with the four clinical remodeling modes examined. This implies that these modes play an important role in cardiac remodeling. These clinical modes partition the variation into shape features which are linked to clinical outcomes and which have an intuitive visual interpretation.

Acknowledgements. This project was supported by award numbers R01HL087773 and R01HL121754 from the National Heart, Lung, and Blood Institute. MESA was supported by contracts N01-HC-95159 through N01-HC-95169 from the NHLBI and by grants UL1-RR-024156 and UL1-RR-025005 from NCRR. DETERMINE was supported by St. Jude Medical, Inc; and the National Heart, Lung and Blood Institute (R01HL91069). A list of participating DETERMINE investigators can be found at http://www.clinicaltrials.gov. David A. Bluemke is supported by the NIH intramural research program. Xingyu Zhang would like to gratefully acknowledge financial support from the China Scholarship Council.

References

1. Sutton, M.G.S.J., Sharpe, N.: Left ventricular remodeling after myocardial infarction pathophysiology and therapy. Circulation **101**, 2981–2988 (2000)
2. Gjesdal, O., Bluemke, D.A., Lima, J.A.: Cardiac remodeling at the population level—risk factors, screening, and outcomes. Nat. Rev. Cardiol. **8**, 673–685 (2011)
3. Lieb, W., Gona, P., Larson, M.G., Aragam, J., Zile, M.R., Cheng, S., et al.: The natural history of left ventricular geometry in the community clinical correlates and prognostic significance of change in LV geometric pattern. JACC Cardiovasc. Imaging **7**, 870–878 (2014)
4. Zile, M.R., Gaasch, W.H., Patel, K., Aban, I.B., Ahmed, A.: Adverse left ventricular remodeling in community-dwelling older adults predicts incident heart failure and mortality. JACC Heart Fail. **2**, 512–522 (2014)
5. Wong, S.P., French, J.K., Lydon, A.-M., Manda, S.O.M., Gao, W., Ashton, N.G., et al.: Relation of left ventricular sphericity to 10-year survival after acute myocardial infarction. Am. J. Cardiol. **94**, 1270–1275 (2004)
6. White, H.D., Norris, R.M., Brown, M.A., Brandt, P.W., Whitlock, R.M., Wild, C.J.: Left ventricular end-systolic volume as the major determinant of survival after recovery from myocardial infarction. Circulation **76**, 44–51 (1987)
7. Di Donato, M., Dabic, P., Castelvecchio, S., Santambrogio, C., Brankovic, J., Collarini, L., et al.: Left ventricular geometry in normal and post-anterior myocardial infarction patients: sphericity index and 'new' conicity index comparisons. Eur. J. Cardio-Thorac. Surg. **29**, S225–S230 (2006)
8. Jolliffe, I.: Principal Component Analysis. Wiley Online Library, New York (2005)
9. Zhang, X., Cowan, B.R., Bluemke, D.A., Finn, J.P., Fonseca, C.G., Kadish, A.H., et al.: Atlas-based quantification of cardiac remodeling due to myocardial infarction. PLoS ONE **9**, e110243 (2014)

10. McLeod, K., Caiazzo, A., Fernández, M.A., Mansi, T., Vignon-Clementel, I.E., Sermesant, M., Pennec, X., Boudjemline, Y., Gerbeau, J.-F.: Atlas-based reduced models of blood flows for fast patient-specific simulations. In: Camara, O., Pop, M., Rhode, K., Sermesant, M., Smith, N., Young, A. (eds.) STACOM 2010. LNCS, vol. 6364, pp. 95–104. Springer, Heidelberg (2010)

11. Bild, D.E., Bluemke, D.A., Burke, G.L., Detrano, R., Roux, A.V.D., et al.: Multi-ethnic study of atherosclerosis: objectives and design. Am. J. Epidemiol. **156**, 871–881 (2002)

12. Kadish, A.H., Bello, D., Finn, J., Bonow, R.O., Schaechter, A., Subacius, H., et al.: Rationale and design for the defibrillators to reduce risk by magnetic resonance imaging evaluation (DETERMINE) trial. J. Cardiovasc. Electrophysiol. **20**, 982–987 (2009)

13. Fonseca, C.G., Backhaus, M., Bluemke, D.A., Britten, R.D., Do Chung, J., Cowan, B.R.: The Cardiac Atlas Project—an imaging database for computational modeling and statistical atlases of the heart. Bioinformatics **27**, 2288–2295 (2011)

14. Medrano-Gracia, P., Cowan, B.R., Bluemke, D.A., Finn, J., Lima, J.A., Suinesiaputra, A., Young, A.A.: Large scale left ventricular shape atlas using automated model fitting to contours. In: Ourselin, S., Rueckert, D., Smith, N. (eds.) FIMH 2013. LNCS, vol. 7945, pp. 433–441. Springer, Heidelberg (2013)

15. Medrano-Gracia, P., Cowan, B., Ambale-Venkatesh, B., Bluemke, D., Eng, J., Finn, J., et al.: Left ventricular shape variation in asymptomatic populations: the multi-ethnic study of atherosclerosis. J. Cardiovasc. Magn. Reson. **16**, 56 (2014)

16. Young, A.A., Cowan, B.R., Thrupp, S.F., Hedley, W.J., Dell'Italia, L.J.: Left ventricular mass and volume: fast calculation with guide-point modeling on MR images 1. Radiology **216**, 597–602 (2000)

17. Izumo, M., Lancellotti, P., Suzuki, K., Kou, S., Shimozato, T., Hayashi, A., et al.: Three-dimensional echocardiographic assessments of exercise-induced changes in left ventricular shape and dyssynchrony in patients with dynamic functional mitral regurgitation. Eur. J. Echocardiogr. **10**, 961–967 (2009)

18. Hoffmann, W.: Iterative algorithms for Gram-Schmidt orthogonalization. Computing **41**, 335–348 (1989)

Models of Mechanics

A Framework for Determination of Heart Valves' Mechanical Properties Using Inverse-Modeling Approach

Ankush Aggarwal[✉] and Michael S. Sacks

Center for Cardiovascular Simulation, Institute for Computational Engineering
and Sciences, Department of Biomedical Engineering,
University of Texas at Austin, Austin, TX 78712, USA
ankush@ices.utexas.edu

Abstract. Heart valves play a very important role in the functioning of the heart and many of the heart failures are related to the valvular dysfunctions, e.g. aortic stenosis and mitral regurgitation. Relationship between the biomechanical properties of valve leaflets and their function has long been established, however, determining these properties in a non-invasive manner remains a challenge. Here we present a framework for such a tool for biomechanical properties determination. We use an inverse-modeling approach, where the only input is through imaging the leaflet tissue as it is loaded naturally during its functional cycle. Using a structural model for the leaflet material behavior allows us to reduce the number of parameters to be determined to only two, which in addition to dramatically reducing the computational time also allows one to visualize the cost function and the minimization process. We close with discussion about the contributions of the current framework and other constituents needed to make it a clinically viable tool.

1 Introduction

Heart valves are critical components of heart because of their role in ensuring the controlled flow of blood synchronized with the contraction of ventricles. The valve leaflets are complex, membrane-structural entities primarily made of collagen, elastin, extracellular matrix and cells. Either due to congenital defects or due to the changes due to remodeling with age, the leaflets functionality is altered sometimes resulting in less efficient heart output. Such problems are hard to diagnose at an early stage and may lead to heart failure if left untreated. Therefore, a tool for determining the functional properties of heart valves from non-invasive imaging modalities will be of utmost clinical importance. Such a tool will help diagnose valve diseases at an early stage as well as provide tools for monitoring the performance of replaced prosthetic valves. The three major components of such a tool are shown in Fig. 1. In order to define the correct reference configuration, we need the pre-strain information about leaflets. For defining the unknown fiber architecture, we need the population average architectures that can be used to constrain parameter optimization. Also, the proper choice of

© Springer International Publishing Switzerland 2015
H. van Assen et al. (Eds.): FIMH 2015, LNCS 9126, pp. 285–294, 2015.
DOI: 10.1007/978-3-319-20309-6_33

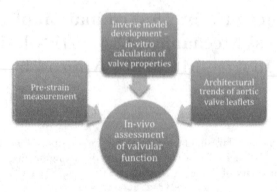

Fig. 1. Three components for an in-vivo assessment tool are pre-strain measurement, inverse model and population average fiber architecture of valve leaflets

constitutive model is important to best exploit fibrous structural information. Finally, the most critical component is development of a numerical inverse-model, which can be applied to in-vivo imaging data. In this paper, we focus on the development of inverse model using an in-vitro dataset and briefly discuss the other components.

2   Methods

To demonstrate the ideal capabilities of our approach, we utilized archival quality in-vitro experimental data from pericardial bioprosthetic heart valve to develop the parameter estimation tool based upon inverse modeling. The leaflets were imaged at three different static transvalvular pressures using surface markers and dual-camera setup [1] (Fig. 2). Leaflets were then removed and their fiber architecture was determined using light scattering technique [2]. Then the leaflets were dissected and put under biaxial test to determine their stress-strain relationship. This highly comprehensive data set with high resolution, marker positions at multiple pressures, valve specific fiber architecture and biaxial data was ideal for this study – to design the inverse model, validate it and calculate its sensitivity to various input parameters and optimization constraints. There is no existing method for inverse modeling of heart valves as per authors' knowledge. This is due to the difficulty in modeling them as well as their small size and thickness and fast movement in-vivo, leading to poor quality of imaging.

2.1   The Forward Model

Each of the valve leaflet was modeled as a thin shell and the contact between leaflets was treated using the augmented Lagrange multiplier method. The basal attachment was fixed as in the experiments and a normal pressure on the leaflets was applied linearly increasing upto 120 mm of Hg. The forward model was implemented in FEBio [3] using quadrilateral shell elements using quasi-static solver.

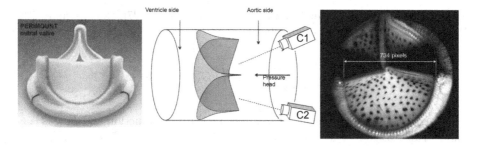

Fig. 2. The experimental setup to obtain valve shape in-vitro at different transvalvular pressure levels

The shape of the valve was provided by Edwards in the form of a quadrilateral meshed geometry file, where each leaflet was made up of 1025 cells and 1082 nodes. To improve the refinement as well as quality of elements, the geometry was imported into Hypermesh and remeshed with each leaflet made up of 2789 cells and 2880 nodes. Leaflets were discretized using four node quadrilateral elements, with each node assigned six degrees of freedom (three for translation and three for rotation). Therefore, the complete discrete version of the model had 51,840 degrees of freedom. A constant thickness was assigned to each leaflet as measured during the experiment. The stresses were integrated through the shell thickness using 3 point Gauss quadrature rule to obtain bending moments. To map the fiber architecture from two-dimensional output of the above-mentioned experimental method onto the three-dimensional valve structure, a spline-based technique previously developed was used [4].

2.2 The Material Model

The valve tissues are made up of multiple layers (fibrosa, ventricularis etc.) with different types of fibers embedded in a matrix for each of them [5,6]. However it is very hard to take into account the contribution of each layer separately. Therefore, here we take one of the most realistic material models developed – where tissue is idealized as a planar network of collagen fibers (main load bearing component) embedded in a ground substance (glycosaminoglycans, elastin, water) [7]. Thus, the strain energy is assumed to be a sum of the contributions from fibrillar (anisotropic) and non-fibrillar (isotropic) components expressed in terms of the tissue level Green-Lagrange strain $\mathbf{E}$:

$$\psi = \psi_{\text{iso}}^{\text{M}}(\mathbf{E}) + \psi_{\text{aniso}}^{\text{F}}(\mathbf{E}). \tag{1}$$

Here $\psi_{\text{iso}}^{\text{M}}$ and $\psi_{\text{aniso}}^{\text{F}}$ are strain energy functions representing ground-matrix and fiber contributions respectively. The ground-matrix function describes low-strain behavior and provides stiffness in the unloaded state. Tissue response to large strains is accommodated by the anisotropic term characterizing the fibrillar microstructure averaged through the thickness. The isotropic contribution is assumed to be neo-Hookean. Following the formulation in [7], the ensemble 2nd Piola-Kirchhoff stress in a single fiber is

$$S_{ens}(E_{ens}) = \frac{\partial \psi_{ens}}{\partial E_{ens}} = \eta \int_0^{E_{ens}} D(x) \frac{E_{ens} - x}{(1 + 2x)^2} dx \qquad (2)$$

where, η is the fiber stiffness, D is the fiber recruitment function and E_{ens} is the uniaxial Lagrangian strain in single fiber aligned in direction N so that $E_{ens} = N \cdot E \cdot N$. However, in present formulation, to avoid extremely high computational times, a simplified model was used [8], where the integral is approximated as an exponential function:

$$S_{ens}(E_{ens}) = A(e^{BE_{ens}} - 1) \qquad (3)$$

The important aspect of this approach is that it allows us to reduce the number of unknown material parameters in fiber component to only two, thus, reducing the computational time in determining them. The matrix component was modeled as a compressible neo-Hookean material and the properties were borrowed from flexural testing results previously published. For a distribution of fibers in a planar biaxial state, we get the planar stress by integrating over the distribution

$$\mathbf{S}_{aniso}^{F}(\mathbf{E}) = \frac{\partial \psi_{aniso}^{F}}{\partial \mathbf{E}} = \int_{-\pi/2}^{\pi/2} R(\theta) S_{ens}(N \cdot \mathbf{E} \cdot N) N \otimes N d\theta \qquad (4)$$

The fiber distribution $R(\theta)$ was determined at every point of the tissue using SALS setup [2] and approximated as a normalized Gaussian distribution. For the matrix part ψ_{iso}^{M}, a neo-Hookean model was used for which there parameters were estimated from flexural studies previously conducted.

2.3   The Algorithm

The above forward model was put in an optimization loop where the parameters were initialized with a guess and solution was iterated until convergence was achieved. The objective cost function was defined as the difference in shape of the deformed mesh and that of the experimental data points. The difference was calculated along the normal direction of deformed mesh. Thus, it can be defined mathematically as

$$\mathcal{F} = \sqrt{\sum_{i,\alpha} (x_i^\alpha - \tilde{x}_i^\alpha(c_m))^2} \qquad (5)$$

where, x_i^α are the input points from experiment and $\tilde{x}_i^\alpha$ are their projection on deformed valve surface obtained from the forward model with material parameters c_m. It should be noted that the above cost function does not use the information about experimental data points being the material points. Instead, the points were used only to describe valve shape. This was done to be consistent with the clinical imaging modalities (e.g. ultrasound) which provides only the shape without any information about material points. The minimization of above defined cost function is a highly non-linear function because of the contact constrained between leaflets. The non-linear least squares problem of minimizing the cost function was solved using Levenberg-Marquardt algorithm:

$$\mathbf{J}^T\mathbf{J} + \lambda \text{diag}(\mathbf{J}^T\mathbf{J}))\Delta c_m = \mathbf{J}^T \left(x_i^\alpha - \tilde{x}_i^\alpha (c_m) \right), \tag{6}$$

where derivatives of the cost function $\mathbf{J} = \partial \tilde{x}_i^\alpha (c_m) / \partial c_m$ were calculated numerically using central difference method by perturbing each of the parameters 1 % above and below its current value. The value of λ was kept constant a value of 5 since it provides a balance between the stability and convergence speed of the algorithm. The simulations were run on Texas advanced computing center (TACC) on a single node with 12 processors. The finite difference cases for cost function derivatives were run in parallel using multiprocessing module of python wrapper. For n parameters, this setup required a total of $2n + 1$ simulations in parallel. Each iteration of the inverse model approximately took 35–40 min.

3   Results

Our approach allows the entire problem to be reduced to optimization of just two parameters; with the cost function shape for various cases as a function of the two material parameters A and B is shown in Fig. 3. The results demonstrate the importance of using higher number of frames from in-vivo imaging for inverse modeling input. It also shows that imaging at higher pressures would give a better confidence in the results. It is also noted that the cost function is almost convex everywhere. However, the lowest part of the cost function is a long banana-shaped region instead of a single point. This is a property of the exponential form of the ensemble response function and can be analyzed analytically (see Appendix). Similar behavior of the exponential function has been previously reported in inverse modeling of myocardium [9]. The closed-form equation for the lowest region Eq. 9, which we will refer to as the "iso-stress-strain minima" region, thus obtained can also be used to reduce the number of parameters in the optimization process in the future. Furthermore, it is noted that the minima of the cost function resembles the error function from biaxial testing. This signifies that although the valve is loaded under normal pressure, locally it still behaves like a planar membrane under biaxial load.

Using the current approach, it is also possible to evaulate sensitivity of results to various input parameters. The change in cost function when average fiber architecture is utilized is shown in Fig. 4a and when it is varied the result is shown in Fig. 4b. This result demonstrates that the inverse modeling results are dependent upon the fiber architecture in an average sense. The results were found insensitive to the Poisson's ratio of the matrix, but highly sensitive to the noise in input point cloud from imaging (Fig. 5). Statistics of the final fit are shown in Table 1. Comparison of predicted stress-strain relationship vs. measured are

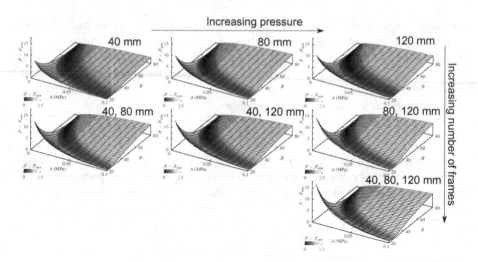

Fig. 3. Cost function using various pressure levels and various number of frames

shown in Fig. 6a and the final fit of valve's deformed shape to point cloud is shown in Fig. 6b. Our framework was able to obtain an excellent fit of the final deformed shape to marker positions. However, there is a small difference in the effect stress-strain response when compared with the biaxial data. The differences could be attributed to multiple factors, but we believe the discrepancy is predominantly due to the effect of preconditioning. Preconditioning is a part of the standard biaxial experiment protocol and was also used for obtaining the experimental results in the current setup. However, its effect is not entirely understood and one could argue that the leaflets were not preconditioned before loading the valve under static pressure head. Additionally, the material models for soft tissues are usually not sufficient for extrapolation of results. Since the strain path employed during biaxial tests is sufficiently different from those during the natural loading of valve, some differences can be expected. The importance of present results and its possible extension to become a viable clinical tool are discussed next.

4   Discussion

Biomechanical properties of heart valve leaflets play an important role in determining their durability as well as susceptibility to disease development. Abnormal mechanical behavior leads to changes in the stress state of VICs and, eventually causing them to adopt a pathological phenotype. Therefore, changes in the mechanical properties of valve leaflets during their lifetime could be used an indicator for pathological functionality. However, the current methods for determining biomechanical properties are limited by their need for an excised valve leaflet. Therefore, a tool that could be used for going from in-vivo imaging to mechanical properties of the valve will be very useful. Additionally, such a

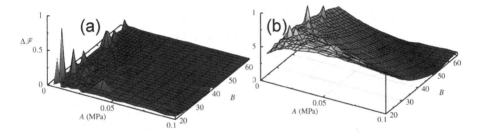

Fig. 4. Changes in cost function due to variation in fiber splay: (a) using average value of 0.58 rad and (b) using a value of 0.5 rad

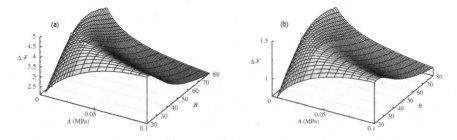

Fig. 5. Changes in the cost function due to noise in the input data – uniform/systematic noise (a) and random Gaussian noise with standard deviation of 0.2 mm (b)

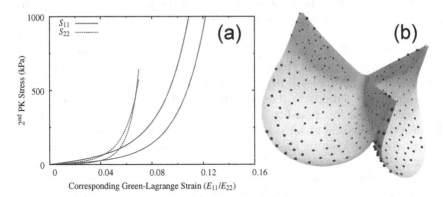

Fig. 6. (a) Comparison of stress-strain response from experiment and inverse model. (b) The final fit to the point cloud using current framework.

tool will also be useful in determining the causes of failure of prosthetic valves, and help us improve their designs in the long run.

The framework developed in this paper presents an approach for inverse model development for heart valves exploiting mapped leaflet structures. Herein each leaflet is considered to have homogenous elastic parameters. The novelty of the present work lies in the incorporation of mapped fiber data of each leaflet

Table 1. Statistics of the final fit

Pressure	Leaflet	mean($|d|$)	std($|d|$)	max($|d|$)
40 mm	1	0.195	0.150	0.718
40 mm	2	0.238	0.196	0.977
40 mm	3	0.182	0.124	0.619
80 mm	1	0.214	0.169	0.786
80 mm	2	0.294	0.232	1.150
80 mm	3	0.199	0.140	0.574
120 mm	1	0.270	0.216	0.938
120 mm	2	0.270	0.238	1.155
120 mm	3	0.196	0.137	0.611

and utilizing it through simplified structural model [8]. Differences in fiber archi-tectures of normal and pathological valve leaflets previously determined [4] can be combined with this tool to predict abnormal biomechanical behaviors. The sensitivity and convergence studies provide guidelines for creating this tool in an effective manner, e.g. the imaging resolution is a critical component of accurate determination, and higher number of frames leads to higher confidence.

To develop a method for evaluating the material parameters from in-vivo imaging, we are also working on other required components. The average fiber architecture of valve leaflets was calculated using a novel spline technique [4] and the pre-strain in leaflets was estimated by comparing the in-vivo shape and explanted leaflets' shape [10]. This information when combined with the inverse model presented in this work will lead to an in-vivo assessment tool for heart valves. Such a tool will help determine the mechanical properties of heart valves and find relations to disease development and help diagnose problems in the mechanical functionality at an early stage. Additionally, this approach will have the potential to serve as a general-purpose in-vivo assessment tool for heart valves for evaluating the performance of replaced prosthetic valves as well as monitoring the progression of valve diseases. In future, the method will be extended for the parameters to vary spatially. The information thus produced over time will help us understand the mechanical behavior of valve leaflets and design better surgical tools to repair and replace them.

Acknowledgment. We gratefully acknowledge the help from Will Zhang with para-meter estimation and analysis of biaxial data. This work was supported by NIH Grant HL108330 and Moncrief Chair funds (M.S.S.) and American Heart Association Post-doctoral Fellowship 14POST18720037 (A.A.). The authors acknowledge the Texas Advanced Computing Center at The University of Texas at Austin for providing HPC resources.

Appendix

Exponential function of the form $f(A, B, x) = A(e^{Bx} - 1)$, where $x \in \mathbb{R}^+$ is the independent variable, is very common in soft tissue mechanics to describe the stress-strain behavior so that its stiffness increases linearly with stress. Here $A \in \mathbb{R}^+$ and $B \in \mathbb{R}^+$ are the material parameters that determine the quantitative nature of this function or stress-strain relationship. However, many different pairs of (A, B) values can give very similar responses. To analyze this aspect, we first construct a functional that calculates the difference between two functions with different set of parameters:

$$\mathcal{F}(A, B, \bar{A}, \bar{B}, \epsilon) = \int_0^\epsilon \left[A(e^{Bx} - 1) - \bar{A}(e^{\bar{B}x} - 1) \right]^2 dx. \tag{7}$$

$\epsilon \in \mathbb{R}^+$ is the upper strain limit for a given application. To find the curve in $A - B$ parameter space along which the function $f(A, B, x)$ is "closest" to $f(\bar{A}, \bar{B}, x)$, we minimize the functional $\mathcal{F}$ for a given B, i.e.

$$\underset{A \in \mathbb{R}^+}{\arg \min} \mathcal{F}(A, B, \bar{A}, \bar{B}, \epsilon) = \underset{A \in \mathbb{R}^+}{\arg \min} \int_0^\epsilon \left[A(e^{Bx} - 1) - \bar{A}(e^{\bar{B}x} - 1) \right]^2 dx = A^{\min}(B, \bar{A}, \bar{B}, \epsilon) \tag{8}$$

One can obtain a closed form solution under reasonable conditions (which are satisfied if $\epsilon \leq 1$, something usually true for strain):

$$A^{\min} \left[g(2B) - 2g(B) + 1 \right] = \bar{A} \left[g(B + \bar{B}) - g(B) - g(\bar{B}) + 1 \right], \tag{9}$$

where,

$$g(B) = \frac{e^{B\epsilon} - 1}{B\epsilon}. \tag{10}$$

References

1. Sun, W., Abad, A., Sacks, M.S.: Simulated bioprosthetic heart valve deformation under quasi-static loading. J. Biomech. Eng. **127**(6), 905–914 (2005)
2. Sacks, M., Smith, D., Hiester, E.: A small angle light scattering device for planar connective tissue microstructural analysis. Ann. Biomed. Eng. **25**(4), 678–689 (1997)
3. Maas, S.A., Ellis, B.J., Ateshian, G.A., Weiss, J.A.: Febio: finite elements for biomechanics. J. Biomech. Eng. **134**(1), 011005 (2012)
4. Aggarwal, A., Ferrari, G., Joyce, E., Daniels, M.J., Sainger, R., Gorman III, J.H., Gorman, R., Sacks, M.S.: Architectural trends in the human normal and bicuspid aortic valve leaflet and its relevance to valve disease. Ann. Biomed. Eng. **42**(5), 986–998 (2014)
5. Sacks, M., Yoganathan, A.: Heart valve function: a biomechanical perspective. Philos. Trans. Royal Soc. B: Biol. Sci. **362**(1484), 1369–1391 (2007)

6. Sacks, M., David Merryman, W., Schmidt, D.: On the biomechanics of heart valve function. J. Biomech. **42**(12), 1804–1824 (2009)
7. Sacks, M., et al.: Incorporation of experimentally-derived fiber orientation into a structural constitutive model for planar collagenous tissues. J. Biomech. Eng. **125**(2), 280 (2003)
8. Fan, R., Sacks, M.S.: Simulation of planar soft tissues using a structural constitutive model: finite element implementation and validation. J. Biomech. **47**(9), 2043–2054 (2014)
9. Xi, J., Lamata, P., Niederer, S., Land, S., Shi, W., Zhuang, X., Ourselin, S., Duckett, S.G., Shetty, A.K., Rinaldi, C.A., Rueckert, D., Razavi, R., Smith, N.P.: The estimation of patient-specific cardiac diastolic functions from clinical measurements. Med. Image Anal. **17**(2), 133–146 (2013)
10. Aggarwal, A., Aguilar, V.S., Lee, C.H., Ferrari, G., Gorman, J.H., Gorman, R.C., Sacks, M.S.: Patient-specific modeling of heart valves: from image to simulation. In: Ourselin, S., Rueckert, D., Smith, N. (eds.) FIMH 2013. LNCS, vol. 7945, pp. 141–149. Springer, Heidelberg (2013)

Patient-Specific Biomechanical Modeling of Cardiac Amyloidosis – A Case Study

D. Chapelle[1]([⊠]), A. Felder[1], R. Chabiniok[1,3], A. Guellich[2], J.-F. Deux[2], and T. Damy[2]

[1] Inria Saclay Ile-de-France, MΞDISIM Team, Palaiseau, France
dominique.chapelle@inria.fr
[2] Henri Mondor Hospital, Créteil, France
[3] St. Thomas' Hospital, King's College London, London, UK

Abstract. We present a patient-specific biomechanical modeling framework and an initial case study for investigating cardiac amyloidosis (CA). Our patient-specific heartbeat simulations are in good agreement with the data, and our model calibration indicates that the major effect of CA in the biophysical behavior lies in a dramatic increase of the passive stiffness. We also conducted a preliminary trial for predicting the effects of pharmacological treatments – which is an important clinical challenge – based on the model combined with a simple venous return representation. This requires further investigation and validation, albeit provides some valuable preliminary insight.

Keywords: Cardiac modeling · Patient-specific · Amyloidosis · Heart failure

1 Introduction

Amyloidosis is a systemic disease with extracellular deposition of insoluble aggregates of beta-fibrillar proteins in various organs, leading to disturbances of normal tissue architecture and function [12]. Cardiac involvement is associated with poor prognosis. Morphological and functional abnormalities combine increase in cardiac wall thickness due to amyloid infiltration, and left ventricle (LV) diastolic and systolic dysfunction.

Global LV systolic function evaluated by ejection fraction is often preserved until advanced stages of the disease [4]. More recently, several studies have shown the ability of myocardial strain derived from echocardiography to detect subtle changes in systolic function in various diseases [6]. This approach seems to be useful in detecting cardiac involvement and was used to evaluate the prognosis of cardiac amyloidosis (CA) [8,9,18]. Moreover, magnetic resonance imaging (MRI) provides clinically valuable information beyond systolic function such as characterizing the myocardial tissue, and MRI late gadolinium enhancement

---

A. Felder—Inria Saclay Ile-de-France, MΞDISIM Team, Palaiseau, France

© Springer International Publishing Switzerland 2015
H. van Assen et al. (Eds.): FIMH 2015, LNCS 9126, pp. 295–303, 2015.
DOI: 10.1007/978-3-319-20309-6_34

(LGE) has been shown to be helpful for assessing the severity of cardiac amyloid deposition [16].

Cardiological treatment of CA is also challenging as the usual drugs used in heart failure – such as those with negative inotropic and/or negative chronotropic effects (calcium channel blockers, beta-blockers) – are contraindicated. Furthermore, inotropic drugs may not be tolerated either due to the restrictive type of the cardiopathy.

Therefore, being able to formulate a biophysical model integrating the essential mechanisms of CA is expected to be valuable in order to better understand the pathophysiology of CA and the effect of drugs with different pharmacological properties. Moreover, setting up this type of model on a patient-specific basis [2] can be envisioned as a means to ultimately allow for optimizing therapeutic strategies based on predictive personalised modeling tools.

In the present article we sought to (1) build a personalised cardiac biophysical model of a patient with cardiac amyloidosis, (2) validate the model, (3) explore the effects of different pharmacological interventions using the model.

2   Data

Patient description – 72 year old man, with a CA due to a punctual mutation on TTR gene (Val122Ile). The patient gave his informed consent for the study.

MR Imaging Protocol and Processing – Cardiac MRI was performed with a 1.5-T scanner (Magnetom Avanto, Siemens Healthcare). Contiguous short-axis sections encompassing the ventricles were acquired by the k-space segmented Steady State Free Precession Cine sequence, with the following parameters: TR/TE, 2.8/1.4 (apparent TR, 31.4 ms; 11 segments); flip angle, 82°; matrix size, 192 × 192; FOV, 300 × 270 mm; slice thickness, 8 mm. Retrospective ECG gating was used with 25 phases per section. We then segmented the two ventricles in the end-diastolic images using the CardioViz3D software [17], and the Image Registration Toolkit (IRTK, under license from Ixico Ltd., see [13]) was used – with the motion tracking algorithm of [15] – to propagate the segmentation in all other time frames, hence, to compute LV volumes.

Echocardiography Data Acquisition and Analysis – Echocardiography was performed using a Vivid 7 system (GE Vingmed). Data were analysed offline using the ECHOPac software (GE Healthcare). Acquisitions were digitally recorded at a high frame rate (>60 frames/s) in three consecutive cycles. Longitudinal strain was computed from the standard LV apical views (2, 3 and 4 cavities) using 2D speckle tracking echocardiography analysis by automated function imaging (AFI, ECHO-Pac, GE). For strain processing, the region of interest (ROI) was automatically positioned to track the LV speckles frame-by-frame throughout the cardiac cycle. The endocardial contour and width were manually adjusted when necessary to provide optimal tracking. LV global (GLS) and regional longitudinal strain values of the 17 AHA classification segments were obtained. Average longitudinal strain was also calculated at the three LV levels (apical, mid-cavity and basal).

Pressure Measurements – The catheterization procedure was performed in the fasting state without sedation. Systemic blood pressure, central venous pressure, pulmonary artery pressure, and pulmonary capillary wedge pressure (PCWP) were determined over an average of 5 end-expiratory cardiac cycles, at steady state, with the patient in the supine position. The 0 mmHg reference level was the mid-axillary line with the patient in the supine position. The pressure transducer was carefully maintained at the same level throughout the procedure. PCWP was assessed in West's zone 3, as verified by fluoroscopy, and found to be around 25 mmHg, a clear symptom of chronic heart failure.

The above 3 sets of data were acquired separately, albeit all within 24 h, under the exact same medication, and with similar cardiac frequencies (around 80 bpms).

3 Methods

3.1 Hierarchical Modeling Strategy

We employ the same cardiac biomechanical model as in [1], for which the fundamentals have been presented and analysed in [3]. This model was also experimentally validated in [1], while some clinical validations of its predictive capabilities were given in [14].

As proposed in [1], in combination with a detailed 3D continuum mechanics model, we also consider a simplified – albeit consistent – model of the left ventricle. This simplified model is derived from the 3D model by assuming some spherical symmetry properties while retaining the full complexity of the constitutive properties, based on multi-scale considerations and compatible with the fundamental thermo-mechanical laws. Of course, spherical symmetry is not quite present in an actual left ventricle, but the resulting model – being in the form of a set of ordinary differential equations – is mostly used for fast simulation and calibration purposes. As explained in [1] we can very effectively calibrate the biophysical parameters using this simplified model, and then carry over the parameters to the 3D model in a straightforward manner.

3.2 Personalisation Strategy

The patient-specific anatomical model was produced by using the segmentation of the end-systolic MR frame. The resulting surface mesh was further processed using the Yams software, and then the Ghs3D software was used to create a 3D tetrahedral mesh for a two-cavity (ventricular) model [5]. All the subsequent steps – for preparing the model and running simulations – were performed with the HeartLab software[1]. First, specific regions of interest were designated on the outer surfaces to provide for adequate boundary conditions [14], then the tetrahedral elements were tagged according to the AHA regions specifications. Furthermore, fiber directions were prescribed based on a given variation of the elevation angle across the thickness of the walls, see Fig. 1.

---

[1] Proprietary software of Inria (main author P. Moireau).

Fig. 1. Personalised anatomical model: AHA regions (left) and fiber directions (right)

Concerning the personalisation of the biophysical parameters, in this preliminary study we did not resort to automatic estimation methods as in [2], and limited ourselves to – mostly global – manual calibration based on the simplified model and then carried over to the complete 3D model. Nevertheless, this calibration process is quite complex and requires some progressive systematic strategy as explained in [1].

3.3   Venous Return Model

A major challenge for this type of patient lies in determining an adequate therapeutic strategy – usually by a combination of drugs and pacing device – specifically adapted to the patient in his/her current state, susceptible to quite rapid variations. Nevertheless, the actual cardiac state – e.g. as observed in the pressure-volume loop – is governed not only by the cardiac behavior coupled with the arterial response that determines the afterload, as accounted for in our above model, but also by the venous return that determines the preload. The latter is not included in our modeling components so far, as in fact we *prescribed* the preload. In order to account for variations in venous return as well, we use a description classically employed in cardiac physiology – valid for stationary states – that relies on total blood conservation written in the form [7]

$$C_{vs}P_{vs} + C_d P_d = V_{\text{eff}}, \tag{1}$$

where P_d and P_{vs} respectively denote the distal arteries and venous system pressures, while C_d and C_{vs} denote the corresponding global elastance parameters, and V_{eff} represents the total amount of blood that can fluctuate between the arterial and venous compartments. In addition, the distal flow between the arteries and veins is given by

$$Q_d = \frac{P_d - P_{vs}}{R_d}, \tag{2}$$

where R_d denotes the distal resistance. Of course, in a stationary regime this flow equals the cardiac output Q, hence we can use (1) to eliminate P_d from (2) and obtain

$$Q = \frac{V_{\text{eff}}}{R_d C_d} - \left(1 + \frac{C_{vs}}{C_d}\right)\frac{P_{vs}}{R_d}, \tag{3}$$

namely, a linear relationship between cardiac output and venous pressure.

4   Results

4.1   Model Calibration and Simulation Results

Compared with previous modeling trials, our calibration here was guided by clinical evidence that amyloidosis results into a stiffer passive behavior, while the slope of the end-systolic pressure-volume relationship (ESPVR) is roughly preserved – albeit shifted towards large volumes [11]. As a consequence, we increased the passive coefficients without changing the contractility parameter – compared to a normal heart – while adjusting the Frank-Starling function to operate with larger volumes, see [1] for a definition of this function. We show in Fig. 2 the resulting LV pressure, volume and flow indicators computed with the 3D model and compared with the processed data. We also show in Fig. 3 a comparison of the model boundaries with MRI in short axis view at end-systole and end-diastole. The calibrated biophysical parameters are given in Table 1, compared with those calibrated in [1] – in the same units – for a normal heart. As we immediately observe here, the main difference in these parameters concerns the passive law, for which an increase by a factor 20 or so – for the main two parameters (C_0, C_2) – was found to be consistent with the data.

Having at our disposal peak longitudinal strain data in AHA regions, we proceeded further in the calibration by allowing the contractility parameter to vary over these regions. We can thus account for the non-uniform deposition of

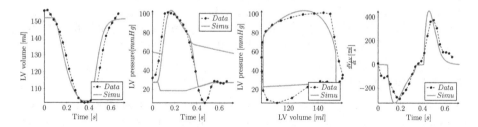

Fig. 2. Comparison of simulation results and data for LV. From left to right: volume; pressure; PV-diagram; outgoing flow. Black dots correspond to MR frames, registered with pressure recordings based on R-waves of ECGs

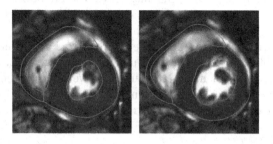

Fig. 3. Comparison of simulated mesh contours with MR data at end-systole (left) and end-diastole (right)

Table 1. Comparison of CA patient-specific model parameters with normal heart from [1]. (C_0, C_2) and (C_1, C_3) denote elastic parameters, η viscous modulus, σ_0 contractility (active behavior), (R_p, C_p) and (R_d, C_d) Windkessel parameters (proximal and distal)

	C_0, C_2	C_1, C_3	σ_0	R_p	C_p	R_d	C_d	η
CA	$1.1\,10^5$	$1.1\,10^{-1}$	$4.2\,10^5$	$1.8\,10^7$	$5.6\,10^{-10}$	$1.1\,10^8$	$2\,10^{-8}$	560
Normal	$5.7\,10^3$	$1.1\,10^{-1}$	$6.2\,10^5$	$2\,10^7$	$2\,10^{-9}$	$2\,10^8$	$0.5\,10^{-8}$	70

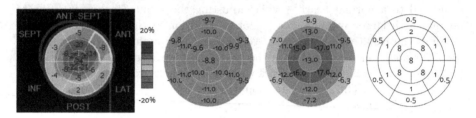

Fig. 4. Comparison of simulation results and processed data for peak longitudinal systolic strain in AHA regions. From left to right: data (AHA region #1 surrounded in bold yellow); simulation with uniform contractility; with non-uniform contractility; non-uniform contractility scaling coefficients (Color figure online)

amyloid fibrils in the myocardium, with a much higher density near the base [10]. We accordingly prescribed a varying contractility by applying a scaling coefficient in the AHA regions. This is shown in Fig. 4 with the resulting longitudinal strains averaged within each AHA region, and compared with the data.

4.2   Prediction of Patient States

We now want to conduct a preliminary study of the predictive capabilities of our modeling, using the simplified cardiac model presented above together with the venous return relation (3). Assuming a fixed difference between the preload and the venous return – prescribed to 7.5 mmHg in our simulations, in accordance with our pressure data – the cardiac function itself determines another relation between the venous pressure and the output, via the Frank-Starling mechanism. In fact, we can very effectively use our above simplified model to characterize this relation by running simulations with a given preload until the cardiac cycle stabilizes. The current cardiac state is then summarized by the intersection of the two curves. We plot these curves for our patient in Fig. 5-left, compared with those obtained for normal hearts. In our case, we inferred the venous return curve (3) by observing that the maximum flow equals $V_{\text{eff}}/R_d C_d$, where there is no reason to believe that V_{eff} differs from a physiological value. Subsequently, as the operating point is known – from the data – for the patient we can draw the whole line, and infer C_{vs}.

With this simplified – albeit global and personalized – model of the cardiovascular system we can investigate the effect of various drugs on the patient.

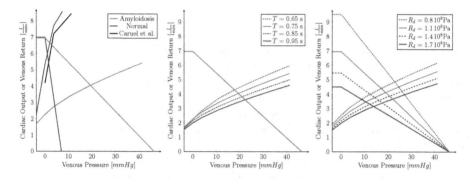

Fig. 5. Computed cardiac function and venous return curves. From left to right: patient in baseline compared to normal heart in [7] (green) and [1] (black); rhythm variations (*T* denotes beat period in s); systemic resistance variations (Color figure online)

Clearly, judging from the differences in (1) cardiac states, and (2) slopes of the cardiac and venous functions in Fig. 5-left, we can presume that the patient's responses will be very different from those of a healthy individual. In Fig. 5-center we show the effect of varying the heart rhythm, which can be produced by using beta-blockers, in particular. In our patient's case, lowering the rhythm – an effect of beta-blockers – would decrease the output and increase the – already dangerously high – venous pressure, hence this should be avoided. In Fig. 5-right we display the effect of varying the systemic resistance, as could be obtained with ACE-inhibitors. The effect is more complex, as both cardiac function and venous return curves are affected by this parameter. We see that a decrease in systemic resistance – as induced by ACE-inhibitors – would improve the output, but again at the cost of increasing the venous pressure, quite importantly in fact due to the low slope of the cardiac curve.

5 Discussion

Concerning the numerical simulations of cardiac cycles, the main difference appearing in Fig. 2 in the global indicators between the numerical results and the data pertains to the initial part of the rapid filling phase, during which the measured LV pressure decreases significantly below the simulated value. In fact, we know from the ECG that the patient suffers from a strong delay in atrio-ventricular conduction, hence, during that initial filling phase the left atrium is already contracting. Therefore, the simulation-data discrepancy can only be due to a strong heterogeneity between venous, atrial and ventricular pressures during that phase, whereas our model is designed to instead equilibrate these pressures. This could be changed, but we do not consider this improvement as very important from the point of view of the overall cardiac function.

Strain results obtained with varying contractility have already shown some good adequacy with the data, recall Fig. 4. In order to obtain more accurate results, a more detailed estimation of the parameter values would be needed, see below.

6   Conclusions

We have presented a patient-specific modeling framework and an initial case study for investigating cardiac amyloidosis. Our patient-specific heartbeat simulations show very good agreement with the data as regards the pressure-volume indicators, and reasonable agreement with the strain measurements when allowing the physical parameters to spatially vary. Better accuracy in the strains would require resorting to automatic parameter estimation as in [2], which is a natural perspective of this study.

We also conducted a preliminary trial for predicting the effects of pharmacological treatments on the patient based on the model. This of course requires further investigation and validation, including with retrospective assessments.

Our case study indicates that the major effect of CA in the biophysical behavior lies in a dramatic increase of the passive stiffness. This needs to be confirmed by including other similar cases in a further study, albeit provides some valuable preliminary insight. Also, beyond the specific type of CA considered here – quite important in terms of prevalence with 4% of African-American people carrying such Val122Ile gene mutations – other types of CA warrant similar investigations.

Acknowledgments. The authors are very grateful to Philippe Moireau for valuable discussions and for the use of his numerical simulation software HeartLab, and also wish to thank Gabriel Valdes Alonzo (intern) for some helpful numerical verifications.

References

1. Caruel, M., et al.: Dimensional reductions of a cardiac model for effective validation and calibration. Biomech. Model. Mechan. **13**(4), 897–914 (2014)
2. Chabiniok, R., et al.: Estimation of tissue contractility from cardiac cine-MRI using a biomechanical heart model. Biomech. Model. Mechanobiol. **11**(5), 609–630 (2012)
3. Chapelle, D., et al.: Energy-preserving muscle tissue model: formulation and compatible discretizations. Int. J. Multiscale Comput. **10**(2), 189–211 (2012)
4. Damy, T., et al.: Role of natriuretic peptide to predict cardiac abnormalities in patients with hereditary transthyretin amyloidosis. Amyloid **20**(4), 212–220 (2013)
5. Frey, P., George, P.L.: Mesh Generation, 2nd edn. Wiley, New York (2008)
6. Gray Gilstrap, L., et al.: Predictors of survival to orthotopic heart transplant in patients with light chain amyloidosis. J. Hear. Lung Transplant. **33**(2), 149–156 (2014)
7. Koeppen, B.M., Stanton, B.A.: Berne and Levy Physiology, 6th edn. Mosby, Philadelphia (2009)
8. Koyama, J., Falk, R.H.: Prognostic significance of strain Doppler imaging in light-chain amyloidosis. J. Am. Coll. Cardiol. Imging **3**(4), 333–342 (2010)
9. Koyama, J., et al.: Longitudinal myocardial function assessed by tissue velocity, strain, and strain rate tissue Doppler echocardiography in patients with AL (primary) cardiac amyloidosis. Circulation **107**(19), 2446–2452 (2003)

10. Liu, D., et al.: Echocardiographic evaluation of systolic and diastolic function in patients with cardiac amyloidosis. Am. J. Cardiol. **108**(4), 591–8 (2011)
11. McCarthy, R.E., Kasper, E.K.: A review of the amyloidoses that infiltrate the heart. Clin. Cardiol. **21**(8), 547–552 (1998)
12. Roig, E., et al.: Outcomes of heart transplantation for cardiac amyloidosis: sub-analysis of the Spanish registry for heart transplantation. Am. J. Transplant. **9**(6), 1414–1419 (2009)
13. Rueckert, D., et al.: Nonrigid registration using free-form deformations: application to breast MR images. IEEE Trans. Med. Imaging **18**(8), 712–721 (1999)
14. Sermesant, M., et al.: Patient-specific electromechanical models of the heart for the prediction of pacing acute effects in CRT: a preliminary clinical validation. Med. Image Anal. **16**(1), 201–215 (2012)
15. Shi, W., et al.: A comprehensive cardiac motion estimation framework using both untagged and 3D tagged MR images based on non-rigid registration. IEEE Trans. Med. Imaging **31**(6), 1263–1275 (2012)
16. Syed, I.S., et al.: Role of cardiac magnetic resonance imaging in the detection of cardiac amyloidosis. J. Am. Coll. Cardiol. Imging **3**(2), 155–164 (2010)
17. Toussaint, N., et al.: An integrated platform for dynamic cardiac simulation and image processing: application to personalised tetralogy of Fallot simulation. In: Proceedings, Eurographics Workshop on Visual Computing for Biomedicine (2008)
18. Vikram, C.S.: Removing the diffraction halo effect in speckle photography of sinusoidal vibration. Appl. Opt. **29**(25), 3572–3573 (1990)

Relationship Between Cardiac Electrical and Mechanical Activation Markers by Coupling Bidomain and Deformation Models

Piero Colli-Franzone[1], Luca F. Pavarino[2]($^{(\boxtimes)}$), and Simone Scacchi[2]

[1] Dipartimento di Matematica, Istituto di Matematica Applicata e Tecnologie Informatiche, Università di Pavia and IMATI-CNR, Via Ferrata 1, 27100 Pavia, Italy
colli@imati.cnr.it
[2] Dipartimento di Matematica, Università di Milano,
Via Saldini 50, 20133 Milano, Italy
{luca.pavarino,simone.scacchi}@unimi.it

Abstract. The aim of this study is to simulate the electromechanical behavior of a cardiac wedge following an endo- or epicardial stimulation, and to study different markers of mechanical contraction times. We investigate how tissue anisotropy affects the performance of the mechanical markers and we evaluate their delay distributions with respect to the electrical activation time. The main results of this study show that: the electrical and mechanical activation sequences are very well correlated; the electromechanical delay displays heterogeneous distributions even if the electrical and mechanical cellular properties are assumed homogeneous; the electromechanical delay is larger in the regions where depolarization proceeds along fiber than across fiber.

Keywords: Cardiac electromechanical markers · Cardiac excitation and contraction · Orthotropic bidomain model · 3D parallel simulations

1 Introduction

In the intact heart, local myocyte contraction follows local electrical activation by some tens of milliseconds. This phenomenon is referred to in the literature as *electromechanical delay*. Previous experimental and simulation studies have shown that the distribution of electromechanical delay is not homogeneous throughout the tissue and depends on the electrical activation sequence, see e.g. [5,16,25]. The aim of the present work is to investigate how the orthotropic architecture of the ventricular tissue affects the spatial distribution of the electromechanical delay. To this end, we simulate the excitation-contraction coupling of a cardiac wedge following an endo- or epicardial stimulation, and we compute different markers of mechanical contraction times and the associated distributions of electromechanical delay with respect to the electrical activation time. Our simulation study is based on the cardiac electromechanical model consisting of four coupled systems of differential equations: (a) a nonlinear quasi-static

© Springer International Publishing Switzerland 2015
H. van Assen et al. (Eds.): FIMH 2015, LNCS 9126, pp. 304–312, 2015.
DOI: 10.1007/978-3-319-20309-6_35

finite elasticity system describing the mechanical deformation of the cardiac tissue, based on an orthotropic strain energy function, see [14,26]; (b) a system of nonlinear ordinary differential equations, known as active tension model, describing the dynamics of biochemically generated active force, proposed in [21]; (c) a nonlinear system of reaction-diffusion partial differential equations, known as Bidomain system, describing the electrical excitation and recovery of the cardiac tissue, based on [10]; (d) a system of nonlinear ordinary differential equations, known as ionic membrane model, describing the dynamics of the ionic currents through the myocyte membrane; here we use the ten Tusscher et al. membrane model [28] for human ventricular cells, augmented with stretch-activated channels. Previous studies of cardiac electromechanical models can be found in [1,2,7,8,10,12–15,19,20,27,29–31]. In order to describe the sequence of mechanical contraction, some markers of myocardial shortening have been studied in the experimental investigations on spontaneous beating or pacing hearts [3,4,18,25,32]. In this study, we quantitatively evaluate different markers of electrical activation and mechanical contraction, their correlations and spatial distributions, as well as the associated electromechanical delays.

2   Cardiac Electromechanical Models

(a) **Mechanical Model of Cardiac Tissue.** We denote by $\mathbf{X} = (X_1, X_2, X_3)^T$ the material coordinates of the undeformed cardiac domain $\widehat{\Omega}$, by $\mathbf{x} = (x_1, x_2, x_3)^T$ the spatial coordinates of the deformed cardiac domain $\Omega(t)$ at time t, by $\mathbf{F}(\mathbf{X}, t) = \frac{\partial \mathbf{x}}{\partial \mathbf{X}}$ the deformation gradient.

   The cardiac tissue is modeled as a nonlinear hyperelastic material satisfying the steady-state force equilibrium equation

$$\text{Div}(\mathbf{FS}) = \mathbf{0}, \qquad \mathbf{X} \in \widehat{\Omega}. \tag{1}$$

The second Piola-Kirchoff stress tensor $\mathbf{S} = \mathbf{S}^{pas} + \mathbf{S}^{vol} + \mathbf{S}^{act}$ is the sum of passive, volumetric and active components. The passive and volumetric components are defined as

$$S_{ij}^{pas,vol} = \frac{1}{2} \left(\frac{\partial W^{pas,vol}}{\partial E_{ij}} + \frac{\partial W^{pas,vol}}{\partial E_{ji}} \right) \quad i,j = 1,2,3,$$

where $\mathbf{E} = \frac{1}{2}(\mathbf{C} - \mathbf{I})$ is the Green-Lagrange strain tensor, W^{pas} is an exponential strain energy function modeling the myocardium as an orthotropic hyperelastic material (see [14]), and $W^{vol} = K(J-1)^2$ is a volume change penalization term, with K a positive bulk modulus and $J = det\mathbf{F}$, added in order to model the myocardium as nearly incompressible.

(b) **Mechanical Model of Active Tension.** The active component $\mathbf{S}^{act}$ is given in terms of the active tension, that we assume developed along the myofiber direction only, $\mathbf{S}^{act} = T_a \frac{\widehat{\mathbf{a}}_l \otimes \widehat{\mathbf{a}}_l}{\widehat{\mathbf{a}}_l^T \mathbf{C} \widehat{\mathbf{a}}_l}$, with $\widehat{\mathbf{a}}_l$ the fiber direction in the reference configuration and the biochemically generated active tension $T_a = T_a\left(Ca_i, \lambda, \frac{d\lambda}{dt}\right)$ is calcium, stretch ($\lambda = \sqrt{\widehat{\mathbf{a}}_l^T \mathbf{C} \widehat{\mathbf{a}}_l}$), and stretch-rate ($\frac{d\lambda}{dt}$)

dependent, with the dynamics described by a system of ODEs proposed in Land et al. [21].

(c) **Electrical Model of Cardiac Tissue: The Bidomain Model.** We will use the following parabolic-elliptic formulation of the modified Bidomain model on the reference configuration $\widehat{\Omega} \times (0, T)$,

$$
\begin{cases}
c_m J \frac{\partial \widehat{v}}{\partial t} - \mathrm{Div}(J \, \mathbf{F}^{-1} D_i \mathbf{F}^{-T} \mathrm{Grad}\, (\widehat{v} + \widehat{u}_e)) + J \, i_{ion}(\widehat{v}, \widehat{\mathbf{w}}, \widehat{\mathbf{c}}, \lambda) = 0 \\
-\mathrm{Div}(J \, \mathbf{F}^{-1} D_i \mathbf{F}^{-T} \mathrm{Grad}\, \widehat{v}) - \mathrm{Div}(J \, \mathbf{F}^{-1}(D_i + D_e) \mathbf{F}^{-T} \mathrm{Grad}\, \widehat{u}_e) = J \, \widehat{i}^{\,e}_{app} \quad (2) \\
\frac{\partial \widehat{\mathbf{w}}}{\partial t} - \mathbf{R}_w(\widehat{v}, \widehat{\mathbf{w}}) = 0, \qquad \frac{\partial \widehat{\mathbf{c}}}{\partial t} - \mathbf{R}_c(\widehat{v}, \widehat{\mathbf{w}}, \widehat{\mathbf{c}}) = 0.
\end{cases}
$$

for the transmembrane potential $\widehat{v}$ extracellular potential $\widehat{u}_e$ gating and ionic concentrations variables $(\widehat{\mathbf{w}}, \widehat{\mathbf{c}})$. This system is completed by prescribing initial conditions, insulating boundary conditions, and applied current $\widehat{i}^{\,e}_{app}$, see [11] for further details. The conductivity tensors are given by

$$
D_{i,e}(\mathbf{x}) = \sigma_l^{i,e} \, \mathbf{a}_l(\mathbf{x}) \mathbf{a}_l^T(\mathbf{x}) + \sigma_t^{i,e} \, \mathbf{a}_t(\mathbf{x}) \mathbf{a}_t^T(\mathbf{x}) + \sigma_n^{i,e} \, \mathbf{a}_n(\mathbf{x}) \mathbf{a}_n^T(\mathbf{x}), \qquad (3)
$$

where $\sigma_l^{i,e}$, $\sigma_t^{i,e}$, $\sigma_n^{i,e}$ are the conductivity coefficients in the intra- and extracellular media measured along and across the fiber direction $\mathbf{a}_l, \mathbf{a}_t, \mathbf{a}_n$.

(d) **Ionic Membrane Model and Stretch-Activated Channel Current.** The functions $I_{ion}(v, \mathbf{w}, \mathbf{c}, \lambda)$ ($i_{ion} = \chi I_{ion}$), $R_w(v, \mathbf{w})$ and $R_c(v, \mathbf{w}, \mathbf{c})$ in the Bidomain model (2) are given by the ionic membrane model by ten Tusscher et al. [28], available from the cellML depository (models.cellml.org/cellml). The ionic current is the sum $I_{ion}(v, \mathbf{w}, \mathbf{c}, \lambda) = I_{ion}^m(v, \mathbf{w}, \mathbf{c}) + I_{SAC}$ of the ionic term $I_{ion}^m(v, \mathbf{w}, \mathbf{c})$ given by the ten Tusscher model and a stretch-activated channel current I_{SAC}. This last current is modeled as in [22] as the sum of non-specific and specific currents $I_{SAC} = I_{SAC,n} + I_{Ko}$.

3   Methods

Space Discretization. We discretize the cardiac domain with an hexahedral structured grid T_{h_m} for the mechanical model (1) and T_{h_e} for the electrical Bidomain model (2), where T_{h_e} is a refinement of T_{h_m}, i.e. h_m is an integer multiple of h_e. We then discretize all scalar and vector fields of both mechanical and electrical models by isoparametric Q_1 finite elements in space.

Time Discretization. The time discretization is performed by a semi-implicit splitting method, where the electrical and mechanical time steps could be different. At each time step,

(a) given v^n, w^n, c^n at time t_n, solve the ODE system of the membrane model with a first order IMEX method to compute the new w^{n+1}, c^{n+1}.

(b) given the calcium concentration Ca_i^{n+1}, which is included in the concentration variables c^{n+1}, solve the mechanical problems (1) and the active tension system to compute the new deformed coordinates $\mathbf{x}^{n+1}$, providing the new deformation gradient tensor $\mathbf{F}_{n+1}$.

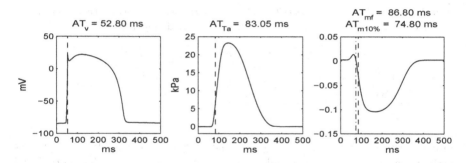

Fig. 1. Examples of transmembrane potential waveform v with markers AT_v (left), active tension T_a with AT_{Ta} (middle) and E_{ll} with AT_{mf} and $\text{AT}_{m10\%}$ (right).

(c) given w^{n+1}, c^{n+1}, $\mathbf{F}_{n+1}$ and $J_{n+1} = \det(\mathbf{F}_{n+1})$, solve the Bidomain system (2) with a first order IMEX method and compute the new electric potentials v^{n+1}, u_e^{n+1} with an operator splitting method.

We refer to [9] for more details.

Computational Kernels. Due to the employed space and time discretization strategies, at each time step, the main computational efforts consist of:

(i) solving the nonlinear system deriving from the discretization of the mechanical problem by a Newton-GMRES-Algebraic Multigrid method, see [10];
(ii) solving the two linear systems associated with the elliptic and parabolic Bidomain equations using the Conjugate Gradient method preconditioned by a Multilevel Additive Schwarz preconditioner studied in [23]. Our parallel simulations have been performed on a Linux cluster using the parallel library PETSc [6] from the Argonne National Laboratory.

Domain Geometry and Parameter Calibration. The cardiac domain $\widehat{\Omega} = \Omega(0)$ is the image of a cartesian slab using ellipsoidal coordinates, yielding a portion of truncated ellipsoid. The values of the orthotropic conductivity coefficients (see (3)) used in all the numerical tests are $\sigma_l^i = 3$, $\sigma_t^i = 0.31525$, $\sigma_n^i = 0.031525$, $\sigma_l^e = 2$, $\sigma_t^e = 1.3514$, $\sigma_n^e = 0.6757$, all expressed in $\text{m}\Omega^{-1}\text{cm}^{-1}$. The parameters of the orthotropic strain energy function are given in [14]. The bulk modulus is $K = 200\,\text{kPa}$.

Stimulation Site, Initial and Boundary Conditions. The depolarization process is started by applying a cathodal extracellular stimulus of $i_{app}^e = -250\,\text{mA/cm}^3$ lasting $1\,\text{ms}$ on a small volume of $0.4 \times 0.4 \times 0.2\,\text{mm}^3$ located in the center of the endocardial or epicardial surface. The initial conditions are at resting values for all the potentials and gating variables of the ten Tusscher model, while the boundary conditions are for insulated tissue. For what concerns the mechanical boundary conditions, to prevent rigid body motion, the displacement is set to zero at the base of the wedge, similarly to previous studies [16,17].

Table 1. Correlation coefficients (CC) between the three mechanical activation markers and the electrical activation marker, for endocardial and epicardial stimulations.

	$CC(\text{AT}_{Ta},\text{AT}_v)$	$CC(\text{AT}_{mf},\text{AT}_v)$	$CC(\text{AT}_{m10\%},\text{AT}_v)$
ENDO STIM	0.97	0.93	0.81
EPI STIM	0.94	0.90	0.80

Moreover, we assume that the wedge is not subject to any external loading, therefore homogeneous Neumann boundary conditions are prescribed elsewhere. In all the electromechanical simulations, the electrical mesh size is $h_e = 0.01\ cm$, while the mechanical mesh size is $h_m = 0.08\ cm$, and the electrical time step size is $\Delta_e t = 0.05\ ms$, while the mechanical times step is $\Delta_m t = 0.25\ ms$.

Time Markers. At selected nodes $\mathbf{x} \in \Omega$, we computed (see Fig. 1):

- the electrical activation time, defined as the unique instant $\text{AT}_v(\mathbf{x})$ of maximum time derivative of the transmembrane potential $v(\mathbf{x}, t)$ during the upstroke phase of the action potential;
- three markers of mechanical activation time (see [3,4,12,16,18,25,29,32]), defined as:
  - (i) the instant AT_{Ta} of maximum time derivative of the active tension $T_a(\mathbf{x}, t)$;
  - (ii) the instant AT_{mf} of maximum rate of the myofiber shortening, i.e. the minimum time derivative of the fiber strain $E_{ll} = \hat{\mathbf{a}}_l^T \mathbf{E} \hat{\mathbf{a}}_l = (\lambda(\mathbf{x}, t)^2 - 1)/2$;
  - (iii) the instant $\text{AT}_{m10\%}$ when, starting from the onset of myofiber shortening, the fiber strain E_{ll} decreases below the threshold $M - 10\%(M - m)$, where $M = \max_t E_{ll}(\mathbf{x}, t)$ and $m = \min_t E_{ll}(\mathbf{x}, t)$ are both computed during the contraction phase.

4   Results

Figure 2 shows the spatial distributions of the electrical activation marker AT_v and of the mechanical activation markers AT_{Ta}, AT_{mf} and $\text{AT}_{m10\%}$, on the epicardial, midmyocardial and endocardial surfaces, in case of epicardial stimulation. The three patterns of the mechanical activation markers match well with the electrical activation sequence, with an average delay ranging between 15 and 35 ms, exhibiting a faster mechanical contraction along fibers than across similarly to the electrical propagation. The strong qualitative agreement between the mechanical and the electrical activation sequences is confirmed by the correlation coefficients reported in Table 1, which range from 0.97 for the pair AT_{Ta}, AT_v to 0.8 for the pair $\text{AT}_{m10\%}$, AT_v. The correlation coefficients of the pair $\text{AT}_{m10\%}$, AT_v on each epi- endocardial layer (not shown) decrease toward the breakthrough layer.

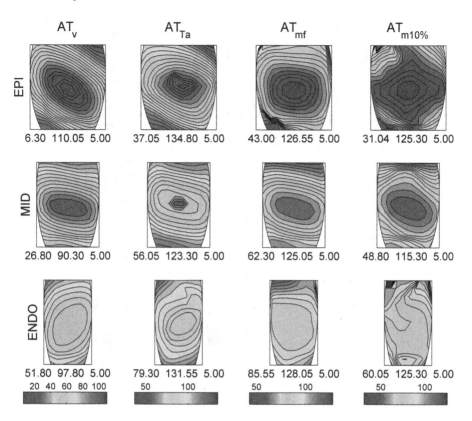

Fig. 2. Epicardial stimulation. Distributions of the electrical activation marker AT_v (first column) and mechanical activation markers AT_{Ta} (second column), AT_{mf} (third column), $AT_{m10\%}$ (fourth column), on epicardium (top row), midwall (central row) and endocardium (bottom row). Below each contour plot are reported the minimum, maximum, step size of the corresponding map.

Figure 3 reports the epicardial, midwall and endocardial distributions of the delay of the mechanical activation times AT_{Ta}, AT_{mf} and $AT_{m10\%}$ with respect to the electrical activation time AT_v. The delays are computed as the differences $AT_{Ta} - AT_v$, $AT_{mf} - AT_v$ and $AT_{m10\%} - AT_v$. The results show that the delay of mechanical activation is large in the regions where the depolarization wavefront proceeds along fiber ($-45°$ at epicardium, $0°$ at midwall, $+72°$ at endocardium). Moreover, we observe that the cellular mechanical activation time AT_{Ta} is always delayed with respect to the electrical activation time AT_v. On the other hand, the delay of AT_{mf} and $AT_{m10\%}$ is negative in the regions where the propagation is across fibers, indicating that the contraction of these regions, driven by both the passive deformation and the contraction of the neighboring intramurally activated regions, precedes the local electrical activation. Similar considerations on mechanical activation maps and delay distributions apply also in case of an endocardial stimulation (not shown).

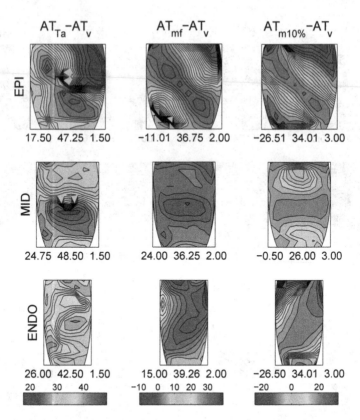

Fig. 3. Epicardial stimulation. Distributions of electromechanical delays $\mathrm{AT}_{Ta}-\mathrm{AT}_v$ (first column), $\mathrm{AT}_{mf}-\mathrm{AT}_v$ (second column), $\mathrm{AT}_{m10\%}-\mathrm{AT}_v$ (third column), on epicardium (top row), midwall (central row) and endocardium (bottom row). Below each contour plot are reported the minimum, maximum, step size of the corresponding map.

5   Discussion and Conclusions

The main findings of this study are: (i) the electrical and mechanical activations sequences are very well correlated ($CC \geq 0.8$), particularly for the AT_{mf}, AT_v pair ($CC \geq 0.9$) and, as expected, for the "local" AT_{Ta}, AT_v pair ($CC \geq 0.94$); (ii) the spatial distributions of electromechanical delay depends on the mechanical activation marker considered; (iii) the electromechanical delay distributions display heterogeneous patterns, even if the electrical and mechanical cellular properties are assumed homogeneous; (iv) in the regions where depolarization proceeds along fiber, the electromechanical delay is larger than in the regions where propagation is mainly across fiber. Understanding the relationship between electrical and mechanical activation times is important for the design of effective therapeutical pacing strategies aimed at resynchronize the contraction of ventricles. Since large electromechanical delays have been observed in diseased and/or dyssynchronous hearts, our findings might lead to a better insight into the mechanisms and distribution of electromechanical delays and into the optimization of cardiac resynchronization therapies.

Limitations. We have used an idealized ellipsoidal geometry instead of a realistic biventricular geometry and homogeneous cellular properties, in order to separate the study of electromechanical delay heterogeneity from the effects of complex cardiac geometries and intrinsic cellular heterogeneities. We have also disregarded the effects of intracavitary blood pressure on the endocardial surface. In a future study, we plan to extend this work to include: heterogeneous transmural and apex-base cellular properties, the repolarization/relaxation phase, active tension developed also across fiber (see e.g. [7,24]), realistic cardiac anatomies and the presence of Windkessel pressure boundary conditions on the endocardial surface.

References

1. Adeniran, I., Hancox, J.C., Zhang, H.: Effect of cardiac ventricular mechanical contraction on the characteristics of the ECG: a simulation study. J. Biomed. Sci. Eng. **6**, 47–60 (2013)
2. Ambrosi, D., et al.: Electromechanical coupling in cardiac dynamics: the active strain approach. SIAM J. Appl. Math. **71**, 605–621 (2011)
3. Ashikaga, H., et al.: Transmural dispersion of myofiber mechanics. J. Am. Coll. Cardiol. **49**, 909–916 (2007)
4. Augustijn, C.H., et al.: Mapping the sequence of contraction of the canine left ventricle. Pflug. Arch. **419**, 529–533 (1991)
5. Badke, F.R., Boinay, P., Covell, J.W.: Effects of ventricular pacing on regional left ventricular performance in the dog. Am. J. Physiol. **238**, H858–H867 (1980)
6. Balay, S., et al.: PETSc users manual. Technical report, ANL-95/11 - Revision 3.3, Argonne National Laboratory (2012)
7. Bovendeerd, P.H.M., Kroon, W., Delhaas, T.: Determinants of left ventricular shear strain. Am. J. Physiol. HCP **297**, H1058–H1068 (2009)
8. Carapella, V., et al.: Quantitative study of the effect of tissue microstructure on contraction in a computational model of rat left ventricle. PLoS One **9**, e92792 (2014)
9. Colli Franzone, P., Pavarino, L.F., Scacchi, S.: Mathematical Cardiac Electrophysiology. MSA, vol. 13. Springer, New York (2014)
10. Colli Franzone, P., Pavarino, L.F., Scacchi, S.: Parallel multilevel solvers for the cardiac electro-mechanical coupling. Appl. Numer. Math. (To appear) (2014)
11. P. Colli Franzone, L. F. Pavarino, and S. Scacchi, A numerical simulation study of the influence of mechanical feedbacks on the bioelectrical activity in a cardiac electro-mechanical model, Math. Mod. Meth. Appl. Sci. (2015) (Submitted)
12. Constantino, J., Hu, Y., Trayanova, N.A.: A computational approach to understanding the cardiac electromechanical activation sequence in the normal and failing heart, with translation to the clinical practice of CRT. Progr. Biophys. Mol. Biol. **110**, 372–379 (2012)
13. de Oliveira, B.L., et al.: Effects of deformation on transmural dispersion of repolarization using in silico models of human left ventricular wedge. Int. J. Numer. Meth. Biomed. Eng. **29**, 1323–1337 (2013)

14. Eriksson, T.S.E., et al.: Influence of myocardial fiber/sheet orientations on left ventricular mechanical contraction. Math. Mech. Solids **18**, 592–606 (2013)

15. Fritz, T., et al.: Simulation of the contraction of the ventricles in a human heart model including atria and pericardium. Biomech. Mod. Mechanobiol. **13**(3), 627–641 (2014)

16. Gurev, V., et al.: Distribution of electromechanical delay in the heart: insights from a three-dimensional electromechanical model. Biophys. J. **99**, 745–754 (2010)

17. Kerckhoffs, R.C.P., et al.: Homogeneity of cardiac contraction despite physiological asyncrony of depolarization: a model study. Ann. Biomed. Eng. **31**, 536–547 (2003)

18. Kerckhoffs, R.C.P., et al.: Timing of depolarization and contraction in the paced canine left ventricle: model and experiment. J. Cardiovasc. Electrophysiol. **14**, S188–S195 (2003)

19. Kerckhoffs, R.C.P., et al.: Intra- and interventricular asynchrony of electromechanics in the ventricularly paced heart. J. Eng. Math. **47**, 201–216 (2003)

20. Kerckhoffs, R.C.P., et al.: Electromechanics of paced left ventricle simulated by straightforward mathematical model: comparison with experiments. Am. J. Physiol. HCP **289**, H1889–H1897 (2005)

21. Land, S., et al.: An analysis of deformation-dependent electromechanical coupling in the mouse heart. J. Physiol. **590**, 4553–4569 (2012)

22. Niederer, S.A., Smith, N.P.: A mathematical model of the slow force response to stretch in rat ventricular myocites. Biophys. J. **92**, 4030–4044 (2007)

23. Pavarino, L.F., Scacchi, S.: Multilevel additive Schwarz preconditioners for the Bidomain reaction-diffusion system. SIAM J. Sci. Comput. **31**, 420–443 (2008)

24. Pluijmert, M., et al.: Effects of activation pattern and active stress development on myocardial shear in a model with adaptive myofiber reorientation. Am. J. Physiol. HCP **306**, H538–H546 (2014)

25. Prinzen, F.W., et al.: The time sequence of electrical and mechanical activation during spontaneous beating and ectopic stimulation. Eur. Heart. J. **13**, 535–543 (1992)

26. Rossi, S., et al.: Thermodynamically consistent orthotropic activation model capturing ventricular systolic wall thickening in cardiac electromechanics. Eur. J. Mech. A-Solids **48**, 129–142 (2014)

27. Sainte-Marie, J., et al.: Modeling and estimation of cardiac electromechanical activity. Comput. Struct. **84**, 1743–1759 (2006)

28. ten Tusscher, K.H.W.J., et al.: A model for human ventricular tissue. Am. J. Phys. HCP **286**, H1573–H1589 (2004)

29. Usyk, T.P., McCulloch, A.D.: Relationship between regional shortening and asynchronous electrical activation in a three-dimensional model of ventricular electromechanics. J. Cardiovasc. Electrophysiol. **14**, S196–S202 (2003)

30. Wall, S.T., et al.: Electromechanical feedback with reduced cellular connectivity alters electrical activity in an infarct injures left ventricle: a finite element model study. Am J. Physiol. HCP **302**, H206–H214 (2012)

31. Weise, L.D., Nash, M.P., Panfilov, A.V.: A discrete model to study reaction-diffusion-mechanics systems. PLos One **6**(7), e21934 (2011)

32. Wyman, B., et al.: Mapping propagation of mechanical activation in the paced heart with MRI tagging. Am. J. Physiol. HCP **45**, H881–H891 (1999)

Influence of Polivinylalcohol Cryogel Material Model in FEM Simulations on Deformation of LV Phantom

Szymon Cygan[1(✉)], Jakub Żmigrodzki[1], Beata Leśniak-Plewińska[1], Maciej Karny[2], Zbigniew Pakieła[2], and Krzysztof Kałużyński[1]

[1] Institute of Metrology and Biomedical Engineering, Warsaw University of Technology, Warsaw, Poland
{s.cygan,j.zmigrodzki,b.lesniakplewinska,k.kaluzynski}@mchtr.pw.edu.pl
[2] Faculty of Materials Science and Engineering, Warsaw University of Technology, Warsaw, Poland
maciej_karny@wp.pl, zpakiela@inmat.pw.edu.pl

Abstract. One of the available tools for validation of strain imaging methods are physical phantoms, most frequently produced of polivinylalcohol cryogel (PVA). This material was often assumed to exhibit elastic properties, but it has more complex nature. In this work we examine the influence of the applied material model – elastic vs hyperelastic – on the strains within the numerical model of the phantom obtained from FEM. This influence appeared significant – hyperelastic model provides lower strain contrasts and also the ratios between radial, circumferential and longitudinal strains differ for both models.

Keywords: LV phantom · Strain imaging · Polivinylalcohol · PVA · Finite Element Method

1 Introduction

Fast development of cardiac strain imaging methods without clear benchmarking guidelines for evaluating their performance calls for development of reliable methods that will allow for quantitative assessment of strain imaging methods performance. These developments include two main directions of studies – application of physical models with known properties and numerical models for creating synthetic medical data [1]. Both approaches have their advantages. Physical models – phantoms or animal, make use of real ultrasound scanner systems and therefore provide more realistic data, but the ground truth data reliability and extent depends on many factors, including other measurement methods (sonomicrometry, mechanical testing etc.). Synthetic data based on numerical models provides extensive reference data, but on the other hand some simplifications concerning. e.g. the mechanics of the deformation or the ultrasound data generation may result in unrealistic data [1].

The most frequently used material for physical phantoms is the polyvinylalcohol (PVA) cryogel, question arises how its properties should be taken into account in Finite Element Method (FEM) model to make the latter a useful reference to physical experiments.

© Springer International Publishing Switzerland 2015
H. van Assen et al. (Eds.): FIMH 2015, LNCS 9126, pp. 313–320, 2015.
DOI: 10.1007/978-3-319-20309-6_36

This material is often assumed to be elastic when applied to LV phantoms [2, 3], but it has been shown that its behavior should be described as hyperelastic or even visco-hyperelastic [4]. This might not be of crucial importance when the "stiffness" is used to roughly give an idea of the properties and another method is applied as reference as was done in [3], but if the properties are to be used to model the behavior of the phantom, then the applied model should be chosen carefully.

2   Materials and Methods

2.1   Material Characterization

To obtain realistic data for modeling the PVA cryogel, mechanical examinations of a sample were carried out. A cylindrical sample with 34.26 mm in diameter and 28.85 mm in height was made from a liquid solution of 10 mass% (m%) PVA (Sigma Aldrich, St. Louis, USA) and 10 m% glycerin (Chempur, Piekary Slaskie, Poland). This solution was solidified by 2 cycles of freezing and thawing (24/24 h) at temperature of −25 °C, similarly as described in [3], where the addition of glycerin was justified.

The examinations were carried out on a MTS Qtest/10 material testing machine (MTS, USA) with deformation ranging up to 30 % of the original sample size. Strain-stress curves were registered. The measurements provided experimental data on one sample of PVA cryogel. Stiffer materials can be obtained through appropriate processing [3] – using higher PVA concentration and/or applying more freezing-thawing cycles. For the purpose of this study behavior of stiffer materials was simulated by linearly scaling the experimental strain-stress curves by 1.5, 2 and 3. This approach may not correspond to the complex way the properties of PVA change when composed or processed differently, but was assumed sufficient for the purpose of this study.

2.2   Numerical Model of a Phantom

Numerical model of the LV phantom, corresponding to a physical phantom, was designed as a half of a thick-walled ellipsoid with internal dimensions of 35 mm (cross-section diameter) and length of 80 mm, with uniform 15 mm wall thickness (Fig. 1).

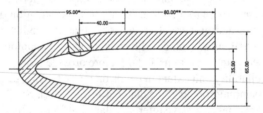

Fig. 1. Schematics of the left ventricle phantom, where * marks the part simulating the LV and ** the fixing collar.

Proposed geometry resulted in 52 ml volume which is in agreement with physiological end-systole volume of left ventricle [5]. A tubular collar was added for mounting physical phantoms in the laboratory setup. Stiffer inclusions were created as cylinders with main axes normal to the surface of the phantom, crossing it 40 mm from the base. This study concerns a homogenous phantom and a phantom with a transmural inclusion with diameter of 20 mm.

2.3   FEM Modeling

Numerical model of the phantom geometry, created using the Autodesk Inventor 2012 software (Autodesk Inc., USA) was exported in CAD format to the Finite Element simulation software – Abaqus 6.13-3 (Simulia, USA). The phantom was meshed into linear hexaheadral elements (type C3D8H). The collar was divided into 7020 elements via 3 mm seeding, the phantom into 7176 elements (seed 3 mm) and the inclusion into 2080 elements (seed 1, 5 mm).

The PVA cryogel was modeled as an elastic and as a hyperelastic material based on strain-stress relation derived from experiments. Elastic material was defined as isotropic with Poisson's ratio 0, 45 and with Youngs modulus obtained by fitting a linear function to the experimental data using linear regression. The hyperelastic material was defined as isotropic, with 3-rd order Yeoh strain energy potential definition, with Poisson's ratio 0, 45 and with stress-strain relation taken directly from measurements [4].

In the simulations the phantom was fixed by immobilizing both internal and external surfaces of the mounting collar. Deformation was forced by applying pressure load to the inner surface of the elliptical part of the phantom. The pressure linearly increased in 20 steps reaching 12 kPa. Except to the standard global coordinate system, a cylindrical system was defined, with radial direction corresponding to the radial direction of the phantom and longitudinal direction aligned with the phantoms' main axis.

As a result of finite element analysis values of logarithmic strains at all nodes were returned by the FEM software. A set of nodes defining a line-of-interest crossing the inclusion in the direction of phantom long axis was defined (Fig. 2). Three normal strain components for points of this set were exported from Abaqus for further processing in

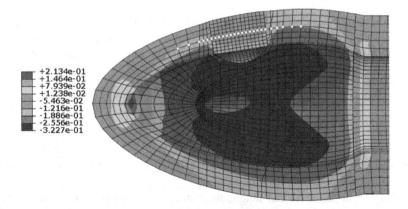

Fig. 2. Radial strain color map for the hyperelastic phantom with the inclusion (stiffness × 3). Nodes defining the line-of-interest are marked white.

Matlab 2012b (Mathworks, USA). The logarithmic strains $\varepsilon_{\log}$ have been transformed to the engineering strains ε_{eng} [6]:

$$\varepsilon_{eng} = exp\left(\varepsilon_{log}\right) - 1 \tag{1}$$

3   Results

3.1   PVA Characteristics

Measured relation between strain and stress is shown in the Fig. 3 in blue color. "Stiffer" variations of the PVA are a result of scaling of this relation by 1.5, 2 and 3. A linear function fitted to the data resulted in Young's modulus values presented in the legend box in the Fig. 3.

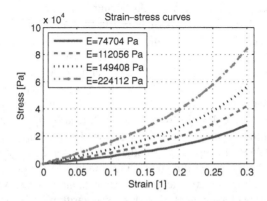

Fig. 3. Stress-strain curves. Solid line – from measurements, dashed, dotted and with markers as a result of scaling by factors of 1.5, 2 and 3 respectively (Color figure online).

3.2   Modeling Results

Series of 3D logarithmic strain maps were obtained in Abaqus software (Fig. 2). Plots of strains as a function of distance from the first node, counting from the base, were created to assess the strain profile across the inclusion. For more compact presentation of results a measure of strain contrast C has been proposed:

$$C = \frac{\left|\overline{\varepsilon_{ph}} - \overline{\varepsilon_{inc}}\right|}{\left|\overline{\varepsilon_{ph}}\right|} \cdot 100 \tag{2}$$

Where $\overline{\varepsilon_{ph}}$ is the mean strain in the phantom material along the line-of-interest and $\overline{\varepsilon_{inc}}$ is the mean strain in the inclusion.

Radial, circumferential and longitudinal strains were examined. Figure 4 shows selected strain profiles – radial and longitudinal for the homogenous phantom (a and b) and for the phantom with an inclusion of highest stiffness (c and d). Each plot shows a

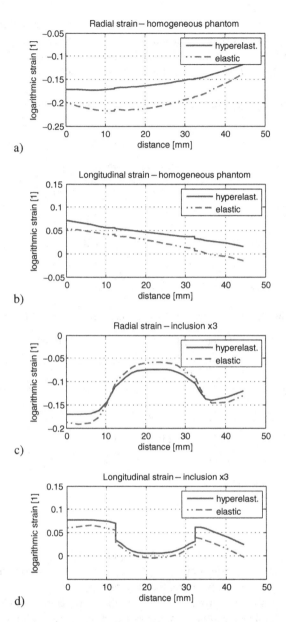

Fig. 4. Radial and longitudinal strain plots along the line-of-interest for homogenous phantom (a) and (b) and for the phantom with an inclusion 3x stiffer than surrounding material (c) and (d) (Color figure online).

profile for elastic (dashed red) and hyperelastic (blue) material model. Relative root mean squared differences (RRMSD) between strain profiles for the two material models are presented in the Table 1.

Strain contrasts calculated using the proposed formula (2) for both material models and all relative stiffness values proposed in Sect. 2.1 are given in Table 2. Contrasts obtained for the hyperelastic model are also presented as plots in Fig. 5, that shows additionally how the contrasts increase with increase of relative stiffness of the inclusion.

Table 1. Root mean squared differences between strain profiles for elastic and hyperelastic material model.

	Relative RMSD		
	Radial	Circ	Long.
Homogenous	27.3 %*	34.3 %	29.3 %*
Stiffness × 1.5	11.1 %	17.2 %	24.7 %
Stiffness × 2.0	7.2 %	11.1 %	21.8 %
Stiffness × 3.0	7.8 %*	10.2 %	19.6 %*

*marks cases plotted in Fig. 4

Table 2. Strain contrast values calculated for the line-of-interest.

	Hyperelastic			Elastic		
	Radial	Circ	Long.	Radial	Circ	Long.
Homogenous	0.13 %*	0.47 %	1.0 %*	2.1 %*	2.2 %	7 %*
Stiffness × 1.5	16 %	12 %	45 %	21 %	18 %	66 %
Stiffness × 2.0	29 %	23 %	63 %	38 %	36 %	80 %
Stiffness × 3.0	45 %*	41 %	78 %*	57 %*	57 %	88 %*

*marks cases plotted in Fig. 4

4   Discussion

Obtained stress-strain relationship for PVA cryogel is in agreement with other published results [3], but exact properties depend on the material composition and solidification procedure (freezing time, temperature, thawing rate etc.).

FEM simulations results provided complete data on phantom model behavior, but as with every model these results are an approximation of the real process. This can be observed for example in the map shown in Fig. 2 as irregularities of the radial strain in regions, where it should be homogenous. Also the plots of strain profiles for the homogenous phantom (Fig. 4a and b) exhibit small discontinuities at the borders of the inclusion which is defined as a material identical with the surrounding area.

Obtained strain profiles and strain contrasts (Fig. 4, Table 2 and Fig. 5) show that the difference of strains within the inclusion and the surrounding phantom increases as the inclusion material becomes stiffer. This occurs for both applied material models,

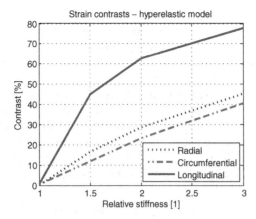

Fig. 5. Strain contrast change with increase in relative stiffness difference for the hyperelastic material model for the three strain components (Color figure online).

which is in agreement with common sense and other research. The differences between the elastic and the hyperelastic models are clearly visible. Magnitude of those differences is presented in Table 1. Strain contrasts, as defined here, are higher for the elastic model as compared to the hyperelastic in all cases (Table 2).

Inspecting the longitudinal strain profiles for the stiffest inclusion case (Fig. 4d) it can be noticed that the difference between strains outside and in the inclusion is higher for the hyperelastic model (strain difference of 0.051) than for the elastic model (strain difference of 0.041), but table Table 1 shows higher strain contrast for the latter (88 % for the elastic vs. 78 % for the hyperelastic model). This results from the assumed definition of contrast – lower value of ε_{ph} in the denominator of the formula (2) produces higher relative contrast value for the elastic model in case of longitudinal strains.

Figures 4a and b show that the phantom with hyperelastic material model deforms less then elastic in radial and circumferential[1] direction but more in longitudinal direction. Those different ratios of strain components result from the fact, that the stiffness of the material in each direction depends on the amount of deformation present. For this reason a phantom with hyperelastic material model deforms differently than an elastic, also on global scale. The shape of the deformed phantom depends on the material model applied.

5   Conclusions

The selection of material model for simulation affect all the strains and the strain contrasts. Both the "stiffness" and the material type play here a role.

The strain contrast values vary strongly with material model adopted (Table 1). There is no linear relation between the same kinds of strain for both material models.

---

[1] This was not visualized in this paper, but inspected during our research.

It may be of interest to use FEM results as a reference data for physical experiments, as they are more complete than results obtained from sonomicrometry. The material model used in FEM simulations plays a crucial role and simple elastic approximations may lead to erroneous results. For more complete reference, mechanical examinations should be carried out on samples with different properties produced exactly as the phantom material taking into account also the viscous properties of the PVA cryogel.

Relation between relative stiffness of the material used and the strain contrast (Fig. 5) can be useful in physical experiments planning. It provides valuable information for phantom design. As the hyperelastic phantom deforms differently than the elastic one, further investigation of how the hyperelasticity can be influenced by manufacturing process may improve the control of the ratios of strains in physical models.

Acknowledgements. This study was supported by the Polish National Science Centre by the decision DEC-2012/07/B/ST7/01441.

Characterization of PVA material was supported by Polish-Norwegian Research Programme under grant agreement Pol-Nor/209584/19/2013.

References

1. De Craene, M., Marchesseau, S., Heyde, B., Gao, H., Alessandrini, M., Bernard, O., Piella, G., Porras, A.R., Tautz, L., Hennemuth, A., Prakosa, A., Liebgott, H., Somphone, O., Allain, P., Makram Ebeid, S., Delingette, H., Sermesant, M., D'hooge, J., Saloux, E.: 3D strain assessment in ultrasound (Straus): a synthetic comparison of five tracking methodologies. IEEE Trans. Med. Imaging **32**(9), 1632–1646 (2013)
2. Jia, C., Kim, K., Weitzel, W.F., Jia, C., Rubin, J.M., Kolias, T.J.: P1E-1 left ventricular phantom with pulsatile circulation for ultrasound strain rate imaging. In: IEEE Ultrasonics Symposium, 2006, pp. 1317–1320 (2006)
3. Heyde, B., Cygan, S., Choi, H.F., Lesniak-Plewinska, B., Barbosa, D., Elen, A., Claus, P., Loeckx, D., Kaluzynski, K., D'hooge, J.: Regional cardiac motion and strain estimation in three-dimensional echocardiography: a validation study in thick-walled univentricular phantoms. IEEE Trans. Ultrason. Ferroelectr. Freq. Control **59**(4), 668–682 (2012)
4. Karimi, A., Navidbakhsh, M., Beigzadeh, B.: A visco-hyperelastic constitutive approach for modeling polyvinyl alcohol sponge. Tissue Cell **46**(1), 97–102 (2014)
5. Carlsson, M., Ugander, M., Mosén, H., Buhre, T., Arheden, H.: Atrioventricular plane displacement is the major contributor to left ventricular pumping in healthy adults, athletes, and patients with dilated cardiomyopathy. Am. J. Physiol. Heart Circ. Physiol. **292**(3), H1452–H1459 (2007)
6. D'hooge, J., Heimdal, A., Jamal, F., Kukulski, T., Bijnens, B., Rademakers, F., Hatle, L., Suetens, P., Sutherland, G.R.: Regional strain and strain rate measurements by cardiac ultrasound: principles, implementation and limitations. Eur. J. Echocardiogr. J. Work. Group Echocardiogr. Eur. Soc. Cardiol. **1**(3), 154–170 (2000)

Image-Derived Human Left Ventricular Modelling with Fluid-Structure Interaction

Hao Gao[1]([✉]), Colin Berry[2], and Xiaoyu Luo[1]

[1] School of Mathematics and Statistics, University of Glasgow, Glasgow, UK
hao.gao@glasgow.ac.uk
[2] Institute of Cardiovascular and Medical Science,
University of Glasgow, Glasgow, UK

Abstract. In this study, we have developed a human left ventricular model using a hybrid immersed boundary- finite element description. The left ventricle model is built based on clinical cardiac magnetic resonance images, and completed with the inflow (left atrium) and outflow (aorta) tracts. The model is used to simulate the left ventricular dynamics with fully-coupled fluid structure interaction, with parameters optimised, and the results are in reasonably good agreement with the in vivo measurements.

1 Introduction

Heart diseases are the leading causes of death worldwide, claiming millions of lives every year. There is a growing recognition that computational approach for ventricular biomechanics, when used in line with clinical images, can provide insights into heart function and dysfunction. For example, carefully designed models have been used to infirm various improvement therapies post-myocardial infarction [15].

Cardiac dynamics are complex multiphysics problems that involve dynamic blood flow, electrophysiology of the excitable myocardium, anisotropic nonlinear deformation of the ventricular wall, and mostly importantly, the interactions between them [1]. Substantial effort has been devoted to developing computational models of the heart [1,2,7,9,13]. However, three-dimensional modelling of the heart that fully accounts for fluid-structure and electro-mechanical interactions still present significant computational challenges. To date only a couple of models have been developed to simulate the fluid structure interaction (FSI) of ventricular dynamics, in which the arbitrary lagrangian-Eulerian approach is adopted [11,14].

The immersed boundary (IB) method is an alternative approach for FSI simulation [12]. One of the recent extensions of IB is the hybrid finite difference-finite element method (IB/FE) [5], which is capable of simulating left ventricular (LV) dynamics with hyperelastic representation of the fibre-reinforced myocardium [4]. However, these previous IB/FE LV models are truncated, and hence cannot assess the FSI properly. In this study, we develop a more completed IB/FE human LV model with aorta and atrium tracts, based on clinical

© Springer International Publishing Switzerland 2015
H. van Assen et al. (Eds.): FIMH 2015, LNCS 9126, pp. 321–329, 2015.
DOI: 10.1007/978-3-319-20309-6_37

cardiac magnetic resonance (CMR) imaging, so that the flow field inside the LV is more realistic. In this model, both the passive and active parameters of the myocardium are inversely determined so that the predicted LV dynamics agree quantitatively with the image-based measurements. Preliminary results of the LV dynamics throughout cardiac cycles will be shown with fully coupled FSI.

2   Methodology

Imaging-Derived LV Model. CMR imaging was performed on a young healthy volunteer (28 years). The study was approved by the local ethics committee, and written informed consent was obtained before the study. Steady-state free precession cine imaging was used for functional assessment, including the short-axis, horizontal long-axis, and vertical long-axis planes. Typical cine imaging parameters: slice thickness: 10 mm; in-plane pixel size: $1.3 \times 1.3\,\mathrm{mm}^2$; frame rate: 25 per cardiac cycle. Phase contrast imaging was also performed to measure the flow rate across the aortic valve.

Ventricular wall boundaries were extracted at early diastole when the LV pressure is lowest. Seven slices of short-axis views from base to apex were used for the LV chamber reconstruction, long-axis slices were selected for reconstructing the inflow (left atrium) and outflow (aorta) tracts. Following a manual segmentation, the wall boundaries were imported into SolidWorks (SolidWorks Corp., Waltham, MA USA) to generate the three dimensional geometry via B-spline surface fitting.

The rebuilt LV geometry is shown in Fig. 1. Notice that the geometries above the two valvular rings are not derived from the CMR images; they are the artificial extensions for applying flow boundary conditions. Since there is limited information on the organization of myocytes in the regions between the base and the two valvular rings (denoted as valvular region), only the region bellow the basal plane (referred to as the LV region) is modelled to have the active contractility. The valvular region, the inflow and outflow tracts, are assumed to only passively bear the load. A rule-based myocardial fibre generation method [13] was employed to model the fibre and sheet architecture of the myocardium inside the LV region. The fibre angle rotates from $-60°$ to $60°$ from endocardium to epicardium, and the sheet angle rotates from $-45°$ to $45°$.

IB/FE Formulation. The IB/FE description of the fluid-structure coupled system is given by the following equations [5].

$$\rho\left(\frac{\partial \mathbf{u}}{\partial t}(\mathbf{x},t) + \mathbf{u}(\mathbf{x},t)\cdot\nabla\mathbf{u}(x,t)\right) = -\nabla p(\mathbf{x},t) + \mu\nabla^2\mathbf{u}(\mathbf{x},t) + \mathbf{f}^{\mathrm{s}}(\mathbf{x},t),$$

$$\nabla\cdot\mathbf{u}(\mathbf{x},t) = 0,$$

$$\mathbf{f}^{\mathrm{s}}(\mathbf{x},t) = \int_U \nabla_{\mathbf{X}}\cdot\mathbb{P}^{\mathrm{s}}(\mathbf{X},t)\,\delta(\mathbf{x} - \chi(\mathbf{X},t))\,\mathrm{d}\mathbf{X}$$

$$- \int_{\partial U} \mathbb{P}^{\mathrm{s}}(\mathbf{X},t)\,\mathbf{N}(\mathbf{X})\,\delta(\mathbf{x} - \chi(\mathbf{X},t))\,\mathrm{d}A,$$

$$\frac{\partial\chi}{\partial t}(\mathbf{X},t) = \int_\Omega \mathbf{u}(\mathbf{x},t)\,\delta(\mathbf{x} - \chi(\mathbf{X},t))\,\mathrm{d}\mathbf{x}$$

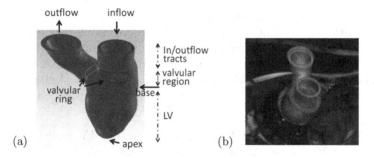

Fig. 1. Reconstructed LV model with inflow and outflow tracts (a) and superimposed with one short-axis CMR image (b). The geometry is divided into three parts: the LV region, the valvular region, and the inflow and outflow tracts.

where $\mathbf{x} = (x_1, x_2, x_3)$ denotes the Eulerian coordinates, $\mathbf{X} = (X_1, X_2, X_3)$ denotes the Lagrangian coordinates attached to the immersed solid. ρ is the fluid density, μ is the fluid viscosity, $\mathbf{u}(\mathbf{x}, t)$ represents the Eulerian velocity, and $p(\mathbf{x}, t)$ is the Eulerian pressure field. $\delta(\mathbf{x}) = \delta(x_1)\,\delta(x_2)\,\delta(x_3)$ is a smoothed three-dimensional Dirac delta function, $\mathbf{f}^{\mathrm{s}}(\mathbf{x}, t)$ represents the Eulerian force density derived from the first Piola-Kirchoff stress tensor $\mathbb{P}^{\mathrm{s}}(\mathbf{X}, t)$ of the immersed structure. These equations express the conservation of the momentum and mass in the Eulerian form while using a Lagrangian description for the structural deformation and stress tensor.

The total Cauchy stress for the fluid-structure coupled system is

$$\boldsymbol{\sigma}(\mathbf{x}, t) = -p\,\mathbb{I} + \mu\left[\nabla\mathbf{u} + (\nabla\mathbf{u})^T\right] + \begin{cases} \boldsymbol{\sigma}^{\mathrm{s}}(\mathbf{x}, t) & \text{for } \mathbf{x} \in \text{immersed solid,} \\ \mathbf{0} & \text{for } \mathbf{x} \in \text{otherwise,} \end{cases} \tag{1}$$

in which $\boldsymbol{\sigma}^{\mathrm{s}} = J^{-1}\,\mathbb{P}^{\mathrm{s}}\,\mathbb{F}^T$, is the structure-like Cauchy stress, and $\mathbb{F} = \partial\boldsymbol{\chi}/\partial\mathbf{X}$ is the deformation gradient associated with the immersed structure and $J = \det(\mathbb{F})$. To use standard C^0 FE methods for nonlinear elasticity, a weak formulation for the structure domain is employed [5].

Myocardial Mechanics. We model the myocardial stress tensor as the summation of the active and the passive stress tensors, i.e. $\boldsymbol{\sigma}^{\mathrm{s}} = \boldsymbol{\sigma}^{\mathrm{p}} + \boldsymbol{\sigma}^{\mathrm{a}}$. The myocardial passive response is described using an invariant-based strain energy function

$$W = \frac{a}{2b}e^{[b(I_1 - 3 - 2\log(J))]} + \sum_{i=\mathrm{f,s}} \frac{a_i}{2b_i}\{e^{[b_i(I_{4i}^\star - 1)^2]} - 1\} + \frac{a_{\mathrm{fs}}}{2b_{\mathrm{fs}}}\{e^{[b_{\mathrm{fs}}(I_{8\mathrm{fs}})^2]} - 1\}, \tag{2}$$

in which the invariants are $I_1 = \mathrm{tr}(\mathbb{C})$, $I_{4\mathrm{f}} = \mathbf{f}_0^T\mathbb{C}\mathbf{f}_0 = \mathbf{f}\cdot\mathbf{f}$, $I_{4\mathrm{s}} = \mathbf{s}_0^T\mathbb{C}\mathbf{s}_0 = \mathbf{s}\cdot\mathbf{s}$, $I_{8\mathrm{fs}} = \mathbf{f}_0^T\mathbb{C}\mathbf{s}_0 = \mathbf{f}\cdot\mathbf{s}$, and $I_{4i}^\star = \max(I_{4i}, 1)$. Here $\mathbf{f}_0$ and $\mathbf{s}_0$ are the initial fibre and sheet directions, and $\mathbb{C} = \mathbb{F}^T\mathbb{F}$ is the right Cauchy-Green deformation tensor. From which we derive $\boldsymbol{\sigma}^{\mathrm{p}} = \frac{1}{J}\frac{\partial W}{\partial\mathbb{F}}\mathbb{F}^T - \frac{\beta_s}{J}\log(J^2)$, where β_s is a constant to reinforce the incompressibility.

The active tension is computed as $\sigma^{\mathrm{a}} = T_{\mathrm{scale}} T \mathbf{f} \otimes \mathbf{f}$, in which T is determined from the active contraction model of Niederer et al. [10], and is a function of the fibre stretch λ_f, stretch rate $\partial \lambda_f / \partial t$, intracellular calcium transient, and contractility T_{ref}. The active stress is scaled using a constant T_{scale}, which enables the model to produce the realistic systolic dynamics. A simple model of intracellular calcium dynamics [8] is used to trigger the myocardial contraction uniformly and simultaneously inside the LV region.

Boundary Conditions and Implementation. The LV model is immersed in a 16.5 cm × 16.5 cm × 16.5 cm fluid box, with the inflow and outflow tracts attached to the top plane as shown in Fig. 2(a). The inflow and outflow tracts are fixed during the simulations using a tethering force, defined as $k(\mathbf{x}_1 - \mathbf{x}_0)$, where k is the penalty parameter (taken to be 1.0e7 dyne/cm), $\mathbf{x}_0$ and $\mathbf{x}_1$ are the desired and boundary positions, respectively. Since only the LV region can actively contracts, the longitudinal and circumferential displacements of the LV base are set to be zero, with the radial motion allowed. The reminder of the LV region is left free.

An isotropic hyperelastic material model is used to describe the valvular region, with much lower stiffness than that of the LV region. The valvular and the LV regions overlap slightly ($\approx$1 mm), which allows the valvular region to follow the LV basal motion. During the diastolic filling, we also impose a constrain so that the valvular region does not deform more than 6 mm, using a tethering force. When the LV starts to generate active tension (beginning of the isovolumetric contraction), the valvular region above the LV base is gradually pulled back to the original position within 40 ms using a tethering force, this will keep the valvular region in the place until the end of systole. This is because the valvular region as modelled is unable to generate active force to resist the systolic ventricular pressure.

To simulate the ventricular flow, different boundary conditions are applied in the four different phases of the cardiac cycle: (1) diastolic filling, (2) isovolumetric contraction, (3) systolic ejection and (4) isovolumetric relaxation. During phase 1, we apply the pressure directly to the endocardial surface of the LV region until it reaches the end-diastolic value (8 mmHg: population-based average value), and ensure that volumetric flow rate $Q_{\mathrm{out}} = 0$ in the outflow boundary $\partial \Omega_{\mathrm{out}}$, and zero pressure in the inflow boundary $\partial \Omega_{\mathrm{in}}$ ($P_{\mathrm{in}} = 0$), as shown in Fig. 2(a). These are simplified boundary conditions to mimic the sucking effect of the rapid diastolic filling phase. During phase 2, we let $Q_{\mathrm{in}} = Q_{\mathrm{out}} = 0$. When the LV pressure is greater than the diastolic aortic pressure, phase 3 begins, and we set the diastolic aortic pressure to be the measured diastolic cuff pressure (85 mmHg) (taken from the brachial artery of the volunteer). During the systolic ejection, a three-element Windkessel circulation model [6] is connected to the outflow boundary as shown in Fig. 2(b). When Q_{out} approaches zero, indicating the closure of the aortic valve, we set $Q_{\mathrm{out}} = 0$, and disconnect the circulation model from the LV model. There is no flow at $\partial \Omega_{\mathrm{in}}$ in the systole. During phase 4, $Q_{\mathrm{in}} = Q_{\mathrm{out}} = 0$.

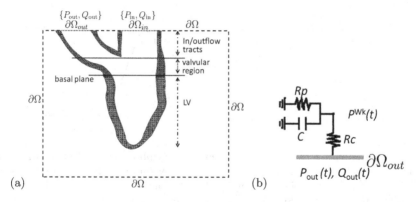

Fig. 2. Schematic illustration of boundary conditions of the IB/FE LV model (a) and the lumped three-element Windkessel circulation model (b) with characteristic resistance R_c, peripheral resistance R_p and arterial compliance C. $\partial\Omega_{in}$ is the inflow boundary, P_{in} and Q_{in} are the pressure and volumetric flow rate at $\partial\Omega_{in}$. $\partial\Omega_{out}$ is the outflow boundary, P_{in} and Q_{in} are the associated pressure and volumetric flow rate. $\partial\Omega$ is the fluid box boundary excluding $\partial\Omega_{in}$ and $\partial\Omega_{out}$. In the simulations, $R_c = 0.033$ (mmHg ml^{-1} s), $R_p = 0.79$ (mmHg ml^{-1} s) and $C = 1.75$ (ml mmHg^{-1}).

The model is then run for another cardiac cycle. Before starting the next cardiac cycle, the LV model is allowed to reach a fully relaxed state with $P_{in} = 0$, $Q_{out} = 0$ and $\sigma^a = 0$. During the whole cardiac cycle, zero pressure is set at the reminder boundary of the fluid box.

The passive material parameters are optimized by matching the end-diastolic volume of the healthy volunteer and the regional circumferential strain estimated from the cine images using a multi-step optimization scheme [3]. T_{scale} in the active contraction model is chosen to match the measured ejection fraction and the flow rate across the aortic valve. The simulations are carried out using the open-source IBAMR software framework (http://ibamr.googlecode.com).

3   Results

The measured end-diastolic volume from the short axis images is 143 mL with an ejection fraction 57 %; the computed end-diastolic volumes from the optimized model is 142.6 mL with an ejection fraction 52 %, which is slightly lower than the measured ejection fraction. The optimized parameters are found to be $a = 0.19$ kPa, $b = 5.08$, $a_f = 1.2$ kPa, $b_f = 4.15$, $a_s = 0.70$ kPa, $b_s = 1.6$, $a_{fs} = 0.24$ kPa, $b_{fs} = 1.3$, and $T_{scale} = 4$.

The computed LV pressure-volume loop is shown in Fig. 3(a). The diastolic filling starts from the left bottom corner with linearly ramped pressure until it reaches the end-diastole. The isovolumetric contraction is the second phase where the pressure increases sharply. When the ventricular pressure exceeds the diastolic aortic pressure (85 mmHg), the blood is ejected towards the aorta, and ventricular ejection phase then begins. It takes about 51 ms for the LV model to

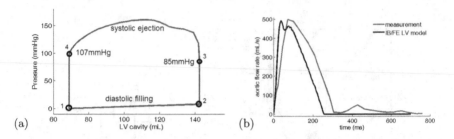

Fig. 3. (a): Simulated LV pressure-volume loop. Symbols 'o'1 and 'o'2 indicate the diastolic filling; symbols 'o'2 and 'o'3 indicate the isovolumtric contraction; symbols 'o'3 and 'o'4 indicate the systolic ejection; symbols 'o'4 and 'o'1 indicate the isovolumtric relaxation. (b): the aortic flow rates comparison between the IB/FE LV model (black) and the CMR measurements (red). Note that 0 ms is the beginning of the systolic ejection, or the end of the isovolumetric contraction (Color figure online).

develop a high enough ventricular pressure to open the aortic valve from end-diastole. Ejection continues until the end-systole when the ventricular pressure at end-systole is 107 mmHg. Then isovolumetric relaxation occurs, during which time the ventricular pressure drops dramatically.

Overall the model outcome agrees very well with the CMR measurements, with only small discrepancies. The peak LV pressure computed in systole is 161 mmHg, this is slightly higher than the measured systolic cuff pressure (150 mmHg). Figure 3(b) compares the simulated aortic flow rate with the CMR measurements across the aortic valve during systole. The peak flow rate from the IB/FE model is 491 mL/s, very close to the measured peak value, 498 mL/s. However, the peak in the IB/FE LV model arrives earlier, which is partially because of the lack of the aortic valve. The systolic duration (256 ms) is also shorter compared to the in vivo measurement (300 ms), which may be caused by the prescribed intracellular calcium transient [8].

Figures 4(a, b, c, d) show the deformed endocardial wall at early-diastole, end-diastole, middle-systole and end-systole superimposed with a three-chamber long axis view from the CMR images. The simulated endocardial wall motion agrees well with the CMR images throughout the cardiac cycle. Figures 4(e, f, g, h) show the streamlines at middle-diastole, end-diastole, middle-systole and end-systole. We can see that vortices are formed inside the LV cavity during diastolic filling, especially at middle-diastole (Fig. 4(e)). During systole, the blood is pushed out towards the aorta with a strong jet, as shown in Fig. 4(g).

Following the AHA definition, we divide the basal/middle-ventricular short-axis slices into 6 segments, and the apical slices into 4 segments. The corresponding regions in the LV geometry for each slice are defined by the regions lying in the axial range $(-5, 5)$ mm of that slice. Based on these definitions, the active tension in different regions of the LV is shown in Fig. 5(a) when the peak LV pressure is developed. Except for the most basal and apical regions, the myocardium works equally hard in all other regions to generate the active tension (mean value: 102 kPa).

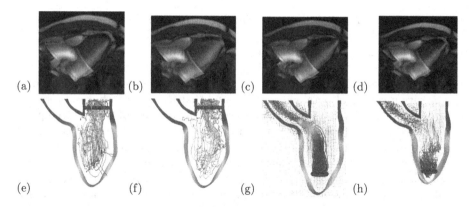

Fig. 4. Wall deformations at early-diastole (a), end-diastole (b), middle-systole (c) and end-systole (d); Ventricular flow patterns at middle-diastole (e), end-diastole (f), middle-systole (g) and end-systole (h), coloured by velocity magnitude. The colour figure can be found in the online version (Color figure online).

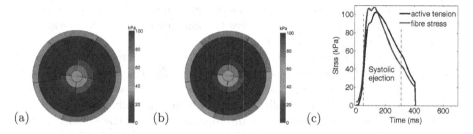

Fig. 5. Bullseye plots of active tension (a) and myofibre stress (b) when the LV cavity pressure peaks, and the average active tension and myofibre stress in the LV region from the beginning of the isovolumtric contraction to the end-systole (c). The two dash vertical lines in (c) indicates the systolic ejection phase, 0 ms is the beginning of the isovolumetric contraction, after 400 ms, the LV model is in the fully relaxed state with zero active tension. The colour figure can be found in the online version (Color figure online).

Similar results can be found for the myofibre stress distribution in Fig. 5(b), with a mean value of 104 kPa. Figure 5(c) shows the average active tension and fibre stress in the LV region in systole. In the early-systole, the myofibre stress is slightly higher than the active tension, indicating the stretching of the myofibres, while in the late-systole, the active tension is higher than the total fibre stress, since the myofibres become compressed when the LV cavity volume is decreased.

4   Discussion and Conclusion

In this study, we have developed a contractile human LV model with inflow and outflow tracts, using a hybrid IB/FE formulation. It is derived based on clinical in vivo measurement of a healthy subject, and incorporates fluid structure interaction and a rule-based myocardial fibre structure. In the model we have adopted a simplified electro-mechanical coupling model for active contraction, and a population-based end-diastolic pressure (8 mmHg).

We followed the approach in [3] using the pressure-volume data and the regional circumferential strain estimated from cine images to inversely determine the passive myocardial parameters. We found that this approach could robustly determine the mechanical response, however the uniqueness of the material parameters can not be ensured. This is partially because of the correlations among these parameters, and partially due to the limited in vivo measurements we could obtain. In our previous work [3], we applied our multi-step parameter fitting technique to estimate the passive parameters based on a known model, and found that it is possible to obtain a unique set of parameters. However, we had to use the LV pressure and volume at end diastole and end systole, along with sufficient number of regional circumferential strains. The challenge here is that we need to personalize the LV model using the in vivo imaging data, which are not sufficient to provide all the required information for all the parameters. For example the strain data, which have been used for fitting the material properties, are far fewer compared to that the multi-step parameter fitting would need in [3]. Furthermore, parameters from the active tension cannot be easily measured or estimated from the clinical images.

Despite of these limitations, it seems that the myocardial mechanical response has been estimated robustly. The simulated LV deformations agree well with that from the CMR imaging, and the computed aortic flow rate is also close to the measurement. The intracellular calcium transient profile, however, is not personalised, which may explain the discrepancy between the simulation and measurement in terms of the systolic ejection duration. Work is in progress to insert the heart valves into the LV model so that more realistic flow field can be obtained. Effects of varied values of the end-diastolic pressure on the model results are also being investigated.

Acknowledgement. We are grateful for funding provided by the UK EPSRC (EP/I1029990), the British Heart Foundation (PG/14/64/31043, PG/11/2/28474).

References

1. Baillargeon, B., Rebelo, N., Fox, D.D., Taylor, R.L., Kuhl, E.: The living heart project: a robust and integrative simulator for human heart function. Eur. J. Mech.-A/Solid. **48**, 38–47 (2014)
2. Gao, H., Carrick, D., Berry, C., Griffith, B.E., Luo, X.Y.: Dynamic finite-strain modelling of the human left ventricle in health and disease using an immersed boundary-finite element method. IMA J. Appl. Math. **79**, 978–1010 (2014)
3. Gao, H., Li, W.G., Cai, L., Berry, C., Luo, X.Y.: Parameter estimation of holzapfel-ogden law for healthy myocardium. J. Eng. Math. (2015, in press). doi:10.1007/s10665-014-9740-3
4. Gao, H., Wang, H., Berry, C., Luo, X., Griffith, B.E.: Quasi-static image-based immersed boundary-finite element model of left ventricle under diastolic loading. Int. J. Numer. Methods Biomed. Eng. **30**, 1199–1222 (2014)

5. Griffith, B., Luo, X.: Hybrid finite difference/finite element version of the immersed boundary method (submitted)
6. Griffith, B.E.: Immersed boundary model of aortic heart valve dynamics with physiological driving and loading conditions. Int. J. Numer. Methods Biomed. Eng. **28**(3), 317–345 (2012)
7. Guccione, J., Moonly, S., Moustakidis, P., Costa, K., Moulton, M., Ratcliffe, M., Pasque, M.: Mechanism underlying mechanical dysfunction in the border zone of left ventricular aneurysm: a finite element model study. Ann. Thorac. Surg. **71**(2), 654–662 (2001)
8. Hunter, P., McCulloch, A., Ter Keurs, H.: Modelling the mechanical properties of cardiac muscle. Prog. Biophys. Mol. Biol. **69**(2–3), 289–331 (1998)
9. Nash, M., Hunter, P.: Computational mechanics of the heart. J. Elast. **61**(1), 113–141 (2000)
10. Niederer, S., Hunter, P., Smith, N.: A quantitative analysis of cardiac myocyte relaxation: a simulation study. Biophys. J. **90**(5), 1697–1722 (2006)
11. Nordsletten, D., Niederer, S., Nash, M., Hunter, P., Smith, N.: Coupling multiphysics models to cardiac mechanics. Prog. Biophys. Mol. Biol. **104**(1), 77–88 (2011)
12. Peskin, C.: Flow patterns around heart valves: a numerical method. J. Comput. Phys. **10**(2), 252–271 (1972)
13. Wang, H., Gao, H., Luo, X., Berry, C., Griffith, B., Ogden, R., Wang, T.: Structure-based finite strain modelling of the human left ventricle in diastole. Int. J. Numer. Methods Biomed. Eng. **29**, 83–103 (2012)
14. Watanabe, H., Sugiura, S., Kafuku, H., Hisada, T.: Multiphysics simulation of left ventricular filling dynamics using fluid-structure interaction finite element method. Biophys. J. **87**(3), 2074–2085 (2004)
15. Zhang, Z., Sun, K., Saloner, D., Wallace, A., Ge, L., Baker, A., Guccione, J., Ratcliffe, M.: The benefit of enhanced contractility in the infarct borderzone: a virtual experiment. Front. Physiol. **3**, 86 (2012)

Fluid-Structure Interaction Model of Human Mitral Valve within Left Ventricle

Hao Gao[1]($^{\boxtimes}$), Nan Qi[1], Xingshuang Ma[2], Boyce E. Griffith[3],
Colin Berry[4], and Xiaoyu Luo[1]

[1] School of Mathematics and Statistics, University of Glasgow, Glasgow, UK
hao.gao@glasgow.ac.uk
[2] School of Aerospace Engineering, Chongqing University, Chongqing, China
[3] Department of Mathematics, University of North Carolina, Chapel Hill, NC, USA
[4] Institute of Cardiovascular and Medical Science,
University of Glasgow, Glasgow, UK

Abstract. We present an integrated model of mitral valve coupled with
the left ventricle. The model is derived from clinical images and takes
into account of the important valvular features, left ventricle contraction, nonlinear soft tissue mechanics, fluid structure interaction, and the
MV-LV interaction. This model is compared with a corresponding mitral-
tube model, and differences in the results are discussed. Although the
model is a step closer towards simulating physiological realistic situation,
further work is required to ensure that the highly complex valvular-
ventricular interaction, and the fluid-structure interaction, can be reliably represented.

1 Introduction

Moderate or severe mitral valve (MV) dysfunction remains a major medical problem. It can be caused, among others, by leaflet prolapse or is secondary to left
ventricular diseases [15]. Computational modelling of the MV mechanics, particularly within the context of the left ventricle (LV) environment, can enhance
our understanding of the valvular-ventricular interaction, and potentially lead
to more efficient MV repairs and replacement.

Research on developing biomechanical MV models can be dated back to
1990s [10]. Recent development involves finite strain deformational kinematics,
realistic anatomical geometries and advanced hyperelastic constitutive models
[2,16,17]. However, most of the studies are based on purely structural analysis with applications to a statically or dynamically pressurized closed valve in
isolated situation [9]. Since the structure of the MV is closely tied to the left
ventricle through the chordae connection, it is important to simulate the dynamics of MV by taking into account of LV dynamics, as well as the fluid-structure
interaction (FSI) between the MV, LV and the blood.

Kunzelman, Einstein and co-workers first started to simulate normal and
pathological mitral function [3,4,11] with FSI. Over the last few years there
have been a number of FSI valvular models using the immersed boundary (IB)

© Springer International Publishing Switzerland 2015
H. van Assen et al. (Eds.): FIMH 2015, LNCS 9126, pp. 330–337, 2015.
DOI: 10.1007/978-3-319-20309-6_38

approach [6,7,13,18]. However, none of these MV models included the effect of
the LV motion, hence the flow field is not physiological. Indeed, Lau et al. [12]
compared the MV dynamics in a straight tube and a U-shaped ventricle, and
found that when the MV is mounted into a LV, the transvalvular velocity is
slower compared to the one estimated using a tubular geometry. Yin et al. [19]
modelled a chordaed MV inside a LV and identified fluid vortices associated
with the LV motion. However, their LV motion is modelled as a set of prescribed
moving boundary, and the MV model is simply constructed using a network of
linear elastic fibres. Chandran and Kim [1] recently reported a prototype FSI
MV dynamics in a simplified LV chamber model using an immersed interface-like
approach. To date, there has been no work reported that included both the MV
and LV models and the fluid-structure interaction properly.

In this study, we have developed a fully integrated MV-LV model, which is
image-derived and simulated using a hybrid immersed boundary-finite element
framework (IB/FE) [8]. We will also compare the differences of the MV dynamics
when the MV is mounted in the LV and in a tube.

2 Methodology

The hybrid IB/FE method employs an Eulerian description of the viscous incompressible fluid, described by $\mathbf{x} = (x_1, x_2, x_3) \in \Omega$, along with a Lagrangian
description of the structure that is immersed in the fluid, described by $\mathbf{X} = (X_1, X_2, X_3) \in U$. Interactions between the Lagrangian ($\mathbf{X}$) and Eulerian ($\mathbf{x}$)
fields are achieved by integral transforms with a Dirac delta function kernel $\delta(\mathbf{x})$.
In brief, the IB form of the equations of motion is:

$$\rho\left(\frac{\partial \mathbf{u}}{\partial t}(\mathbf{x}, t) + \mathbf{u}(\mathbf{x}, t) \cdot \nabla \mathbf{u}(\mathbf{x}, t)\right) = -\nabla p(\mathbf{x}, t) + \mu \nabla^2 \mathbf{u}(\mathbf{x}, t) + \mathbf{f}^s(\mathbf{x}, t),$$

$$\nabla \cdot \mathbf{u}(\mathbf{x}, t) = 0,$$

$$\mathbf{f}^s(\mathbf{x}, t) = \int_U \nabla \cdot \mathbb{P}^s(\mathbf{X}, t) \, \delta(\mathbf{x} - \boldsymbol{\chi}(\mathbf{X}, t)) \, d\mathbf{X}$$
$$- \int_{\partial U} \mathbb{P}^s(\mathbf{X}, t) \, \mathbf{N}(\mathbf{X}) \, \delta(\mathbf{x} - \boldsymbol{\chi}(\mathbf{X}, t)) \, dA(\mathbf{X}),$$

$$\frac{\partial \boldsymbol{\chi}}{\partial t}(\mathbf{X}, t) = \int_\Omega \mathbf{u}(\mathbf{x}, t) \, \delta(\mathbf{x} - \boldsymbol{\chi}(\mathbf{X}, t)) \, d\mathbf{x},$$

in which ρ is the mass density, μ is the viscosity, $\mathbf{u}(\mathbf{x}, t)$ is the Eulerian velocity
filed of the system, $p(\mathbf{x}, t)$ is the Eulerian pressure field, $\mathbf{f}^s(\mathbf{x}, t)$ is the Eulerian
elastic force density. $\boldsymbol{\chi}(\mathbf{X}, t) \in \Omega$ denotes the physical position of material point
$\mathbf{X}$ at time t, and $\mathbf{N}(\mathbf{X})$ is the exterior unit normal at ∂U. $\mathbb{P}^s$ is the first Piola-
Kirchoff stress tensor, which is related to the Cauchy stress σ^s of the immersed
structure by $\mathbb{P}^s = \det(\mathbb{F})\sigma^s\mathbb{F}^{-T}$, in which $\mathbb{F} = \partial\boldsymbol{\chi}/\partial\mathbf{X}$ is the deformation gradient. $\mathbb{P}^s$ is determined from the material properties by strain energy functional W
by $\mathbb{P}^s = \partial W/\partial \mathbb{F}$. Readers may refer to [8] for more details of the IB/FE method.

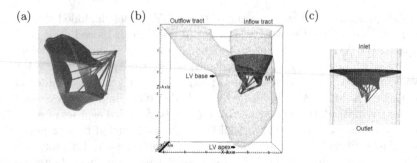

Fig. 1. The MRI-derived MV model (a), the MRI-derived MV-LV model (b), and the MV-tube model (c).

A cardiac magnetic resonance (CMR) study was performed on a healthy young volunteer. The study was approved by the local NHS Research Ethics Committee, and written informed consent was obtained before the CMR scan. Twelve imaging planes along the LV outflow tract view were imaged to cover the MV with cine images. Typical parameters were: slice thickness: 3 mm; in-plane pixel size: 0.7×0.7 mm^2, field of view: 302×400 mm^2; frame rate: 25 per cardiac cycle. The MV geometry was reconstructed from a stack of cine images at early-diastole when it first opens, as shown in Fig. 1(a), with sixteen chordae distributed evenly and running through the leaflet free edges to the annulus ring. Details on the subject-specific MV reconstruction from CMR images are similar as in [6,13]. Short-axial and long-axial cine images were also performed to cover the LV region. Typical parameters were: slice thickness: 10 mm; in-plane pixel size: 1.3×1.3 mm^2; frame rate: 25 per cardiac cycle. The LV geometry including the inflow and outflow tracts was reconstructed from the same volunteer at early of diastole just before the MV opens (Fig. 1(b)) [5]. The MV geometry is mounted to the inflow tract according to the relative positions derived from the CMR images. This forms the integrated MV-LV model. Following the work in [6], the MV is also fixed to a housing disc and mounted to a rigid outer tube (Fig. 1(c)), this forms the MV-tube model.

Simulations of the MV-LV and MV-tube models are run using the open-source IBAMR software framework[1], which provides an adaptive and distributed-memory parallel infrastructure for developing the IB/FE models. In the MV-LV model, the structure below LV base is contractile, the regions above the LV, including the MV and its apparatuses, are made to only bear the load passively. The LV base is allowed to have radial expansion, but fixed in the long axial and circumferential directions. During diastole, zero flow boundary condition is imposed in the outflow tract. The pressure that is linearly ramped to the end-diastolic pressure (assumed to be 8 mmHg) is applied in the inflow tract in 0.8 s. After the end-diastole, the LV region simultaneously contracts, triggered by a spatially homogeneous intracellular calcium transient [5].

---

[1] https://ibamr.googlecode.com.

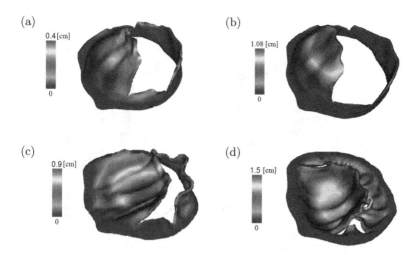

Fig. 2. The fully opened MV in (a) the MV-LV model, (b) the MV-tube model (b). The closed MV in (c) the MV-LV model, and (d) the MV-tube model (d). Coloured by the displacement. The colour figure can be found in the online version.

The increased LV pressure will close the MV and open the aortic valve when the LV pressure exceeds the diastolic aortic pressure, assumed to be the measured diastolic cuff pressure from the healthy volunteer (85 mmHg). Because the aortic valve is not included in the MV-LV model, the aortic tract is either fully open or fully closed, determined by the pressure difference between the value inside LV and the aorta. During the systolic ejection, a three-element Windkessel model [7] is connected to the outflow tract to provide a physiological pressure-flow boundary condition, the systolic phase ends when the LV no longer pumps blood out. The end-diastolic pressure (8 mmHg) is maintained in the inflow boundary until the end-systole. The chordae are not directly attached to the LV wall, but modelled similarly as in [6]. In the MV-tube model, typical pressure profiles are applied on the inlet and outlet boundaries, as described in [6].

3   Results

Because of the increased pressure in the inflow tract during diastole, the volumetric flow rate across the MV linearly increases until end of diastole, with a maximum value of 90 mL/s. The total inflow volume across the MV in diastole is 40 mL, which is less than the real cardiac output (around 80 mL). Figure 2(a) shows the deformed MV leaflets of the MV-LV model when the MV is fully open. This is compared to that of the MV-tube model (Fig. 2(b)). We can see that the orifice in the MV-LV model is smaller, which suggests that the LV wall provides more resistance to the blood flow that fills the LV cavity during the MV opening. This does not happen in the MV-tube model, hence for the same pressure drop, the flow rate is greater. In systole, it takes 80 ms for the LV model to develop

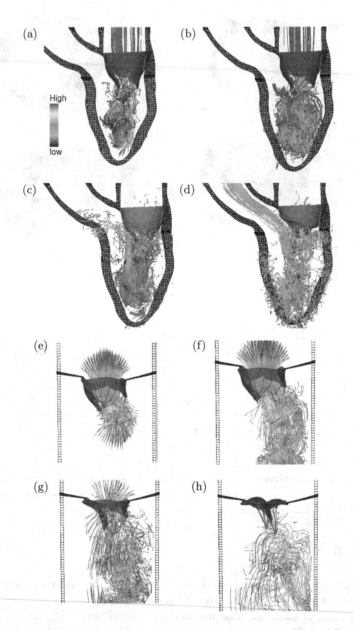

Fig. 3. Streamlines in the MV-LV model at the early diastolic filling (a), at the late diastolic filling (b), when the MV is closing (c), and at the middle of the systolic ejection (d). Streamlines in the MV-tube model at the early diastolic filling (e), at the late diastolic filling (f), when the MV is closing (g) and when the MV is fully closed (h). Coloured by the velocity magnitude, and the colour figure can be found in the online version.

high enought pressure in order to eject the blood through the outflow tract. The closing regurgitation flow across the MV is estimated to be 7.2 mL.

Figure 2(c) shows the deformed MV leaflets when the LV starts to eject blood. Notice that the MV in the MV-LV model is only partially closed, compared to that of the MV-tube model in Fig. 2(d). The MV regurgitation in the MV-LV model persists in systole, which prevents the ventricular pressure to increase efficiently. The peak value of the pressure is 117 mmHg, which is much less than the peak value of 160 mmHg, corresponding to no leak at the inflow tract in systole.

Figure 3 (a, b, c, d) show, respectively, the streamlines in the MV-LV model during the early diastolic filling, the late diastolic filling, when the MV is closing, and when the LV is ejecting blood through the outflow tract. The corresponding streamlines of the MV-tube model is shown in Fig. 3 (e, f, g, h). In the MV-LV model, the flow moves directly towards the LV apex in the early filling, large vortices form in the late diastolic filling in the whole LV cavity, mixing the fresh blood from left atrium with the remaining blood from the previous heart beat. These flow features are absent in the MV-tube model.

4   Discussion and Conclusion

In this study, we have built an integrated MV-LV model based on in vivo CMR images of a healthy volunteer. This model incorporates a MV, a contractile LV, and the fluid structure interaction for the first time. Results are compared with that of a corresponding MV-tube model, and a number of differences are noticed. First, the flow patterns are very different in the MV-LV model. When the MV is opened, the blood passes through the MV and is directed to the posterior side in the MV-tube model before hitting the tubular wall. However, in the MV-LV model, the flow goes directly towards the LV apex, and then bends toward LV base to form large vortices. The MV-tube model is unable to model the flow in systole due to the absence of the aortic part. Those differences suggest that the incorporation of the left ventricle (the valvular-ventricle interaction) is necessary for modelling the flow patterns around the MV. However, if one is only interested in the MV closure, then a reduced-model (MV only) may be sufficient [6,17].

During diastole, the MV-LV model seems to produce a smaller orifice compared to the corresponding MV-tube model. This is because of the extra resistance offered by the LV wall, which is absent in the MV-tube model. Consequently, the total inflow volume through the MV is reduced. According to [14], the diastolic phase could be divided into three phases as the rapid filling, slow filling and atrial contraction. In the rapid filling, the transvalvular flow is resulted from the relaxation of the LV (the sucking effect), and 80 % transvalvular flow occurs [14]. During the slow filling and atrial contraction, the left atrium needs to generate higher pressure for further filling. In the MV-LV model, the ramped pressure in the inflow tract during diastole is similar to the slow filling and atrial contraction. However, the model is less accurate in simulating the rapid filling phase. This suggest that our boundary conditions need to be improved. We also notice that during systole, the MV does not close as well as in the MV-tube model, suggesting that the deformation of the annulus ring and the papillary muscles may have to be included in order to produce a physiological fluid mechanics around the MV.

The heart function is well coordinated through the electrical-mechanical coupling, the valvular closing and opening events, etc., thus the combination of the MV and LV models is more challenging compared to the isolated MV or LV models. Furthermore due to the large number of parameters involved in the integrated MV-LV model, we have not achieved a fully personalized MV-LV model (much less stroke volume compared to the measurements), though the MV and LV models are from the same volunteer. Currently we are working towards optimizing model parameters, estimating pressure boundaries, and incorporating more detailed valvular-ventricular interactions.

Acknowledgement. This work is funded by the UK EPSRC (EP/I1029990), and the British Heart Foundation (PG/14/64/31043, PG/11/2/28474). B.E.G. acknowledges research support from the American Heart Association (AHA award 10 SDG4320049), the National Institutes of Health (award HL117063), and the National Science Foundation (awards DMS 1016554 and ACI 1047734).

References

1. Chandran, K.B., Kim, H.: Computational mitral valve evaluation and potential clinical applications. Ann. Biomed. Eng. 1–15 (2014). doi:10.1007/s10439-014-1094-5

2. Conti, C.A., Stevanella, M., Maffessanti, F., Trunfio, S., Votta, E., Roghi, A., Parodi, O., Caiani, E.G., Redaelli, A.: Mitral valve modelling in ischemic patients: finite element analysis from cardiac magnetic resonance imaging. In: Computing in Cardiology, pp. 1059–1062. IEEE (2010)

3. Einstein, D.R., Kunzelman, K.S., Reinhall, P.G., Nicosia, M.A., Cochran, R.P.: Non-linear fluid-coupled computational model of the mitral valve. J. Hear. Valve Dis. **14**(3), 376–385 (2005)

4. Einstein, D.R., Reinhall, P., Nicosia, M., Cochran, R.P., Kunzelman, K.: Dynamic finite element implementation of nonlinear, anisotropic hyperelastic biological membranes. Comput. Meth. Biomech. Biomed. Eng. **6**(1), 33–44 (2003)

5. Gao, H., Carrick, D., Berry, C., Griffith, B.E., Luo, X.: Dynamic finite-strain modelling of the human left ventricle in health and disease using an immersed boundary-finite element method. IMA J. Appl. Math. **79**, 978–1010 (2014)

6. Gao, H., Ma, X., Qi, N., Berry, C., Griffith, B.E., Luo, X.: A finite strain nonlinear human mitral valve model with fluid-structure interaction. Int. J. Numer. Meth. Biomed. Eng. **30**(12), 1597–1613 (2014)

7. Griffith, B.E.: Immersed boundary model of aortic heart valve dynamics with physiological driving and loading conditions. Int. J. Numer. Meth. Biomed. Eng. **28**(3), 317–345 (2012)

8. Griffith, B.E., Luo, X.: Hybrid finite difference/finite element version of the immersed boundary method (2012). Submitted, preprint available from http://www.cims.nyu.edu/griffith

9. Kamensky, D., Hsu, M.C., Schillinger, D., Evans, J.A., Aggarwal, A., Bazilevs, Y., Sacks, M.S., Hughes, T.J.: An immersogeometric variational framework for fluid-structure interaction: Application to bioprosthetic heart valves. Comput. Meth. Appl. Mech. Eng. **284**, 1005–1053 (2014)

10. Kunzelman, K.S., Cochran, R.: Stress/strain characteristics of porcine mitral valve tissue: parallel versus perpendicular collagen orientation. J. Card. Surg. **7**(1), 71–78 (1992)

11. Kunzelman, K.S., Einstein, D.R., Cochran, R.P.: Fluid-structure interaction models of the mitral valve: function in normal and pathological states. Philos. Trans. Royal Soc. B: Biol. Sci. **362**(1484), 1393–1406 (2007)

12. Lau, K., Diaz, V., Scambler, P., Burriesci, G.: Mitral valve dynamics in structural and fluid-structure interaction models. Med. Eng. Phys. **32**(9), 1057–1064 (2010)

13. Ma, X., Gao, H., Griffith, B.E., Berry, C., Luo, X.: Image-based fluid-structure interaction model of the human mitral valve. Comput. Fluids **71**, 417–425 (2013)

14. Nishimura, R.A., Tajik, A.J.: Evaluation of diastolic filling of left ventricle in health and disease: Doppler echocardiography is the clinician's Rosetta Stone. J. Am. Coll. Cardiol. **30**(1), 8–18 (1997)

15. Ray, R., Chambers, J.: Mitral valve disease. Int. J. Clin. Prac. **68**(10), 1216–1220 (2014)

16. Sun, W., Abad, A., Sacks, M.S.: Simulated bioprosthetic heart valve deformation under quasi-static loading. J. Biomech. Eng. **127**(6), 905–914 (2005)

17. Wang, Q., Sun, W.: Finite element modeling of mitral valve dynamic deformation using patient-specific multi-slices computed tomography scans. Ann. Biomed. Eng. **41**(1), 142–153 (2013)

18. Watton, P.N., Luo, X.Y., Yin, M., Bernacca, G.M., Wheatley, D.J.: Effect of ventricle motion on the dynamic behaviour of chorded mitral valves. J. Fluids Struct. **24**(1), 58–74 (2008)

19. Yin, M., Luo, X., Wang, T., Watton, P.: Effects of flow vortex on a chorded mitral valve in the left ventricle. Int. J. Numer. Meth. Biomed. Eng. **26**(3–4), 381–404 (2010)

Finite Element Simulations Explore a Novel Strategy for Surgical Repair of Congenital Aortic Valve Insufficiency

Peter E. Hammer[✉] and Pedro J. del Nido

Department of Cardiac Surgery, Boston Children's Hospital, Boston, MA 02115, USA
peter.hammer@childrens.harvard.edu

Abstract. For aortic valve reconstruction in the child, techniques are favored that minimize the introduction of foreign material and graft tissues that do not grow with the child. In this study we use computer simulation to study a potential method for conservatively repairing an aortic valve that is regurgitant due to a congenitally undersized leaflet. The surgical approach consists of resecting portions of the aortic root in order to allow the valve leaflets to close completely. We use a structural finite element model of the aortic valve to simulate valve closure following different strategies for resecting portions of the aortic root (e.g., triangular versus rectangular resection). Results show that rectangular resection of the root is able to eliminate regurgitation and produce a viable repair.

Keywords: Aortic valve · Root remodeling · Simulation · Coaptation

1 Introduction

The normal aortic valve consists of three leaflets, approximately equal in size, that open freely during cardiac ejection and close tightly during cardiac filling. In the congenitally malformed aortic valve, the well-described symmetry of the normal valve [1] is lost, and this can have detrimental effects on valve function. At our center, a common finding in children with aortic regurgitation is an undersized leaflet in the right coronary (RC) position. The resulting regurgitation, if severe, can lead to heart failure and death if left untreated.

In adults, regurgitant aortic valves are often replaced with either mechanical valves or bioprosthetic valves. In children, however, these devices have significant disadvantages. The most important is that replacement valves remain fixed in size while the child's heart grows, eventually restricting cardiac output and requiring reoperation. In addition, mechanical valves require lifelong anticoagulant therapy, while bioprosthetic valves suffer from accelerated degeneration and calcification in children. As a result, surgical reconstruction of the diseased valve in children is preferred over valve replacement. The current approach to treating a regurgitant valve due to a single undersized leaflet typically involves augmenting or replacing the undersized leaflet with a graft. This repair typically has good short- to mid-term results, but, in the long-term it is likely

© Springer International Publishing Switzerland 2015
H. van Assen et al. (Eds.): FIMH 2015, LNCS 9126, pp. 338–345, 2015.
DOI: 10.1007/978-3-319-20309-6_39

that the graft, made from autologous or animal tissue that has been chemically fixed, will degrade and calcify requiring reoperation.

A better solution would eliminate the regurgitation without introducing graft material. This results in a reconstructed valve in which all of the structures could develop and grow normally with the child. Based on surgical experience, we have seen that it is possible to bring regurgitant valve leaflets into better approximation by resecting portions of the aortic root and closing the excision with sutures. Specifically, we suspect that it is possible to eliminate regurgitation due to an undersized RC leaflet by resecting a portion of the root bounded by the noncoronary (NC) leaflet. This could potentially increase the mobility of the NC leaflet, allowing it to contact and form a seal with the RC leaflet, eliminating the regurgitation.

To explore this approach without putting patients at risk, we use a computational model of the aortic valve to test different root resection strategies. Computational models have been used to study normal aortic valve function [2–5] as well as surgical repair [6–8]. We have been developing computational modeling tools to study aortic valve function and surgical repair, focusing on structural finite element models aimed at simulating the closed, loaded state of the valve under diastolic pressure [8]. In this study, we apply these modeling tools to study how the insufficiency of an aortic valve with one congenitally undersized leaflet can be improved by resecting tissue from the wall of the aortic root adjacent to one of the other two (normal) leaflets. In the remainder of the paper, we summarize our computational model and present our method for simulating closure of a valve with portions of the aortic root surgically resected. We then present the results of these simulations, focusing on quantitative descriptions of the closed configuration of the valve leaflets. Finally we discuss the significance of the findings and outline future work.

2   Methods

To create a model of a generalized human aortic valve, we scaled the average valve leaflet shape determined from studies of 18 porcine hearts (Fig. 1A) to appropriate relative leaflet sizes for a normal human aortic valve [9]. Then, to model a valve with a congenitally undersized RC leaflet, we uniformly scaled that leaflet to achieve an aortic valve cross-section in which the portion of the aortic root circumference bounded by the RC leaflet was reduced from approximately one-third to one-quarter, in accordance with our clinical observations. This corresponds to scaling the RC leaflet dimensions by a factor of two-thirds. The scaled leaflet outlines were meshed with triangles (Fig. 1B) and joined at their endpoints. This planar, 3-leaflet mesh was wrapped into a cylinder (Fig. 1C) based on anatomical studies which showed that the points of attachment of the leaflets to the aortic root lie on a cylinder [10]. This geometry is further supported by our own experimental observations. For the computational model, each triangle is treated as a 3-node membrane element, with the material properties governed by an orthotropic, hyperelastic constitutive equation fit to biaxial test results from porcine valve leaflets [8]. The principal material direction, corresponding to the local collagen fiber direction, was assigned to mesh elements to reflect the average fiber pattern that we measured in porcine aortic valve

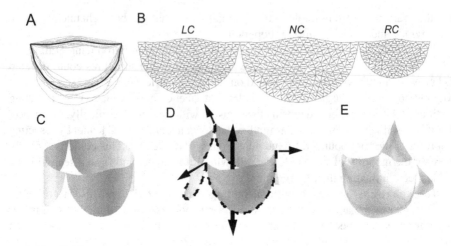

Fig. 1. (A) Aortic valve leaflet shape is determined by averaging the leaflet outlines from 18 porcine hearts. (B) Left coronary (LC), noncoronary (NC), and right coronary (RC) leaflets are scaled to model a human valve with a congenitally undersized RC leaflet, and a mesh of unstructured triangles is generated within each of the three average leaflet outlines. The collagen fiber direction (principal material axis) is also shown. (C) The planar leaflet meshes are wrapped into a cylinder. Closure and loading of the valve is simulated in two steps. (D) First loads are applied to the mesh vertices that lie on the aortic root to simulate the radial and axial dilation of the aortic root in diastole. (E) Then surface normal forces are applied to all leaflet elements to simulate peak diastolic transleaflet pressure.

leaflets that had been excised and stained for collagen. The constitutive equation and collagen pattern have been published previously [11].

In simulating aortic valve repair strategies, our primary concern is whether the repaired valve will close without regurgitation under diastolic pressure load. We simulate the state of diastolic loading in two steps. First, loads are applied to the boundary points where the leaflet mesh meets the aortic root to simulate the state of radial and axial stresses in the aortic root in diastole, modeling the aortic root implicitly as an orthotropic cylinder with elastic moduli taken from the literature [6]. Then surface normal forces are applied to leaflet elements to simulate peak diastolic transleaflet pressure.

To simulate different strategies for resecting portions of the aortic root, we move the leaflet boundary points in order to achieve the modified aortic root shape prior to simulating valve closure. We are interested in the effect of resecting patches of various shapes from the aortic root adjacent to the NC leaflet: (1) a triangular patch oriented such that no material is removed at the level of the annulus (i.e., the nadir of the leaflet attachment) and the maximum is removed at the level where the tops of the leaflets attach (i.e., the commissures), (2) a rectangular patch where a strip of uniform width is resected from the level of the annulus to that of the commissures, (3) a triangular patch oriented such that no material is removed at the level of the commissures and the maximum is removed at the level of the annulus, and (4) a diamond-shaped patch where no material is removed at the level of the annulus or commissures and the maximum is removed at the level halfway between the two (Fig. 2). We refer to these resection shapes as *TriUp*, *Rect*,

TriDown, and *Diamond*, respectively. Note that for the *Rect* and *TriDown* shapes, a portion of the valve annulus (where the leaflet attaches) is removed. In these cases, the NC leaflet meshes are modified to remove a triangular-shaped portion of the leaflet mesh that shares the resected annulus. The mesh is stitched back together by applying spring forces to approximate the cut edges during simulated resection and closure.

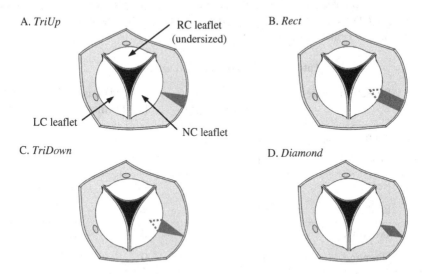

Fig. 2. Drawing of transected aorta showing top view of the aortic valve. Four strategies are shown for resecting a portion (shown here in red) of the NC aortic root to try to compensate for an undersized RC leaflet: (A) triangular resection with base at the level of the commissures, (B) rectangular resection, (C) triangular resection with base at the level of the annulus, and (D) diamond-shaped resection.

A normal aortic valve is characterized by several millimeters of overlap (coaptation) between adjacent leaflets when closed, and it has been shown that achieving this redundancy following valve repair is associated with good long-term outcomes [12]. Accordingly, we determine the adequacy of simulated valve closure by computing the minimum extent of coaptation of adjacent leaflets in the closed state as well as the total area of coaptation for a given leaflet.

We simulate the equations of motion, including contact mechanics, and solve for the final deformed state of the valve using semi-implicit numerical integration with adaptive time-step control. All simulation and analysis software was written in the Matlab programming language (Mathworks, Natick, MA, USA). See our previous work for details of the modeling method [8].

3   Results

The simulated closed state for a valve with a congenitally undersized RC leaflet exhibited failure to close completely under typical diastolic transvalvular pressure of 80 mmHg (Fig. 3A). A map of the portion of the RC leaflet that coapts with the adjacent leaflets shows a large central portion of the RC leaflet that does not contact the other leaflets (Fig. 3B). For the five strategies for resection of the NC root that we simulated, only three (*Rect5*, *Rect10*, and *TriDown*) resulted in at least marginal central coaptation (Fig. 3B). Of those three, only the strategy of resecting a 10 mm wide rectangular section (*Rect10*) of the NC root resulted in a repair with sufficient central coaptation to be considered acceptable. This strategy also resulted in the repair with the largest total area of the RC leaflet involved in coaptation, and the results for total coaptation area generally echoed those for central coaptation height (Table 1).

Table 1. Metrics of simulated valve closure

Resection Strategy	Coaptation area on RC leaflet (mm^2)	Central coaptation height (mm)
No resection	31	0
TriUp	31	0
Rect5	59	0.8
Rect10	**63**	**2.2**
TriDown	58	0.3
Diamond	31	0

4   Discussion

Despite simple anatomy and elegant function, the structures of the aortic valve interact in complex ways. Computational modeling, by codifying the relationships between the anatomy, tissue properties, and pressure, allows the effects of a single variable - like the geometry of a surgically remodeled aortic root - to be isolated and studied. Our simulation results show that it is possible to repair an aortic valve that is regurgitant due to one congenitally undersized leaflet by resecting part of the aortic root adjacent to an opposing leaflet. Results indicate that only the root resection strategies that include the valve annulus and part of the adjoining leaflet effectively abolish the regurgitation. Furthermore, only the case of a 10 mm wide rectangular resection produces a repair with adequate central coaptation of the closed valve.

Examination of the simulated closed valves helps to explain the effects of resecting the aortic root at different levels. When resection primarily reduces the diameter of the valve at the level of the commissures (e.g., *TriUp*), the free edges of all three leaflets drop farther. For the undersized RC leaflet, dropping of the free edge toward the lower LC and NC leaflets is desirable. However, the LC and NC leaflets also drop, and no gains in leaflet coaptation are achieved. In fact, the dropping of the LC and NC leaflets introduces some billowing of the leaflets toward the ventricle, and this moves the net force vector on the leaflets due to transvalvular pressure to be less radially inward and more axially downward. In other words, pulling the commissures together does not result

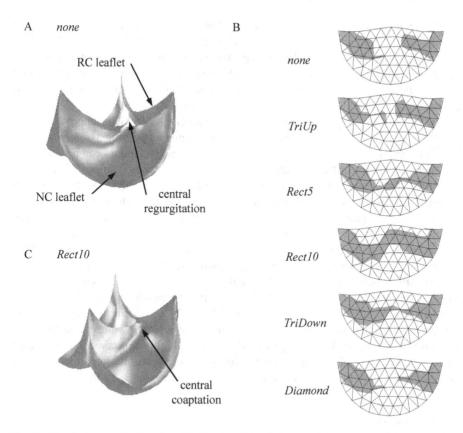

Fig. 3. Simulation results showing (A) the closed, loaded state of the valve model representing the case of the pathological valve with no resection of the NC aortic root. Note the regurgitant orifice formed due to inability of the three leaflets to meet in the valve center. (B) The region of the RC leaflet that coapts with the other two leaflets (shaded gray) for various root resection shapes. *Rect5* and *Rect10* refer to the width of the resected rectangle in millimeters. The width of the triangle and diamond regions was approximately 10 mm. (C) The closed, loaded state of the valve model representing the *Rect10* resection shape. Note the central coaptation where the three leaflets meet.

in the leaflets being pushed into tighter closure but rather in their being pushed more downward toward the ventricle. When resection primarily reduces the diameter of the valve at the level of the annulus (e.g., *TriDown*), the heights of the leaflets along their midlines become more in proportion to the smaller annulus diameter. By resecting or folding the portion of the NC leaflet that had been attached to the now resected section of annulus, billowing is avoided, and the direction of the net force vector on the leaflet due to pressure retains its strong radially inward component. However, without the reduction of the root at the level of the commissures, the dropping of the RC leaflet is not present, and the valve still does not close with significant coaptation in the center. Rectangular resection reduces the root diameter at both the commissural and annular levels, and both effects described above combine to achieve a fully-closed valve with substantial coaptation. The efficacy of this strategy depends on the amount of root wall that is resected. Our simulations showed that resecting a 5 mm wide strip did result in full valve closure, but the central coaptation height – which can be thought of as a factor of safety for the repaired valve – is marginal. Raising the width of the rectangle to 10 mm (corresponding to a ~12 % reduction in root diameter for the valve size that we simulated) gives a more acceptable amount of central coaptation.

Results of our simulation study, while encouraging, should be interpreted with caution. First, the modeling methods are based on many simplifying assumptions, and careful experimental validation is mandatory before these techniques can be applied in the operating room. In fact, the tangible benefit of our results is to provide specific, quantitative hypotheses to test experimentally, reducing the otherwise prohibitively large parameter space that would have to be explored. Second, our metric for assessing adequacy of the repaired valve was based on diastolic function, i.e., the competence of the closed valve under load. However, the resection strategies suggested by our simulations reduce the caliber of the aortic root and, consequently, can increase outflow resistance during systole. Many of the patients who present with the type of congenital aortic insufficiency considered here have enlarged aortic roots, so resecting part of the root is unlikely to limit cardiac output. However, in patients whose aortic root is normal or small preoperatively, root resection could be problematic. Further modeling and experimental studies will be necessary to quantify post-repair pressure gradients across the valve in systole. While our hope would be that the highly conserved, reconstructed valve would undergo physiological remodeling to restore valve size to normal and keep pace with growth of the child, we do not yet have direct evidence for this.

5   Conclusion

We have shown that the method of localized aortic root reduction holds promise as a method for aortic valve repair for aortic regurgitation due to one congenitally undersized leaflet. These results suggest that this technique might warrant further investigation, and we have begun conducting experiments to verify and extend these simulation results in an animal model. Work on producing computational valve models from 3d ultrasound is also underway, both in our group and in others [13, 14], and promises to enable patient-specific surgical planning.

Acknowledgments.   This work was supported by NIH grant R01 HL110997.

References

1. Thubrikar, M.: The Aortic Valve, pp. 8–10. CRC Press, Boca Raton (1990)
2. Grande, K.J., Cochran, R.P., Reinhall, P.G., Kunzelman, K.S.: Stress variations in the human aortic root and valve. Ann. Biomed. Eng. **26**(4), 534–545 (1998)
3. Nicosia, M.A., Cochran, R.P., Einstein, D.R., Rutland, C.J., Kunzelman, K.S.: A coupled fluid-structure finite element model of the aortic valve and root. J. Heart Valve Dis. **12**(6), 781–789 (2003)

4. De Hart, J., Peters, G.W.M., Schreurs, P.J.G., Baaijens, F.P.T.: Collagen fibers reduce stresses and stabilize motion of aortic valve leaflets during systole. J. Biomech. **37**, 303–311 (2004)
5. Labrosse, M.R., Lobo, K., Beller, C.J.: Structural analysis of the natural aortic valve in dynamics. J. Biomech. **43**, 1916–1922 (2010)
6. Grande-Allen, K.J., Cochran, R.P., Reinhall, P.G., Kunzelman, K.S.: Finite element analysis of aortic valve-sparing. IEEE Trans. Biomed. Eng. **48**, 647–659 (2001)
7. Labrosse, M.R., Boodhwani, M., Sohmer, B., Beller, C.J.: Modeling leaflet correction techniques in aortic valve repair: a finite element study. J. Biomech. **44**, 2292–2298 (2011)
8. Hammer, P.E., Chen, P.C., del Nido, P.J., Howe, R.D.: Computational model of aortic valve surgical repair using grafted pericardium. J. Biomech. **45**, 1199–1204 (2012)
9. Sim, E.K.W., Muskawad, S., Lim, C.-S., Yeo, J.H., Lim, K.H., Grignani, R.T., Durrani, A., Lau, G., Duran, C.: Comparison of human and porcine aortic valves. Clin. Anat. **16**, 193–196 (2003)
10. Swanson, W.M., Clark, R.E.: Dimensions and geometric relationships of the human aortic valve as a function of pressure. Circ. Res. **35**, 871–882 (1974)
11. Hammer, P.E., Pacak, C.A., Howe, R.D., del Nido, P.J.: Straightening of curved pattern of collagen fibers under load controls aortic valve shape. J. Biomech. **47**, 341–346 (2014)
12. Augoustides, J.G., Szeto, W.Y., Bavaria, J.E.: Advances in aortic valve repair. J. Cardiothorac. Vasc. Anesth. **24**, 1016–1020 (2010)
13. Mansi, T., Voigt, I., Georgescu, B., Zheng, X., Mengue, E.A., Hackl, M., Ionasec, R.I., Noack, T., Seeburger, J., Comaniciu, D.: An integrated framework for finite-element modeling of mitral valve biomechanics from medical images. Med. Image Anal. **16**, 1330–1346 (2012)
14. Burlina, P., Sprouse, C., Mukherjee, R., DeMenthon, D., Abraham, T.: Patient-specific mitral valve closure prediction using 3D echocardiography. Ultrasound Med. Biol. **39**(5), 769–783 (2013)

Determining Anisotropic Myocardial Stiffness
from Magnetic Resonance Elastography:
A Simulation Study

Renee Miller[1(✉)], Haodan Jiang[2], Ria Mazumder[3], Brett R. Cowan[1], Martyn P. Nash[4,5],
Arunark Kolipaka[3], and Alistair A. Young[1,4]

[1] Department of Anatomy with Radiology, University of Auckland, Auckland, New Zealand
renee.miller@auckland.ac.nz
[2] Ohio Supercomputing Center, Columbus, OH, USA
[3] Department of Radiology, Ohio State University, Columbus, OH, USA
[4] Auckland Bioengineering Institute, University of Auckland, Auckland, New Zealand
[5] Department of Engineering Science, University of Auckland, Auckland, New Zealand

Abstract. Although magnetic resonance elastography (MRE) has the potential to non-invasively measure myocardial stiffness, myocardium is known to be anisotropic and it is not clear whether all material parameters can be uniquely determined from MRE data. In this study, we examined the determinability of anisotropic stiffness parameters using finite element analysis (FEA) simulations of harmonic steady state wave behavior. Two models were examined: (i) a cylindrical and (ii) a canine left ventricular geometry with realistic fiber architecture. A parameter sweep was carried out, and the objective function, which summed the error between reference displacements and displacements simulated from MRE boundary data and material parameters, was plotted and determinability assessed from the Hessian. Then, given prescribed boundary displacements from the ground truth simulation with added Gaussian noise, an anisotropic material parameter optimization was run 30 times with different noise in order to investigate the effect of noise on determination of material parameters. Results show that transversely isotropic material parameters can be robustly determined using this method.

Keywords: Magnetic resonance elastography · Myocardial stiffness · Cardiac anisotropy · Finite element modelling

1 Introduction

Myocardial stiffness is an important determinant of cardiac function with significant increases in global stiffness associated with diastolic heart failure [1]. However, myocardial stiffness is not widely measured clinically or in research studies due to the invasiveness of measurements. Magnetic resonance elastography (MRE) is a recent, non-invasive technique for measuring myocardial stiffness. MRE is a three stage process in which (1) non-invasive, external actuators are used to generate waves in the myocardium [2], (2) the wave is visualized using a phase-contrast gradient echo MRI sequence [3], and (3) the

© Springer International Publishing Switzerland 2015
H. van Assen et al. (Eds.): FIMH 2015, LNCS 9126, pp. 346–354, 2015.
DOI: 10.1007/978-3-319-20309-6_40

displacement information is converted into stiffness maps using an inversion algorithm [4]. This process shows considerable promise for non-invasive assessment of global myocardial stiffness.

A large portion of MRE research has focused on developing an inversion algorithm which accurately and robustly quantifies stiffness from the MRI displacement images. The literature notes that shear stiffness determined in this way can only be seen as "effective" stiffness and does not represent the true stiffness of the material [5–8]. "Effective" tissue stiffness can be important for relative comparison of adjacent tissues in a region (e.g. tumor detection). However, when assessing disease progression in a single patient, or comparing tissue stiffness within a cohort of patients, true stiffness values are preferable. The "effective" nature of the resulting stiffness is a result of the underlying assumptions made by a number of the inversion algorithms, principally, that the material is isotropic, and that the medium is infinite.

This paper is an initial work towards the development of a novel method for determining homogeneous anisotropic stiffness in myocardium by integrating finite element analysis (FEA) and DTI fiber data with MRE displacement data. The identifiability of the anisotropic material parameters, based only on the modelled displacement field, is assessed to demonstrate that this is a well-posed inverse problem, such that the material parameters can be uniquely determined.

2   Methods

2.1   Finite Element Models

Abaqus 6.13 (Dassault Systèmes Simulia Corp., Providence, USA) was used to model and solve both a cylindrical phantom and a canine LV geometry. The cylindrical phantom model contained 50061 nodes and 46592 first-order hexahedral elements. Fibers were oriented along the long-axis of the cylinder. A harmonic deformation of 0.2 mm was applied to the lower flat surface of the model to simulate deformation caused by pressure from a pneumatic MRE driver. The magnitude of the applied deformation was chosen in order to obtain displacement values similar to those commonly seen in cardiac MRE experiments [7]. The curved surface of the phantom was fixed with a no-displacement boundary condition. The top surface of the cylinder was unconstrained (Fig. 1).

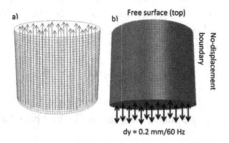

Fig. 1. (a) Cylindrical fibers (direction of highest stiffness); (b) boundary conditions

The LV model contained 5490 nodes and 4320 first-order hexahedral elements. Physiologically realistic helical fibers from histology were imbedded and interpolated at each nodal location. Fiber angles measured with respect to the short axis plane varied from approximately −60 to +60 degrees transmurally from epicardium to endocardium, respectively. The epicardial basal nodes were fixed and a displacement of 0.2 mm at 60 Hz was prescribed at 41 epicardial nodes at the apex (Fig. 2).

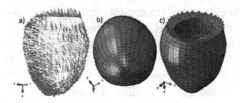

Fig. 2. (a) LV fibers; (b) apical nodes in red which were displaced 0.2 mm/60 Hz; (c) epicardial basal nodes in red that were fixed (no-displacement) (Color figure online)

A linearly elastic, transversely isotropic material model was used to describe the passive mechanics of the material in the cylindrical and LV models. This is considered to be a good assumption at the small strains induced by MRE. To generate the ground truth for later comparison, the fiber direction ($E3$) was assigned a Young's modulus of 60 kPa and the transverse directions ($E1$, $E2$) were set to 18 kPa. A Poisson's ratio of 0.495 was applied as cardiac tissue is largely incompressible and a density of 1.06 g/cm^3 was assumed.

In Abaqus, the steady-state response due to harmonic loading is solved using the harmonic equation of motion where P^* is the complex forcing term, u^* is the complex displacement, M is the mass matrix, C is the damping matrix and K is the stiffness matrix. The stiffness matrix is the inverted elasticity matrix, which is a function of the elasticity constants ($E1$, $E2$ and $E3$) and Poisson's ratios.

$$-\omega^2 M u^* + i\omega C u^* + K u^* = P^* \tag{1}$$

Structural damping was applied in order to provide a means of extracting energy from the model as would be expected for biological tissue. Structural damping forces are proportional to forces caused by stressing the structure and oppose the velocity. One common representation of structural damping is as a complex stiffness (K^*) involving a real component (K_S) and an imaginary component ($K_L = \omega C$); the latter acts as a loss modulus which accounts for damping.

$$K^* = iK_L + K_S = K_S(is + 1) \tag{2}$$

where s is the structural damping coefficient, which relates the real and imaginary components of stiffness to one another. Equation (1) then becomes:

$$-\omega^2 M u^* + (1 + is) K_S u^* = P^* \tag{3}$$

2.2   Objective Function

To assess the determinability of the transversely isotropic stiffness parameters from the MRE displacement field, an objective function of the unknown parameters must be defined. Four parameters need to be defined: $E_{1/2}$, E_3, v_{12}, and $v_{13/23}$ to describe a transversely isotropic material. A simplification was made where the Poisson's ratios, which account for the incompressibility of the tissue, were assumed to be the same in the transverse-transverse (1-2, 2-1) and fiber-transverse directions (3-1, 3-2) thereby reducing the number of parameters to three. The Poisson's ratios in the transverse-fibre directions (1-3, 2-3) are related to the Young's moduli and the Poisson's ratios in the fibre-transverse direction according to the following equation:

$$v_{ij}/E_i = v_{ji}/E_j \tag{4}$$

In this study we wanted to also identify the damping coefficient (s) resulting in the following four parameters to be identified: s, E_1, E_3 and v_{12}.

The error, $E(u^*)$, was calculated as the root mean square of the Euclidian distance between nodal displacements generated from the estimated parameters and ground truth displacements generated from the true parameters and expressed as a percentage of the root mean square ground truth displacement.

$$E(u^*) = \frac{\sqrt{\frac{1}{3N} \sum^N \sum^{x,y,z} \left[\left(U_R^C - U_R^M \right)^2 + \left(U_I^C - U_I^M \right)^2 \right]}}{\sqrt{\frac{1}{3N} \sum^N \sum^{x,y,z} \left[\left(U_R^M \right)^2 + \left(U_I^M \right)^2 \right]}} \tag{5}$$

where subscripts R and I indicate the real and imaginary components of displacement; superscripts C and M represent the calculated (or simulated) and the ground truth displacement, respectively; and N denotes the number of nodes. In both the cylindrical and LV models, only the internal nodes, which were not subject to displacement boundary constraints, were used to calculate the error function.

2.3   Determinability of Material Properties

Three criteria were evaluated in order to assess the determinability of the material parameters from the Hessian matrix computed at the minimum of the objective function, [9, 10].

D-Optimality Criterion. This is related to the volume of the indifference region, defined by the p-dimension hyper ellipsoid with size determined by the eigenvalues of the Hessian matrix. The indifference region is named thus since a small change in the parameter space within this region doesn't affect the error function significantly. Since the volume is inversely proportional to the determinant of the Hessian, a D-optimal design would maximize det(H).

Eccentricity Criterion. The eccentricity, or ratio of the longest to shortest axis, of the ellipsoid describing the region of indifference is a measure of the discrepancy between the least- and best-determined linear combination of parameters. The ratio of the largest eigenvalue to the smallest eigenvalue reflects the ellipsoidal eccentricity. An eccentric-optimal design would minimize cond(H). Since the four parameters evaluated in this study are of different scales (by three orders of magnitude), the eccentricity criterion was evaluated from a Hessian computed using parameter vectors which were normalized to have equal magnitude, H_{norm}.

M-Optimality Criterion. The third criterion relates to the interaction between design parameters. An ellipsoid axis which lies at some angle with respect to parameter axes indicates a correlation between parameters. Interaction between parameters is minimized when the determinant of a scaled Hessian matrix, $M = det\left(\tilde{H}\right)$, is equal to one. The (i,j) entry of $\tilde{H}$ can be written as:

$$\tilde{h}_{ij} = h_{ij}/\sqrt{h_{ij}h_{jj}}, i,j \in \{1,2,3,4\} \quad \text{(No summation)} \tag{6}$$

2.4 Material Parameter Optimization with Gaussian Noise

In both models, Gaussian noise was added to the real and imaginary components of the nodal displacements. In the LV model, epicardial displacements from the ground truth case with added noise were prescribed to the outer surface. The noisy displacements from the side and bottom of the cylindrical ground truth simulation were used as boundary conditions for the cylindrical model. The Gaussian distribution of noise was computed for the x, y and z directions independently to have a mean of zero and standard deviation computed as:

$$\sigma_{noise}^{x,y,z} = 15\% \cdot \sigma_{disp}^{x,y,z} \tag{7}$$

where $\sigma_{disp}^{x,y,z}$ is the standard deviation of the displacement in each direction. 15 % was chosen based on previous MRE simulation studies [11].

An optimization algorithm, provided in the Scipy package, was used to determine the material parameters that minimized the objective function when noise was added to the model. The algorithm (*fmin_cobyla*) is a nonlinearly constrained optimization that does not require knowledge of derivatives [12]. Each iteration forms a linear approximation to the objective and constraint functions by interpolation at the vertices of a simplex. The optimization was run 30 times with random noise re-generated at each run.

3   Results

3.1   Cylindrical Phantom Model

Figure 3 shows the ground truth displacement maps in the cylindrical model and Fig. 4 shows objective function plots generated by a parameter sweep. Based on the Hessian at the minimum of the objective function, the D-optimality criterion (det(H)) for identifiability was 5600; the eccentricity criterion (cond(H_{norm})) was 19; and the M-optimality (det($\tilde{H}$)) criterion was 0.84. Although the eccentricity criterion is large (>1), which indicates that some parameters are more identifiable than others, the D-optimality criterion is quite high indicating that the volume of the indifference region is extremely small. In addition, an M-optimality value of 0.838 indicates a reasonable degree of independence between parameters. Parameter values resulting from the 30 material parameter optimizations for the cylindrical model with noise added to the boundary conditions were: $s = 0.249 \pm 0.0003$, $E1 = 17.778 \pm 0.0417$ kPa, $E3 = 60.288 \pm 0.0980$ kPa and $v_{12} = 0.4955 \pm 0.00031$. Overall, these results indicate that the reference material parameters are identifiable based on the displacement field and for this cylindrical model.

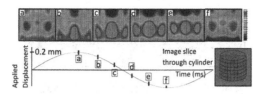

Fig. 3. Displacement maps at six points during one harmonic cycle in the ground truth FE cylindrical model. The figure is a 2D view passing through the central axis of the cylinder (see diagram) and represents displacement in the vertical direction. Red: +0.2 mm, blue -0.2 mm (Color figure online).

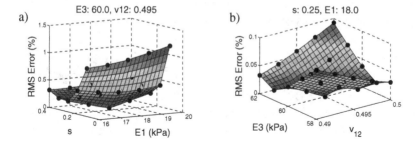

Fig. 4. Percent RMSE for (a) s and $E1$ and (b) $E3$ and v_{12}. Plots are shown with additional interpolated data points; black spheres indicate points where error values were calculated (Color figure online).

3.2 LV Model

Figure 5 shows the ground truth displacement maps and Fig. 6 shows the objective function plots. Based on the Hessian at the minimum of the objective function for the LV brute force parameter sweep, the D-optimality criterion (det(H)) for identifiability was 0.032; the eccentricity criterion (cond(H_{norm})) was 9.61; and the M-optimality (det($\tilde{H}$)) criterion was 0.94. The eccentricity criterion was lower than the cylindrical model but still greater than one. The D-optimality criterion was much lower than in the cylindrical model. This marked difference was due to the magnitude of the displacement being significantly lower on average in the LV model. The magnitude of the displacement errors directly affects the magnitude of values in the Hessian, and consequently the determinant. Despite the small magnitude of the D-optimality criterion, plots of the objective function show a clear descent to a single global minimum. An M-optimality value close to one indicates high independence between parameters in the LV model. Parameter values resulting from the 30 material parameter optimizations with simulated noise were: $s = 0.250 \pm 0.00058$, $E1 = 18.032 \pm 0.048$ kPa, $E3 = 60.023 \pm 0.051$ kPa and $v_{12} = 0.495 \pm 0.00005$. The identified values were close to the reference values and the standard deviations were small. Overall, these results indicate that the reference material parameters are identifiable in the LV model.

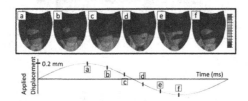

Fig. 5. Displacement maps at six points during one harmonic cycle in the ground truth FE LV model. The figure is a 2D view passing through the long axis of the LV and represents magnitude of displacement ($\sqrt{x^2 + y^2 + z^2}$); red: +0.2 mm, blue: 0 mm (Color figure online).

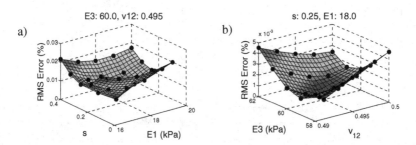

Fig. 6. Percent RMSE for (a) s and $E1$ and (b) $E3$ and v_{12}. Plots are shown with additional interpolated data points; black spheres indicate points where error values were calculated (Color figure online).

4   Discussion and Conclusions

The results from the cylindrical and LV simulations demonstrate the determinability of anisotropic material parameters based solely on the displacement field of a propagating wave as in MRE. The D-optimality criterion was greater for the cylindrical model than in the LV model. This was due to the difference in magnitude of the error calculated by the objective function. The cylindrical model resulted in greater percentage RMSE over the parameter sweep than the LV model. Since the parameter range was the same for both models, the resulting Hessian, and subsequently the determinant of the Hessian, was affected by the difference in the size of the error.

Although the high eccentricity criteria for both models indicate that some parameters are more identifiable than others, this criterion is secondary in assessing determinability of parameters [9] where the D-optimality criterion is the primary indicator of identifiability. The M-optimality criterion showed that, for these problems, there is relatively little interaction between the four parameters.

In both the cylindrical and LV models, the transversely isotropic material parameter optimizations resulted in values close to the reference values and showed little variance. In future work, the method will be applied to MRE data acquired from the cylindrical phantom and in vivo hearts. Overall, this initial work indicates that using displacement data from MRE to identify anisotropic material properties is a well-posed problem even in the presence of noise.

Some limitations of this study include the limited range of the parameter sweep and the resolution of the models compared with MRE images. Although the objective function shows one global minimum, a larger sweep of the parameter space should be performed in order to rule existence of any other minima. Additionally, the models have higher resolution than MRE images. For example, the cylindrical model has approximately 3.5 nodes per mm compared to one set of experimental MRE phantom data which has approximately one pixel per mm. The lower resolution of MRE images could provide an additional challenge when applying this methodology to experimental data.

Acknowledgements.   This research was supported by an award from the National Heart Foundation of New Zealand and by the NeSI high performance computing facilities.

References

1. Zile, M.R., Baicu, C.F., Gaasch, W.H.: Diastolic heart failure–abnormalities in active relaxation and passive stiffness of the left ventricle. N. Engl. J. Med. **350**, 1953–1959 (2004)
2. Kolipaka, A., et al.: Magnetic resonance elastography as a method for the assessment of effective myocardial stiffness throughout the cardiac cycle. Magn. Reson. Med. **64**, 862–870 (2010)
3. Muthupillai, R., et al.: Magnetic resonance elastography by direct visualization of propagating acoustic strain waves. Science **269**(5232), 1854–1857 (1995)
4. Manduca, A., et al.: Magnetic resonance elastography: non-invasive mapping of tissue elasticity. Med. Image Anal. **5**(4), 237–254 (2001)
5. Huwart, L., et al.: Liver fibrosis: non-invasive assessment with MR elastography. NMR Biomed. **19**, 173–179 (2006)
6. Kolipaka, A., et al.: Evaluation of a rapid, multiphase MRE sequence in a heart-simulating phantom. Magn. Reson. Med. **62**, 691–698 (2009)
7. Kolipaka, A., et al.: Magnetic resonance elastography as a method for the assessment of effective myocardial stiffness throughout the cardiac cycle. Magn. Reson. Med. **64**, 862–870 (2010)
8. Kolipaka, A., et al.: Magnetic resonance elastography: inversions in bounded media. Magn. Reson. Med. **62**, 1533–1542 (2009)

9. Nathanson, M., Saidel, G.: Multiple-objective criteria for optimal experimental design: application to ferrokinetics. Am. J. Physiol. **248**, 378–386 (1985)
10. Babarenda Gamage, T.P., et al.: Identification of mechanical properties of heterogeneous soft bodies using gravity loading. Int. J. Numer. Meth. Biomed. Eng. **27**, 391–407 (2011)
11. Van Houten, E.E.W., et al.: Subzone based magnetic resonance elastography using a rayleigh damped material model. Med. Phys. **38**, 1993 (2011)
12. Powell, M.: A direct search optimization method that models the objective and constraint functions by linear interpolation. Adv. Optim. Numer. Anal. **275**, 51–67 (1994)

Myocardial Stiffness Estimation: A Novel Cost Function for Unique Parameter Identification

Anastasia Nasopoulou[1]([✉]), Bojan Blazevic[1], Andrew Crozier[1], Wenzhe Shi[4],
Anoop Shetty[1,3], C. Aldo Rinaldi[1,3], Pablo Lamata[1,2], and Steven A. Niederer[1]

[1] Division of Imaging Sciences and Biomedical Engineering,
King's College London, London, UK
`anastasia.nasopoulou@kcl.ac.uk`
[2] Department of Computer Science, University of Oxford, Oxford, UK
[3] Department of Cardiology, Guy's and St. Thomas' NHS Trust, London, UK
[4] Department of Computing, Imperial College London, London, UK

Abstract. Myocardial stiffness is a clinical biomarker used to diagnose
and stratify diseases such as heart failure. This biomechanical property
can be inferred from the personalisation of computational cardiac models
to clinical measures. Nevertheless, previous attempts have been unable
to determine a unique set of material constitutive parameters. In this
study we address this shortcoming by proposing a new cost function
that allows us to uncouple key parameters and uniquely describe passive
material properties in patients from available clinical data.

Keywords: Cardiac mechanics · Myocardial stiffness · Parameter estimation · Myocardial characterisation · Diastolic biomarkers

1 Introduction

Ischemic heart disease is the most common pathology underlying heart failure
(HF), affecting more than half a million people in the UK[1]. It is characterised
by adverse myocardial remodelling, which alters the myocardial microstructure
leading to an increased material stiffness at the sites of scar formation [14].
Moreover, ventricular stiffness plays a major role in HF with normal ejection
fraction [18]. The reliable estimation of myocardial stiffness can provide essential
insight into the pathophysiology of HF as well offer a biomechanically relevant
tool for its diagnosis and monitoring.

Passive mechanical parameters can be estimated by the personalisation of
computational models to available clinical data [1,20]. Nevertheless, the process
involves uncertainties associated with noise in the clinical measurements, the
cost function (CF) used in the optimisation, as well as the optimisation technique

---

P. Lamata—Equal contribution senior authors.
S.A. Niederer—Equal contribution senior authors.
[1] http://www.nice.org.uk/guidance/cg108/resources/guidance-chronic-heart-failure-pdf.

© Springer International Publishing Switzerland 2015
H. van Assen et al. (Eds.): FIMH 2015, LNCS 9126, pp. 355–363, 2015.
DOI: 10.1007/978-3-319-20309-6_41

itself. In addition, the majority of the constitutive equations employed to model the myocardium physiology with enough fidelity involve parameter coupling, which results in multiple parameter sets predicting the same model solution [4]. The uncertainty in stiffness parameter values is demonstrated by cardiac stiffness estimates of the same species differing by two orders of magnitude [11].

To shed light on this critical issue for the field of myocardial parameter estimation the current study evaluates the performance of several widely employed CFs in reducing the material parameter coupling and proposes a novel CF that uniquely identifies one of the parameters, allowing the remainder of the parameters to be uniquely defined. The proposed methodology is then applied to models derived from human datasets to assess its performance in a clinical framework.

2   Materials and Methods

2.1   Evaluated Cost Functions

Geometry Based Metrics. In this section we evaluate the parameter identifiability of existing geometry based metrics, which range from measures of cardiac deformation from across the myocardial domain Ω to simpler metrics that can be directly derived from imaging modalities such as 3D echo (wall thickness and cavity volume) or cine MRI (apicobasal distance).

The evaluated geometry based CFs are the L^2 error norm of displacements ($|\Delta u|$) [22], a metric which expresses the average error in the displacements (u) between data (u_{dat}) and simulations (u_{sim}):

$$|\Delta u| = \sqrt{\frac{\int_\Omega (u_{sim} - u_{dat})^2 \, d\Omega}{\int_\Omega d\Omega}} \qquad (1)$$

and the absolute differences between the data and the simulation in cavity volume, wall thickness (averaged over equatorial mesh nodes) and apico-basal distance (averaged over endo- or epicardial nodes of the base and the apex).

Energy Based Metric. The main contribution of this study is the introduction of a novel, energy based CF. This CF is based on the principle of energy conservation, which under the assumption of quasistatic loading and purely elastic behaviour [1,4,20], can be expressed as the external work being stored as elastic strain energy inside a deformable body (Eq. 2). Previous work has demonstrated the need to account for residual active tension (AT) in diastole [2,20]. But focusing on the late diastolic window the residual AT can be assumed to be zero and its contribution to the work estimation negligible. Expressing the energy conservation (Eq. 2) over any two late diastolic frames, denoted as diastolic frame 1 (DF_1) and diastolic frame 2 (DF_2) (and stating that the ratio of the external work at these two frames should equal the ratio of the elastic energy at the same frames) the energy based functional f is formulated (Eq. 3), where V is the domain of the undeformed (reference) configuration and $\Psi|_{DF_i}$ and $W_{ext}|_{DF_i}$

denote the strain energy density function and the external work respectively at the diastolic frame DF_i $(i = 1, 2)$.

$$\int_V \Psi \, dV = W_{\text{ext}} \tag{2}$$

$$f = \frac{\int_V \Psi|_{DF_1} \, dV}{\int_V \Psi|_{DF_2} \, dV} - \frac{W_{\text{ext}}|_{DF_1}}{W_{\text{ext}}|_{DF_2}} \tag{3}$$

The elastic energy estimation relies on calculating the Green-Lagrange strains derived from the data as a result of the deformation field between the reference and the deformed geometries at a given frame (DF_1 or DF_2).

The external work is estimated as the work performed from the LV cavity pressure to increase the volume of the LV chamber. It should be noted that, unlike the geometry based CFs, which rely on the comparison of a geometric quantity between the data and the model simulation, the energy based CF is entirely data driven (see Fig. 1).

2.2 Diastolic Mechanical Model of the LV

To evaluate the performance of the CFs a model of the left ventricle (LV) in late diastole using clinical data is constructed, where diastolic behaviour is simulated as passive inflation under hydrostatic pressure.

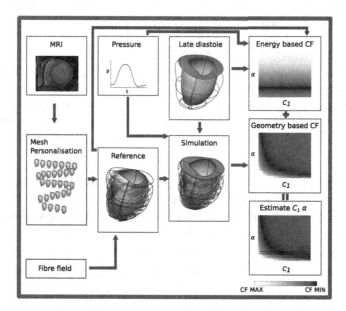

Fig. 1. Summary of the followed pipeline, from FE model generation to CF evaluation and paramter estimation.

Clinical Data. The clinical data used in this study consist of LV cavity pressure recordings used for the loading BCs and images from cine MRI for the deformation field and mechanical reference configuration in the model. The data were acquired from 2 Cardiac Resyncronisation Therapy (CRT) patients according to the clinical protocols followed for CRT patients in St. Thomas' Hospital, London and one healthy volunteer. Imaging data consist of 2-D cine MRI stacks with ECG gating. The pressure data are derived from pressure catheterisation procedure before pacing protocols begin.

Mesh Personalisation. Finite element (FE) meshes of the LV geometry at each frame were created using an automatic pipeline combining mesh personalisation [8] and image registration [13] techniques. The mesh personalisation requires solely a binary mask representing the LV domain as input, and performs an automatic fitting of an idealised template mesh to the anatomy of the patient [8]. This technique is robust to the sparse resolution of cine sequences, and therefore the myocardial domain was defined by the manual segmentation of the end diastolic (ED) frame of each sequence. A temporal sparse free form registration technique extracts the warping field from the same sequence [13]. This field, that is continuous in time and space, is then fitted to the domain of the ED mesh using a variational technique [7], obtaining the sequence of meshes. Finally, a postprocessing step improves mesh quality and simulation stability [8,9]. As in previous studies [20,21], a relatively coarse mesh is chosen (12 elements and 1308 degrees of freedom with cubic Lagrange interpolation for the displacement field and linear for the pressure), in order to reduce the computational cost of simulations and regularise the deformation field from clinical data.

Myocardial Material Description. In this study the myocardium was modelled as a homogeneous, transversely isotropic material. The reason for choosing a transversely isotropic over an orthotropic material behaviour was two-fold: firstly, transvesely isotropic material laws utilise fewer material parameters and are thus more tractable within a parameter estimation framework (as opposed to more complex orthotropic material laws [5]). Secondly, this choice provides the opportunity to use a very popular transversely isotropic material law proposed by Guccione [3]. For the purpose of this study, the reformulated version of this equation (as proposed by [20]) will be adopted, since it helps illustrate the major direction of parameter coupling (between parameters C_1 and α), describing the stored elastic energy density Ψ in each material point of the myocardium as:

$$\Psi = \frac{1}{2}C_1(e^Q - 1) \tag{4}$$

$$Q = \alpha(r_{ff}E_{ff}^2 + r_{fs}(2E_{fs}^2 + 2E_{fn}^2) + r_{sn}(E_{ss}^2 + E_{nn}^2 + 2E_{sn}^2)) \tag{5}$$

$$\alpha = b_{ff} + b_{fs} + b_{sn} \tag{6}$$

$$r_{ff} = b_{ff}/\alpha, r_{fs} = b_{fs}/\alpha, r_{sn} = b_{sn}/\alpha \tag{7}$$

where $Eij(i,j = f,s,n)$ are the Green-Lagrange strain tensor components in fiber coordinates [12]. The parameter ratios (r_{ff}, r_{fs}, r_{sn}) are assumed constant throughout this study and were assigned values that correspond to the average reported values in the literature.

FE Model of the Reference Configuration. The finite strain framework employed in cardiac mechanics involves the use of a reference and a deformed state configuration. The reference configuration represents an idealised geometry where the domain is unloaded and there are zero stresses and strains. However, finding this idealised geometry in practice is an elusive task which is associated with many uncertainties. In this work this problem is circumvented by considering only geometries available in the clinical data as possible reference frame candidates. More specifically after examining several possible reference states the early-diastolic frame with the minimum volume was chosen as the reference configuration in all cases. The FE model of the LV is completed with the fitting of a standard fiber field [15] to the reference frame mesh (Fig. 1).

Mechanical Simulations. For the estimation of the geometry based CFs mechanical simulations were performed with parameter sweeps of C_1 and α parameters. These involve the passive inflation of the LV reference geometry to ED pressure with prescribed ED basal nodes displacements using the *CHeart*[2] nonlinear FE solver.

The meshes and simulation outputs were visualised with *cmGui*[3].

3   Results

3.1   Geometry Based Metrics

The parameter combinations that minimise the geometry based CFs are shown in Fig. 2, where it can be observed that they represent similar regions in the parameter space. Therefore these metrics can not improve the identifiability of the stiffness model. The exception to this observation is the landscape of the wall thickness CF in Patient case B, which demonstates the inability of this simpler metric to provide a reliable CF.

3.2   Energy Based Metric

The formulation of the energy based CF is such that f does not depend on C_1 (C_1 is cancelled out, see Eq. 8), allowing for the unique identification of the α parameter.

$$f = \frac{\int_V (e^Q - 1)|_{DF_1}\, \mathrm{d}V}{\int_V (e^Q - 1)|_{DF_2}\, \mathrm{d}V} - \frac{W_{\text{ext}}|_{DF_1}}{W_{\text{ext}}|_{DF_2}} \qquad (8)$$

The typical landscape of this CF in parameter space is shown in the contour plot titled 'Energy based CF' in Fig. 1. The clear straight line where the CF is

---

[2] http://cheart.co.uk/.

[3] http://www.cmiss.org/cmgui.

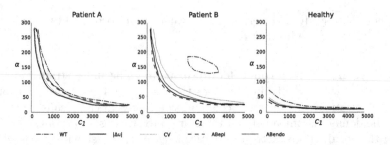

Fig. 2. The parameter combinations where the various geometry based CFs (wall thickness (WT), average displacement error ($|\Delta u|$), cavity volume (CV), apicobasal displacements at endo and epi (ABendo, ABepi)) are minimised are shown for all three cases. Using the geometry based metrics alone the parameter space can not be confined.

Table 1. Estimated material parameters for two CRT patient and one healthy case and comparison to previous published results.

	C_1 (kPa)	b_{ff}	b_{sn}	b_{fs}	$|\Delta u|$ (mm)
Healthy	1.5	8	3	4	1.7
Patient A	3.5	14	5	6	1.3
Patient B	2.5	25	9	11	2
healthy, Xi et al. [20]	2	20	10	12	1.78
Case 1, Xi et al. [20]	2	53	22	29	1.58
Case 2, Xi et al. [20]	2	51	17	27	1.39
Wang (dog, [17])	1.7	14	4	0.8	1.81

minimised indicates that the CF is independent of C_1, and that it allows the unique estimation of α. The C_1 parameter is then identified from the minimisation of one of the geometry based CFs (here the $|\Delta u|$ error norm was used) for the known α value (see Fig. 1 under the title "Estimate C_1, α").

3.3   Parameter Estimation Results

Using the proposed methodology (see Sect 3.2) a unique set of (C_1, α) parameters was identified and forward simulations with this parameter set were performed. The discrepancy between the simulation and the data is presented in terms of the $|\Delta u|$ error norm as in previous studies [17,20] to facilitate comparison of the proposed method against other published approaches (Table 1).

4   Discussion

In this study a novel, energy based CF for unique myocardial stiffness parameter identification is proposed.

Removing the caveat of parameter coupling inherent in the current parameter estimation approaches [20] will increase their validity and allow for more reliable comparisons between complex FE stiffness estimations and standard clinical stiffness metrics [23]. To our knowledge this is the first time that a solution is proposed for the problem of the material parameter coupling.

The performance of the proposed CF combination is assessed on patient specific models. The use of comprehensive datasets of anatomical geometry and recorded pressure throughout the cardiac cycle increases the reliability of the parameter estimation results [21]. The results show that the identified parameters managed to capture sufficiently well the mechanical deformation of the LV, as suggested by the discrepancy between the data and the resulting LV model from the parameter estimation. As can be seen in Table 1 the quantified error is smaller or equivalent to that reported in previous studies using similar datasets and modelling approaches, despite the fact that our LV model is significantly less constrained (in this study only basal nodes were prescribed, as opposed to both basal and apical ones in [20] and [17]). Our results also confirm the reported coupling in the material parameters using a variety of popular CFs, and show that using geometry based CFs alone can not confine the solution parameter space (Fig. 2).

However, there are several limitations included in this work. The assumed myocardial material homogeneity renders the approach applicable to patients with dilated cardiomyopathy, but less so for cases where infarcts are present. Additionally, the impact of the constant parameter ratios (r_{ff}, r_{fs}, r_{sn}) and the generic fiber field on the proposed CF need to be analysed (although the impact of using anatomical fiber fields on bulk geometry based metrics has been reported to be nominal [10]). Despite evidence of myocardial tissue acceleration during early diastole [6], the mechanical analysis was based on a quasistatic approach as in previous studies [4,17,20]. This assumption has been shown to hold reasonably well in a 1-D model of cardiac fibres [19] but its effect on whole tissue cardiac mechanics remains to be examined. The cavity pressure was assumed to be homogeneous which is a reasonable approximation since pressure distribution on the endocardial boundary at end diastole has limited spatial variation [16]. Another limitation in our model is the assumption of incompressibility, which is violated by local strain concentrations due to capillary and coronary flow throughout the cycle. Finally, the proposed method has been demonstrated to uniquely identify the material parameters in a limited set of three clinical cases and needs to be validated more extensively in the future.

5   Conclusions

In this work a novel and clinically tractable pipeline for passive myocardial material estimation is proposed, which manages to decouple the material constitutive law parameters and guarantee reliable material estimation. This is an important step towards the use of myocardial stiffness as a reliable tool for understanding of cardiac pathophysiology and the development of biomechanically relevant biomarkers.

Acknowledgements. The authors would like to acknowledge financial support from the NIHR Biomedical Research Centre at Guy's and St. Thomas' NHS Foundation Trust and KCL, and support from the Wellcome Trust and EPSRC Centre of Excellence in Medical Engineering. S.A.N is supported by BHF PG/11/101/29212. PL holds a Sir Henry Dale Fellowship funded jointly by the Wellcome Trust and the Royal Society (grant no. 099973/Z/12/Z).

References

1. Augenstein, K.F., Cowan, B.R., LeGrice, I.J., Nielsen, P.M.F., Young, A.A.: Method and apparatus for soft tissue material parameter estimation using tissue tagged magnetic resonance imaging. J. Biomech. Eng. **127**(1), 148 (2005)
2. Bermejo, J., Yotti, R., Pérez del Villar, C., del Álamo, J.C., Rodríguez-Pérez, D., Martínez-Legazpi, P., Benito, Y., Antoranz, J.C., Desco, M.M., González-Mansilla, A., Barrio, A., Elízaga, J., Fernández-Avilés, F.: Diastolic chamber properties of the left ventricle assessed by global fitting of pressure-volume data: improving the gold standard of diastolic function. J. Appl. Physiol. (Bethesda, Md.: 1985) **115**(4), 556–568 (2013)
3. Guccione, J.M., McCulloch, A.D., Waldman, L.K.: Passive material properties of intact ventricular myocardium determined from a cylindrical model. J. Biomech. Eng. **113**, 42–55 (1991)
4. Hadjicharalambous, M., Chabiniok, R., Asner, L., Sammut, E., Wong, J., Carr-White, G., Lee, J., Razavi, R., Smith, N., Nordsletten, D.: Analysis of passive cardiac constitutive laws for parameter estimation using 3D tagged MRI. Biomech. Model. Mechanobiol. 1–22 (2014)
5. Holzapfel, G.A., Ogden, R.W.: Constitutive modelling of passive myocardium: a structurally based framework for material characterization. Philos. Trans. Ser. A, Math. Phys. Eng. Sci. **367**, 3445–3475 (2009)
6. Karwatowski, S.P., Brecker, S.J.D., Yang, G.Z., Firmin, D.N., Sutton, M.S.J., Underwood, S.R.: Cardiovascular physiology: a comparison of left ventricular myocardial velocity in diastole measured by magnetic resonance and left ventricular filling measured by doppler echocardiography. Eur. Heart J. **17**(5), 795–802 (1996)
7. Lamata, P., Niederer, S., Nordsletten, D., Barber, D.C., Roy, I., Hose, D.R., Smith, N.: An accurate, fast and robust method to generate patient-specific cubic Hermite meshes. Med. Image Anal. **15**(6), 801–813 (2011)
8. Lamata, P., Roy, I., Blazevic, B., Crozier, A., Land, S., Niederer, S.A., Hose, D.R., Smith, N.P.: Quality metrics for high order meshes: analysis of the mechanical simulation of the heart beat. IEEE Trans. Med. Imaging **32**(1), 130–138 (2013)
9. Lamata, P., Sinclair, M., Kerfoot, E., Lee, A., Crozier, A., Blazevic, B., Land, S., Lewandowski, A.J., Barber, D., Niederer, S., Smith, N.: An automatic service for the personalization of ventricular cardiac meshes. J. Royal Soc. Interface/Royal Soc. **11**(91), 20131023 (2014)
10. Land, S., Niederer, S., Lamata, P.: Estimation of diastolic biomarkers: sensitiviy to fibre orientation. In: Camara, O., Mansi, T., Pop, M., Rhode, K., Sermesant, M., Young, A. (eds.) STACOM 2014. LNCS, vol. 8896, pp. 105–113. Springer, Heidelberg (2015)

11. Nordbø, O., Lamata, P., Land, S., Niederer, S., Aronsen, J.M., Louch, W.E., Sjaastad, L., Martens, H., Gjuvsland, A.B., Tøndel, K., Torp, H., Lohezic, M., Schneider, J.E., Remme, E.W., Smith, N., Omholt, S.W., Vik, J.O.: A computational pipeline for quantification of mouse myocardial stiffness parameters. Comput. Biol. Med. **53**, 63–75 (2014)
12. Nordsletten, D.A., Niederer, S.A., Nash, M.P., Hunter, P.J., Smith, N.P.: Coupling multi-physics models to cardiac mechanics. Progr. Biophys. Mol. Biol. **104**(1–3), 77–88 (2011)
13. Shi, W., Jantsch, M., Aljabar, P., Pizarro, L., Bai, W., Wang, H., O'Regan, D., Zhuang, X., Rueckert, D.: Temporal sparse free-form deformations. Med. Image Anal. **17**(7), 779–789 (2013)
14. Sutton, M.G.S.J., Sharpe, N.: Left ventricular remodeling after myocardial infarction: pathophysiology and therapy. Circulation **101**(25), 2981–2988 (2000)
15. Usyk, T., Mazhari, R., McCulloch, A.: Effect of laminar orthotropic myofiber architecture on regional stress and strain in the canine left ventricle. J. Elast. Phys. Sci. Solids **61**(1–3), 143–164 (2000)
16. de Vecchi, A., Gomez, A., Pushparajah, K., Schaeffter, T., Nordsletten, D.A., Simpson, J.M., Penney, G.P., Smith, N.P.: Towards a fast and efficient approach for modelling the patient-specific ventricular haemodynamics. Prog. Biophys. Mol. Biol. **116**(1), 3–10 (2014)
17. Wang, V.Y., Lam, H.I., Ennis, D.B., Cowan, B.R., Young, A.A., Nash, M.P.: Modelling passive diastolic mechanics with quantitative MRI of cardiac structure and function. Med. Image Anal. **13**, 773–784 (2009)
18. Westermann, D., Kasner, M., Steendijk, P., Spillmann, F., Riad, A., Weitmann, K., Hoffmann, W., Poller, W., Pauschinger, M., Schultheiss, H.P., Tschöpe, C.: Role of left ventricular stiffness in heart failure with normal ejection fraction. Circulation **117**(16), 2051–2060 (2008)
19. Whiteley, J.P., Bishop, M.J., Gavaghan, D.J.: Soft tissue modelling of cardiac fibres for use in coupled mechano-electric simulations. Bull. Math. Biol. **69**(7), 2199–2225 (2007)
20. Xi, J., Lamata, P., Niederer, S., Land, S., Shi, W., Zhuang, X., Ourselin, S., Duckett, S.G., Shetty, A.K., Rinaldi, C.A., Rueckert, D., Razavi, R., Smith, N.P.: The estimation of patient-specific cardiac diastolic functions from clinical measurements. Med. Image Anal. **17**(2), 133–146 (2013)
21. Xi, J., Shi, W., Rueckert, D., Razavi, R., Smith, N.P., Lamata, P.: Understanding the need of ventricular pressure for the estimation of diastolic biomarkers. Biomech. Model. Mechanobiol. **13**(4), 747–757 (2014)
22. Zienkiewicz, O., Taylor, R., Zhu, J.: The Finite Element Method: Its Basis and Fundamentals. Butterworth-Heinemann, UK (2013)
23. Zile, M.R.: New concepts in diastolic dysfunction and diastolic heart failure: part i: diagnosis, prognosis, and measurements of diastolic function. Circulation **105**(11), 1387–1393 (2002)

Hemodynamics in Aortic Regurgitation Simulated Using a Computational Cardiovascular System Model

G. Palau-Caballero[1]([⊠]), J. Walmsley[1], P. Rudenick[2],
A. Evangelista[3], J. Lumens[1], and T. Delhaas[1]

[1] Department of Biomedical Engineering, Cardiovascular Research Institute Maastricht
(CARIM), Maastricht University, Maastricht, The Netherlands
g.palaucaballero@maastrichtuniversity.nl
[2] Physense, Universitat Pompeu Fabra (UPF), Barcelona, Spain
[3] Hospital General Universitari Vall d'Hebron, Barcelona, Spain

Abstract. The influence of left ventricular and aortic tissue properties on hemodynamics in patients with aortic regurgitation (AR) is unclear. In this study we
aim: (1) to assess the capability of the CircAdapt model of the heart and circulation
to simulate hemodynamics in AR; (2) to determine the interaction between aortic
compliance and AR using CircAdapt. We simulated three degrees of AR by
changing the aortic regurgitant orifice area (ROA) with normal and low aortic
compliance. The higher the ROA is, the higher the systolic left ventricular and
aortic pressures, the lower the diastolic aortic pressures and the higher the
diastolic left ventricular pressures are. For low aortic compliance, those effects
are exacerbated, but the regurgitant blood volume is decreased. These simulation
data show the capability of CircAdapt to simulate hemodynamics in AR, and
suggest that patient-to-patient variability in aortic compliance should be taken
into account when assessing AR severity using imaging-based hemodynamic
metrics.

Keywords: Aortic insufficiency · Compliance · Computational modeling ·
Retrograde flow

1 Introduction

In normal cardiac function, the aortic valve prevents retrograde flow from the aorta back
into the left ventricle (LV) during diastole. The aortic valve can fail to close properly
due to congenital defects or acquired valve pathology. The resulting regurgitant orifice
area (ROA) in the aortic valve gives rise to diastolic backward flow, known as aortic
regurgitation (AR). The associated hemodynamic changes with AR can cause cardiac
remodeling. In severe AR patients, this remodeling may result in heart failure and high
mortality, as reported in several patient studies [1].

In clinical practice, the severity of AR is evaluated using quantitative and qualitative
imaging-based metrics that provide information about the aortic valve anatomy and
function, as well as on the presence and severity of cardiac and vascular structural
remodeling [2, 3]. However, the influence of LV and aortic function and tissue properties

© Springer International Publishing Switzerland 2015
H. van Assen et al. (Eds.): FIMH 2015, LNCS 9126, pp. 364–372, 2015.
DOI: 10.1007/978-3-319-20309-6_42

on hemodynamics in AR, and, therefore, in the assessment of AR severity, still remains unclear [4, 5].

Computational models on cardiac and vascular function can contribute to better understanding of cardiovascular function under AR as they allow precise control of individual parameters. With computational models it is therefore possible to investigate the combined effect of system properties in a controlled way, which is difficult or impossible to achieve in a clinical or experimental setting.

The goal of this study is, firstly, to assess the capability of the CircAdapt model of the human cardiovascular system to simulate hemodynamics when AR is present [6, 7]; and secondly, to determine the interaction between aortic compliance and AR using hemodynamics simulated in CircAdapt.

2   Methods

2.1   Model Description

CircAdapt [6, 7] is a lumped-parameter model that simulates beat-to-beat cardiovascular mechanics and hemodynamics in both healthy and pathological conditions. The model is based on modules representing elements of the cardiovascular system, including atrial and ventricular cavities, the pericardium, and systemic and pulmonary circulations (see Fig. 1). The cardiac valve module is used to simulate blood flow across the ventriculo-arterial and atrio-ventricular valves. The blood vessel module simulates pressure and flow in the aorta, the pulmonary artery, and the systemic and pulmonary veins. The hemodynamics and mechanics of every element are governed by physics and physiological principles. When connected together the modules form a system of ordinary differential equations describing the dynamics of the closed-loop cardiovascular system [6].

2.2   Aortic Valve

The aortic valve is modeled as described previously [6], and can be used for healthy and pathological valve functions. The aortic valve is a narrow and varying orifice area where blood flow, $q_{AoValve}$, passes through assuming unsteady, incompressible and non-viscous plug flow. Blood flow is considered forward flow ($q_{AoValve} \geq 0$) when it travels from the LV into the aorta, and backward flow ($q_{AoValve} < 0$) in the opposite direction. The aortic valve mechanics is governed by the differential equation:

$$\Delta p(t) = \rho l_{AoValve} \frac{dv_{AoValve}}{dt} + \frac{1}{2}\rho \left\{ \begin{array}{l} v_{max}^2(t) - v_{LV}^2(t), q_{AoValve}(t) \geq 0 \\ v_{Ao}^2(t) - v_{max}^2(t), q_{AoValve}(t) < 0 \end{array} \right. \tag{1}$$

The blood flow across the valve at time t, $q_{AoValve}(t)$, causes a transvalvular pressure drop, $\Delta p(t)$, which is the sum of inertia and Bernoulli effects (first and second right hand terms in Eq. 1, respectively). Briefly, the inertia term in Eq. 1 adds the effect of blood acceleration in change of pressure between the LV and the aorta by considering the valve as a small channel (with length $l_{AoValve}$) where blood flow has a velocity change in time $dv_{AoValve}/dt$. The second right hand term adds the kinetic energy that must be conserved

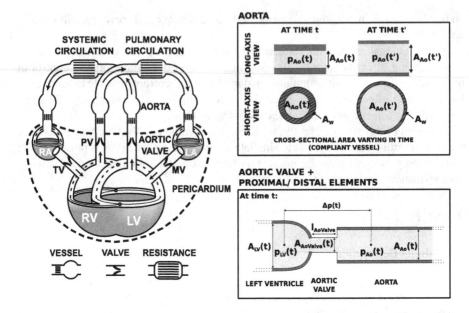

Fig. 1. CircAdapt model representation (left) with aorta and aortic valve parameters (right). RA: right atrium, LA: left atrium, RV: right ventricle, LV: left ventricle, TV: tricuspid valve, PV: pulmonary valve, MV: mitral valve, p_{Ao}: aortic pressure, A_{Ao}: cross-sectional area of the aorta, A_w: cross-sectional area of the aortic wall, A_{LV}: cross-sectional area of the left ventricle, p_{LV}: left ventricular pressure, $A_{AoValve}$: cross-sectional area of the aortic valve, $l_{AoValve}$: length of the aortic valve, Δp: pressure drop across the aortic valve.

along a streamline of flow according to Bernoulli's law (neglecting gravitation and pressure loss due to friction or turbulence). Flow velocities at the aortic valve, at the LV and at the aorta are $v_{AoValve}(t)$, $v_{LV}(t)$ and $v_{Ao}(t)$, respectively, and $v_{max}(t)$ is:

$$v_{max}(t) = max\left\{v_{LV}(t), v_{AoValve}(t), v_{Ao}(t)\right\} \tag{2}$$

All flow velocities are computed as the flow passing across the aortic valve $q_{AoValve}(t)$ divided by the corresponding cross-sectional area at t. Blood density is ρ. The orifice area of the aortic valve at instant t, $A_{AoValve}(t)$, is defined by an opening orifice area (OOA) and a regurgitant orifice area (ROA), which corresponds to the cross-sectional area of the opened and closed (or regurgitant) valve, respectively. The signs of pressure drop $\Delta p(t)$ and blood flow $q_{AoValve}(t)$ determine whether the valve is opened or closed at t. Therefore, an ROA value greater than 0 allows simulating AR. In the healthy situation, the OOA and ROA of the aortic valve are 5 cm^2 and 0 cm^2, respectively.

2.3 Aortic Compliance

The aorta is modeled as a compliant large blood vessel, whose proximal element is the aortic valve and whose distal element is the systemic circulation resistance. The vessel hemodynamics are governed by a pressure-cross-sectional area relationship arising from stress and strain in the vessel walls [6] (see Eq. 3). $p_{Ao}(t)$ and $p_{Ao,Ref}$ are the current and the reference transmural pressure of the aorta, respectively; and $A_{Ao}(t)$ and $A_{Ao,Ref}$ are the current and the reference cross-sectional area of the aorta, respectively. $A_{Ao,Ref}$ is defined as the cross-sectional area such that the aortic pressure is the reference transmural pressure $p_{Ao,Ref}$. A_W is the cross-sectional area of the aortic wall.

$$p_{Ao}(t) = p_{Ao,Ref} \left(\frac{A_w + 3A_{Ao}(t)}{A_w + 3A_{Ao,Ref}} \right)^{\frac{k-3}{3}} \tag{3}$$

The parameter k in Eq. 3 determines the non-linearity in the stiffness of the aortic wall. The higher k is, the higher the pressure will be at the same cross-sectional area. Therefore, increasing k decreases aortic compliance (increases aortic stiffness).

2.4 Simulations

We simulated three degrees of AR by changing the aortic ROA in a cardiovascular system with either normal or low aortic compliance, resulting in six simulations (see Fig. 2). Normal aortic compliance is obtained with a parameter $k = 8$ in Eq. 3, representing a younger population [8]. A low aortic compliance is simulated using $k = 15$, representing an older population. The ROA values for the three simulated degrees of AR are 0 cm^2, 0.15 cm^2 and 0.40 cm^2. According to current clinical guidelines [3], these values correspond to no AR, moderate AR and severe AR, respectively. All other model parameters were the same in all simulations. No structural remodeling of the cardiac and vascular tissue was simulated. Resting conditions were simulated under pressure-flow control with a mean arterial pressure of 91 mmHg, a cardiac output of 5 L/min and a heart rate of 71 bpm [6].

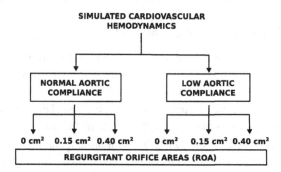

Fig. 2. Simulated protocol. Three AR degrees (0 cm^2, 0.15 cm^2 and 0.40 cm^2) with normal and low aortic compliance. The parameters k and ROA in CircAdapt are changed to simulate low/normal aortic compliance and different AR degrees, respectively.

3 Results

3.1 Effect of Aortic Compliance on Pressure

Figure 3 shows the isolated effect of decreasing aortic compliance on LV and aortic pressures with no AR (ROA of 0 cm^2). Systolic aortic and LV pressures increase from 118 mmHg with normal compliance to 141 mmHg with low compliance. Aortic diastolic pressures decrease from 70 mmHg with normal compliance to 53 mmHg with low compliance. As a result, the aortic pulse pressure increases from 48 mmHg with normal compliance to 88 mmHg with low compliance.

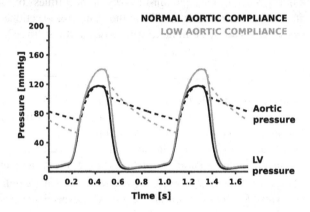

Fig. 3. Effect of aortic compliance on pressures at no AR. Simulated pressures of the aorta (dashed line) and left ventricle (solid line) for normal aortic compliance (black) and low aortic compliance (gray) both at ROA of 0 cm^2 (no AR).

3.2 Effect of AR and Aortic Compliance on Pressure

The effect of aortic regurgitation and aortic compliance on aortic and LV pressures is displayed in Fig. 4. Increasing the ROA while aortic compliance is normal, results in an increase of systolic aortic and LV pressures increase from 118 mmHg to 149 mmHg at ROA of 0 cm^2 and 0.40 cm^2, respectively (see Fig. 4A). Diastolic aortic pressures decreased under these circumstances from 70 mmHg to 46 mmHg at ROA of 0 cm^2 and 0.40 cm^2, respectively. Also, increased ROA with normal aortic compliance results in increased diastolic LV pressures and increased end-diastolic LV volumes (132 mL vs. 214 mL at ROA of 0 cm^2 and 40 cm^2, respectively). Decreasing aortic compliance in AR results in much higher systolic pressures compared to the normal aortic compliance case (see Figs. 4A and 4B). Similarly, the diastolic aortic pressures decrease more with low compliance than with normal compliance. In Fig. 4B, the increase in diastolic LV pressures with low compliance is similar to the one in Fig. 4A with normal compliance.

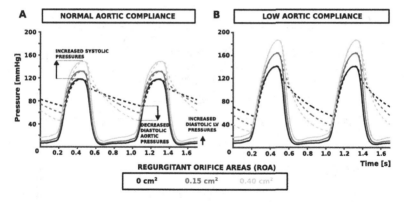

Fig. 4. Effect of AR and aortic compliance on pressure. Aortic pressures (dashed line) and left ventricular pressures (solid line) for normal (A) and low (B) aortic compliance displayed for each AR degree: ROA of 0 cm² (black), 0.15 cm² (gray) and 0.40 cm² (light gray) (Color figure online).

3.3 Effect of AR and Aortic Compliance on Flow

The effect of aortic regurgitation and aortic compliance on flow across the aortic valve is displayed in Fig. 5. Increasing the ROA with normal compliance results in an increased of peak forward blood flow from 424 mL/s to 719 mL/s for 0 cm² and 0.40 cm² of ROA, respectively (see Fig. 5A). It also causes an increase in peak backward blood flow from 0 mL/s to -184 mL/s for 0 cm² and 0.40 cm² of ROA, respectively. Also, comparing Figs. 5A and 5B, decreasing aortic compliance with the same ROA causes a decrease in peak forward and backward blood flow. Additionally, there is an earlier aortic valve opening by increasing the ROA.

3.4 Simulated and Measured Blood Flow Velocities

Figure 6A shows the simulated blood flow velocities with normal and low aortic compliance both at ROA of 0.15 cm² (moderate AR). Regarding to the steepness of both backward blood flow velocity curves in Fig. 6A, the decrease in aortic compliance causes a steeper velocity curve. Regarding to the forward blood flow velocity curves, the peak velocity is reached earlier for low than for normal compliance. This results in a more asymmetric shape of the forward flow velocity curve for low compliance compared to normal compliance. Figure 6B shows a continuous-wave (CW) Doppler velocity across the aortic valve recorded from a patient (78-year-old) with AR. Comparing Figs. 6A and 6B, the simulated velocities show consistent results with measured clinical data.

4 Discussion

The CircAdapt model enables simulation of hemodynamics when AR is present, and subsequently allows investigation of the changes in cardiovascular function that take place. Moreover, the model allows us to decrease the aortic compliance and predict its influence on hemodynamics with AR. Though the model does not contain a detailed,

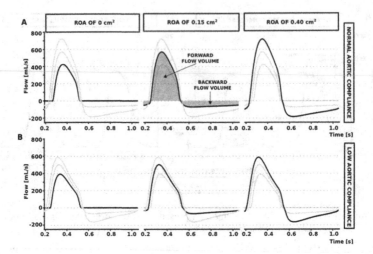

Fig. 5. Effect of AR and aortic compliance on flow. Simulated flows with normal (A) and low (B) aortic compliance are displayed for ROA of 0, 0.15 and 0.40 cm^2 (*). Flows for the corresponding ROA and the other ROA values with the same compliance are in black and gray, respectively.

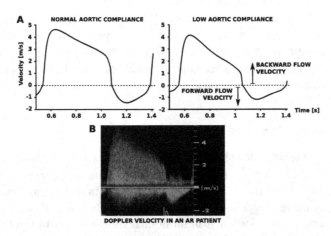

Fig. 6. Simulated and measured blood flow velocities. A: Simulated blood flow velocities at ROA of 0.15 cm^2 with normal (left) and low (right) aortic compliance; B: CW Doppler velocity in a 78-year-old patient with AR. Positive axis correspond to backward blood flow velocity.

image derived geometry of the aortic valve, the lumped-model simulations plausibly describe hemodynamic responses and ventricular-arterial interaction in AR.

Increasing ROA in the model leads to an increase of the regurgitant blood volume flowing backwards across the aortic valve during diastole. Consequently, this regurgitant blood volume results in a decrease in simulated diastolic aortic pressure, which is in agreement with clinical data [9]. Moreover, the increased regurgitant blood volume into the LV causes an increase in forward blood volume during systole. Therefore, the combined effects in the simulations are an increase in pulse pressure, an increase in the LV preload (higher diastolic LV pressure and end-diastolic LV volume), and an increase in the LV afterload (higher systolic LV pressure) assuming mean arterial pressure to be maintained through homeostatic regulation. Note that the hemodynamics are simulated assuming pressure-flow control of the cardiovascular system. This may explain the large increase of LV afterload, which may be exaggerated.

In CircAdapt, decreasing the aortic compliance in AR causes a larger increase in systolic aortic pressure accompanied by a decrease in diastolic aortic pressure. Consequently, low aortic compliance combined with AR simulates a decrease in pressure drop across the aortic valve during diastole, as shown in experimental studies [5]. Therefore, decreased aortic compliance in AR results in a decrease in simulated backward blood volume through the aortic valve. The simulated flow velocities with normal and low aortic compliance (Fig. 6A) are both consistent with velocities measured in patients. Although patient-specific simulation of AR is currently beyond the capabilities of our model, our results may contribute to improved understanding of cardiovascular response in AR and, therefore, to better assessment of disease severity.

5   Conclusions

We successfully simulated hemodynamics in AR using CircAdapt. Simulations predicted that hemodynamics in AR depend on both ROA and aortic compliance. The simulated changes in left ventricular and aortic pressures, and aortic flow velocity were exacerbated by lowering aortic compliance, while the regurgitant blood volume was decreased. Therefore, these simulation data suggest that patient-to-patient variability in aortic compliance should be taken into account during clinical assessment of AR severity using imaging-based metrics.

References

1. Enriquez-Sarano, M., Tajik, A.J.: Aortic regurgitation. N. Eng. J. Med **351**, 1539–1546 (2004)
2. Bonow, R.O.: Aortic Regurgitation: Time to Reassess Timing of Valve Replacement? JACC Cardiovasc. Imaging **4**, 223–230 (2011)
3. Nishimura, R.A., Otto, C.M., Bonow, R.O., Carabello, B.A., et al.: 2014 AHA/ACC guideline for the management of patients with valvular heart disease. JACC **63**, e57–e185 (2014)
4. Detaint, D., Messika-Zeitoun, D., Maalouf, J., Tribouilloy, C., et al.: Quantitative echocardiographic determinants of clinical outcome in asymptomatic patients with aortic regurgitation. JACC Cardiovasc. Imaging **1**, 1–11 (2008)
5. Griffin, B.P., Flachskampf, F.A., Siu, S., Weyman, A.E., et al.: The effects of regurgitant orifice size, chamber compliance, and systemic vascular resistance on aortic regurgitant velocity slope and pressure half-time. Am. Heart J. **122**, 1049–1056 (1991)
6. Arts, T., Delhaas, T., Bovendeerd, P., Verbeek, X., Prinzen, F.W.: Adaptation to mechanical load determines shape and properties of heart and circulation: the CircAdapt model. Am. J. Physiol. Heart Circ. Physiol. **288**, H1943–H1954 (2005)

7. Lumens, J., Delhaas, T., Kirn, B., Arts, T.: Three-wall segment (TriSeg) model describing mechanics and hemodynamics of ventricular interaction. Ann. Biomed. Eng. **37**, 2234–2255 (2009)
8. Arts, T., Reesink, K., Kroon, W., Delhaas, T.: Simulation of adaptation of blood vessel geometry to flow and pressure: implications for arterio-venous impedance. Mech. Re. Commun. **42**, 15–21 (2012)
9. Teague, S.M., Heinsimer, J.A., Anderson, J.L., Sublett, K., et al.: Quantification of aortic regurgitation utilizing continuous wave doppler ultrasound. JACC **8**, 592–599 (1986)

How to Choose Myofiber Orientation
in a Biventricular Finite Element Model?

Marieke Pluijmert[1,2]([✉]), Frits Prinzen[1], Adrián Flores de la Parra[2],
Wilco Kroon[3], Tammo Delhaas[1], and Peter H.M. Bovendeerd[2]

[1] Cardiovascular Research Institute Maastricht,
Departments of Biomedical Engineering/Physiology, Maastricht University,
P.O. Box 616, 6200 MD Maastricht, The Netherlands
marieke.pluijmert@maastrichtuniversity.nl
[2] Department of Biomedical Engineering, Eindhoven University of Technology,
P.O. Box 513, 5600 MB Eindhoven, The Netherlands
[3] Institute of Computational Science, University of Lugano, Lugano, Switzerland

Abstract. Biventricular (BiV) finite element (FE) models of cardiac
electromechanics are evolving to a state where they can assist in clinical
decision making. Carefully designed patient-specific geometries are combined with generic myofiber orientation data, because of lack of accurate
techniques to measure myofiber orientation. However, it remains unclear
to what extent the assumption of a generic myofiber orientation influences predictions on cardiac function from BiV FE models. As an alternative approach, it was suggested to let the myofiber orientation adapt
in response to fiber cross-fiber shear. The aim of this study was to investigate to what extent variations in myofiber orientation as induced by
adaptive myofiber reorientation caused variations in global stroke work
in a BiV FE model and whether the adaptation model could be used
as an alternative approach to prescribe the myofiber orientation in these
models. An average change in myofiber orientation over an angle of about
8°, predominantly in transmural direction, resulted in a 91 % increase of
LV and 20 % increase of RV stroke work. These findings indicate the
importance for a more thorough effort to address a realistic myofiber
orientation. The currently used model for adaptive myofiber reorientation seems a useful approach to prescribe the myofiber orientations in
BiV FE models.

1 Introduction

Biventricular (BiV) finite element (FE) models eventually could be used to assist
in clinical decision making in cardiac disease [1–6]. The strength of these models is
their ability to describe the relation between local tissue function and global pump
function. Therefore, the heart's geometry, myofiber orientation, and myocardial
tissue properties are taken into account. Whenever possible, the model input is
based on *in vivo* measurements. For example, geometrical properties are taken
from imaging data and included into models using meshing algorithms [7]. However, usually data from *in vitro* measurements is used to prescribe local myofiber

© Springer International Publishing Switzerland 2015
H. van Assen et al. (Eds.): FIMH 2015, LNCS 9126, pp. 373–381, 2015.
DOI: 10.1007/978-3-319-20309-6_43

orientation [8,9] and myocardial tissue properties. While the morphed myofiber orientation data is fixed, material properties are tuned for maximum agreement between model predicted and clinically measured cardiac function [1,4].

The accuracy of MRDTI is on the order of $\pm 10°$ [10], even in the *in vitro* case [11,12], making it hard to assess the true variation of fiber orientation in between hearts. In addition, in models of only the left ventricle (LV) it has been shown that variations in the spatial distribution of myofiber orientations of $8°$, introduced variations in myofiber stress of 10% [13]. In addition, shear deformation could only be realistically predicted when a transmural component in myofiber orientation of $\pm 10°$ was included [14]. As an alternative approach for estimating myofiber orientation, it has been hypothesized that myofiber orientation is the result of an adaptation process. This hypothesis was successfully tested in an LV model that described the adaptive change of myofiber orientation in response to fiber cross-fiber shear [15,16]. It remains unclear to what extent the assumption of a generic myofiber orientation influences predictions on cardiac function from BiV FE models and whether adaptive myofiber reorientation could also serve as a method for predicting the myofiber orientation in these models.

The aim of this study was to investigate to what extent variations in myofiber orientation induced by adaptive reorientation of myofibers according to the model in [15] cause variations in cardiac function in a BiV FE model of cardiac mechanics and whether the adaptation model could be used to refine the myofiber orientation in these models.

2   Methods

Geometry. In the unloaded state, the BiV mesh was described by two intersecting, truncated ellipsoids (Fig. 1A). The 8 parameters listed and explained in Table 1 defined the whole geometry. Values were set such that the LV and RV inner radii were 20 mm and 53 mm, LV and RV inner length were 74 mm and 70 mm, and LV and RV wall volumes were 160 ml and 40 ml, respectively, and both LV and RV cavity volume were 60 ml.

Material Properties. Myocardial tissue Cauchy stress $\boldsymbol{\sigma}$ was composed of a passive component $\boldsymbol{\sigma}_p$ and an active component σ_a. Passive material behavior was assumed nonlinearly elastic, transversely isotropic, and nearly incompressible. It was described by a strain energy density function that consisted of an isotropic, an anisotropic, and a volume part as described in [17]. Active stress σ_a acted in the myofiber direction only. The magnitude of σ_a depended on time elapsed since activation t_a, sarcomere length l_s, and sarcomere shortening velocity as described in [14]. Active stress was initiated simultaneously throughout both ventricles with a cycle time of 800 ms.

Adaptive Myofiber Reorientation. Myofiber orientation in the unloaded state $\boldsymbol{e}_{f,0}$ was defined in a local cardiac coordinate system $\{\boldsymbol{e}_{t,0},\ \boldsymbol{e}_{l,0},\ \boldsymbol{e}_{c,0}\}$ (Fig. 1C). The transmural direction $\boldsymbol{e}_{t,0}$ was defined perpendicular to the endo- and epicardial surfaces, the longitudinal direction $\boldsymbol{e}_{l,0}$ was defined parallel to these surfaces from apex to base, and the circumferential direction $e_{c,0}$ was defined in clockwise direction when viewing the geometry in apex-to-base direction. $e_{f,0}$ was quantified by two angles. The helix angle $\alpha_{h,0}$ was defined as the angle between $e_{c,0}$ and the projection of $e_{f,0}$ on the circumferential-longitudinal plane ($e_{c,0}$, $e_{l,0}$). The transverse angle $\alpha_{t,0}$ was defined as the angle between $e_{c,0}$ and the projection of $e_{f,0}$ on the circumferential-transmural plane ($e_{c,0}$, $e_{t,0}$). To ensure a smooth transition of the coordinate system and myofiber orientation in all regions, including the RV-LV attachment, the bi-directional spherical linear interpolation method of Bayer et al. [18] was used.

During simulation of the cardiac cycle, $e_{f,0}$ was subject to adaptation that caused a structural change in myofiber orientation in the unloaded reference state. Adaptation of $e_{f,0}$ was simulated with the model published by Kroon et al. [15] and explained in detail in [16]. The mechanism behind the incremental update of $e_{f,0}$ is explained in Fig. 1B and described by the following equation:

$$\frac{\partial e_{f,0}}{\partial t} = \frac{1}{\kappa}(e_f^* - e_{f,0}) \tag{1}$$

with an adaptation time constant κ (set to 3200 ms). e_f^* represents the actual myofiber orientation in the deformed tissue corrected for rigid body rotation.

Governing Equations and Boundary Conditions. In the model, the quasi-static equations of conservation of linear momentum were solved numerically with a Galerkin type finite element method using 27-noded hexahedral elements with a tri-quadratic interpolation of the displacement field. Rotational movement of LV endocardial base and longitudinal displacement of both LV and RV base were suppressed. LV and RV endocardial surfaces were subjected to a left and right ventricular pressure, p_{lv} and p_{rv}, respectively, computed from the interaction of the LV and RV with a lumped parameter model of both systemic and pulmonary circulation, respectively (Fig. 1A). To capture characteristic of the human circulation, parameter values for resistances and compliances were based on [19, 20].

Simulations Performed. The simulation started with an initial distribution of $\alpha_{h,0}$ that varied throughout the whole myocardium nonlinearly with the transmural position from 65° at the endocardium to -50° at the epicardium [14]. The initial distribution of $\alpha_{t,0}$ was set to zero.

First, the model ran for 8 consecutive cardiac cycles to approximate a hemodynamic steady state while myofiber reorientation was disabled. In the following cardiac cycles myofiber orientation was adapted per node according to the adaptation model in [15].

Global function was quantified as cardiac output (CO), LV and RV ejection fraction (EF), and stroke work W, which was calculated from the area within the pressure-volume loops for both ventricles. The change in myofiber orientation between the initial and adapted state was quantified by the angle γ (Fig. 3). This orientation change was further quantified in terms of $\alpha_{h,0}$ and $\alpha_{t,0}$.

Table 1. Biventricular geometry input parameter values and explanation.

Parameter	Definition	Value	Unit	Explanation
V_{lvw}		160	[ml]	LV wall volume
f_V	$V_{lv,0}/V_{lvw}$	0.375	[-]	LV cavity-to-wall volume ratio in the stress-free state
f_{VRL}	$V_{rv,0}/V_{lv,0}$	1.0	[-]	RV-to-LV cavity volume ratio
f_R	$R_1/(h+Z_1)$	0.55	[-]	Endocardial LV radius-to-length ratio
f_h	h/Z_1	0.40	[-]	Truncation height above equator-to-LV endocardial length ratio
f_{sep}	R_{1s}/R_1	0.70	[-]	Septal endocardial radius (R_{1s})-to-LV endocardial radius (R_1) ratio
f_T	$\frac{R_2-R_1}{R_4-R_3}$	0.33	[-]	LV-to-RV wall thickness ratio
θ_A		0.85π	[rad]	Lowest attachment angle of RV to LV

3   Results

Global function variables in the initial state were averaged over cycles 5 through 8 (Fig. 2). Average values were 0.54 J for LV stroke work (W_{lv}), 0.20 J for RV stroke work (W_{rv}), 3.8 l/min for cardiac output (CO), 0.43 for LV ejection fraction (EF), and 0.54 for RV EF. Adaptive myofiber reorientation was enabled during cycles 9 through 30. Pump function increased directly (Fig. 2, right). From the initial to the adapted state, after 22 cardiac cycles of adaptation, W_{lv} increased by $\sim$91 % to 1.03 J, W_{rv} by $\sim$20 % to 0.24 J, CO by $\sim$35 % to 5.1 L/min, LV EF by $\sim$23 % to 0.53, and RV EF by $\sim$7 % to 0.57.

In the left panel of Fig. 3, the change in orientation of $e_{f,0}$ between the initial and adapted state is visualized as the angle γ. The mean ($\pm$SD) change was 7.9° (6.3). Larger values for γ were found in the apical, basal and attachment regions (white regions in the figure). Model predicted data on $\alpha_{h,0}$ and $\alpha_{t,0}$ were taken from the LV free wall (LVfw) for comparison with experimental data (Fig. 3 middle and right). The change in orientation took place especially in the transmural direction: distributions of $\alpha_{h,0}$ remained similar after reorientation, while the distribution of $\alpha_{t,0}$ developed a gradient with negative values near the apex and positive values near the base. Figure 3 shows that model predicted values on $\alpha_{h,0}$ and $\alpha_{t,0}$ were very well within the range of experimental data.

4   Discussion

In this study, it was investigated whether variations in myofiber orientation as induced by adaptive reorientation would affect the pump function as predicted from a BiV FE model. An average change in myofiber orientation over an angle of about 8° resulted in a 91 % increase of LV and 20 % increase of RV stroke work.

The increase of LV stroke work was caused by a decrease in LV stiffness. The decrease in stiffness was caused by the change in myofiber orientation $e_{f,0}$ and

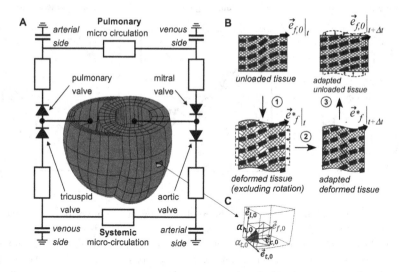

Fig. 1. A: The BiV finite element (FE) mesh (684 elements) is incorporated in a lumped parameter model of the circulation. The valves are modeled as ideal diodes. Both pulmonary and systemic circulation are modeled using compliances and resistances. **B:** Mechanism behind incremental update of myofiber orientation $e_{f,0}$ [15]. (1) Shear deformations during the cardiac cycle are assumed to damage the connections between extra-cellular matrix (ECM, raster) and myofibers (tick black lines). (2) Rigid body rotations are considered irrelevant for adaptation. New connections are formed between the ECM and rotation-corrected myofiber orientation e_f^*. (3) The adapted myofiber orientation is used as the orientation in the unloaded state at the next time step $t + \Delta t$. **C:** Myofiber orientation vector $e_{f,0}$ is quantified by helix angle $\alpha_{h,0}$ and transverse angle $\alpha_{t,0}$ using a local cardiac coordinate system $\{e_{t,0}, e_{l,0}, e_{c,0}\}$.

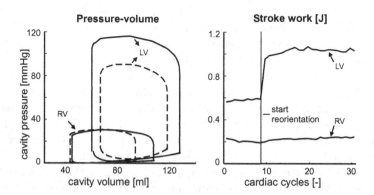

Fig. 2. Left: Pressure-volume loops before (dashed), and after 22 cardiac cycles with myofiber reorientation (solid). **Right:** Changes of LV and RV stroke work as a function of time expressed in cardiac cycles. Myofiber reorientation was enabled during cycles 9 through 30.

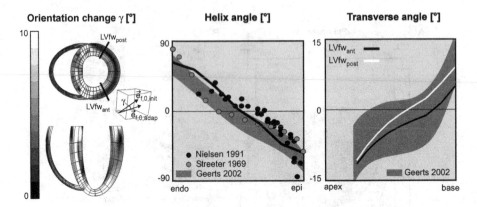

Fig. 3. Orientation change defined as the angle γ between the myofiber orientation in the initial state $e_{f,0,init}$ and in the final state $e_{f,0,adap}$ in a short-axis cross-sections at equatorial level (left, top) and long-axes cross-sections in geometrical plane of symmetry (left, bottom). Adapted myofiber orientation angles ($\alpha_{h,0}$ middle and $\alpha_{t,0}$ right) from the simulation (lines) and experimental data taken from [8, 21, 22]. The data on $\alpha_{h,0}$ (middle) was averaged from the anterior and posterior cross-sections of the LV free wall at equatorial level (LVfw$_{ant}$ and LVfw$_{post}$, respectively). The data on $\alpha_{t,0}$ is presented from LVfw$_{ant}$ and LVfw$_{post}$ and taken from the midwall at 7 locations between apex and base.

not by tuning parameters in the material model. The development of a transmural component in $e_{f,0}$, quantified by the transverse angle $\alpha_{t,0}$ explains the decrease of stiffness through a decrease in the contribution of the stiff myofibers in longitudinal and circumferential direction. As can be seen from the left panel of Fig. 2, a lower LV stiffness resulted in a higher LV cavity volume at a lower end-diastolic pressure. Increased filling leads to increased wall thinning and increased longitudinal and circumferential stretching. While the former effect tends to shorten fibers, the latter two effects tend to lengthen fibers. The net effect is an increased stretch of the fibers and the sarcomeres therein, and subsequently, a more forceful contraction that leads to improved pump function ('Frank-Starling effect').

In the adapted state, a value of 1.3 J for total cardiac stroke work ($W_{lv}+W_{rv}$) is well within the physiological range for a healthy human heart that has a total wall volume of 200 ml [22]. The physiological value for global work indicates that also local mechanical work is within the physiological range, because total global stroke work should be in equilibrium with the sum of local mechanical work that could be computed from the myofiber stress-natural strain loop.

Predicted distributions for $\alpha_{h,0}$ and $\alpha_{t,0}$ were very well within the physiological range of experimental data (Fig. 3). However, critical testing of these distributions is difficult because of the lack of accurate measurements on the spatial distribution of myofiber orientation. A difference in myofiber orientation of 8° is hard to detect with *in vitro*, let alone with *in vivo* MRDTI [23]. New imaging techniques with considerable higher resolution than MRDTI, such as X-ray microtomoraphy [24, 25], might improve imaging of the true variation in myofiber orientation.

At the moment, the model of shear-induced myofiber reorientation seems a useful approach to refine myofiber orientations in BiV FE models.

Study Assumptions and Limitations. In this study, a parameterized description of the BiV was used, while the real heart is not a perfect ellipsoid. However, it has been shown for the LV that large variations in geometry were needed for a considerable change in local mechanics [13]. Still, in a next step, it should be investigated whether adaptive myofiber reorientation induces similar changes in cardiac pump function result in different parameterized geometries and also in patient-specific geometries.

5 Conclusion

The results from the BiV FE model in this study showed that changes in myofiber orientation of about 8° caused a considerable change in total pump work of about 72 %. The findings suggest that implementing a realistic myofiber orientation is at least as important as defining a patient-specific geometry. The currently used model for adaptive myofiber reorientation seems a useful approach to first refine myofiber orientations in BiV FE models, before tuning myocardial parameters to arrive at a patient-specific model.

References

1. Aguado-Sierra, J., Krishnamurthy, A., Villongco, C., Chuang, J., Howard, E., Gonzales, M.J., Omens, J., Krummen, D.E., Narayan, S., Kerckhoffs, R.C.P., McCulloch, A.D.: Patient-specific modeling of dyssynchronous heart failure: a case study. Prog. Biophys. Mol. Biol. **107**(1), 147–155 (2011)
2. Constantino, J., Hu, Y., Trayanova, N.A.: A computational approach to understanding the cardiac electromechanical activation sequence in the normal and failing heart, with translation to the clinical practice of crt. Prog. Biophys. Mol. Biol. **110**(2–3), 372–379 (2012)
3. Kerckhoffs, R.C.P., Healy, S.N., Usyk, T.P., McCulloch, A.H.: Computational methods for cardiac electromechanics. Proc. IEEE **94**(4), 769–783 (2006)
4. Niederer, S.A., Plank, G., Chinchapatnam, P., Ginks, M., Lamata, P., Rhode, K.S., Rinaldi, C.A., Razavi, R., Smith, N.P.: Length-dependent tension in the failing heart and the efficacy of cardiac resynchronization therapy. Cardiovasc. Res. **89**(2), 336–343 (2011)
5. Saint-Marie, J., Chapelle, D., Cimrman, R., Sorine, M.: Modeling and estimation of the cardiac electromechanical activity. Comp. Struc. **84**, 1743–1759 (2006)
6. Xia, L., Huo, M., Wei, Q., Liu, F., Crozier, S.: Analysis of cardiac ventricular wall motion based on a three-dimensional electromechanical biventricular model. Phys. Med. Biol. **50**(8), 1901–1917 (2005)
7. Lamata, P., Sinclair, M., Kerfoot, E., Lee, A., Crozier, A., Blazevic, B., Land, S., Lewandowski, A.J., Barber, D., Niederer, S., Smith, N.: An automatic service for the personalization of ventricular cardiac meshes. J. R. Soc. Interface. **11**(91), 20131023 (2014)

8. Nielsen, P.M., LeGrice, I.J., Smaill, B.H., Hunter, P.J.: Mathematical model of geometry and fibrous structure of the heart. Am. J. Physiol. Heart Circ. Physiol. **260**, H1365–H1378 (1991)
9. Sermesant, M., et al.: Personalised electromechanical model of the heart for the prediction of the acute effects of cardiac resynchronisation therapy. In: Ayache, N., Delingette, H., Sermesant, M. (eds.) FIMH 2009. LNCS, vol. 5528, pp. 239–248. Springer, Heidelberg (2009)
10. Reese, T.G., Weisskoff, R.M., Smith, R.N., Rosen, B.R., Dinsmore, R.E., Wedeen, V.J.: Imaging myocardial fiber architecture in vivo with magnetic resonance. Magn. Reson. Med. **34**(6), 786–791 (1995)
11. Hsu, E.W., Muzikant, A.L., Matulevicius, S.A., Penland, R.C., Henriquez, C.S.: Magnetic resonance myocardial fiber-orientation mapping with direct histological correlation. Am. J. Physiol. **274**(5), H1627–H1634 (1998)
12. Lombaert, H., Peyrat, J.M., Croisille, P., Rapacchi, S., Fanton, L., Cheriet, F., Clarysse, P., Magnin, I., Delingette, H., Ayache, N.: Human atlas of the cardiac fiber architecture: study on a healthy population. IEEE Trans. Med. Imaging **31**(7), 1436–1447 (2012)
13. Geerts-Ossevoort, L., Kerckhoffs, R., Bovendeerd, P., Arts, T.: Towards patient specific models of cardiac mechanics: a sensitivity study. In: Magnin, I.E., Montagnat, J., Clarysse, P., Nenonen, J., Katila, T. (eds.) FIMH 2003. LNCS, vol. 2674, pp. 81–90. Springer, Heidelberg (2003)
14. Bovendeerd, P.H.M., Kroon, W., Delhaas, T.: Determinants of left ventricular shear strain. Am. J. Physiol. Heart Circ. Physiol. **297**(3), H1058–H1068 (2009)
15. Kroon, W., Delhaas, T., Arts, T., Bovendeerd, P.: Computational analysis of the myocardial structure: adaptation of myofiber orientations through deformation in three dimensions. Med. Imag. Anal. **13**, 346–353 (2009)
16. Pluijmert, M., Bovendeerd, P., Kroon, W., Delhaas, T.: The effect of active cross-fiber stress on shear-induced myofiber reorientation. In: Ourselin, S., Rueckert, D., Smith, N. (eds.) FIMH 2013. LNCS, vol. 7945, pp. 35–45. Springer, Heidelberg (2013)
17. Kerckhoffs, R.C.P., Bovendeerd, P.H.M., Kotte, J.C.S., Prinzen, F.W., Smits, K., Arts, T.: Homogeneity of cardiac contraction despite physiological asynchrony of depolarization: a model study. Ann. Biomed. Eng. **31**(5), 536–547 (2003)
18. Bayer, J.D., Blake, R.C., Plank, G., Trayanova, N.A.: A novel rule-based algorithm for assigning myocardial fiber orientation to computational heart models. Ann. Biomed. Eng. **40**(10), 2243–2254 (2012)
19. Shoukas, A.A., Sagawa, K.: Control of total systemic vascular capacity by the carotid sinus baroreceptor reflex. Circ. Res. **33**(1), 22–33 (1973)
20. Shoukas, A.A.: Pressure-flow and pressure-volume relations in the entire pulmonary vascular bed of the dog determined by two-port analysis. Circ. Res. **37**(6), 809–818 (1975)
21. Geerts-Ossevoort, L., Bovendeerd, P., Nicolay, K., Arts, T.: Characterization of the normal cardiac myofiber field in goat measured with mr-diffusion tensor imaging. Am. J. Physiol. Heart Circ. Physiol. **283**(1), H139–H145 (2002)
22. Guyton, A.C., Hall, J.E.: Textbook of Medical Physiology. Elsevier Saunders, Philadelphia (2006)
23. Scollan, D.F., Holmes, A., Winslow, R., Forder, J.: Histological validation of myocardial microstructure obtained from diffusion tensor magnetic resonance imaging. Am. J. Physiol. **275**(6), H2308–H2318 (1998)

24. Stender, B., Schlaefer, A.: Detecting rat heart myocardial fiber directions in x-ray microtomography using coherence-enhancing diffusion filtering. In: Ourselin, S., Rueckert, D., Smith, N. (eds.) FIMH 2013. LNCS, vol. 7945, pp. 63–70. Springer, Heidelberg (2013)

25. Stephenson, R.S., Boyett, M.R., Hart, G., Nikolaidou, T., Cai, X., Corno, A.F., Alphonso, N., Jeffery, N., Jarvis, J.C.: Contrast enhanced micro-computed tomography resolves the 3-dimensional morphology of the cardiac conduction system in mammalian hearts. PLoS One **7**(4), e35299 (2012)

Microstructural Remodelling and Mechanics of Hypertensive Heart Disease

Vicky Y. Wang[1(✉)], Alexander J. Wilson[1,2], Gregory B. Sands[1,2], Alistair A. Young[1,3], Ian J. LeGrice[1,2], and Martyn P. Nash[1,4]

[1] Auckland Bioengineering Institute, University of Auckland, Auckland, New Zealand
{vicky.wang,alexander.wilson,g.sands,a.young,i.legrice,
martyn.nash}@auckland.ac.nz
[2] Department of Physiology, University of Auckland, Auckland, New Zealand
[3] Department of Anatomy with Radiology, University of Auckland, Auckland, New Zealand
[4] Department of Engineering Science, University of Auckland, Auckland, New Zealand

Abstract. Changes in heart geometry and microstructural remodelling of myocardium are pathological processes that occur in heart failure (HF). The combination of effects often leads to compromised mechanical function of the left ventricle (LV). We have developed an image-driven constitutive modelling framework, which integrates subject-specific geometric data and microstructural organisation of the myocardial tissue, in order to investigate the mechanisms of heart function in health and disease. In this study, we investigated the effect of myocardial microstructural remodelling on passive mechanical function of the LV during the progression of hypertension-induced HF. We constructed finite element models of the LV using geometric data derived from *in vivo* magnetic resonance images of healthy and failing rat hearts at both 14-months and 24-months of age. By incorporating subject-specific LV geometries and parameters that directly reflect the tissue microstructure, we demonstrated that a single set of fitted constitutive parameters could reproduce the observed LV chamber compliance curves in all four cases (healthy and diseased hearts at 14-months and 24-months). This study highlights the important role that remodelling of myocardial microstructure plays in the mechanics of the failing heart.

Keywords: Microstructural remodelling · *In vivo* MRI · Chamber compliance · Hypertensive heart disease

1 Introduction

In many forms of heart failure (HF), left ventricular (LV) remodelling has been identified as the leading cause of ventricular dysfunction. Altered haemodynamic loading can lead to changes in LV size, wall thickness, and remodelling of myocardial tissue constituents (e.g. fibrosis) [1]. Myocardial fibrosis involves an accumulation of connective tissue (collagen) in the interstitial space, which has been shown to disrupt the organisation of the laminar microstructure within the heart wall [2]. This can lead to altered ventricular

© Springer International Publishing Switzerland 2015
H. van Assen et al. (Eds.): FIMH 2015, LNCS 9126, pp. 382–389, 2015.
DOI: 10.1007/978-3-319-20309-6_44

chamber mechanical properties (e.g. elevation of chamber stiffness [2, 3]), and thus the mechanical function of the heart.

The extent of myocardial fibrosis in HF has been investigated by several groups using the spontaneously hypertensive rat (SHR) model [3–5]. The SHR develops hypertension with age, which induces HF with similar characteristics to those observed in failing human hearts, making it an effective animal model for studies of HF [3, 6, 7]. The Wistar-Kyoto rat (WKY) serves as an effective control for comparative purposes with the SHR. Extensive proliferation of endomysial and perimysial collagen has been observed using high resolution confocal imaging [2, 3], which demonstrates that perimysial collagen "fills up" the cleavage space [2]. Several groups have shown that SHR hearts exhibit elevated myocardial stiffness inferred from the *ex vivo* stress-strain relationship of trabeculae [3] and/or the *ex vivo* pressure-volume curve of the whole heart [2]. Histological studies have speculated that interstitial fibrosis may be the key cause of LV chamber stiffening [2, 3, 6, 8]. However, the mechanistic link between microstructural and functional remodelling is poorly understood.

We have developed an image-driven constitutive modelling framework to differentiate the roles that geometric alteration and microstructural remodelling play in determining the passive mechanical function of the failing heart. This framework incorporates a new constitutive model of failing myocardium based on knowledge of the microstructural remodelling that occurs. Using subject-specific *ex vivo* LV compliance curves from normal (WKY) and diseased (SHR) hearts, we demonstrate that the differences in LV geometry and proliferation of endomysial and perimysial collagen must both be accounted for in order to reproduce the observed differences in LV chamber compliance with age and disease.

2   Methods

2.1   Multi-scale Structural Data on the Heart

The experimental study was approved by the Animal Ethics Committee of the University of Auckland and conforms to the National Institutes of Health Guide for the Care and Use of Laboratory Animals (NIH Publication No. 85-23).

To test our modelling framework, WKY rats and SHRs, at each of 14-months and 24-months of age were selected. The rats were anaesthetised and *in vivo* cine MRI of the heart was acquired using a Varian 4.7T MRI scanner. Each *in vivo* dataset consisted of images at six evenly spaced short-axis locations between the LV base and apex, and three radially-oriented long-axis locations. The deformation of the heart was captured at 18 time points throughout the cardiac cycle. The MR images were analysed offline using in-house software to reconstruct the 3D LV geometry at all frames of the cardiac cycle. The 3D geometric information at diastasis was used to construct FE models of the LV (Fig. 1(a)).

Subject-specific microstructural imaging data were not available for the animals used in this study, but LeGrice et al. [2] have previously performed extensive confocal imaging studies to characterise the progression of tissue remodelling in SHRs compared to age-matched WKY rats. They reported that the SHRs exhibited

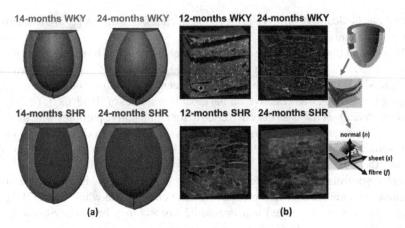

Fig. 1. (a) Anterior views of LV FE models constructed from subject-specific *in vivo* MRI data at diastasis. (b) Microstructural images of hearts from 12-months and 24-months WKY and SHR (Adapted from [2, 11]). In the confocal images, brighter intensity reveals the collagen microstructure, while dark gaps indicate the cleavage planes.

substantial reorganisation of the laminar microstructure (i.e. fusion of sheetlets of myocytes with cleavage space filled with perimysial collagen, and increased endomysial collagen surrounding the myocytes), and that these effects progressed with age (Fig. 1(b)). On the other hand, Pope (2011) reported insignificant differences in the transmural gradients of myocyte fibre orientations both within and between WKY and SHR groups [9]. We, therefore, assumed the fibre angle to vary from -60° on the epicardial surface to +70° on the endocardial surface, and the sheet angle to be +30° throughout the LV models, based on previous histological studies [10, 11].

Following the *in vivo* MRI experiments, rats were sacrificed and the excised hearts were prepared for *ex vivo* passive pressure-volume inflation experiments. Each isolated heart was perfused on a Langendorff rig and a balloon was inserted into the LV through the mitral valve. A series of inflation and deflation cycles of the balloon were carried out using a syringe driven at a controlled rate via a pump. The balloon was inflated to reach a specified series of maximum pressures (i.e. 10, 20, and 30 mmHg) in order to pre-condition the balloon. Once this was achieved, an average pressure-volume (PV) curve with maximum LVP up to 30 mmHg was extracted from recordings over several loading cycles. From the averaged curves, we defined the compliance as a function of LV pressure by computing the inverse of the derivative of the PV curve.

2.2   Modelling Passive LV Inflation

To simulate the *ex vivo* passive pressure-volume inflation experiments, we used the subject-specific 3D LV FE models described in Sect. 2.1. LV mechanics was simulated by solving equations of finite deformation elasticity. The LV base was kinematically constrained, while LV pressure of up to 15 mmHg was incrementally applied to the endocardial surfaces of the models to match the experimental protocols.

2.3   A Constitutive Model of Microstructural Remodelling

To describe the passive mechanical response of the LV, we developed an orthotropic constitutive model (Eq. (1)) of a similar form to that published by Holzapfel and Ogden [12]. In our formulation, we omitted the isotropic term and added an exponential term to independently describe the mechanical response in the sheet-normal (n) direction, and to allow each of the material axes (fibre (f), sheet (s) and sheet-normal (n)) to be separately parameterised. The parameters of this new model (a_f, a_s, a_n, a_{fs}, b_f, b_n, b_s, and b_{fs}) were initially fitted to *ex vivo* experimental data describing simple shear of myocardial tissue blocks [13].

To account for the microstructural differences observed between the WKY and SHR myocardium (Fig. 1(b)), we introduced two microstructural parameters (t_{endo} and t_{peri}) to represent the degree of endomysial and perimysial collagen thickening due to disease progression. Changes in endomysial collagen were postulated to affect the tissue response in the fibre and sheet directions, whereas perimysial collagen was thought to modulate the response in the sheet-normal direction.

$$\overline{W} = \frac{t_{endo}a_f}{2b_f}\left\{\exp\left[b_f\left(I_{4f} - 1\right)^2\right] - 1\right\} + \frac{t_{endo}a_s}{2b_s}\{\exp[b_s(I_{4s} - 1)^2] - 1\} + \frac{t_{peri}a_n}{2b_n}\{\exp[b_n(I_{4n} - 1)^2] - 1\} + \frac{a_{fs}}{2b_{fs}}\{\exp[b_{fs}I_{8fs}^2] - 1\} \tag{1}$$

2.4   Myocardial Constitutive Parameters

To investigate the relationships between geometric, microstructural and functional changes in the failing hearts, we estimated the passive constitutive parameters by matching the simulated compliance-pressure curves to the *ex vivo* experimental data for the 14-months animals. At this age, analyses of high-resolution microstructural images of WKY and SHR myocardium [2] indicated that the endomysial and perimysial collagen were both thicker in the SHR by approximately 5-fold and 10-fold, respectively. Therefore, we set the values of t_{endo} and t_{peri} for the 14-months SHR to be 5 and 10, respectively (note that for the 14-months WKY case, t_{endo} and t_{peri} were both set to 1 as a reference). From the model predicted pressure-volume curves, compliance-pressure relations were derived, and nonlinear optimisation was used to estimate a single set of constitutive parameters (a_f, a_s, a_n, a_{fs}, b_f, b_n, b_s, and b_{fs}). A nonlinear least-squares algorithm (*lsqnonlin* in MATLAB[1]) was used to simultaneously minimise the differences between predicted and experimental compliance data for all subjects.

High resolution microstructural images were not available for the rat hearts at 24-months of age. Therefore, t_{endo} and t_{peri} could not be directly quantified for these older animals. Instead, we tested whether modification of just the microstructural parameters (t_{endo} and t_{peri}), while keeping the other constitutive parameters at values fitted to data from the 14-months animals, could reproduce the observed LV chamber compliance curves for the 24-months animals. The goal was to determine whether the modelling

---

[1]   The MathWorks, Inc., Natick, Massachusetts, United States.

framework could yield consistently accurate fits to the functional data from each of the four animals based on the use of subject-specific LV geometries and microstructural parameters (t_{endo} and t_{peri}), but a single common set of the remaining constitutive parameters in Eq. 1 as described above.

3 Results

3.1 Geometrical and Functional Characteristics

Analyses of the 3D *in vivo* MRI data revealed some substantial differences in LV geometry and function between the four animals (Table 1). LV mass and ejection fraction (EF) for the 24-months WKY were substantially larger than the values for the 14-months WKY, which is likely due to normal growth. The increase in LV mass was accompanied by a decrease in LV end-diastolic volume (EDV), suggesting the development of concentric hypertrophy with age. At 14-months, the SHR had a slightly smaller LV mass, LV EDV, and EF compared to the WKY control. LV mass for the 24-months SHR was nearly double that of the 14-months SHR. This substantial myocardial hypertrophy was associated with a slightly greater LV EDV, and a slightly lower EF compared to the younger SHR. In comparison to the age-matched control, 24-months SHR had a substantially greater LV mass, but a markedly smaller EF, indicating the likely presence of systolic HF.

Table 1. *In vivo* LV geometric and functional measurements for two WKY and two SHR hearts at 14-months and 24-months of age.

Animal	LV Mass (mg)	EDV (µl)	EF (%)
14-months WKY	794	750	63
24-months WKY	941	631	80
14-months SHR	734	702	52
24-months SHR	1415	789	45

To characterise the differences in passive mechanical function of the LV, we compared the *ex vivo* chamber compliance-pressure data (Fig. 2). At 14-months of age, the SHR exhibited reduced compliance compared to the age-matched control. The WKY rats exhibited a decrease in LV compliance with age, whereas the compliance data for the SHRs at 14-months and 24-months of age were similar.

3.2 Constitutive Parameter Estimation

To investigate whether geometric remodelling alone could explain the observed differences in LV compliance between the normal (WKY) and hypertensive (SHR) hearts, a single set of constitutive parameters were estimated (data not shown) to provide the best simultaneous match between the observed compliance data and the subject-specific model predicted compliance curves for the WKY and SHR at 14-months of age. Figure 3(a) demonstrates that accounting only for the geometrical differences could not reproduce the observed differences in LV compliance.

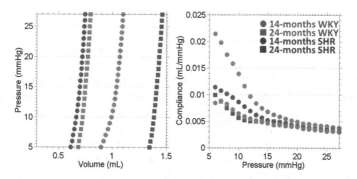

Fig. 2. Subject-specific *ex vivo* pressure-volume curves (left) and derived LV compliance-pressure relationships (right) for each of the cases.

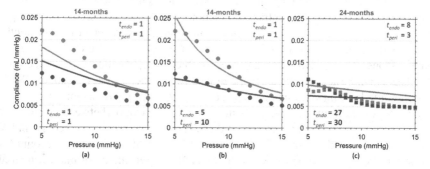

Fig. 3. Model-predicted LV compliance-pressure curves (lines) versus experimental data (symbols) for normal (WKY, orange) and hypertensive (SHR, blue) hearts. (a) Model predictions for the 14-months WKY/SHR cases, accounting only for geometric (but not microstructural) remodelling. Fits to the WKY and SHR animals at (b) 14-months and (c) 24-months of age, accounting for geometric and microstructural remodelling. The subject-specific microstructural parameters (t_{endo} and t_{peri}) are indicated (Color figure online).

Using a similar approach, but also accounting for microstructural changes as described in Sect. 2.3, a new single set of constitutive parameters were fitted (Table 2). In this case, more accurate matches to the observed compliance curves were achieved using this new set of constitutive parameters together with the appropriate set of micro-structural parameters (t_{endo} and t_{peri}) as indicated in Figs. 3(b) and 3(c).

Table 2. Fitted constitutive parameters when accounting for geometric and microstructural remodelling for the WKY and SHR hearts at 14-months of age.

a_f (kPa)	a_s (kPa)	a_n (kPa)	a_{fs} (kPa)	b_f	b_n	b_s	b_{fs}
4.5	3.0	3.2	0.17	15.6	0.2	6.5	10.7

4 Discussion

We have developed an image-driven constitutive modelling framework to investigate the passive mechanics of normal and failing LVs. We used FE models to integrate LV structural and functional data obtained from *in vivo* MRI, *ex vivo* microstructural imaging, and chamber compliance measurements on a subject-specific basis. This individualised mechanics modelling approach allowed us to differentiate the roles of geometric shape variation and microstructural remodelling during HF.

Ex vivo LV cavity compliance-pressure relations were used to characterise the passive mechanical responses of the hearts. Using this protocol, LV cavity pressures could be increased beyond the normal physiological range, and this provided more information with which to characterise the nonlinear myocardial mechanical response. The use of LV cavity compliance curves allowed us to discern changes in passive chamber mechanical function due to microstructural remodelling, independent of changes in LV size.

At 14-months of age, the contractile function of the SHR was similar to the age-matched WKY. However, the LV chamber compliance was lower in the 14-months SHR. This difference could not be explained by the observed differences in the LV geometry alone. By accounting for the observed thickening of endomysial and perimysial collagen, a more accurate match to the *ex vivo* compliance curves for both the WKY and SHR could be achieved using a single set of fitted constitutive parameters.

Functional data from the WKY rats indicated an association between age and the development of concentric hypertrophy of the LV. We also observed a decrease in LV cavity compliance with ageing. The increase in LV mass alone could not account for the lower LV compliance observed in the older WKY rat, therefore microstructural remodelling (collagen proliferation) needed to be taken into account. Our estimated values of t_{endo} and t_{peri} were consistent with this, and with observations from previous studies [2]. The 24-months SHR had a significantly dilated LV compared to both the 24-months WKY and the 14-months SHR. Without accounting for microstructural remodelling, LV dilation predicted a greater LV cavity compliance (data not shown) compared to the experimental data. Again, microstructural remodelling, in the form of proliferation of endomysial and perimysial collagen together with the reorganisation of the extracellular matrix network, is likely to be the main contributor to the stiffening of the myocardium in the failing heart.

The remaining error between measured compliances and the model predictions, particularly for the 24-month animals, indicates that there are additional factors influencing cardiac function that have not been captured using this modelling framework. To address this, it is likely that the reorganisation in the distribution of collagen will also need to be taken into account in the constitutive model. Individualised measurements of local fibre and sheet orientations can also potentially improve the goodness-of-fit. Once measurements of the regional myocardial deformation become available, this framework will be extended to match local myocardial motion, which provides more modes of deformation for characterising material properties.

5   Conclusions

An image-driven modelling framework that can explain the observed differences in passive mechanical function between normal and failing hearts was proposed. Using subject-specific *ex vivo* LV compliance curves, we demonstrated that alterations in LV geometry alone cannot account for the observed differences in LV cavity compliance. To reproduce these differences, we needed to account for the mechanical consequences of microstructural remodelling that have been observed in failing myocardial tissue. This study highlights the important role that tissue reorganisation plays in the passive mechanics of the failing heart.

References

1. Leonard, B.L., Smaill, B.H., LeGrice, I.J.: Structural remodeling and mechanical function in heart failure. Microsc. Microanal. **18**, 50–67 (2012)
2. LeGrice, I.J., Pope, A.J., Sands, G.B., Whalley, G., Doughty, R.N., Smaill, B.H.: Progression of myocardial remodeling and mechanical dysfunction in the spontaneously hypertensive rat. Am. J. Physiol. - Heart Circulatory Physiol. **303**, H1353–H1365 (2012)
3. Conrad, C.H., Brooks, W.W., Hayes, J.A., Sen, S., Robinson, K.G., Bing, O.H.L.: Myocardial fibrosis and stiffness with hypertrophy and heart failure in the spontaneously hypertensive rat. Circulation **91**, 161–170 (1995)
4. Pfeffer, M.A., Pfeffer, J.M., Frohlich, E.D.: Pumping ability of the hypertrophying left ventricle of the spontaneously hypertensive rat. Circulation Res. **38**, 423–429 (1976)
5. Wise, R.G., Huang, C.L.H., Gresham, G.A., Al-Shafei, A.I.M., Carpenter, T.A., Hall, L.D.: Magnetic resonance imaging analysis of left ventricular function in normal and spontaneously hypertensive rats. J. Physiol. **513**, 873–887 (1998)
6. Bing, O.H., Conrad, C.H., Boluyt, M.O., Robinson, K.G., Brooks, W.W.: Studies of prevention, treatment and mechanisms of heart failure in the aging spontaneously hypertensive rat. Heart Fail. Rev. **7**, 71–88 (2002)
7. Brilla, C.G., Matsubara, L., Weber, K.T.: Advanced hypertensive heart disease in spontaneously hypertensive rats: Lisinopril-mediated regression of myocardial fibrosis. Hypertension **28**, 269–275 (1996)
8. Weber, K.T., Brilla, C.G., Janicki, J.S.: Myocardial fibrosis: functional significance and regulatory factors. Cardiovasc. Res. **27**, 341–348 (1993)
9. Pope, A.J.: Characterising myocardial remodelling in hypertensive heart disease. Structural and functional changes in the spontaneously hypertensive rat. Department of Physiology, vol. Ph.D. University of Auckland (2011)
10. Streeter, D.D., Spotnitz, H.M., Patel, D.P., Ross, J., Sonnenblick, E.H.: Fiber orientation in the canine left ventricle during diastole and systole. Circulation Res. **24**, 339–347 (1969)
11. LeGrice, I.J., Smaill, B.H., Chai, L.Z., Edgar, S.G., Gavin, J.B., Hunter, P.J.: Laminar structure of the heart: ventricular myocyte arrangement and connective tissue architecture in the dog. Am. J. Physiol.- Heart Circulatory Physiol. **269**, H571–H582 (1995)
12. Holzapfel, G.A., Ogden, R.W.: Constitutive modelling of passive myocardium: a structurally based framework for material characterization. Philos. Trans. R. Soc. A Math. Physical Eng. Sci. **367**, 3445–3475 (2009)
13. Dokos, S., Smaill, B.H., Young, A.A., LeGrice, I.J.: Shear properties of passive ventricular myocardium. Am. J. Physiol. - Heart Circulatory Physiol. **283**, H2650–H2659 (2002)

Models of Electrophysiology

Inverse Problem of Electrocardiography: Estimating the Location of Cardiac Ischemia in a 3D Realistic Geometry

Carlos Eduardo Chávez[1][✉], Nejib Zemzemi[2,3], Yves Coudière[2,3],
Felipe Alonso-Atienza[4], and Diego Álvarez[1]

[1] University Carlos III of Madrid, Leganés, Spain
cchavez@ing.uc3m.es
[2] INRIA Bordeaux - Soud-Ouest, Bordeaux, France
[3] Electrophysiology and Heart Modeling Institute (IHU LIRYC), Bordeaux, France
[4] University Rey Juan Carlos, Fuenlabrada, Spain

Abstract. The inverse problem of electrocardiography (IPE) has been formulated in different ways in order to non invasively obtain valuable informations about the heart condition. Most of the formulations solve the IPE neglecting the dynamic behavior of the electrical wave propagation in the heart. In this work we take into account this dynamic behavior by constraining the cost function with the monodomain model. We use an iterative algorithm combined with a level set formulation and the use of a simple phenomenological model. This method has been previously presented to localize ischemic regions in a 2D cardiac tissue. In this work, we analyze the performance of this method in different 3D geometries. The inverse procedure exploits the spatiotemporal correlations contained in the observed data, which is formulated as a parametric adjust of a mathematical model that minimizes the misfit between the simulated and the observed data. Numerical results over 3D geometries show that the algorithm is capable of identifying the position and the size of the ischemic regions. For the experiments with a realistic anatomical geometry, we reconstruct the ischemic region with roughly a 47 % of false-positive rate and a 13 % false-negative rate under 10 % of input noise. The correlation coefficient between the reconstructed ischemic region and the ground truth exceeds the value of 0.70).

1 Introduction

Cardiac arrhythmias are disturbances in the normal rhythm of the heart produced by an abnormal electrical activity. They are one of the major causes of death worldwide [19]. Cardiac arrhythmias can be analyzed by means of non-invasive procedures aiming to characterize the cardiac electric sources (membrane potential, epicardial or endocardial potentials, activation times) and/or the cardiac substrate (ischemia regions, post-infarct scars) from voltages recorded by electrodes systems which are not in direct contact with the cardiac tissue [2,17]. From a mathematical point of view, the problem of recovering the

© Springer International Publishing Switzerland 2015
H. van Assen et al. (Eds.): FIMH 2015, LNCS 9126, pp. 393–401, 2015.
DOI: 10.1007/978-3-319-20309-6_45

heart electrical activity from remotes voltages can be formulated as an inverse problem, known as the inverse problem of electrocardiography (IPE) [8]. The IPE is a hard technological challenge since in its general formulation is an ill-posed problem so a number of regularization approaches have been developed over the years to obtain stable and realistic solutions [3,7,9].

In the present study we analyze the IPE in terms of localizing cardiac ischemic regions from remote voltage measurements. Cardiac ischemia is a pathology produced by the lack of blood supplied to the heart muscle that generates electrophysiologically-abnormal substrate which may lead to life threatening arrhythmias, and ultimately to a heart attack [10]. In the clinical practice, determining the size and location of cardiac ischemic regions has shown limited accuracy, specially in severe damaged hearts [20]. Ischemia alters the propagation properties of the electrical impulse through the cardiac muscle and this results in alterations in the associated voltage measurement recordings. These alteration patterns are temporally stable, making it feasible to inversely reconstruct ischemic regions from voltage measurements recordings [22].

Previous works have analyzed the localization of cardiac ischemia from remote voltage measurements [1,5,11,12,22]. In [5,11] the size and position of myocardial infarction is estimated by minimizing the difference between real voltage measurements and model-simulated ones. In [12,15,18] ischemic regions are assessed by reconstructing the epicardial potentials at a single time-instant during the plateau phase of the cardiac membrane potential by using a level-set formulation. Wang et al. [22] formulated the IPE as a constraint minimization problem, in which the size and position of ischemic regions were estimated by inversely computing the membrane potential at a single time instant during the plateau phase. While showing promising results towards clinical validation [15,22], in these studies time instants are treated independently from the others, thus ignoring the spatiotemporal correlation information contained in the voltage measurements. On the other hand, Álvarez et al. [1] presented an algorithm being able to reconstruct disconnected cardiac ischemic regions with a limited number of recording sites by exploiting the spatiotemporal correlation contained in the measured data.

Here, we extend the work presented in [1] to 3D geometries. We analyze the performance of the method in a model of spheres and then we study its use on 3D realistic anatomical model. Our results show that the algorithm is capable of identifying the position and, in most of the cases, approximates the size of the ischemic regions.

2   Forward Calculation

2.1   Action Potential Model

Cellular electrical activity was simulated using a modified version of the Two-Current model (TC), proposed by Mitchell and Schaeffer [14]. The TC consists of just two ordinary differential equations for two variables: the transmembrane potential $v(t)$, and the inactivation gate variable $h(t)$. The voltage, which is dimensionless and varies between zero and one, is defined as follows

$$\frac{dv}{dt} = J_{TC} = J_{stim}(t) + J_{in}(v,h) + J_{out}(v), \tag{1}$$

where J_{stim} represents the initial stimulus, J_{in} and J_{out} denotes the sum of all inward and outward currents, respectively, which are defined as

$$J_{in}(v,h) = \frac{h(1-v)(v-v_{rest})^2}{\tau_{in}}, \tag{2}$$

$$J_{out}(v,h) = -\frac{v - v_{rest}}{\tau_{out}}. \tag{3}$$

where v_{rest} is the resting potential. This parameter was not considered in the original formulation. It was incorporated in [1] to simulate the increase of the resting potential during ischemia.

The gating variable $h(t)$ is dimensionless and varies between zero and one. This variable regulates inward current flows and obeys the following equation

$$\frac{dh}{dt} = \begin{cases} (1-h)/\tau_{open}, & v < v_{crit} \\ -h/\tau_{close}, & v \geq v_{crit} \end{cases} \tag{4}$$

where v_{crit} is the change-over voltage. This model contains four time constants (τ_{in}, τ_{out}, τ_{open} and τ_{close}) which correspond to the four phases of the cardiac action potential: initiation, plateau, decay and recovery (see [1] for details).

The effects of ischemia are simulated by modifying the values of τ_{in} and v_{rest} [4]. For ischemic cells, we set the parameter τ_{in} and v_{rest} equal to 0.8 ms and 0.1, respectively. For healthy cells τ_{in} and v_{rest} are set to 0.2 ms and 0, respectively.

2.2   Cardiac Tissue Model

Let $\Omega_H \in \mathcal{R}^3$ be the cardiac tissue, and $v = v(r_H, t)$ the membrane potential with $r_H \in \Omega_H$. The propagation of v is described according to the monodomain formalism,

$$\frac{\partial v}{\partial t} = \nabla D \cdot (\nabla v) + J_{TC} \tag{5}$$

where J_{TC} is the ion current term current provided by TC model, and D is the intracellular conductivity tensor (assumed constant $D = 1.4\mathcal{I}\,(\text{mm}^2/\text{ms})$). Equation (5) is solved by imposing the initial conditions $v = v_{rest}$ at $t = 0$, and no-flux boundary conditions.

To simulate the effects of ischemia at the tissue level, we consider a regional model of ischemia where the parameters $\tau_{in}(r_H)$, $v_{rest}(r_H)$ vary linearly between healthy and ischemic values [1,21].

2.3   Model of Remote Recordings

According to the Volumen Conductor Theory [13,16], the resulting potential distribution at position $r_T \in \Omega_T$ outside the cardiac tissue Ω_H, is calculated as

$$\varphi(r_T, t) = \frac{1}{4\pi\sigma_0} \int_{\Omega_H} \frac{\nabla D \cdot (\nabla v(r_H, t))}{R(r_H, r_T)} d\Omega_H \tag{6}$$

where $R(r_H, r_T) = \|r_T - r_H\|$ represents the distance from the source location point r_H to the observation point r_T, σ_0 is the medium conductivity (assumed homogeneous and set to $1\mathrm{Sm}^{-1}$), and $v(r_H, t)$ is solution of (5). Equations (1)–(6) comprise a complete description of the forward problem.

3 Inverse Procedure

Let Ω_H be a synthetic cardiac tissue which contains an ischemic region, and $\varphi_R^i(t)$ be a set of associated remote measurements. Note that $\varphi_R^i(t) = \varphi(r_T^i, t)$ with $t \in [0, T]$ and $i = 1, 2, \ldots, N$, being T the total recording time, and N the number of recording points. Hereinafter, we assume that the regional ischemia constitute our true ischemic configuration, and $\varphi_R^i(t)$ is our observed data.

The aim of the inverse procedure is to estimate the shape and the location of the ischemic areas by adjusting the spatial distribution of τ_{in} and v_{rest}. This is achieved by minimizing the misfit between the observed data $\varphi_R^i(t)$, and the data associated to a guess configuration: $\varphi_S^i(t; \tau_{in}, v_{rest}) = \varphi(r_T^i, t; \tau_{in}, v_{rest})$. We use an iterative scheme in which the following cost functional

$$J(\tau_{in}, v_{rest}) = \frac{1}{2} \int_0^T \sum_{i=1}^N \left\| \varphi_R^i(t) - \varphi_S^i(t; \tau_{in}, v_{rest}) \right\|^2 dt \qquad (7)$$

is reduced at each step of the reconstruction process. We use an adjoint formulation to find a direction in the parameter space (τ_{in}, v_{rest}) such that the cost functional decreases. The ischemic region is defined by a level set function [1].

Since both parameters, τ_{in} and v_{rest}, define the same region, we only consider the variation of τ_{in} for the gradient computation. Following [1], the gradient of the function J over the parameter τ_{in} is given by

$$grad_{\tau_{in}} J = \int_0^T w \frac{h(v - v_{rest})^2 (v - 1)}{\tau_{in}^2} dt \qquad (8)$$

where adjoint state $w = w(r_H, t)$ is the solution of the following problem

$$\begin{cases} \dfrac{\partial w}{\partial t} + \nabla D \cdot \nabla w - \dfrac{\partial J_{TC}}{\partial v} w = -\dfrac{1}{4\pi\sigma_0} \sum_{i=1}^N \mathcal{M}_i \ \nabla^2 \left(\dfrac{1}{R_i}\right) \quad \text{in} \quad \Omega_H \\[2ex] \dfrac{\partial w}{\partial n} = \dfrac{1}{4\pi\sigma_0} \sum_{i=1}^N \mathcal{M}_i \ \dfrac{\partial}{\partial n}\left(\dfrac{1}{R_i}\right) \quad \text{on} \quad \partial\Omega_H \\[2ex] w = 0 \quad \text{at} \quad t = T, \end{cases} \qquad (9)$$

here $\mathcal{M}_i = \| \varphi_R^i(t) - \varphi_S^i(t, \tau_{in}, v_{rest})\|$, $R_i = \|r_T^i - r_H\|$ and

$$\frac{\partial J_{TC}}{\partial v} = \frac{h}{\tau_{in}}(v - v_{rest})(2 - 3v + v_{rest}) - \frac{1}{\tau_{out}}. \qquad (10)$$

4   Results

We conduct three experiments in order to analyze the scope of the proposed algorithm on 3D geometries: (1) we initialize the method with different initial guesses using a two concentric spheres geometry; (2) we analyze the approach with different noise levels using the spherical geometry; (3) we study the performance of the algorithm in a real anatomical geometry. For all the experiments, the reconstruction algorithm is applied to a single cardiac cycle of length $T = 240$ ms in the steady state. The inverse procedure stops when the functional cost becomes stationary. We evaluate the inverse solutions using three metrics: (1) the correlation coefficient (CC) between the real and the reconstructed τ_{in} configuration; (2) the sensitive error (SN), which is defined as the rate of true ischemia that is not detected by the algorithm (false negative); and (3) the specificity error (SP) , which is defined as the rate of the misjudged ischemic region out the reconstructed ischemic region (false negative) [22].

4.1   Model of Spherical Surface

We consider the cardiac tissue Ω_H as spherical surface of radius 50.0 mm, which was discretized by a triangular finite element mesh of 4470 nodes generated with Gmsh [6]. For the remote measurement electrodes Ω_T, we set an arrangement of 232 points located on the surface of a concentric sphere of radius 65.0 mm.

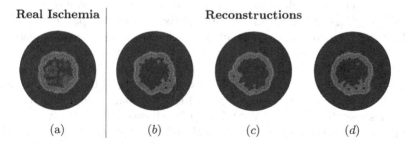

Real Ischemia	Reconstructions

(a)   (b)   (c)   (d)

Fig. 1. Reconstruction of a ischemic area considering different initial guesses in a spherical geometry with random fluctuation of 10 % to the parameter values. (a) Represents the spatial distribution of the real ischemia: red ischemia, green border zone and blue health tissue. Panels (b), (c) and (d) represent final reconstruction of the ischemic region considering different initial guesses ($\tau_{in}(r_H)$) configuration. (b) Case 1, a healthy cardiac tissue. (c) Case 2, initializing with an small ischemic region close to the real one. (d) Case 3, considering an ischemic region elsewhere in the cardiac tissue. (Color figure online)

In this experiment a true circular ischemic region of radius 14 mm was considered, as is shown in Fig. 1 panel (a). We added random fluctuations of 10 % to the model parameters values. The aim of this experiment is to verify if the inverse procedure is able to reconstruct a single ischemia independently of the initial guess. We simulated three different cases: Case 1, a healthy cardiac tissue

Table 1. Comparative quality metrics for the model of spheres when considering different initializations of the reconstruction method with 10 % of noise. Case 1, a healthy cardiac tissue is assumed, i.e., there is not an ischemic region. Case 2, initialize the method with a ischemic region close to the real ischemic area. Case 3, considering an ischemic guess elsewhere in the cardiac tissue.

Experiment	CC	SN×10	SP×10
Case 1	0.92	0.16	0.62
Case 2	0.93	0.22	0.56
Case 3	0.91	0.14	0.53

Table 2. Comparative quality metrics for the model of sphere when a healthy cardiac tissue is assumed as initial condition and adding different noise levels: 15 %, 20 % and 25 %.

Noise	CC	SN×10	SP×10
15 %	0.89	0.18	0.54
20 %	0.85	0.29	0.75
25 %	0.73	0.33	1.04

is assumed, i.e., there is not an ischemic region, Case 2, initialize the method with a ischemic region close to the real ischemic area and Case 3, considering an ischemic guess elsewhere in the cardiac tissue. Figure 1 panel (b), (c) and (d) show reconstructed $\tau_{in}(r_H)$ profile for each case, respectively. A qualitative comparison between the obtained reconstruction and the real ischemic region shows that position and size of the ischemia was reconstructed successfully for all cases. For all cases (see Table 1), CC exceeds a value of 0.89 which is consistent with the reconstructions observed in Fig. 1. Similarly, we obtain a SN and a SP ratios less than 0.3×10 and 0.7×10 for all cases, respectively.

Following with the model of a spherical surface, we study the behavior of the method at different noise levels. We consider a circular ischemia and we added random fluctuations of: 15 %, 20 % and 25 % to the parameters values. Results are shown in Table 2.

4.2   Realistic Anatomical Model

We use a 3D mesh of torso and heart surfaces obtained from a CT scan of a 43 years old woman (Fig. 2 panel (a)). We use the medical imaging software Osirix to segment the heart and the torso from the CT scan DICOM files. We then construct the meshes using the CardioViz3D software. The heart and torso domains are discretized using triangular meshes of nodes 5842 and 2873, respectively.

In this experiment a circular ischemic region of radius 13 mm on the cardiac surface was considered, as is shown in Fig. 2 panel (b). We consider random fluctuations of 10 % to the model parameters values. The remote measurement

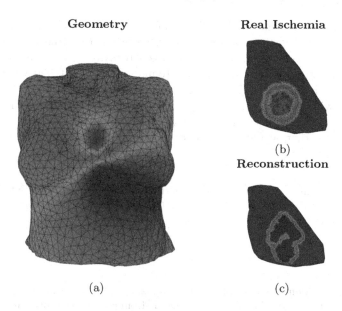

Geometry   Real Ischemia

(b)
Reconstruction

(a)   (c)

Fig. 2. Reconstruction of an ischemic region in a real anatomical geometry with 10 % of noise. (a) A snapshot of the torso potential at time 170 ms. (b): Representations of the true ischemic region $\tau_{in}(r_H)$ configuration). (c): The reconstructed ischemia region.

are located at the torso surface. In contrast to the methodology used in [1], in this experiment we do not assume an ischemic initial guess. We start the iterative method assuming healthy conditions for the entire tissue. At each iteration, the functional cost gradient is computed and both parameters, $\tau_{in}(r_H)$ and $v_{rest}(r_H)$, are updated. After 32 iterations, stop criteria is accomplished obtaining the final reconstruction. For this scenario, we obtain a correlation coefficient up to CC= 0.72 between the reconstructed ischemic region and the ground truth, for all cases. Additionally, we compute the false-negative and false-positive error ratio obtaining a value of $SN = 1.39 \times 10$ and $SN = 4.75 \times 10$.

5   Conclusions

A number of methodologies have been proposed to address the problem of localizing ischemic regions using voltage measurements recorded at the body surface. Most of these methods consider inverse procedures at a single time instant, thus ignoring the spatio-temporal information contained in the recorded body signals. Alternatively, the model proposed in [1] provides a regularization method that exploits the spatio-temporal correlations of cardiac signals.

Here, we extend the methodology proposed in [1] to different 3D geometries. First, we analyze its performance in a geometric model of concentric spheres. We found that the inverse procedure does not depend on the initial guess. Then, we test the algorithm using noisy data by adding different noise levels, showing in all cases high performance. Finally, we analyze the methodology on a realistic geometry, considering that data was corrupted by a 10 % of noise level. Also in this case, the ischemic region is located satisfactorily.

In a future work, we aim to incorporate more realistic settings using an electrophysiology detailed model, with a physiological representation of the ischemia model. We will also investigate the performance of this method in localizing ischemia from real clinical measurements. This assumes that the electrophysiological model that would be used to constrain the minimization problem is sufficiently accurate.

Acknowledgments. This work was supported in part by Spanish MINECO grants TEC-2013-46067-R, FIS2013-41802-R and by Carlos III of Madrid University PIF grant to Carlos E. Chavez.

References

1. Álvarez, D., Alonso-Atienza, F., Rojo-Álvarez, J.L., García-Alberola, A., Moscoso, M.: Shape reconstruction of cardiac ischemia from non-contact intracardiac recordings: a model study. Math. Comput. Model. **55**, 1770–1781 (2012)
2. Berger, T., Fischer, G., Pfeifer, B., Modre, R., Hanser, F., Trieb, T., Roithinger, F.X., Stuehlinger, M., Pachinger, O., Tilg, B., Hintringer, F.: Single-beat noninvasive imaging of cardiac electrophysiology of ventricular pre-excitation. J. Am. Coll. Cardiol. **48**(10), 2045–2052 (2006). Focus issue: Cardiac. Imaging
3. Brooks, D.H., Ahmad, G.F., MacLeod, R.S., Maratos, G.M.: Inverse electrocardiography by simultaneous imposition of multiple constraints. IEEE Trans. Biomed. Eng. **46**(1), 3–18 (1999)
4. Chávez, C., Alonzo-Atienza, F., Alvarez, D.: Avoiding the inverse crime in the inverse problem of electrocardiography: estimating the shape and location of cardiac ischemia. In: Computing in Cardiology Conference (CinC), pp. 687–690, September 2013
5. Farina, D., Dossel, O.: Model-based approach to the localization of infarction. In: Computers in Cardiology, pp. 173–176, 30 September 2007–3 October 2007
6. Geuzaine, C., Remacle, J.F.: Gmsh: a 3-D finite element mesh generator with built-in pre- and post-processing facilities. Int. J. Numer. Methods Eng. **79**(11), 1309–1331 (2009). http://dx.doi.org/10.1002/nme.2579
7. Greensite, F., Huiskamp, G.: An improved method for estimating epicardial potentials from the body surface. IEEE Trans. Biomed. Eng. **45**(1), 98–104 (1998)
8. Gulrajani, R.M.: The forward and inverse problems of electrocardiography. IEEE Eng. Med. Biol. Mag. **17**(5), 84–101 (1998)
9. Huiskamp, G., Van Oosterom, A.: The depolarization sequence of the human heart surface computed from measured body surface potentials. IEEE Trans. Biomed. Eng. **35**(12), 1047–1058 (1988)
10. Lazzara, R., El-Sherif, N., Hope, R.R., Scherlag, B.J.: Ventricular arrhythmias and electrophysiological consequences of myocardial ischemia and infarction. Circ. Res. **42**(6), 740–749 (1978). http://circres.ahajournals.org/content/42/6/740.short
11. Li, G., He, B.: Non-invasive estimation of myocardial infarction by means of a heart-model-based imaging approach: a simulation study. Med. Biol. Eng. Comput. **42**(1), 128–136 (2004)

12. MacLachlan, M.C., Nielsen, B.F., Lysaker, M., Tveito, A.: Computing the size and location of myocardial ischemia using measurements of ST-segment shift. IEEE Trans. Biomed. Eng. **53**(6), 1024–1031 (2006)
13. Malmivuo, J., Plonsey, R.: Bioelectromagnetism: Principles and Applications of Bioelectric and Biomagnetic Fields, 1st edn. Oxford University Press, Oxford (1995)
14. Mitchell, C.C., Schaeffer, D.G.: A two-current model for the dynamics of cardiac membrane. Bull. Math. Biol. **65**(5), 767–793 (2003)
15. Nielsen, B., Lysaker, M., Grottum, P.: Computing ischemic regions in the heart with the bidomain model; first steps towards validation. IEEE Trans. Med. Imaging **32**(6), 1085–1096 (2013)
16. van Oosterom, A., Jacquemet, V.: Genesis of the P wave: atrial signals as generated by the equivalent double layer source model. Europace **7**(s2), S21–S29 (2005)
17. Rudy, Y.: Noninvasive electrocardiographic imaging of arrhythmogenic substrates in humans. Circ. Res. **112**(5), 863–874 (2013)
18. Ruud, T., Nielsen, B., Lysaker, M., Sundnes, J.: A computationally efficient method for determining the size and location of myocardial ischemia. IEEE Trans. Biomed. Eng. **56**(2), 263–272 (2009)
19. Shah, A.J., Hocini, M., Pascale, P., Roten, L., Komatsu, Y., Daly, M., Ramoul, K., Denis, A., Derval, N., Sacher, F., Dubois, R., Bokan, R., Eliatou, S., Strom, M., Ramanathan, C., Jais, P., Ritter, P., Haissaguerre, M.: Body surface electrocardiographic mapping for non-invasive identification of arrhythmic sources. Arrhythm. Electrophysiol. Rev. **2**(1), 16–22 (2013)
20. Tokuda, M., Tedrow, U.B., Inada, K., Reichlin, T., Michaud, G.F., John, R.M., Epstein, L.M., Stevenson, W.G.: Direct comparison of adjacent endocardial and epicardial electrograms: implications for substrate mapping. J. Am. Hear. Assoc. **2**(2), e000215 (2013). http://jaha.ahajournals.org/content/2/5/e000215.abstract
21. Trénor, B., Romero, L., Ferrero Jr., J.M., Sáiz, J., Moltó, G., Alonso, J.M.: Vulnerability to reentry in a regionally ischemic tissue: a simulation study. Ann. Biomed. Eng. **35**(10), 1756–1770 (2007)
22. Wang, D., Kirby, R.M., MacLeod, R.S., Johnson, C.R.: Inverse electrocardiographic source localization of ischemia: an optimization framework and finite element solution. J. Comput. Phys. **250**, 403–424 (2013)

Sequential State Estimation for Electrophysiology Models with Front Level-Set Data Using Topological Gradient Derivations

A. Collin, D. Chapelle[⊠], and P. Moireau

Inria Saclay Ile-de-France, MΞDISIM Team, Palaiseau, France
dominique.chapelle@inria.fr

Abstract. We propose a new sequential estimation method for making an electrophysiology model patient-specific, with data in the form of level sets of the electrical potential. Our method incorporates a novel correction term based on topological gradients, in order to track solutions of complex patterns. Our assessments demonstrate the effectiveness of this approach, including in a realistic case of atrial fibrillation.

Keywords: Electrophysiology modeling · Data assimilation · Estimation · Observer · Bidomain equations · Topological gradient · Shape derivative

1 Introduction

Our objective in this paper is to propose an effective strategy for performing estimation in an electrophysiology model with data in the form of level sets of the electrical potential, as e.g. with isochrones of the depolarization front reconstructed in electrocardiographic imaging [9, 15]. Estimation is a crucial step for obtaining a personalized model – typically for a patient – that can be used for predictive purposes, with a view to providing diagnosis and prognosis assistance in the clinics [17].

Previous works have already considered the issue of state and/or parameter estimation in electrophysiology models, see e.g. [9,12,14] and references therein. Here, we depart from most existing works by proposing an estimation method of so-called *sequential* type, namely, consisting in a dynamical system obtained by incorporating in the original model a correction taking into account the discrepancy between the current simulation and the data. Sequential estimation – or data assimilation – methods have already been shown to be extremely effective for a variety of models and data, including in cardiac modeling applications with real data [2], albeit must be adapted to each category of model and data of concern. In our case, we will consider a bidomain model [5], without resorting to

---

A. Collin—Current affiliation: Università di Pavia, Italy.

© Springer International Publishing Switzerland 2015
H. van Assen et al. (Eds.): FIMH 2015, LNCS 9126, pp. 402–411, 2015.
DOI: 10.1007/978-3-319-20309-6_46

any model reduction or transformation – e.g. into eikonal-curvature equations – and address the issue of devising a sequential estimation method well-suited to the above type of data, including in cases where multiple fronts with little structure are present in the solution, as typically in fibrillation scenarii.

In the next sections we present in sequence the problem setting (Sect. 2), the proposed estimation methodology (Sect. 3), and detailed assessment results including a realistic complex fibrillation case (Sect. 4), before providing some conclusions.

2 Problem Setting

2.1 Model

We consider the surface bidomain model proposed in [4], particularly well-suited to the atria very thin walls [7]. This model is posed on the midsurface of the wall $\mathcal{S}$, and the primary unknowns are the extracellular potential u_e and the transmembrane potential u. Hence, the intracellular potential reads $u_i = u + u_e$. For all $t > 0$, we seek (u, u_e) with $\int_{\mathcal{S}} u_e \, dS = 0$ such that, for all (ϕ, ψ) with $\int_{\mathcal{S}} \psi \, dS = 0$,

$$
\begin{cases}
A_m \int_{\mathcal{S}} \left(C_m \dfrac{\partial u}{\partial t} + I_{\text{ion}}(u)\right)\phi \, dS + \int_{\mathcal{S}} \left(\boldsymbol{\sigma}_i \cdot (\boldsymbol{\nabla} u + \boldsymbol{\nabla} u_e)\right) \cdot \boldsymbol{\nabla}\phi \, dS \\
\qquad\qquad\qquad\qquad\qquad\qquad = A_m \int_{\mathcal{S}} I_{\text{app}}\phi \, dS, \qquad (1) \\
\int_{\mathcal{S}} \left((\boldsymbol{\sigma}_i + \boldsymbol{\sigma}_e) \cdot \boldsymbol{\nabla} u_e\right) \cdot \boldsymbol{\nabla}\psi \, dS + \int_{\mathcal{S}} \left(\boldsymbol{\sigma}_i \cdot \boldsymbol{\nabla} u\right) \cdot \boldsymbol{\nabla}\psi \, dS = 0.
\end{cases}
$$

Here, the positive constant A_m denotes the ratio of membrane area per unit volume, C_m is the membrane capacitance per unit surface, $I_{\text{ion}}(u)$ a reaction term representing the ionic current across the membrane and also depending on local ionic variables satisfying additional ODEs – in our case we use the ionic model of [8] – and I_{app} a given prescribed stimulus current, when applicable. We define the intra- and extra-cellular diffusion tensors $\boldsymbol{\sigma}_i$ and $\boldsymbol{\sigma}_e$ by

$$
\boldsymbol{\sigma}_{i,e} = \sigma_{i,e}^t \boldsymbol{I} + (\sigma_{i,e}^l - \sigma_{i,e}^t)[I_0(\theta)\boldsymbol{\tau}_0 \otimes \boldsymbol{\tau}_0 + J_0(\theta)\boldsymbol{\tau}_0^\perp \otimes \boldsymbol{\tau}_0^\perp], \qquad (2)
$$

where $\boldsymbol{I}$ denotes the identity tensor in the tangential plane – also sometimes called the surface metric tensor – $\boldsymbol{\tau}_0$ is a unit vector associated with the local fiber direction on the atria midsurface, and $\boldsymbol{\tau}_0^\perp$ such that $(\boldsymbol{\tau}_0, \boldsymbol{\tau}_0^\perp)$ gives an orthonormal basis of the tangential plane. The functions $I_0(\theta) = \frac{1}{2} + \frac{1}{4\theta}\sin(2\theta)$ and $J_0(\theta) = 1 - I_0(\theta)$ represent the effect of an angular variation 2θ of the fiber direction across the wall. A typical physiological simulation of the model is presented in Fig. 1 in a healthy case, with the parameters given in Table 1. For details on the modeling formulation and parameter calibration we refer to [4,7], and also to [16] where this model was used in the atria for numerical simulations of complete realistic electrocardiograms.

Table 1. Conductivity parameters (all in S.cm^{-1}) and maximal conductance g_{Na} in the different atrial areas (all in nS.pF^{-1}) with RT = regular tissue, PM = pectinate muscles, CT = crista terminalis, BB = Bachman's bundle, FO = fossa ovalis

σ_e^t	σ_e^l	σ_i^t	σ_i^l	g_{Na} (RT)	g_{Na} (PM)	g_{Na} (CT)	g_{Na} (BB)	g_{Na} (FO)
$9.0\,10^{-4}$	$2.5\,10^{-3}$	$2.5\,10^{-4}$	$2.5\,10^{-3}$	7.8	11.7	31.2	46.8	3.9

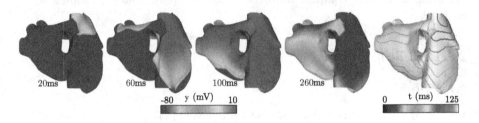

Fig. 1. Atrial electrical depolarization and corresponding synthetic front data

2.2 Data of Interest

We assume in this work that the patient-specific depolarization front is measured, as is the case when isochrones are available. From a mathematical standpoint, the measurement procedure can be modeled by considering that, for a particular solution of (1) denoted by $(\breve{u}, \breve{u}_e)$ and associated with patient-specific parameters and initial conditions, we have at our disposal the time evolution of the front

$$\Gamma_{\breve{u}}(t) = \{\boldsymbol{x} \in \mathcal{S}, \, \breve{u}(\boldsymbol{x}, t) = c_{\text{th}}\}, \tag{3}$$

defining c_{th} as a threshold value characterizing the front, and the already traveled-through region is given by

$$\mathcal{S}_{\breve{u}}^{\text{in}}(t) = \{\boldsymbol{x} \in \mathcal{S}, \, \breve{u}(\boldsymbol{x}, t) > c_{\text{th}}\}, \tag{4}$$

up to some measurement noise. These data can be represented as a sequence of images – denoted by $z(t)$ – that essentially take two different values inside and outside the traveled-through region. Our objective is to use the image sequence $z(t)$ to reconstruct the target solution $(\breve{u}, \breve{u}_e)$, in a context where the initial conditions and some physical parameters are uncertain.

3 Estimation Methodology

3.1 Sequential Estimation Principles

We consider a general dynamical system $\dot{y} = A(y, \theta, t)$, where y is the state variable, A the model operator, and θ some parameters of interest. In this abstract setting, we consider a specific target trajectory $\{\breve{y}(t), t > 0\}$ solution of the model with initial condition $\breve{y}(0) = y_\diamond + \breve{\zeta}_y$, where $y_\diamond$ is a known *a priori* whereas $\breve{\zeta}_y$

is unknown, and assuming the same type of decomposition for the parameters $\theta = \theta_\diamond + \check{\zeta}_\theta$. We further assume that we have at our disposal some indirect measurements of the target trajectory represented by the observation variable $z(t)$. Our estimation problem consists in reconstructing the solution $\{\check{y}(t), t > 0\}$ and possibly identifying the parameters $\check{\theta}$ from the data $z(t)$.

As a prerequisite to any estimation strategy, we must be able to define – at each time – a similarity/discrepancy measure $\mathcal{D}(y, z)$ between the data z and the state variable y. When $\mathcal{D}(y, z)$ vanishes, the state is exactly compatible with the data. By contrast, when $\mathcal{D}(y, z)$ is non-zero, the data indicate that $y(t) \neq \check{y}(t)$.

To achieve our estimation objective, we adopt a so-called *sequential strategy* where we define an *observer* system – also known as *sequential estimator* system – as a new dynamical system of the form

$$\begin{cases} \dot{\hat{y}} = A(\hat{y}, \hat{\theta}) + B_y(\mathcal{D}, \nabla_y \mathcal{D}), \quad y(0) = y_\diamond, \\ \dot{\hat{\theta}} = B_\theta(\mathcal{D}, \nabla_y \mathcal{D}), \quad \hat{\theta}(0) = \theta_\diamond. \end{cases} \tag{5}$$

Here, $B = \begin{pmatrix} B_y \\ B_\theta \end{pmatrix}$ is a feedback operator designed with the objective that $(\hat{y}, \hat{\theta})$ converge in time to $(\check{y}, \check{\theta})$, using only the discrepancy measure available at each time. When assuming that $\check{\zeta}_\theta = 0$, namely, that the model parameters are perfectly known, we choose $B_\theta = 0$, and we then focus on the *state sequential estimator*, also called *state observer*. However, when the parameters must also be identified, the dynamical system (5) defines a *joint state-parameter observer*.

3.2   State Observer Using Shape Derivatives

In this paper, we focus on the design of a state observer using the front data. Our objective is to extend the state observer introduced in [6], where a more extensive development of the general methodology is provided, including an additional parameter estimation component. Here, our extended observer will also be compatible with the same parameter estimation strategy.

In [6], the state observer consists in finding for all $t > 0$ the solution $(\hat{u}, \hat{u}_e)$ with $\int_S \hat{u}_e \, dS = 0$ such that for all (ϕ, ψ) with $\int_S \psi \, dS = 0$ we have

$$\begin{cases} A_m \int_S \left(C_m \frac{\partial \hat{u}}{\partial t} + I_{\text{ion}}(\hat{u}) \right) \phi \, dS + \int_S \left(\sigma_i \cdot (\nabla \hat{u} + \nabla \hat{u}_e) \right) \cdot \nabla \phi \, dS \\ = A_m \int_S I_{\text{app}} \phi \, dS - \lambda \int_{\Gamma_{\hat{u}}} \frac{1}{|\nabla \hat{u}|} \left(\left(z - C_1(\mathcal{S}_{\hat{u}}^{\text{in}}) \right)^2 - \left(z - C_2(\mathcal{S}_{\hat{u}}^{\text{in}}) \right)^2 \right) \phi \, d\Gamma, \\ \int_S \left((\sigma_i + \sigma_e) \cdot \nabla \hat{u}_e \right) \cdot \nabla \psi \, dS + \int_S \left(\sigma_i \cdot \nabla \hat{u} \right) \cdot \nabla \psi \, dS = 0. \end{cases} \tag{6}$$

Here, we have

$$\Gamma_{\hat{u}}(t) = \{ \boldsymbol{x} \in \mathcal{S}, \ \hat{u}(\boldsymbol{x}, t) = c_{\text{th}} \} \text{ and } \mathcal{S}_{\hat{u}}^{\text{in}}(t) = \{ \boldsymbol{x} \in \mathcal{S}, \ \hat{u}(\boldsymbol{x}, t) > c_{\text{th}} \},$$

which is similar to (3) and (4), but computed with the observer solution $\hat{u}$, $\lambda > 0$ is a constant weighing the confidence in the data, and

$$C_1(\mathcal{S}_{\hat{u}}^{\text{in}}) = \frac{\int_{\mathcal{S}_{\hat{u}}^{\text{in}}} z \, dS}{\int_{\mathcal{S}_{\hat{u}}^{\text{in}}} dS}, \quad C_2(\mathcal{S}_{\hat{u}}^{\text{in}}) = \frac{\int_{\mathcal{S} \backslash \overline{\mathcal{S}_{\hat{u}}^{\text{in}}}} z \, dS}{\int_{\mathcal{S} \backslash \mathcal{S}_{\hat{u}}^{\text{in}}} dS}. \tag{7}$$

As discussed in [6] the observer design originates from an analogy with the dynamical behavior of contours tracking an object in a "Mumford-Shah"-based segmentation. Indeed, introducing the data-fitting term of the functional in [3]

$$\mathcal{J}_u = \int_{\mathcal{S}_u^{\text{in}}} \left(z - C_1(\mathcal{S}_u^{\text{in}})\right)^2 dS + \int_{\mathcal{S} \backslash \overline{\mathcal{S}_u^{\text{in}}}} \left(z - C_2(\mathcal{S}_u^{\text{in}})\right)^2 dS,$$

we can show that the observer data correction term in (6), with respect to the original model (1), corresponds to the *shape derivative* of $\mathcal{J}_u$ given by

$$\boldsymbol{\nabla}_{\text{sh}} \mathcal{J}_u = \frac{\delta_{\Gamma_u}}{|\boldsymbol{\nabla} u|} \left(\left(z - C_1(\mathcal{S}_u^{\text{in}})\right)^2 - \left(z - C_2(\mathcal{S}_u^{\text{in}})\right)^2 \right),$$

where δ_{Γ_u} is the single-layer distribution associated with the boundary Γ_u, namely, the distribution such that $\int_{\mathcal{S}} \delta_{\Gamma_u} \psi \, dS = \int_{\Gamma_u} \psi \, d\Gamma$, for all ψ.

In [6], a mathematical justification was given for the convergence of the observer, at least for "reasonably small" errors. Moreover, numerical tests have demonstrated the effectiveness of this approach, in particular when the target trajectory corresponds to a simple depolarization wave. However, this observer is limited in the sense that it can only correct the front $\Gamma_{\hat{u}}$ that is already present in the simulation. This means that, if we do not have a propagation front to be associated with the data, we are not able to compute any observer correction. Our goal here is to circumvent this limitation, in order to track complex types of propagation patterns such as fibrillations.

3.3 A New State Observer Based on the Topological Gradient

In order to improve the observer formulation, we intend to follow a strategy of increasing importance in shape optimization [1] or "level-set"-based image segmentation [10,11]. The central idea is to complement the required shape derivatives, used to modify the shape contours, by a topological derivative that represents the sensitivity of $\mathcal{J}_u$ when removing a small part of the domain. Following [1], we formally define the topological derivative of our functional by

$$d\mathcal{J}_u(\mathcal{S}_u^{\text{in}})(\boldsymbol{x}) = \lim_{\boldsymbol{x} \to 0} \frac{\mathcal{J}_u(\mathcal{S}_u^{\text{in}} \backslash B_{\rho,\boldsymbol{x}}) - \mathcal{J}_u(\mathcal{S}_u^{\text{in}})}{|\mathcal{S}_u^{\text{in}} \cap B_{\rho,\boldsymbol{x}}|}, \quad \forall \boldsymbol{x} \in \mathcal{S}_u^{\text{in}},$$

$B_{\rho,\boldsymbol{x}}$ denoting a ball of radius ρ and center $\boldsymbol{x}$. This gives in our case as in [11]

$$d\mathcal{J}_u(\mathcal{S}_u^{\text{in}})(\boldsymbol{x}) = (z(\boldsymbol{x}) - C_1(\mathcal{S}_u^{\text{in}}))^2 - (z(\boldsymbol{x}) - C_2(\mathcal{S}_u^{\text{in}}))^2. \tag{8}$$

Then, we incorporate this topological derivative in our observer as a new reaction term, when we detect a topological difference between the target solution and the observer solution. In this respect, we follow a simple reasoning inspired from [10]

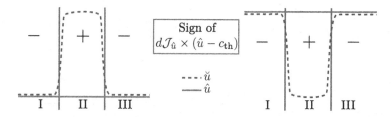

Fig. 2. 1D case example: two opposite topological differences between the observer solution and the data

and summarized by the 1D examples of Fig. 2, where we show two opposite situations of topological differences. We see that in both cases we want to act only on region II, namely, the region where $d\mathcal{J}_u \times (u - c_{\text{th}})$ is positive. We infer that we should consider a *topological gradient* term of the form

$$\nabla_{\text{top}}\mathcal{J}_u = \left(1 + \text{sign}\left(d\mathcal{J}_u \times (u - c_{\text{th}})\right)\right)d\mathcal{J}_u,$$

where the first term selects the region where we want to act

$$\left(1 + \text{sign}\left(d\mathcal{J}_u \times (u - c_{\text{th}})\right)\right) = \begin{cases} 0 & \text{if } x \in I \cup III \\ 2 & \text{if } x \in II \end{cases} \tag{9}$$

while the topological derivative $d\mathcal{J}_u$ provides the direction (creation or destruction, see Fig. 2 left and right, respectively).

When incorporating this term in the previous state observer, our new state observer is defined by: find $(\hat{u}, \hat{u}_e)$ with $\int_S \hat{u}_e \, dS = 0$, such that for all $t > 0$,

$$\begin{cases} A_m \int_S \left(C_m \dfrac{\partial \hat{u}}{\partial t} + I_{\text{ion}}(\hat{u})\right)\phi \, dS + \int_S \left(\sigma_i \cdot \left(\nabla \hat{u} + \nabla \hat{u}_e\right)\right) \cdot \nabla \phi \, dS \\ \qquad = -\int_S \left(\lambda \nabla_{\text{sh}}\mathcal{J}_{\hat{u}} + \mu \nabla_{\text{top}}\mathcal{J}_{\hat{u}}\right)\phi \, dS + A_m \int_S I_{\text{app}}\phi \, dS, \qquad (10) \\ \int_S \left((\sigma_i + \sigma_e) \cdot \nabla \hat{u}_e\right) \cdot \nabla \psi \, dS + \int_S \left(\sigma_i \cdot \nabla u\right) \cdot \nabla \psi \, dS = 0, \end{cases}$$

with λ, μ two strictly positive constants.

4   Results

4.1   1D Test Case

As a first illustration and assessment, we consider a monodomain version of (1) coupled with a simple Mitchell-Schaeffer ionic model, in a 1D domain, see [6]. In Fig. 3 we compare the performances of the observer of [6] with those of our new observer, in a case where the initial condition is shifted in one area (to the left) and entirely misses another depolarized area (right). Moreover, we show in the bottom row of Fig. 3 the results obtained when adding a Gaussian noise of 30 % in the data – namely, directly added to $z(t)$, here.

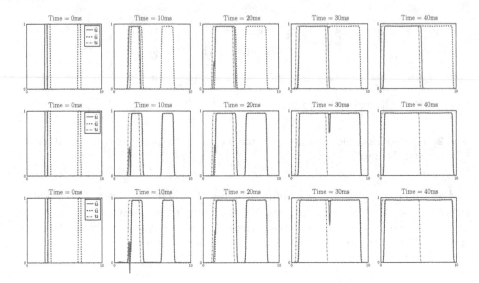

Fig. 3. Comparison of 1D observer based on the shape gradient only (top), and with topological gradient (middle); topological gradient observer with a Gaussian noise of 30 % in the data (bottom)

4.2 Atrial Fibrillation Example

We proceed to consider a more realistic configuration where we purport to track an atrial fibrillation starting from a physiologically "healthy" initial condition. In order to synthetically generate the atrial fibrillation, we use a standard S1-S2 protocol [5]. The location of the stimulation S2 is near the left pulmonary inferior vein. Indeed, the pulmonary veins are known to be prone to frequent re-entries. The S1 stimulus corresponds to a standard sinus stimulation (natural pacemaker of the heart) at t = 0 and t = 700 ms, corresponding to the cardiac cycle period considered here. The S2 stimulus is triggered at t = 290 ms. The results of the simulation of the target atrial fibrillation are displayed in Fig. 4 (left column). The figure shows successive time steps between t = 300 ms (note that before the S2 stimulus the simulation is the same as in the healthy case) and t = 820 ms, and compares the reference solution with the results obtained with our observer, and with a direct model simulation undergoing S1 stimulation only like the observer. We also show in Fig. 5 the results obtained when considering noisy data, in this case Gaussian noise added to the depolarization times, see the corresponding isochrones in the figure.

5 Discussion

Concerning the 1D simulations, we see in Fig. 3 that the original observer of [6] nicely tracks the first depolarized area, but is completely unable to adapt to the other area, hence also when the two areas merge. By contrast, our new observer

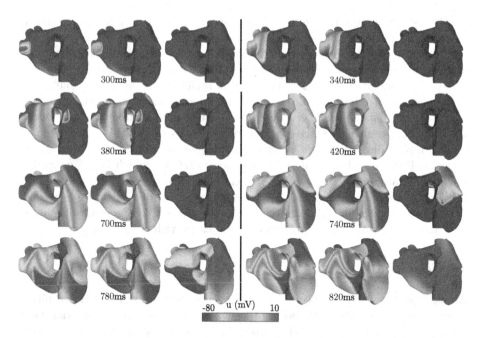

Fig. 4. Atrial fibrillation: target solution (left), observer solution (middle) and direct simulation without data (right)

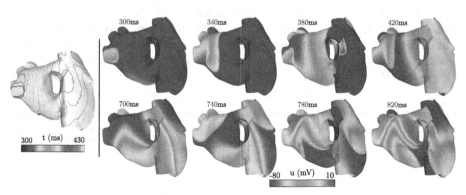

Fig. 5. Atrial fibrillation: synthetic front data with noise (left, note that only the first times of passage after the onset of fibrillation are displayed, for the sake of clarity), and corresponding observer solutions (right)

very effectively retrieves the complete solution. Moreover, the trial with added Gaussian noise shows that our method is very robust to a rather high level of noise, as will be the case with real data.

As regards the atrial fibrillation example, we see in Fig. 4 that the observer very quickly and accurately captures the complex nature of the fibrillation solution, even though the initial condition of the system was devised to generate a healthy depolarization. Here again, the sensitivity to noisy data has been further assessed and confirms the robustness of our approach. Therefore, we can envision in future works to reconstruct the state solution from real isochrone data, thereby complementing classical ECGI approaches, as a second step after the ECGI wavefront inverse problem formulation reviewed e.g. in [5, Sect. 6.5].

Further perspectives include extending the present observer to perform state-parameter estimation, which is straightforward using the strategy originally proposed in [13], see also [6]. Of course, our method is also applicable in principle to 3D electrophysiology models – as would be adequate for ventricular electrophysiology – albeit directly so only with isochrone *surfaces*, which is not generally how actual data are available – more often in the form of lines within outer surfaces (epi- and/or endocardium). Therefore some degree of adaptation is needed to incorporate such clinical data in a 3D observer system. Of course, ultimately some assessments are required in real clinical cases with actual data.

6   Concluding Remarks

We have proposed a new observer for performing estimation in an electrophysiology model with data in the form of level sets of the electrical potential. Our method builds up on a previously-proposed observer based on shape derivatives, and incorporates a new term based on the topological gradient concept, in order to allow the tracking of solutions of complex topological structures, as e.g. in fibrillation cases. Our assessments have demonstrated the effectiveness of this approach, including in a realistic case of atrial fibrillation. We also emphasize that our approach is conceptually general, hence it can be applied to other models of propagative reaction-diffusion type – as e.g. for wild-fire propagation – in a very straightforward manner.

Acknowledgment. The research leading to these results has received partial funding from the European Union's Seventh Framework Programme for research, technological development and demonstration, under grant agreement #611823 (VP2HF Project).

References

1. Burger, M., Hackl, B., Ring, W.: Incorporating topological derivatives into level set methods. J. Comput. Phys. **194**(1), 344–362 (2004)
2. Chabiniok, R., Moireau, P., Lesault, P.-F., Rahmouni, A., Deux, J.-F., Chapelle, D.: Estimation of tissue contractility from cardiac cine-MRI using a biomechanical heart model. Biomech. Model. Mechanobiol. **11**(5), 609–630 (2012)
3. Chan, T.F., Vese, L.A.: Active contours without edges. IEEE Trans. Image Process. **10**(2), 266–277 (1991)
4. Chapelle, D., Collin, A., Gerbeau, J.-F.: A surface-based electrophysiology model relying on asymptotic analysis and motivated by cardiac atria modeling. M3AS **23**(14), 2749–2776 (2013)
5. Colli Franzone, P., Pavarino, L.F., Scacchi, S.: Mathematical Cardiac Electrophysiology. MS&A, vol. XIV. Springer, Switzerland (2014)

6. Collin, A., Chapelle, D., Moireau, P.: A Luenberger observer for reaction-diffusion models with front position data. Preprint available in HAL electronic archive (2014)
7. Collin, A., Gerbeau, J.-F., Hocini, M., Haïssaguerre, M., Chapelle, D.: Surface-based electrophysiology modeling and assessment of physiological simulations in atria. In: Ourselin, S., Rueckert, D., Smith, N. (eds.) FIMH 2013. LNCS, vol. 7945, pp. 352–359. Springer, Heidelberg (2013)
8. Courtemanche, M., Ramirez, R.J., Nattel, S.: Ionic mechanisms underlying human atrial action potential properties: insights from a mathematical model. Am. J. Physiol. **275**, H301–H321 (1998)
9. Doessel, O., Krueger, M.W., Weber, F.M., Schilling, C., Schulze, W.H.W., Seemann, G.: A framework for personalization of computational models of the human atria. In: EMBC, vol. 2011, pp. 4324–4328 (2011)
10. He, L., Kao, C.Y., Osher, S.: Incorporating topological derivatives into shape derivatives based level set methods. J. Comput. Phys. **225**(1), 891–909 (2007)
11. Hintermüller, M., Laurain, A.: Multiphase image segmentation and modulation recovery based on shape and topological sensitivity. J. Math. Imaging Vis. **35**(1), 1–22 (2009)
12. Konukoglu, E., Clatz, O., Menze, B.H., Stieltjes, B., Weber, M.-A., Mandonnet, E., Delingette, H., Ayache, N.: Image guided personalization of reaction-diffusion type tumor growth models using modified anisotropic eikonal equations. IEEE Trans. Med. Imaging **29**(1), 77–95 (2010)
13. Moireau, P., Chapelle, D.: Reduced-order unscented Kalman filtering with application to parameter identification in large-dimensional systems. ESAIM Control Optim. Calc. Var. **17**(2), 380–405 (2011)
14. Moreau-Villeger, V., Delingette, H., Sermesant, M., Ashikaga, H., McVeigh, E., Ayache, N.: Building maps of local apparent conductivity of the epicardium with a 2-D electrophysiological model of the heart. IEEE Trans. Biomed. Eng. **53**(8), 1457–1466 (2006)
15. Ramanathan, C., Ghanem, R.J., Jia, P., Ryu, K., Rudy, Y.: Noninvasive electro-cardiographic imaging for cardiac electrophysiology and arrhythmia. Nat. Med. **10**, 422–428 (2004)
16. Schenone, E., Collin, A., Gerbeau, J.-F.: Numerical simulations of full electrocar-diogram cycle. Accepted for Publication in IJNMBE (2015)
17. Smith, N., de Vecchi, A., McCormick, M., Nordsletten, D., Camara, O., Frangi, A.F., Delingette, H., Sermesant, M., Relan, J., Ayache, N., Krueger, M.W., Schulze, W.H.W., Hose, R., Valverde, I., Beerbaum, P., Staicu, C., Siebes, M., Spaan, J., Hunter, P., Weese, J., Lehmann, H., Chapelle, D., Rezavi, R.: euHeart: person-alized and integrated cardiac care using patient-specific cardiovascular modelling. Interface Focus **1**(3), 349–364 (2011)

The Role of Endocardial Trabeculations in Low-Energy Defibrillation

Adam Connolly[✉] and Martin J. Bishop

Department of Imaging Sciences and Biomedical Engineering,
St. Thomas' Hospital, King's College London, London, UK
adam.connolly@kcl.ac.uk

Abstract. Recent novel low-voltage defibrillation protocols have been shown to terminate arrhythmias in animals with significantly reduced energies compared to standard shocks. Although the importance of fine-scale structural heterogeneity in driving this process through the formation of virtual electrodes has been suggested, a full understanding of these phenomena is still lacking. Here, we perform a detailed computational investigation into how specific geometrical properties of endocardial trabeculations (size, relative curvature, neighbouring proximity) affects the applied electric field strength required to generate a propagated action potential from surface virtual electrodes. We demonstrate that the applied field for propagation is lower for a single compared to multiple neighbouring trabeculations, and decreases as the structure width increases. Initial propagation occurs at trabeculation edges for single and 'tips' for multiple structures. Our findings may help optimise variables associated with low-voltage protocols, advancing clinical application.

Keywords: Bidomain · Defibrillation · Cardiac modelling

1 Introduction

The current strategy for termination of lethal cardiac arrhythmias such as fibrillation, by Implantable Cardioverter Defibrillators (ICDs), is to discharge a high-power capacitor through electrodes implanted in and around the heart. Such high-energy shocks are thought to activate excitable regions of tissue between the fractionated activation wavefronts causing the arrhythmia, removing the path through which these wavefronts propagate, thus terminating the arrhythmia [1]. However, the high-energies required to defibrillate rapidly deplete the ICD batteries, whilst inappropriate therapies cause serious physical pain, damage and psychological stress to the patient and increase mortality rates [2].

Recently, there has been much focus on developing novel strategies for defibrillation that require a fraction of the energy of standard shock treatments [3–6]. It is thought that these techniques are driven by the 'virtual electrode' effect, in which discontinuities in conductivity (in the form of structural discontinuities

© Springer International Publishing Switzerland 2015
H. van Assen et al. (Eds.): FIMH 2015, LNCS 9126, pp. 412–420, 2015.
DOI: 10.1007/978-3-319-20309-6_47

or varying fibre directions) cause re-distribution of current between the intra-/extra-cellular domains giving rise to localised regions of depolarisation (and hyper-polarisation). Action potential (AP) propagation from these numerous activation sites rapidly spread throughout the tissue to close down excitable gaps and extinguish re-entrant wavefronts [7]. Low-energy defibrillation aims to use a series of low intensity pulses to induce depolarisation from virtual electrodes to gradually defibrillate the myocardium. It has been shown that defibrillation can be achieved with a high success rate ($\approx 90\%$) [6] using approximately 1/10th the energy [3] of current shock protocols when low-energy pulses are used.

The exact nature of the fine-scale structural heterogeneity driving virtual-electrode formation during defibrillation has been suggested to be blood vessels [3,4,7] and/or endocardial trabeculae invaginations (insertion site into the endocardial surface) [5,8]. Detailed theoretical analysis has also suggested virtual electrodes may form on the outer 'tip' of a trabecular or papillary muscle where curvature is negative [9]. Endocardial trabeculae have been shown to effectively contain wave-front filaments in 3D high-resolution monodomain simulations of the rabbit ventricles, for relatively longer periods (than other anatomical structures) [10], making these structures important for defibrillation. However, a detailed understanding of these phenomena is still lacking. In particular, how geometrical factors relating to the size, relative curvature and proximity of neighbouring trabeculations affect the formation of virtual electrodes is not well understood. Such knowledge may be key when considering known significant inter-species differences in endocardial anatomy and differences in spatial scales when applying these findings to the human ventricle. In this study, we perform such a detailed investigation into how the specific nature of the endocardial surface geometry affects the field strength required to generate a propagated AP from surface virtual electrodes, and demonstrate the importance of the surrounding conductive media on these effects.

2   Methods

Computational Methods. The bidomain equations, stated in parabolic form for isotropic conductivities, are

$$\frac{\sigma_i \sigma_e}{\sigma_i + \sigma_e} \nabla^2 V_m = \beta \left(C_m \frac{\partial V_m}{\partial t} + I_{ion}(V_m, \boldsymbol{\eta}) \right),$$

$$\sigma_b \nabla^2 \phi_b = -I_{eb} \tag{1}$$

where ϕ_i and ϕ_e are the intra and extra-cellular potentials, $V_m = \phi_i - \phi_e$ is the transmembrane potential, σ_i and σ_e are the intra and extra-cellular conductivities, β is the membrane surface area to volume ratio, ϕ_b is the bath potential, I_{eb} is the extracellular current stimuli applied to the bath-space, C_m is the membrane capacitance per unit area and I_{ion} is the membrane ionic current density, as a function of the transmembrane potential V_m and the vector of state variables $\boldsymbol{\eta}$. The boundary conditions for (1) are

$$\sigma_i \boldsymbol{n} \cdot \nabla \phi_i = 0 \quad \text{on} \quad \partial \Omega_t,$$
$$\sigma_e \boldsymbol{n} \cdot \nabla \phi_e = \sigma_b \boldsymbol{n} \cdot \nabla \phi_b \quad \text{on} \quad \partial \Omega_{tb},$$

(2)

where the subscripts t and tb mean the tissue and tissue-bath interface respectively and $\boldsymbol{n}$ is the unit-normal vector to the tissue. The linear finite element solver Cardiac Arrhythmia Research Package [11] (CARP) was used to solve (1). The harmonic mean $((\sigma_i \sigma_e)/(\sigma_i + \sigma_e))$ conductivity was $0.0176\,\text{S/m}$ [12] and the bath conductivity (σ_b) was $1.0\,\text{S/m}$.

Computational Models. In order to elicit meaningful results, the dimensionality is restricted to 2D and the endocardial surface is assumed to have a highly idealized geometry consisting of periodic ellipsoids (representing endocardial trabeculations) or a single ellipsoid (representing a single trabeculation). In this initial study, geometries are based-on rabbit right ventricular MR data [7]. It is further assumed that the epicardial surface has a straight edge. Figure 1 shows a schematic of the geometries used. In the case of multiple trabeculae, each equivalent point along the surface of an individual trabeculation may be represented by the period (i.e. the angle subtended with the positive horizontal), between π and 2π. Differences in curvature were obtained by varying the semi-minor (b) and semi-major (a) axes of the semi-ellipsoids representing the trabeculations. In this work we keep t and a constant at $2500\,\mu\text{m}$ and $250\,\mu\text{m}$ respectively and vary b in $\{150, 200, 250, 300, 350\}\,\mu\text{m}$. Unstructured triangular meshes of the geometries shown in Fig. 1 were created with a maximum element edge length of $25\,\mu\text{m}$, with local element refinement around cusps and other sharp changes in curvature.

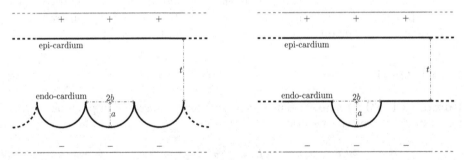

Fig. 1. Computational geometries used to represent multiple, periodic (*left*) and single (*right*) trabeculations. The semi-major and -minor axes are a and b, respectively. Electrode locations are shown in blue. t is tissue thickness (Color figure online).

Extra-cellular fields were applied between the electrodes shown in Fig. 1 of varying strength, of duration $2\,\text{ms}$. A bisection search algorithm was implemented to find the minimum applied electrode potential difference (accurate to $\pm 0.5\,\text{mV}$) to elicit a propagated AP from any point on the curved boundary. Throughout, V_m and ϕ_e were computed at all points within the domain.

3   Results

Figure 2 shows the variation in the minimum applied potential difference with semi-minor axis b for the single trabeculation (top) and multiple trabeculations (bottom) geometries. For all values of curvature, the minimum potential difference for propagation is consistently higher in the case of the single trabeculation, always requiring a field strength of $> 0.1\,$V greater than the case of multiple structures. In both geometries, as b increases, there is a monotonic decrease in the minimum potential difference required for AP propagation from any point on the endocardial surface, with a relatively faster decrease seen in the case of the single trabeculation.

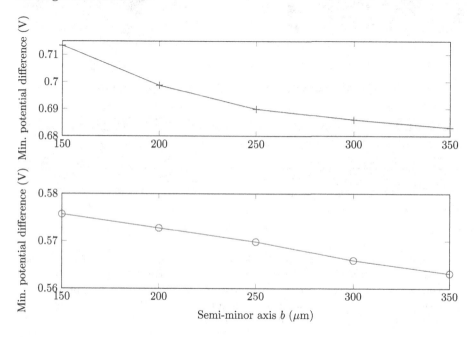

Fig. 2. Legend: —+— periodic, —⊖— single trabeculation. The minimum potential difference between the electrodes required to induce AP propagation from the endocardium following a 2 ms shock for the geometries shown in Fig. 1.

Figure 3 demonstrates the exact location from where AP propagation is first elicited due to the shock. For multiple structures (panel (a)), the very edge of the trabeculation ($\tau = 3\pi/2$) experiences the strongest depolarising effect of the shock; in the case of the single trabeculation (panel (b)), the flat surface away from the structure experiences the strongest depolarising effect. As such, propagation initiates from very different locations in both instances (panels (c) & (d)) resulting in very different initial activation patterns (panels (e) & (f)).

We now examine the relation between the shock-induced V_m levels at different sites along the trabeculation surface (which drives the initial propagation of the AP) and the curvature of the geometry for different geometrical setups. Figure 4

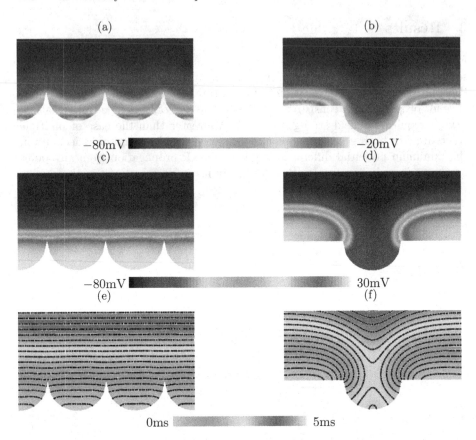

Fig. 3. Initiation of the AP after termination of the minimum potential difference shock for multiple (*left*) and single trabeculation (*right*) geometries ($a = b$). (a) & (b) show V_m distributions at shock-end ($t = 2\,\mathrm{ms}$), with (c) & (d) showing initial propagation of AP at $t = 3\,\mathrm{ms}$. Note different colour-bars. (e) & (f) show the activation time maps (contours drawn at 0.2 ms) (Color figure online).

shows the variation in curvature (top panels) and shock-induced change in V_m (lower panels) along the endocardial surface for different semi-minor axes (b). Note, in the case of the multiple structures (panel (a)), these quantities are plotted as a function of angle along the trabeculation surface due to the periodic nature; in the case of the single trabeculation (panel (b)), these quantities are plotted merely as a function of distance, x, along the endocardial surface.

In the case of the multiple trabeculations (panels (a)), maximum curvature is seen to occur at the centre of each structure ($\tau = 3\pi/2$) for $b < 250\,\mu\mathrm{m}$, with minimum curvature at the edges ($\tau = \pi$ and $\tau = 2\pi$); for $b > 250\,\mu\mathrm{m}$ this trend is reversed. For $b < 250\,\mu\mathrm{m}$, ΔV_m is largest in the centre of the structure, decreasing towards the edges; for $b > 250\,\mu\mathrm{m}$, ΔV_m remains high throughout most of the trabeculation, with a slight peak towards the edge as it approaches

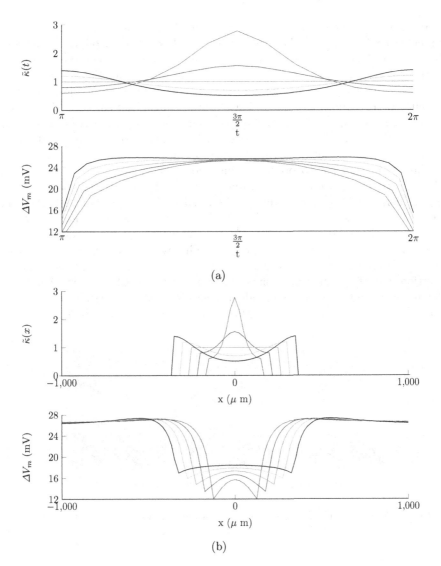

Fig. 4. —— $b = 150\,\mu$m —— $b = 200\,\mu$m —— $b = 250\,\mu$m —— $b = 300\,\mu$m
—— $b = 350\,\mu$m. (a) Variation in normalised curvature ($\tilde{\kappa} = \kappa/\kappa_{circ}$, where $\kappa_{circ} = 1/250\,\mu$m is circular curvature) and rise in V_m (at shock-end for min potential shock) vs period ($\tau \in (\pi, 2\pi)$) along endocardial surface for periodic trabeculation; (b) shows same vs distance from centre of single trabeculation.

the corresponding region of high curvature. Interestingly, absolute values of ΔV_m are similar in all setups. These relatively high values of induced V_m throughout a large central region of the trabeculation explain why propagation initiates from within this region in Fig. 3 and appears to do so as a planar wave.

In the case of a single trabeculation (panels (b)), a similar trend in curvature is seen to that witnessed above within each individual multiple trabeculation, with similar absolute values. (Note, there is no curvature outside the trabeculation). However, in contrast to the case of multiple structures, ΔV_m is consistently highest outside the trabeculation on the flat endocardial surface, close to where the curved trabeculation begins. This location of highest induced V_m is consistent with the site of initial AP propagation found in Fig. 3. Absolute values of V_m (again similar for different setups of b) in these regions, are significantly higher here than at any sites within the trabeculation itself.

Finally, in contrast to previous simulations studie [9], here we used a full bidomain representation of the cardiac tissue. This meant that the resulting electric field $\boldsymbol{E}$ was not constant (as imposed in [9]) throughout the domain. Figure 5 shows the magnitude of the gradient of the extra-cellular potential, $|\nabla\phi_e|$ throughout both geometries. In both cases, $|\nabla\phi_e|$ is low and relatively constant outside of tissue due to the low conductivity of the extra-cellular bath medium. However, $|\nabla\phi_e|$ does slightly increase outside of the tissue close to the insertion sites of the trabeculation with the endocardial surface (light-blue, green regions). Within the tissue, the gradient peaks in a similar location (close to the insertion sites), representing strong localised heterogeneity in $|\nabla\phi_e|$.

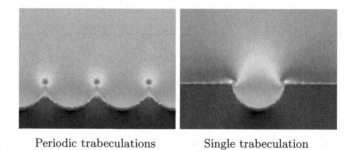

Periodic trabeculations   Single trabeculation

Fig. 5. $|\nabla\phi_e|$ for case $a = b$. Colour scales are normalised (Color figure online).

4 Discussion and Conclusions

In this study, we have used a series of idealised computational models which have allowed us to dissect the fundamental relationships between trabeculation geometry and the formation of virtual electrodes upon application of low-voltage shocks. We have showed that the minimum applied field strength necessary to elicit AP propagation is slightly lower for a single trabeculation than multiple neighbouring structures, and decreases as trabeculation width increases in both cases. When multiple structures are present, propagation was shown to initiate from the trabeculation 'tip' ($\tau = 3\pi/2$) as ΔV_m was relatively large throughout the trabeculation compared to at the edges where it inserts into the endocardium.

In contrast, for a single trabeculation, APs are initially elicited from near these edge insertion sites which in this case show much greater changes in V_m than the central structure itself.

These complex interactions are driven by the use of a full bidomain setup in our simulations which causes the applied electric field to vary along the surface with geometry (shown in Fig. 5). Such a feature underlies the discrepancy with previous work which predicted virtual electrode formation at the 'tip' of a singular trabeculation as a constant applied field was imposed [9].

The existence of a trabeculation actually makes the endocardium effectively thicker, so the same voltage must be dropped over a larger distance of tissue. This explains the higher $|\nabla\phi_e|$ on the flat endocardial surface in Fig. 5 (*right*) and the corresponding larger ΔV_m in this region (Fig. 4). However, local effects of curvature around the trabeculation are also important and cause $|\nabla\phi_e|$ to peak near the insertion site, from where propagation initially originates (Fig. 3). When understanding these bidomain effects, it is important to consider that the current takes the most efficient path between the two electrodes. For multiple trabeculations, the situation is more complex as the current cannot avoid these structures. Despite $|\nabla\phi_e|$ peaking between trabeculations (at the insertion site, Fig. 5), propagation initiates from the 'tip' of the structure (Fig. 3). This is most likely due to the fact that whether an AP succeeds in capturing the local tissue also depends on the local electrotonic loading for wavefront propagation, which is low in the convex shape of the trabeculation [13].

The findings in this study are important when considering the application of the recently-proposed low-voltage protocols to the human ventricle whose endocardial anatomy may be very different to the dogs [4,6] or rabbits [5] upon which these protocols were developed, and which will be the focus of our future investigations. Furthermore, our results may be of relevance in relation to the known inter-ventricular differences in endocardial structures in the human (highly trabeculated right ventricle, fewer, larger structures in the left) to help guide the practical application and optimisation of these methods. We intend to extend this study to 2D and 3D anatomical geometries derived from high-resolution MRI images and validate predictions with future electrophysiological measurements made on rabbit ventricle wedge preparations.

References

1. Zipes, D., Fischer, J., King, R., Nicoll, A., Jolly, W.: Termination of ventricular fibrillation in dogs by depolarizing a critical amount of myocardium. Am. J. Cardiol. **36**, 37–44 (1975)
2. Larsen, G., Evans, J., Lambert, W., Chen, Y., Raitt, M.: Shocks burden and increased mortality in implantable cardioverter-defibrillator patients. Heart Rhythm **8**, 1881–1886 (2011)
3. Fenton, F., Luther, S., Cherry, E., Otani, N., Krinsky, V., Pumir, A., et al.: Termination of atrial fibrillation using pulsed low-energy far-field stimulation. Circulation **120**, 467–476 (2009)

4. Luther, S., Fenton, F., Kornreich, B., Squires, A., Bittihn, P., Hornung, D., et al.: Low-energy control of electrical turbulence in the heart. Nature **475**, 235–239 (2012)

5. Rantner, L., Tice, B., Trayanova, N.: Terminating ventricular tachyarrhythmias using far-field low-voltage stimuli: mechanisms and delivery protocols. Heart Rhythm **10**, 1209–1217 (2013)

6. Janardhan, A., Li, W., Fedorov, V., Yeung, M., Wallendorf, M., Schuessler, R., et al.: A novel low-energy electrotherapy that terminates ventricular tachycardia with lower energy than a biphasic shock when antitachycardia pacing fail. JACC **60**, 2393–2398 (2012)

7. Bishop, M., Plank, G., Vigmond, E.: Investigating the role of the coronary vasculature in the mechanisms of defibrillation. Circ. Arrhyth. Elect. **5**, 210–219 (2012)

8. Bishop, M., Boyle, P., Plank, G., Welsh, D., Vigmond, E.: Modeling the role of the coronary vasculature during external field stimulation. IEEE Trans. Biomed. Eng. **57**, 2335–2345 (2010)

9. Bittihn, P., Horning, M., Luther, S.: Negative curvature boundaries as wave emitting sites for the control of biological excitable media. PLR **109**, 118106 (2012)

10. Bishop, M.J., Plank, G.: The role of fine-scale anatomical structure in the dynamics of reentry in computational models of the rabbit ventricles. J. Physiol. **590**(18), 4515–4535 (2012)

11. Vigmond, E., Hughes, M., Plank, G., Leon, L.J.: Computational tools for modeling electrical activity in cardiac tissue. J. Electrocardiol. **36**, 69–74 (2003)

12. Clerc, L.: Directional differences of impulse spread in trabecular muscle from mammalian heart. J. Physiol. **255**(2), 335–346 (1976)

13. Bishop, M.J., Connolly, A., Plank, G.: Structural heterogeneity modulates effective refractory period: a mechanism of focal arrhythmia initiation. PLoS ONE **9**, e109754 (2014)

Computational Modelling of Low Voltage Resonant Drift of Scroll Waves in the Realistic Human Atria

Sanjay R. Kharche[1], Irina V. Biktasheva[2], Gunnar Seemann[3], Henggui Zhang[4], Jichao Zhao[5], and Vadim N. Biktashev[1(✉)]

[1] College of Engineering, Mathematics and Physical Sciences,
University of Exeter, Exeter, EX4 4QF, UK
V.N.Bitashev@exeter.ac.uk
[2] Department of Computer Science, University of Liverpool, Liverpool L69 3BX, UK
[3] Institute of Biomedical Engineering, Karlsruhe Institute of Technology,
76021 Karlsruhe, Germany
[4] Biological Physics Group, Department of Physics, University of Manchester,
Manchester M13 9PL, UK
[5] Auckland Bioengineering Institute, University of Auckland, Auckland 1142, New Zealand

Abstract. This study evaluated the effects of human atrial anatomy and fibre orientation on the effectiveness of a low voltage resonant defibrillation method. The Courtemanche-Ramirez-Nattel model was modified to simulate scroll wave re-entry that may represent a form of atrial fibrillation. The cell models were incorporated into a 3D anatomical model to simulate re-entry. The single shock threshold to eliminate re-entry in the isotropic and anisotropic 3D models was estimated as the reference point for the low energy defibrillation effectiveness. The low voltage scroll wave termination protocol was based on the resonant drift of stationary scroll waves due to feedback-controlled periodic stimulation. The global resonant feedback stimulation can work in the realistic anatomy model in principle. Further investigation to find optimal parameters for the resonant low energy defibrillation in anatomically realistic models must include optimal location of electrodes as well as stimulation protocol improvement.

Keywords: Atrial arrhythmia · Electrical cardioversion · Computational cardiology · Mathematical modelling

1 Introduction

Atrial arrhythmias cause loss in quality of life and are often driven by scroll waves, which require electrical stimulation for their termination. Low voltage stimulation protocols may potentially improve defibrillation. Modern computational cardiology allows the study of re-entry termination using detailed atrial electrophysiology and atrial anatomy. The locations of pacing electrodes for internal scroll wave termination can be investigated using mathematical modelling [1].

The present work draws upon theoretical developments by Biktashev and Holden [2] where feedback-controlled repetitive low voltage global shocks were proposed

© Springer International Publishing Switzerland 2015
H. van Assen et al. (Eds.): FIMH 2015, LNCS 9126, pp. 421–429, 2015.
DOI: 10.1007/978-3-319-20309-6_48

to be effective. Using a 2D Fitz-Hugh Nagumo model, they showed that a train of small amplitude stimuli applied at appropriate times (i.e. applied "resonantly" with respect to the stationary re-entry) were sufficient to eliminate spiral waves. This work was extended by the same authors to further demonstrate the principles of low voltage defibrillation [3]. Subsequently, Morgan et al. [4] considered two stimulation electrode and point or line registration electrodes providing the feed-back in 2D human atrial sheets. The goal of this study was to verify the effectiveness of a resonant drift low-voltage pacing protocol in a 3D human realistic atrial geometry by estimation of scroll wave termination parameters.

2   Methods

Atrial Excitation Model: The established human atrial action potential (AP) cell model by Courtemanche et al. [5] (CRN) was used in this study. Atrial fibrillation (AF) induced electrophysiological changes were implemented as in our previous study [6]. Thus defined AF condition gave stationary re-entry in isotropic homogeneous sheets of atrial tissue.

Conduction Velocity in Atrium: In the 3D anisotropic cases, the larger eigenvalue of the diffusion tensor D was taken to be 0.21 mm^2/ms to give a conduction velocity of 0.7 mm/ms along the fibres, and the smaller eigenvalue of D was 0.07 mm^2/ms, giving conduction velocity of 0.4 mm/ms across the fibres. In the isotropic case, D was taken to be 0.07 mm^2/ms. The mono-domain reaction-diffusion system was solved using a second order finite-difference approximation of the diffusion operator using 7 point (3D, isotropic) or 19 point (3D, anisotropic) stencils. A 3D human atrium model with rule based fibre orientation in the conduction pathways [7] was used in this study.

Initiation of Scroll Waves: Scroll wave initiation was implemented using the phase distribution method [8]. It involves pacing of the cell model rapidly till steady transients over one period in all state variables are obtained. The transients are then used to pace a 1D tissue with the same D values as in the 3D model. The state variables recorded from the middle of the 1D tissue during one steady propagating pulse are then used to distribute the phase of the electrical excitation to induce a spiral or scroll wave according to an Archimedean spiral formula. Using such a protocol ensures the initiation of a scroll wave at a prescribed location in the 3D models. As the atrial wall in the anatomical model is approximately 3 mm thick with a maximum of 6 computational nodes transmurally, the initiated scroll waves has filaments that were short and stubby.

Tip Tracing: The tips of the transmural filaments, i.e. the centres of the quasi two-dimensions epicardial surface spiral waves, were tracked by the software EZView (http://empslocal.ex.ac.uk/people/staff/vnb262/software/EZView/) based on the Marching Cubes algorithm [9] as implemented by Dowle and Barkley [10].

Evaluation of Single Shock Threshold: To estimate the minimum single shock required to eliminate re-entry, a mono-phasic rectangular waveform constant electrical shock was applied to the whole model for 5 ms. The shocks were applied at 12 different timings to cover an entire 80 ms rotor cycle. A single shock was deemed successful if no re-entry was detectable at 500 ms after termination of the shock.

Pacing Protocol: Global Resonant Drift Pacing: The resonant drift pacing is based on previous theoretical developments [2–4, 11] and involves:

- A registration electrode registers electrical activity, locally at a point or globally over the selected area, to provide a feedback signal.
- A "trigger event" is defined with respect to the registered signal.
- A time delay between the trigger event and a subsequent shock is defined.
- A global rectangular shaped electric shock is issued upon expiry of the time delay.

Each weak electrical shock is supposed to cause a displacement of the scroll wave, and the feedback ensures that these displacements accumulate causing the "resonant drift", which drives the scroll wave towards an inexcitable boundary. The amplitude of the shock may be lower than the threshold of excitation; however, exact parameters of the protocol required for efficient work depend on the model. In our simulations, the protocol was implemented as follows:

- We used point electrodes measuring transmembrane potential. A single registration electrode was chosen near an inexcitable boundary.
- The trigger event was defined as the moment when the signal crossed -40 mV while increasing at the registration electrode site.
- We used fixed time delays of 0 ms, 20 ms, 40 ms, and 60 ms.
- The electric shocks were the same form as for single-shock protocol, only with smaller amplitude.

The resonant drift pacing was deemed successful if the scroll waves were eliminated in 40 s after the start of applying the feedback-driven stimulation.

Cardiac Simulation Environment: BeatBox, a multi-functional High Performance Computing cardiac simulation environment was used in this study [12, 13]. In the 3D human atrium simulations of this study, 256 processors yielded the 40 s of simulated electrical activity within 48 h, reflecting its highly scalable nature [13, 14].

3 Results

Evolution of Scroll Waves in the Isotropic and Anisotropic Atrium: In a previous study [6], we have shown that anatomical features such as pectinate muscles (PMs) (see Fig. 1A and B) serve as perturbation and may cause spontaneous drift of scroll waves. It has also been shown that the drift might be due to a gradient of the surface curvature [15]. In addition, the main conduction pathways in the human atrium have fibre orientation shown in Fig. 1C. An example of the isotropic anatomy induced drift is illustrated in Fig. 1D. Therefore, before the scroll wave termination, the externally unperturbed scroll waves were simulated in the anatomy. We choose 4 scroll wave initiation locations, named L1, L2, L3, L4, shown in Fig. 2, where the isotropic anatomy induced

Human atrium anatomy, fibre orientation, and scroll waves

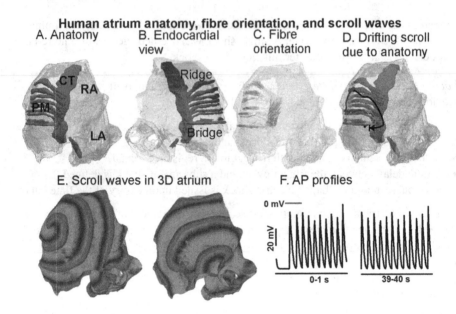

A. Anatomy   B. Endocardial   C. Fibre   D. Drifting scroll
                  view   orientation   due to anatomy

E. Scroll waves in 3D atrium   F. AP profiles

Fig. 1. Human atrial anatomy, fibres and anatomical effect on scroll. A: Epicardial view of the right atrial surface showing pectinate muscles (PM), cristae terminalis (CT), right atrium (RA) and left atrium (LA). B: Endocardial view of the atria showing anatomical heterogeneity of PM and CT that form "bridges" and "ridges" in the tissue. C: Fibre orientation in the human atrial model where the fibres are colour coded by their horizontal axis component. D: Drift of a scroll wave due to anatomical heterogeneity. The scroll wave initiation location is shown by the red dot, trajectory by the green line, and pinning location by the black square after 40 s (Colour figure online).

spontaneous drift was known to be insignificant in a 15 s duration. In the isotropic case, the scroll waves remained localised. L1 and L2 are in the predominantly isotropic regions of the 3D model, far away from the fibre orientation regions of the conduction pathways. In contrast, when the scroll wave was initiated at location L3 with anisotropy "on", it showed a definite spontaneous drift. The drift was in the direction away from the complex regions to the isotropic region. In case of the scroll wave initiated at L4 in the anisotropic model, the scroll wave moved along the pectinate muscle. This movement can be expected as there are anatomical as well as fibre orientation effects. The pectinate muscle is thicker than the atrial wall, see the recent theoretical results that sharp variations in thickness may also cause drift [16]. In addition, conduction along the pectinate muscle is faster due to fibre orientation thus promoting the scroll wave core to move along the pectinate muscle. Note that spatial variations of anisotropy can be viewed as geometric features in terms of Riemannian geometry [17]. Thus, we note that anisotropy also causes a spontaneous drift of the scroll wave. This drift may not be towards any model boundaries, but will affect the effectiveness of the feedback stimulated resonant drift protocol of scroll wave termination.

Single Shock Scroll Wave Termination Threshold: Multiple single shock amplitudes and timings were experimented with in the 2D isotropic model. Small stimulation amplitudes

Tip traces of scroll waves under isotropy and anisotropy

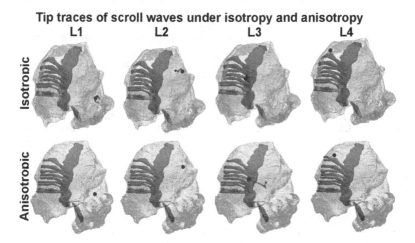

Fig. 2. Evolution of scroll waves when initiated at sites L1 to L4. In all panels, translucent grey shows atrial wall and solid yellow shows conduction pathways. Start of filaments is shown by a red dot (time t = 0), while end is shown by a black square (time t = 40 s). Isotropic filament tip traces (top panels) are shown in green, while anisotropic filament tip traces are shown in blue (bottom panels). In several instances, the drift of the scroll wave was minimal. Therefore, the start and end points, as well as the tip trajectories are virtually indistinguishable (Colour figure online).

(between 0.1 and 1 pA/pF) caused the spirals to displace without termination. Stronger stimuli (1 to 4 pA/pF) eliminated original spiral waves but boundary effects gave rise to secondary wavelets that left the re-entry persistent. A shock of 5 pA/pF was found to terminate the spiral wave activity in the 2D isotropic sheet. In the 3D isotropic anatomy case, the single shock threshold was found to be between 5 and 5.2 pA/pF, while in the anisotropic case it was found to be between 5.2 and 5.6 pA/pF.

Resonant Drift Pacing in 3D Isotropic and Anisotropic Cases: After quantifying the scroll wave behaviour in 2D, resonant drift pacing was applied to scroll waves in the 3D model as illustrated in Fig. 3. A registration electrode at the right AV border (Fig. 3A), registered a signal, after which a global stimulus was applied. In all simulations, isotropic and anisotropic, it should be noted that the small amplitude of the global periodic simulation of 0.5 pA/pF did not give rise to secondary wavelets and stimulus induced AF, but simply caused the scroll wave to incrementally change its location. Four representative delays of 0 ms, 20 ms, 40 ms, and 60 ms were chosen at each of the four locations to span the scrolls unperturbed period of 80 ms. In the isotropic case with initiation at L1 or L2, a correlation between the stimulation delay and the resonant feedback drift of the scroll wave can be seen from Fig. 3 (top panels). At delay 0, the filament of the scroll wave drifts approximately orthogonally to the line joining the initiation site and the registration electrode (Fig. 3A for electrode location, Fig. 3B, for the *isotropic case*). Therefore, in case of L1 and L2, an appropriate choice of delay may be chosen to cause the scroll waves to drift towards a model boundary, a large blood vessel opening. The resonant drift trajectory changes as the delay was increased. When the scroll waves

A. Location of
registration electrode (purple)
close to model boundary
close to AV border.
Model boundary is shown
as a green border.

B. Tip traces of scroll wave filaments under resonant pacing

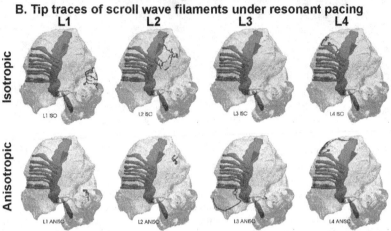

Fig. 3. 3D resonant drift stimulation. A: Location of registration electrode (purple cone) on the model's back opening of the AV border. B: Registration electrode position is designated by the same purple cone, but now it is on the other side and we see its projection through two atrial walls. In each of the panels, the location of the registration electrode was the same. The stimulation amplitude was taken to be 0.5 pA/pF in all simulations. Shown are the drift trajectories for delay values of 0 ms (blue), 20 ms (brown), 40 ms (beige) and 60 ms (green) in the isotropic case (top panels) and anisotropic case (bottom panels). In case of L2 anisotropic case, the tip trajectories are coincidental at the 4 delay values. In case of L3 anisotropic case, there was a large resonant drift at delay 0 (blue) and 60 ms (green). At the delay 60 ms (green trajectory) the drift was markedly in the same direction which then became coincidental with the delay 0 case. In the L3 anisotropic case, the tip traces for delay 20 ms and 40 ms were highly fragmented and whenever a filament tip was detected there, it was almost coincidental with the delay 0 ms or delay 60 ms cases (Colour figure online).

are initiated at L3, the anatomical features (locally altering thickness or ridge structures) inhibit a strong correlation between the feedback parameters and the drift trajectory. The drift trajectories show that the resonant drift pacing may not move the scroll waves in a desired direction. In case of L4, the scroll waves drifted along the pectinate muscle ridge. In this case, the resonant drift stimulation combined with the anatomical perturbation due to the ridge formed by the PM-atrial wall and caused the scroll to drift along the PM ridge.

With anisotropy "on" and scroll wave being initiated at L1, there was only marginal drift due to the feedback stimulation, see L1 bottom row in Fig. 3. In the L2 case (Fig. 3, bottom row), the combination of anatomy and fibre orientation results in all 4 delay cases giving almost identical tip trajectories. In cases of L3 and L4, the scroll wave drift due to the fibre orientation anisotropy was substantially larger. This caused a large scroll wave drift (Fig. 3, L3 and L4 bottom panels). Due to the combined high degree of anatomical and fibre orientation heterogeneity, the scroll wave drifted, led to spiral wave break up, and formation of secondary scroll waves.

4 Conclusions and Discussion

We have demonstrated that feedback-controlled resonant pacing can cause drift of scroll waves in the atrial geometry. The trajectory of the spiral drift depends on the relative position of the spiral core and the registration electrode of the feedback loop, and the delay in the feedback. This trajectory is also affected by the atrial geometry and aniso-tropy. Feedback controlled resonant pacing can push the spiral wave onto an inexcitable boundary, i.e. a "hole", of which in the human atrium geometry there are several, including the right and left AV border and the openings of the blood vessels, and this can be achieved at the amplitude of the electric shocks as small as 1/10 of the single shock scroll wave termination threshold. Thus, we have demonstrated that feedback controlled pacing is capable in principle of eliminating re-entrant activity in the atria.

However, our study also reveals the new difficulties arising due to the complicated geometry of the atria. One difficulty is that pushing a scroll wave onto an inexcitable boundary alone is insufficient to terminate arrhythmia. There could be other spiral waves simultaneously present which is a known complication; as it was shown previously [4, 11] it is not an absolute deterrent: when one re-entrant source has been eliminated, the feedback loop automatically engages on another, still existing re-entrant source, and so can eliminate them one by one. Another, more significant complication for the theory of resonant drift to be applicable is that re-entrant sources in complex media can be anatomical re-entries rather than spiral waves; and the resonant pacing is known to be not effective for anatomical re-entries [18]. Clinically, this outcome itself may be bene-ficial due to the anatomical re-entries have lesser frequencies and are therefore less dangerous. A more daring approach would be to try and avoid occurrence of anatomical re-entries after elimination of spirals, by directing each spiral not to any anatomical obstacle, but choosing those obstacles in such a way that the resulting topological charges of all obstacles are zero. The challenges in terms of mathematical feasibility and hardware implementation are a subject for further investigation.

Our results confirm that the anatomy including fibre orientation has an utmost effect for the feedback-controlled resonant drift pacing. Such pacing is hoped to induce cardiac re-entry to migrate to an inexcitable region or a boundary. Low-voltage protocols very similar or coinciding with ours have been successfully attempted in experiments with heart preparations [19]. There have been also other low-voltage protocols suggested recently [20]. Similar to the strengths used in this study, a recent experimental study found that multiple shocks of strength 10 fold lower than that of the threshold success-fully terminated scroll waves [21]. Further work is needed to compare these methods, and fully understand the mechanisms of scroll wave termination found in experimental studies. It may be a limitation of the present study that the defibrillation shock was applied globally. Another limitation may be that a mono-domain formulation was used in the computations, whereas the effects of an external shock may be better assessed by a bidomain formulation. We are also pursuing an improvement of boundary condition implementation in a structured finite difference mesh paradigm. In any case, this study indicates that anatomy and anatomical fibre orientation of the heart impose significant perturbations on cardiac re-entry and will play a major role in the success of any low energy defibrillation protocol. Therefore, good detailed anatomical models including true fibre orientation might be the key components for further computational simulation tests. Our findings present an experimentally or clinically testable hypothesis that can provide improvements in cardioversion technology.

References

1. Yang, F., Sha, Q., Patterson, R.P.: A novel electrode placement strategy for low-energy internal cardioversion of atrial fibrillation: a simulation study. Int. J. Cardiol. **158**, 149–152 (2012)
2. Biktashev, V.N., Holden, A.V.: Design principles of a low voltage cardiac defibrillator based on the effect of feedback resonant drift. J. Theor. Biol. **169**, 101–112 (1994)
3. Biktashev, V.N., Holden, A.V.: Control of re-entrant activity in a model of mammalian atrial tissue. Proc. Biol. Sci. **260**, 211–217 (1995)
4. Morgan, S.W., Plank, G., Biktasheva, I.V., Biktashev, V.N.: Low energy defibrillation in human cardiac tissue: a simulation study. Biophys. J. **96**, 1364–1373 (2009)
5. Courtemanche, M., Ramirez, R.J., Nattel, S.: Ionic mechanisms underlying human atrial action potential properties: insights from a mathematical model. AJP **275**, H301–321 (1998)
6. Kharche, S., Biktasheva, I.V., Seemann, G., Zhang, H., Biktashev, V.N.: A computer simulation study of anatomy induced drift of spiral waves in the human atrium. BioMed Research International (2015, in press)
7. Seemann, G., Hoper, C., Sachse, F.B., Dossel, O., Holden, A.V., Zhang, H.: Heterogeneous three-dimensional anatomical and electrophysiological model of human atria. Philos. Trans. Math. Phys. Eng. Sci. **364**, 1465–1481 (2006)
8. Biktashev, V.N., Holden, A.V.: Reentrant waves and their elimination in a model of mammalian ventricular tissue. Chaos **8**, 48–56 (1998)
9. Montani, C., Scateni, R., Scopigno, R.: A modified look-up table for implicit disambiguation of marching cubes. Vis. Comput. **10**, 353–355 (1994)
10. Barkley, D., Dowle, M.: EZ-SCROLL: a code for simulating scroll waves (http://www.warwick.ac.uk/~masax/Software/ez_software.html) (2007)
11. Biktashev, V.N., Holden, A.: Resonant drift of autowave vortices in 2D and the effects of boundaries and inhomogeneities. Chaos, Solitons Fractals **5**, 575–622 (1995)
12. McFarlane, R., Biktasheva, I.V.: Beatbox—a computer simulation environment for computational biology of the heart. In: Gelenbe, E., Abramsky, S., Sassone, A. (eds.) Visions of Computer Science— BCS International Academic Conference, pp. 99–109. British Computer Society, London (2008)
13. Biktashev, V.N., Karpov, A.V., Biktasheva, I.V., McFarlane, R., Kharche, S.R., Antonioletti, M., Jackson, A.: BeatBox - HPC environment for biophysically and anatomically realistic cardiac simulations. (http://empslocal.ex.ac.uk/people/staff/vnb262/projects/BeatBox/index.html) (2013)
14. Kharche, S., Biktasheva, I.V., Zhang, H., Biktashev, V.N.: Cardioversion in the human atria: a simulation study. In: CINC Conference (2012)

15. Dierckx, H., Brisard, E., Verschelde, H., Panfilov, A.V.: Drift laws for spiral waves on curved anisotropic surfaces. PRE **88**, 012908 (2013)
16. Biktasheva, I.V., Dierckx, H., Biktashev, V.N.: Drift of scroll waves in thin layers caused by thickness features: asymptotic theory and numerical simulations. PRL (2015, in press)
17. Verschelde, H., Dierckx, H., Bernus, O.: Covariant stringlike dynamics of scroll wave filaments in anisotropic cardiac tissue. PRL **99**, 168104 (2007)
18. Nikolaev, E.V., Biktashev, V.N., Holden, A.: On feedback resonant drift and interaction with the boundaries in circular and annular excitable media. Chaos, Solitons Fractals **9**, 363–376 (1998)
19. Pak, H.N., Liu, Y.B., Hayashi, H., Okuyama, Y., Chen, P.S., Lin, S.F.: Synchronization of ventricular fibrillation with real-time feedback pacing: implication to low-energy defibrillation. AJP **285**, H2704–2711 (2003)
20. Fenton, F.H., Luther, S., Cherry, E.M., Otani, N.F., Krinsky, V., Pumir, A., Bodenschatz, E., Gilmour Jr, R.F.: Termination of atrial fibrillation using pulsed low-energy far-field stimulation. Circulation **120**, 467–476 (2009)
21. Ambrosi, C.M., Ripplinger, C.M., Efimov, I.R., Fedorov, V.V.: Termination of sustained atrial flutter and fibrillation using low-voltage multiple-shock therapy. Heart Rhythm **8**, 101–108 (2011)

Efficient Numerical Schemes for Computing Cardiac Electrical Activation over Realistic Purkinje Networks: Method and Verification

Matthias Lange[1]([✉]), Simone Palamara[2], Toni Lassila[1], Christian Vergara[2], Alfio Quarteroni[3], and Alejandro F. Frangi[1]

[1] CISTIB, Department of Electronic and Electrical Engineering,
The University of Sheffield, Sheffield, UK
{m.lange,t.lassila,a.frangi}@sheffield.ac.uk
[2] MOX, Dipartimento di Matematica, Politecnico di Milano, Milano, Italy
{simone.palamara,christian.vergara}@polimi.it
[3] Chair of Modelling and Scientific Computing, École Polytechnique Fédérale
de Lausanne, Lausanne, Switzerland
alfio.quarteroni@epfl.ch

Abstract. We present a numerical solver for the fast conduction system in the heart using both a CPU and a hybrid CPU/GPU implementation. To verify both implementations, we construct analytical solutions and show that the L^2-error is similar in both implementations and decreases linearly with the spatial step size. Finally, we test the performance of the implementations with networks of varying complexity, where the hybrid implementation is, on average, 5.8 times faster.

1 Introduction

The cardiac Purkinje fibre network is an important contributor to the coordinated contraction of the heart as it can provide a fast conduction system reaching out to large areas of the sub-endocardium. The Purkinje fibres form an extensively branching and rejoining network, which is important for the reliability and fault-tolerance of the propagation of the action potential [1,2]. The ability to simulate propagation in physiological Purkinje networks is essential for studies of the healthy heart to obtain realistic activation patterns [3,4]. It is equally important in the simulation of pathological hearts, where disturbances in the conduction system can alter the activation pattern greatly, e.g. bundle branch blocks and long duration ventricular tachycardia [5].

Typically the simulation of the action potential propagation in a Purkinje network is based on the bidomain equations [6], or on the cable equation with a reaction term [7]. The approach of Vigmond et al. [7] treats the Purkinje conduction system as a branching tree of conducting segments without loops. Our approach also allows current loops in the Purkinje network, which are observed in realistic Purkinje networks [2].

We present first briefly the approach of Vigmond et al., and then explain its implementation on the CPU and on a CPU/GPU hybrid platform. Then, we

© Springer International Publishing Switzerland 2015
H. van Assen et al. (Eds.): FIMH 2015, LNCS 9126, pp. 430–438, 2015.
DOI: 10.1007/978-3-319-20309-6_49

present a simple model with exact known solution and use it for verification of both solvers. Finally, we compare the performance of both implementations.

2 Methods

2.1 Mathematical Model and Solution Method

The electrophysiology of cardiac tissue can be described either by the bidomain or the monodomain model. The former assumes an extracellular and intracellular space with different anisotropic conductivity tensors; if these tensors are linearly dependent the model reduces to the monodomain model [7].

The monodomain equation is considered in one dimension, because the Purkinje network can be approximated by a network of 1-D line segments. Here we assume that the extracellular space is not affected by the Purkinje network, and ignore it in the following. The monodomain equation reads

$$\partial_x(\sigma_i \partial_x V_m) = \beta(C_m \partial_t V_m + I_{\text{ion}}(V_m, \xi)), \tag{1}$$

where x is the local coordinate, V_m is the transmembrane potential, I_{ion} is the current that flows through the ion channels, ξ are the state variables of the membrane model, β is the surface-to-volume ratio of the cell membrane, where σ_i is the intracellular conductivity, and C_m is the membrane capacitance.

To derive a coupling condition between two or more line segments, needed to complete (1), we follow the idea of Vigmond et al. [7]. The equations on each line segment are coupled together by a boundary condition resulting from the enforcement of continuity of the potential and the conservation of charges (Kirchhoff's law). To satisfy the boundary conditions, the transmembrane potential, V_M, and the current, I, are needed. Since $I = \sigma_i \partial_x V_m$, the spatial derivatives of the potential need to be computed. The system is discretized using a cubic Hermite finite element method (FEM), which allows the current I to be recovered as a continuous quantity.

In view of the numerical discretization with the Finite Element Method, each node of the mesh is assumed to be located in the gap junction between two cells, where the unknowns are the intracellular potential ϕ_i and the current I_g through the gap junction. Two ghost nodes are created on both sides of the gap junction, where the transmembrane potential $V_{\pm}$, and ionic channel current I_{ion} are defined. The advantage of the ghost nodes is that with the gap junction modelled as a resistor R, the current I_g can be obtained from Ohm's law

$$V_{\pm} = \phi_i - \phi_e \mp \frac{I_g R}{2}, \tag{2}$$

where ϕ_e is the extracellular potential, taken constant in this work.

To correct for the introduced gap junction resistance, we use the equivalent conductivity $\sigma^* = (\sigma_i l)/(l + \sigma_i R\pi\rho^2)$, where l is the length of the Purkinje cell and ρ the radius. This means, that σ_i is the conductivity in the cell only, while σ^* is the conductivity of the cell and the gap junction. In this notation (1) becomes

$$\partial_x(\sigma^*\partial_x\phi_i^\pm) = \beta(C_m\partial_tV_\pm + I_{ion}(V_\pm,\xi_\pm)), \tag{3}$$

where $\phi_i^\pm$ is the intracellular potential in the ghost nodes. Furthermore, we apply an operator splitting technique to (3):

$$\begin{cases} \partial_tV + L_1(V) = 0 \\ \partial_tV + L_2(V) = 0 \end{cases}, \tag{4}$$

where $L_1 = I_{ion}$ is part of the differential operator that represents the nonlinear term of (3), whereas $L_2 = \partial_x(\sigma^*\partial_x)$ represent the diffusion term of (3). A fractional-step method with a discretization of the temporal derivatives by a first-order approximation is introduced, where the superscript n refers to the numerical solution computed at time t^n:

$$\frac{V^{n+1/2} - V^n}{\Delta t} = -L_1(V^n), \quad \frac{V^{n+1} - V^{n+1/2}}{\Delta t} = -L_2(V^{n+1}). \tag{5}$$

Algorithm 1. To Solve the Cable Equation with a Splitting Scheme

Step 1. Recover the transmembrane potential $V_\pm^n$ with (2) from I_g^n, ϕ_i^n, ϕ_e^n.

Step 2. Solve the first equation of the (5), which is the update of the ionic current in the ghost nodes

$$V_\pm^{n+1/2} = V_\pm^n - \frac{I_{ion}(V_\pm^n,\xi)}{C_m}\Delta t. \tag{6}$$

Step 3. Compute ϕ_i and I_g with the new values of $V_\pm^{n+1/2}$ in the real node:

$$\phi_i^{n+1/2} = \frac{V_+^{n+1/2} + V_-^{n+1/2}}{2} + \phi_e^n, \quad I_g^{n+1/2} = \frac{V_+^{n+1/2} - V_-^{n+1/2}}{R}. \tag{7}$$

Step 4. Use the FEM for the second stage of the operator splitting. By noticing $\phi_i = \frac{\phi_i^+ + \phi_i^-}{2}$ and using the linearity of L_2, we find:

$$\beta C_m\partial_t(\phi_i - \phi_e) = \partial_x(\sigma^*\partial_x\phi_i). \tag{8}$$

Introducing a discretization in time results in:

$$\beta C_m\frac{(\phi_i^{n+1} - \phi_e^{n+1}) - (\phi_i^{n+1/2} - \phi_e^n)}{\Delta t} = \partial_x\sigma^*\partial_x\phi_i^{n+1}, \tag{9}$$

which is solved with the FEM with 1-D cubic Hermite shape functions.

Now the cable equation can be solved in four steps (Algorithm 1). To handle branching and joining of segments in the network, the node where the three segments join is triplicated. The triplicated point is used to enforce the boundary conditions, and thus couple together the solutions obtained for the different line segments. In the case that segment 1 bifurcates into segments 2 and 3, we enforce the continuity of the potential $\phi_1 = \phi_2 = \phi_3$ and the conservation of current $I_1 = I_2 + I_3$. In contrast to Vigmond et al., our implementation covers the case where segments 1 and 2 join to from segment 3, in which case the coupling condition of the currents is $I_1 = I_3 - I_2$. These boundary conditions are introduced in the FEM system matrix associated to (9) and the right hand side.

2.2   Hardware Implementation

We now outline the CPU and the CPU/GPU hybrid implementations. The solver for the cable equation used the FEM in Step 4, and was implemented using the LifeV library (http://www.lifev.org). Parallelism was achieved using OpenMPI. We parallelized only Steps 1–3 of the algorithm and solve the linear system in Step 4 serially. The reason for this is that calculating the ionic model can be done without knowing the mesh geometry and is computationally intensive. On the other hand, it is less trivial to parallelize the solving of the linear system. The resulting computational workflow is shown in Fig. 1.

The Steps 1, 3, and 4 were always implemented on the CPU, only Step 2 is run on the GPU. In the hybrid implementation, between Steps 1 and 2, an additional copy of the transmembrane potential $V_\pm$ from the CPU to the GPU is made. To minimize the time spent copying the data, CUDA streams are used, which allow asynchronous tasks to be queued to the CPU. All computations were performed with Dell a Precision-WorkStation-T7500 featuring two Intel(R) Xeon(R) CPUs E5620 at 2.40GHz and NVIDIA Quadro 4000 GPU with 256 CUDA Cores.

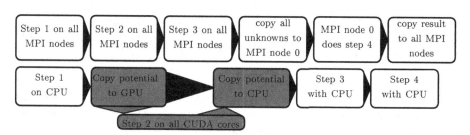

Fig. 1. Workflow for the CPU (above), and CPU/GPU hybrid (below) implementation. The CPU implementation needs to copy the potential in the gap junctions and the current, while the hybrid implementations needs to copy the potential of the ghost nodes. White boxes represent CPU tasks, and grey GPU tasks.

3   Numerical Experiments

To verify the correct and efficient implementation of the solvers, two numerical experiments are performed. The first experiment uses an analytical solution to estimate the absolute error and then to carry out a convergence test. The second experiment compares the performs of the CPU and CPU/GPU hybrid algorithm.

3.1   Numerical Error and Convergence

We first introduce a simplified model and develop two test problems with analytical solutions. The non-physiological model is [8]

$$\partial_t V = pV, \tag{10}$$

where V is the transmembrane potential and p is a model parameter. Depending on the sign of p the cells are stable ($p < 0$) and return exponentially to 0, or are unstable ($p > 0$) and the transmembrane potential increases exponentially.

Next, we introduce two different test cases and derive their analytical solutions. For the first case the domain D_1 considered is an infinite line, which is composed of three subintervals $D_{1,1} = (-\infty, -a)$, $D_{1,2} = [-a, a]$, and $D_{1,3} = (a, \infty)$. In $D_{1,2}$ unstable cells are assumed, while in the surrounding regions $D_{1,1}$, $D_{1,3}$ the cells are stable, with results in a spatial depend parameter of the simplified model

$$p(x) = \begin{cases} p_2 & \text{for } x \in D_{1,2} \\ -p_1 & \text{otherwise} \end{cases}, \tag{11}$$

where $p_1, p_2 > 0$. Inserting the cell model in (1), we need to solve

$$\begin{aligned}
C_m \partial_t V &= \delta \partial_x^2 V - p(x) V \\
V_1(-a) = V_2(-a), \quad V_2(a) &= V_3(a) \\
V_1'(x)|_{x=-a} = V_2'(x)|_{x=-a}, \quad V_2'(x)|_{x=a} &= V_3'(x)|_{x=a} \\
V_1(-\infty) = 0, \quad V_3(\infty) &= 0,
\end{aligned} \tag{12}$$

with $\delta = \sigma^*/\beta$. The solution presented by Artebrant et al. [8] is

$$V = \begin{cases} c_1 e^{\sqrt{p_1/\delta}\, x} & x < -a \\ \cos(\sqrt{p_2/\delta}\, x) & \|x\| \le a \\ c_1 e^{-\sqrt{p_1/\delta}\, x} & x > a \end{cases} \text{ with,} \begin{array}{l} p_1 = p_2 \tan^2(\sqrt{p_2/\delta}\, a), \\ c_1 = \cos(-\sqrt{p_2/\delta}\, a) e^{\sqrt{p_1/\delta}\, a}, \end{array} \tag{13}$$

where a and p_2 are the model parameters.

In the second test case, the domain D_2 is a double-bifurcation with an analytical solution. The domain consist of two rays, $D_{2,1} = (-\infty, -a)$ and $D_{2,2} = (-\infty, -a)$ joining to form a line segment $D_{2,3} = [-a, a]$ in the middle, which then splits again into two rays $D_{2,4} = (a, \infty), D_{2,5} = (a, \infty)$, resulting in a domain of five subintervals in total. The line segment $D_{2,3}$ consists of active cells, while all the rays consist of passive cells. The problem is symmetric with respect to zero, so we will look at the negative domain only. Furthermore, the rays $D_{2,1}$ and $D_{2,2}$ are identical, thus it suffices to solve the following problem for only one of them:

$$\begin{aligned}
\delta V_1'' - p_1 V_1 &= 0, \quad &\forall\, x \in D_{2,1} \\
\delta V_3'' + p_2 V_3 &= 0, \quad &\forall\, x \in D_{2,3} \\
V_1(-a) = V_3(-a), \quad 2\, V_1'(x)|_{-a} &= V_3(x)'|_{-a}, \quad V_1(-\infty) = 0,
\end{aligned} \tag{14}$$

where the two in the derivatives is a result of Kirchhoff's current law. The solution is very similar to the problem on one infinite line, with the ansatz

Table 1. The computational time for different Purkinje networks in the left ventricle (LV) and right ventricle (RV).

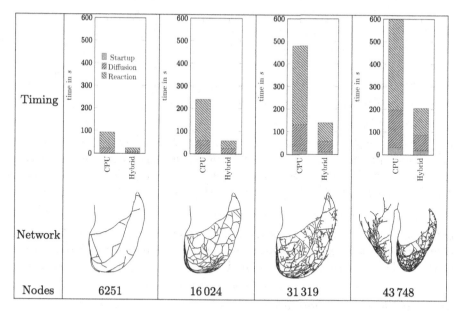

Timing				
Network				
Nodes	6251	16 024	31 319	43 748

functions $V_1 = c_1 e^{k_1 x}$, $V_3 = c_3 \cos(k_2 x)$ the constant c_1 is still given by (13). A relation between p_1 and p_2 follows from

$$2V_1'(-a) = V_3'(-a)$$
$$\overset{(13)}{\Rightarrow} 2k_1(c_3 \cos(-k_2 a)e^{k_1 a})e^{-k_1 a} = -k_2 c_3 \sin(-k_2 a)$$
$$\Rightarrow p_1 = \tfrac{p_2}{4} \tan^2(\sqrt{p_2/\delta}a).$$
(15)

Again, we need to fix V_3 at one point to get a unique solution.

Comparison of the Absolute Error: For numerical simulations, we used the parameters $p_2 = 1\ kS$, $a = 1\ cm$, $C_m = 1\ \mu F$, and $c_2 = 1\ mV$. The cell length has been chosen to $l = 62.5\ \mu m$, and a radius of $\rho = 16.0\ \mu m$, which is within the physiological limits [9]. Furthermore, we make the arbitrary choice $\delta = 1\ [kS/cm^2]$, $R = 0.1\ k\Omega$ and recall, that $\delta = \sigma^*/\beta$, from which we find the conductivity $\sigma_i = 1967\ [kS/cm]$. The spatial discretisation step h is then chosen to be a integer multiple of l, i.e. $h = nl$, $n \in \mathbb{N}$, which means that each finite element contains $n - 1$ virtual gap junctions and only the nth gap junction is explicitly included.

Convergence Test: For the error convergence test, we ran the simulation with the same parameters as before for $n = \{1, 2, 3, 4, 5\}$ in the spatial discretisation and calculated the L^2-Error for each step size (Fig. 3). The CPU and CPU/GPU

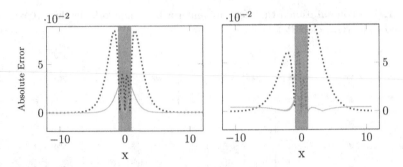

Fig. 2. The absolute error between the analytic solution of the potential and the numerical solution. For the test case on an infinite line (left), and for the simple branching network (right), where the dotted line has a step size of 0.1 cm, while the solid is 0.00625 cm and the error is multiplied by 10. The active cells are in the mark region.

hybrid implementation give the same linear convergence of the error. We conjecture that the fourth order of convergence, which is expected for Hermite bases, is not reached because of the step 1 and 3.

3.2   Performance Comparison

To compare the efficiency of the two implementations, four Purkinje fibre networks were generated with the method presented in [10]. The last two networks are realistic networks for the left ventricle and for both ventricles, respectively. The simulation was performed with a spatial resolution of 0.1 mm and a temporal step size of 0.02 ms. The duration of 45 ms was chosen because all networks were fully depolarised by that time. The membrane model of Di Francesco-Noble was used [11]. The CPU code was run with eight parallel processes, while the hybrid code was run with one CPU. Table 1 shows both the networks and the total computational time spent obtaining the respective solutions. Furthermore, the same figure shows the time spent solving the diffusion problem and the reaction problem separately. In the pure CPU implementation, the majority of the time

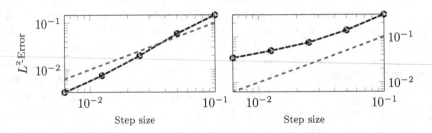

Fig. 3. Linear convergence in h (dashed line) and the convergence rates of the potential computed with the CPU (dotted line), and from the CPU/GPU hybrid (solid line). Results are for the single line case (left) and for the simple branching network (right).

is used to solve the ionic models. This is due to the fact that a detailed ionic model with 15 state variables was used, while the linear system for the diffusion step is comparably simple to solve, as the moving activation front is limited to the vicinity of a few node points. For the hybrid implementation the situation changes, and the time for solving the reaction and diffusion steps are the roughly the same, because the GPU offers a larger number of parallel cores. As a result the solution of the reaction step is ca. 4.7 times faster with the GPU. We also notice a decrease in the time spent solving the diffusion step. This can be a result of several factors, including that there is no memory copy between the CPUs, and the CPU can be used in turbo mode, because of CPU core switches.

4   Conclusion

We presented an extension of the work of Vigmond et al. to solve more realistic Purkinje networks, and implemented it both on a CPU and in a hybrid CPU/GPU architecture. To evaluate the accuracy of both we performed a convergence test of the L^2-Error, and showed that the solver converges linearly with the step size. The branching points introduce a small additional error in the numerical solution. Both implementations had equivalent numerical accuracy.

The performance test indicated that the hybrid implementation using 256 CUDA cores and 1 CPU was in average 5.8 times faster than the CPU implementation run with 8 CPUs. This motivates our future work on developing an implementation, which performs all the remaining steps of the algorithm on the GPU to realize even greater performance gains.

Acknowledgements. Simone Palamara has been funded by "Fondazione Cassa di Risparmio di Trento e Rovereto" (CARITRO) within the project "Numerical modelling of the electrical activity of the heart for the study of the ventricular dyssynchrony". Christian Vergara has been partially supported by the Italian MIUR PRIN09 project no. 2009Y4RC3B_001.

References

1. Cooper, L.L., Odening, K.E., Hwang, M.-S., Chaves, L., Schofield, L., Taylor, C., Gemignani, A.S., Mitchell, G.F., Forder, J.R., Choi, B.-R., Koren, G.: Electromechanical and structural alterations in the aging rabbit heart and aorta. Am. J. Physiol. Heart Circ. Physiol. **302**, H1625–H1635 (2012)
2. Ansari, A., Ho, S.Y., Anderson, R.H.: Distribution of the Purkinje fibres in the sheep heart. Anat. Rec. **254**, 92–97 (1999)
3. Vergara, C., Palamara, S., Catanzariti, D., Nobile, F., Faggiano, E., Pangrazzi, C., Centonze, M., Maines, M., Quarteroni, A., Vergara, G.: Patient-specific generation of the Purkinje network driven by clinical measurements of a normal propagation. Med. Biol. Eng. Comput. **52**(10), 813–826 (2014)
4. Palamara, S., Vergara, C., Catanzariti, D., Faggiano, E., Centonze, M., Pangrazzi, C., Maines, M., Quarteroni, A.: Computational generation of the Purkinje network driven by clinical measurements: the case of pathological propagations. Int. J. Num. Meth. Biomed. Eng. **30**(12), 1558–1577 (2014)

5. Bogun, F., Good, E., Reich, S., Elmouchi, D., Igic, P., Tschopp, D., Dey, S., Wimmer, A., Jongnarangsin, K., Oral, H., Chugh, A., Pelosi, F., Morady, F.: Role of Purkinje fibers in post-infarction ventricular tachycardia. J. Am. Coll. Cardiol. **48**(12), 2500–2507 (2006)
6. Bordas, R.M., Gillow, K., Gavaghan, D., Rodriguez, B., Kay, D.: A bidomain model of the ventricular specialized conduction system of the heart. SIAM J. Appl. Math. **72**, 1618–1643 (2012)
7. Vigmond, E.J., Clements, C.: Construction of a computer model to investigate sawtooth effects in the Purkinje system. IEEE Trans. Biomed. Eng. **54**, 389–399 (2007)
8. Artebrant, R., Tveito, A., Lines, G.T.: A method for analyzing the stability of the resting state for a model of pacemaker cells surrounded by stable cells. Math. Biosci. Eng. **7**, 505–526 (2010)
9. Stankovicová, T., Bito, V., Heinzel, F., Mubagwa, K., Sipido, K.R.: Isolation and morphology of single Purkinje cells from the porcine heart. Gen. Physiol. Biophys. **22**(3), 329–340 (2003)
10. Sebastian, R., Zimmerman, V., Romero, D., Frangi, A.F.: Construction of a computational anatomical model of the peripheral cardiac conduction system. IEEE Trans. Biomed. Eng. **58**(12), 3479–3482 (2011)
11. DiFrancesco, D., Noble, D.: A model of cardiac electrical activity incorporating ionic pumps and concentration changes. Philos. Trans. R. Soc. Lond. B. Biol. Sci. **307**(1133), 353–398 (1985)

Left and Right Atrial Contribution to the P-wave in Realistic Computational Models

Axel Loewe[1](✉), Martin W. Krueger[1,2], Pyotr G. Platonov[3,4],
Fredrik Holmqvist[3,4], Olaf Dössel[1], and Gunnar Seemann[1]

[1] Institute of Biomedical Engineering, Karlsruhe Institute of Technology (KIT),
Kaiserstr. 12, 76128 Karlsruhe, Germany
axel.loewe@kit.edu
publications@ibt.kit.edu
http://www.ibt.kit.edu
[2] ABB Corporate Research, Ladenburg, Germany
[3] Department of Cardiology and The Center for Integrative Electrocardiology
(CIEL), Lund University, Lund, Sweden
[4] Department of Cardiology, Skåne University Hospital, Lund, Sweden

Abstract. ECG markers derived from the P-wave are used frequently
to assess atrial function and anatomy, e.g. left atrial enlargement. While
having the advantage of being routinely acquired, the processes underlying the genesis of the P-wave are not understood in their entirety.
Particularly the distinct contributions of the two atria have not been
analyzed mechanistically. We used an *in silico* approach to simulate
P-waves originating from the left atrium (LA) and the right atrium (RA)
separately in two realistic models.

LA contribution to the P-wave integral was limited to 30 % or less.
Around 20 % could be attributed to the first third of the P-wave which
reflected almost only RA depolarization. Both atria contributed to the
second and last third with RA contribution being about twice as large as
LA contribution. Our results foster the comprehension of the difficulties
related to ECG-based LA assessment.

Keywords: Atrial modeling · Electrocardiogram · P-wave · Mathematical modeling · Left atrial enlargement

1 Introduction

Features of the P-wave measured in the body surface ECG have long been used
to gain insight into atrial anatomy and function [19,24]. Current guidelines for
ECG interpretation recommend the use of morphological P-wave characteristics
to assess atrial abnormalities such as right and left atrial enlargement [8]. Using
these measures, clinicians aim to stratify the risk for a patient to develop diseases
as e.g. atrial fibrillation (AF) [25,27]. In contrast to other techniques like ultrasound, magnetic resonance imaging, electroanatomical mapping, or ECG imaging, these ECG markers have the advantage of being noninvasive, fast and easy to

© Springer International Publishing Switzerland 2015
H. van Assen et al. (Eds.): FIMH 2015, LNCS 9126, pp. 439–447, 2015.
DOI: 10.1007/978-3-319-20309-6_50

acquire and available almost everywhere. However, despite a multitude of empirical studies relating P-wave features to left and right atrial influences [1,3,11], we are still lacking mechanistic understanding of left and right atrial contribution to the P-wave. While some sources state that the peaks corresponding to left and right atrial excitation are normally almost simultaneous, thus fused into a single peak [8], others argue that the second half of the P-wave mainly corresponds to left atrial depolarization [23]. Insight into the question which ECG leads reflect the depolarization of which parts of the atria during the different temporal phases of the P-wave may eventually help to identify patients at risk to develop AF, thus relieving part of the burden from patients and healthcare.

In this study, we used an *in silico* approach giving the unique opportunity to separate left from right atrial depolarization and investigate their contribution to the P-wave in the surface ECG. Such experiments can hardly be conducted *in vivo*. Towards this end, two realistic human models were used to simulate atrial excitation propagation and compute the resulting body surface potentials stemming from the left atrium (LA) and the right atrium (RA) separately as well as in combination.

2   Methods

2.1   Anatomical Modeling

MRI data of two healthy male subjects of age 27 (S1) and 38 (S2) were acquired and processed as described before [16]. In brief, the atria were automatically segmented [7] and converted to a voxel format with isotropic side length of 0.33 mm for electrophysiological simulation. The atrial model of S1 comprised 2.2 million tissue voxels, that of S2 1.1 million. The thoraces were segmented manually and combined with the atrial models in tetrahedral meshes for the calculation of body surface potentials. The mean node distance was 0.8 mm in the atria and 5 mm in the rest of the thorax resulting in 2.9 million torso elements for S1 and 2.1 million elements for S2. The volume of the right atrial cavity including the trunks of the vessels as shown in Fig. 1 was 126.9 ml for S1 and 94.9 ml for S2. LA volumes were 77.8 ml and 83.5 ml, respectively.

The voxel models used for excitation propagation computation were enhanced using a priori knowledge: Crista Terminalis, pectinate muscles, Bachmann's Bundle (BB), inferior isthmus, atrio-ventricular rings and atrial appendages were labeled using a rule-based algorithm [16]. This algorithm also introduced rule-based fiber orientation. Right to left atrial conduction was only possible via 5 discrete interatrial bridges (BB, anterior, 2 posterior, coronary sinus) [10,28].

2.2   Electrophysiological Modeling

The ion kinetics at each of the voxels representing atrial myocytes were computed using the formulation by Courtemanche et al [4]. The different ion currents I_x in this model are described by Hodgkin-Huxley type formulations:

$$I_x = g_x \prod_i \gamma_i (V_m - E_x) \tag{1}$$

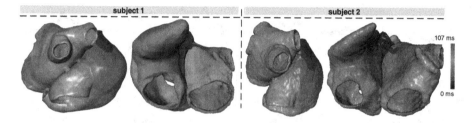

Fig. 1. Geometrical models with color coded activation times resulting from the electrophysiological simulation.

with γ_i being the gating variables describing the open probability, V_m being the transmembrane voltage, and E_x being the Nernst potential of the ion type carrying I_x. The maximum conductivities g_x of several ion channels were adjusted as given in Table 1 to account for regional heterogeneities as described before [15]. The conductivities perpendicular to cardiac fiber orientation and the anisotropy factor for propagation along the fibers are also given in Table 1. Excitation propagation was computed with a monodomain reaction-diffusion solver *acCELLerate* [30]. The finite element method and explicit Euler integration with constant time stepping of $20\,\mu s$ was used after initialization in a single cell environment for 10 cycles to establish steady state conditions.

2.3 Forward Calculation of the ECG

The impressed currents stemming from the gradient of the transmembrane voltage V_m act as the source for the differences of the extracellular potential Φ_e on the body surface that are measured during ECG acquisition:

$$\nabla\left(\sigma_i \nabla V_m\right) = -\nabla\left(\left(\sigma_i + \sigma_e\right)\nabla\Phi_e\right) \tag{2}$$

Table 1. Relative values $\hat{g}_x$ of ion channel conductivities representing regional heterogeneities [15] with respect to the original Courtemanche et al. model of human atrial myocytes [4]. Monodomain conductivity perpendicular to fiber orientation $\sigma_\perp$ and anisotropy factor. Highlighted values differ from normal myocardium.

Anatomical structure	$\hat{g}_{Kr}$	$\hat{g}_{to}$	$\hat{g}_{Ca,L}$	$\sigma_\perp$ (mS/m)	Anisotropy
RA / LA	1.0	1.0	1.0	87	3.75
Crista Terminalis	1.0	1.0	**1.67**	87	**6.56**
Atrial appendages	1.0	**0.68**	**1.06**	87	3.75
Atrio-ventricular rings	1.0	**1.53**	**0.67**	87	3.75
Pectinate muscles	1.0	1.0	1.0	**58**	**10.52**
Bachmann's Bundle	1.0	1.0	1.0	**101**	**9.0**
Inferior isthmus	1.0	1.0	1.0	**75**	**1.0**

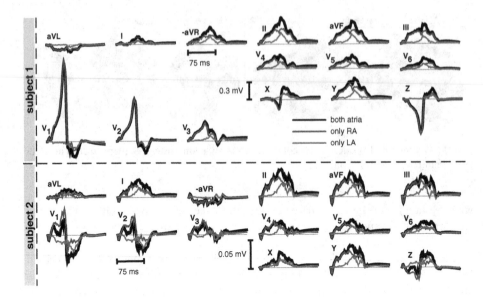

Fig. 2. Simulated 12 lead ECG and VCG leads. The blue traces originate from sources in the RA only, the red traces from sources in the LA. The addition of right and left traces yields the normal P-wave generated by both atria (black traces) (Color figure online).

σ_i represents the intracellular conductivity tensor, σ_e the extracellular conductivity tensor. To compute the ECG, V_m obtained from the electrophysiological simulations was therefore first interpolated onto the tetrahedral torso mesh. A reduced bidomain formulation [14] allowed for the subsequent computation of body surface potentials with a time increment of 1 ms. Tissue conductivities were set as in [14]. Body surface potentials were extracted at the standard 12 lead ECG locations. VCG was computed using the inverse Dower matrix [2].

To evaluate left and right atrial contribution to the P-wave, two separate forward calculations were performed for each set of transmembrane voltages. To obtain only the share of the P-wave caused by the LA, V_m in the RA was set to 0 mV before solving (2), thus muting the sources originating from the RA. Similarly, the LA was muted to compute the right atrial contribution to the P-wave. As the problem is linear, the principle of superposition holds. Hence, the regular P-wave composed of both left and right atrial sources could be obtained by adding the two separate results.

3   Results

Figure 1 shows the activation times resulting from the monodomain simulation. The last voxel was activated at 107 ms after initial stimulation in S1 and after 105 ms in S2. The corresponding ECGs (see Fig. 2) exhibit positive P-waves in leads I, II, and $-aVR$. Furthermore, P-waves in both subjects were monophasic

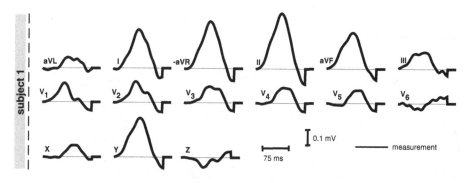

Fig. 3. Measured 12 lead ECG for S1. The VCG leads were computed using the inverse Dower matrix [2].

in II and biphasic in V_1 and the largest amplitude in the limb leads is found in II as expected. For S1, a measured reference ECG (see Fig. 3) was available and showed overall correspondence with the simulated P-waves. However, the polarity in aVL was positive and X was biphasic in the simulation as opposed to the measurement. The electrical axis of the atria α was calculated using the P-wave amplitudes in aVF and I:

$$\alpha = \arctan\left(\frac{2}{\sqrt{3}}\frac{aVF}{I}\right) \tag{3}$$

resulting in $\alpha = 73.22°$ in S1 and $\alpha = 73.27°$ in S2. Maximum P-wave duration (PWd) was 100 ms and 105 ms in S1 and S2. P-wave dispersion amounted to 9 ms and 10 ms, respectively. Left and right atrial contribution to the P-waves are shown in Fig. 2. Table 2 shows that the second third of the P-wave contributes more than 50 % to the integral of absolute values. The contribution of the first third compared to the last is lower in S1 (−9 %) and slightly higher in S2 (+2 %). The contribution of the RA was 72 % in S1 and 70 % in S2 leaving 28 % and 30 % for the LA, respectively. The LA mainly contributed during the second third with integral values being more than twice those of the final third. With

Table 2. Contribution of the two atria to the P-wave integral of absolute values in lead II during the entire P-wave as well as split by temporal thirds. Values were normalized to the integral value in II for each subject.

	S1	S2	S1	S2	S1	S2	S1	S2
both atria	100%	100%	18%	23%	55%	56%	27%	21%
only RA	72%	70%	15%	20%	37%	37%	20%	13%
only LA	28%	30%	3%	3%	17%	19%	8%	8%

the exception of X in S1 and V_1, V_2, V_3, and Z in S2, RA and LA P-waves had the same polarity.

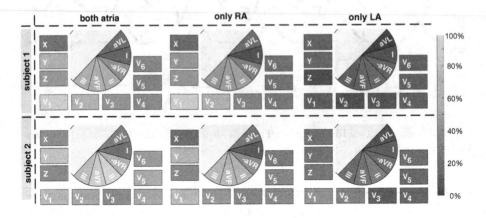

Fig. 4. Relative contributions to the simulated P-waves for both atria separately and combined. Color encodes the integral of absolute values which were normalized to the maximum for each subject (V_1 for S1, II for S2). Values for the RA and the LA add up to the value for both atria.

Figure 4 reveals that P-waves were strongest projected onto leads II, V_1, and Y. While the largest value was observed in V_1 in S1 (26.7 mVms) and II in S2 (9.8 mVms), the overall pattern was consistent across the two subjects — also when looking at RA and LA P-waves separately. In S2, the terminal part of the P-wave was clearly dominated by LA sources in I and V_3 which was not the case for S1 in any of the leads. The first third in most leads in both subjects as well as the entire P-wave in aVL in S1 was dominated by RA sources.

4   Discussion and Conclusion

In this study, we used a set of two realistic models to analyze the genesis of the P-wave with a focus on the distinct contributions of the LA and the RA and their temporal dynamics. Ion kinetics were computed using the Courtemanche model which was selected because of its suitability [32]. The simulated P-waves and the derived indices were within clinically observed ranges [22,31]. However, P-wave dispersion was smaller than in most clinical studies [22]. The time of the latest activation in S1 (107 ms) suggests that the PWd might actually be some milliseconds longer than the visually determined value of 100 ms. The non-zero signal in the PQ segment which we observed particularly in the precordial leads was described before both clinically [6] and in simulations [12]. This phenomenon can be attributed to regional differences in the action potential plateau and heterogeneity of early repolarization. The P-wave amplitude particularly in leads

V_1 and V_2 in S1 were larger in the simulation compared to the measurement. A possible reason for this phenomenon might be an overestimation of the extent of fast-conducting bundles in the RA by the rule-based algorithm used to annotate these anatomical structures [16]. The smaller P-wave amplitude in S2 can be attributed to smaller atria and most importantly to a thinner atrial wall (1.3 mm vs. 2.3 mm with no pronounced interatrial differences).

A limitation of this study is the small number of only two subjects. Thus, our results do not allow to draw universally applicable conclusions. However, basic underlying principles can be understood on the one hand. On the other hand, particularly our results regarding P-wave segments dominated by LA sources show that interindividual differences do play a role. This study used only anatomical models of healthy subjects and an electrophysiological model of healthy atrial substrate. A significant influence of pathologies on the results is unlikely but can not be ruled out. Thus, a sensitivity analysis regarding substrate modifications and other factors such as LA dilation could be addressed in a follow-up study.

Former studies investigated the P-wave *in silico*: Van Oosterom et al. showed that a double layer can serve as a source model for the P-wave [26]. Potse et al. highlighted the importance of considering anisotropic properties [29]. While Lu et al. were able to compute a reasonable P-wave using a generic homogeneous thorax model [20], other studies proved the importance of considering at least the blood and the lungs [5,14]. While all the aforementioned studies used the boundary element method, Krueger et al. conducted finite element computations to investigate P-wave alterations related to hemodialysis [17]. Several computational studies assessed P-waves during AF, however the authors are not aware of any work distinguishing between right and left atrial contribution.

Our results suggest that LA contribution to the P-wave is limited to less than 1/3. Furthermore, we did not see the clear temporal distinction attributing the last third of the P-wave almost exclusively to LA depolarization as described in textbooks [31]. While the LA contribution during the first third was very limited, LA activation translated mostly to P-wave contribution during the middle third rather than the terminal third. These results are in line with the sparse experimental data [18].

One of the most common uses of P-wave analysis regarding the LA is the diagnosis of LA enlargement. While some studies showed good correlation between P-wave markers and LA size [1,11], others showed poor correlation [9,13]. Our results underline the difficulties related to P-wave-based assessment of the LA [8,21] as the LA sources contributing to the P-wave do almost always interfere with RA sources and are even drowned out more often than not. Moreover, the results foster the understanding of P-wave genesis and its spatiotemporal projections.

References

1. Ariyarajah, V., Mercado, K., Apiyasawat, S., et al.: Correlation of left atrial size with P-wave duration in interatrial block. Chest **128**(4), 2615–2618 (2005)

2. Carlson, J., Havmoller, R., Herreros, A., et al.: Can orthogonal lead indicators of propensity to atrial fibrillation be accurately assessed from the 12-lead ECG? Europace **7**(2), 39–48 (2005)

3. Chirife, R., Feitosa, G.S., Frankl, W.S.: Electrocardiographic detection of left atrial enlargement. Correlation of P wave with left atrial dimension by echocardiography. Br. Heart J. **37**(12), 1281–1285 (1975)

4. Courtemanche, M., Ramirez, R.J., Nattel, S.: Ionic mechanisms underlying human atrial action potential properties: insights from a mathematical model. Am. J. Physiol. **275**, H301–321 (1998)

5. van Dam, P.M., van Oosterom, A.: Volume conductor effects involved in the genesis of the P wave. Europace **7**(S2), 30–38 (2005)

6. Debbas, N.M., Jackson, S.H., de Jonghe, D., et al.: Human atrial repolarization: effects of sinus rate, pacing and drugs on the surface electrocardiogram. J. Am. Coll. Cardiol. **33**(2), 358–365 (1999)

7. Ecabert, O., Peters, J., Schramm, H., et al.: Automatic model-based segmentation of the heart in CT images. IEEE Trans. Med. Imaging **27**(9), 1189–1201 (2008)

8. Hancock, E.W., Deal, B.J., Mirvis, D.M., et al.: AHA/ACCF/HRS recommendations for the standardization and interpretation of the electrocardiogram: part V. J. Am. Coll. Cardiol. **53**(11), 992–1002 (2009)

9. Hazen, M.S., Marwick, T.H., Underwood, D.A.: Diagnostic accuracy of the resting electrocardiogram in detection and estimation of left atrial enlargement: an echocardiographic correlation in 551 patients. Am. Heart. J. **122**(3 Pt 1), 823–828 (1991)

10. Ho, S.Y., Sanchez-Quintana, D., Cabrera, J.A., et al.: Anatomy of the left atrium: implications for radiofrequency ablation of atrial fibrillation. J. Cardiovasc. Electrophysiol. **10**(11), 1525–1533 (1999)

11. Hopkins, C.B., Barrett, O.J.: Electrocardiographic diagnosis of left atrial enlargement. Role of the P terminal force in lead V1. J. Electrocardiol. **22**(4), 359–363 (1989)

12. Ihara, Z., van Oosterom, A., Hoekema, R.: Atrial repolarization as observable during the PQ interval. J. Electrocardiol. **39**(3), 290–297 (2006)

13. Josephson, M.E., Kastor, J.A., Morganroth, J.: Electrocardiographic left atrial enlargement. Electrophysiologic, echocardiographic and hemodynamic correlates. Am. J. Cardiol. **39**(7), 967–971 (1977)

14. Keller, D.U.J., Weber, F.M., Seemann, G., et al.: Ranking the influence of tissue conductivities on ECGs. IEEE Trans. Biomed. Eng. **57**(7), 1568–1576 (2010)

15. Krueger, M.W., Dorn, A., Keller, D.U.J., et al.: In-silico modeling of atrial repolarization in normal and atrial fibrillation remodeled state. Med. Biol. Eng. Comput. **51**(10), 1105–1119 (2013)

16. Krueger, M.W., Seemann, G., Rhode, K., et al.: Personalization of atrial anatomy and electrophysiology as a basis for clinical modeling of radio-frequency ablation of atrial fibrillation. IEEE Trans. Med. Imaging **32**(1), 73–84 (2013)

17. Krueger, M.W., Severi, S., Rhode, K., et al.: Alterations of atrial electrophysiology related to hemodialysis session: insights from a multiscale computer model. J. Electrocardiol. **44**(2), 176–183 (2011)

18. Lemery, R., Birnie, D., Tang, A.S.L., et al.: Normal atrial activation and voltage during sinus rhythm in the human heart: an endocardial and epicardial mapping study in patients with a history of atrial fibrillation. J. Cardiovasc. Electrophysiol. **18**(4), 402–408 (2007)

19. Lipman, B.S.: Clinical scalar electrocardiography. Acad. Med. **40**, 815 (1965)

20. Lu, W., Zhu, X., Chen, W., et al.: A computer model based on real anatomy for electrophysiology study. Adv. Eng. Softw. **42**(7), 463–476 (2011)
21. de Luna, A.B., Platonov, P., Cosio, F.G., et al.: Interatrial blocks. A separate entity from left atrial enlargement. J. Electrocardiol. **45**, 445–451 (2012)
22. Magnani, J.W., Williamson, M.A., Ellinor, P.T., et al.: P wave indices: current status and future directions in epidemiology, clinical, and research applications. Circ. Arrhythm. Electrophysiol. **2**(1), 72–79 (2009)
23. Michelucci, A., Bagliani, G., Colella, A., et al.: P wave assessment: state of the art update. Card. Electrophysiol. Rev. **6**(3), 215–220 (2002)
24. Morris, J.J.J., Estes, E.H.J., Whalen, R.E., et al.: P-wave analysis in valvular heart disease. Circulation **29**, 242–252 (1964)
25. Ndrepepa, G., Zrenner, B., Deisenhofer, I., et al.: Relationship between surface electrocardiogram characteristics and endocardial activation sequence in patients with typical atrial flutter. Z. Kardiol. **89**(6), 527–537 (2000)
26. van Oosterom, A., Jacquemet, V.: Genesis of the P wave: atrial signals as generated by the equivalent double layer source model. Europace **7**(S2), 21–29 (2005)
27. Ozdemir, O., Soylu, M., Demir, A.D., et al.: P-wave durations as a predictor for atrial fibrillation development in patients with hypertrophic cardiomyopathy. Int. J. Cardiol. **94**(2–3), 163–166 (2004)
28. Platonov, P.G., Mitrofanova, L., Ivanov, V., et al.: Substrates for intra-atrial and interatrial conduction in the atrial septum. Heart Rhythm **5**(8), 1189–1195 (2008)
29. Potse, M., Dube, B., Vinet, A.: Cardiac anisotropy in boundary-element models for the electrocardiogram. Med. Biol. Eng. Comput. **47**(7), 719–729 (2009)
30. Seemann, G., Sachse, F.B., Karl, M., et al.: Framework for modular, flexible and efficient solving the cardiac bidomain equation using PETSc. Math. Ind. **15**, 363–369 (2010)
31. Wagner, G.S., Strauss, D.G.: Marriott's Practical Electrocardiography. Lippincott Williams & Wilkins, Philadelphia (2013)
32. Wilhelms, M., Hettmann, H., Maleckar, M.M.C., et al.: Benchmarking electrophysiological models of human atrial myocytes. Front. Physiol. **3**, 1–16 (2013)

Propagation of Myocardial Fibre Architecture Uncertainty on Electromechanical Model Parameter Estimation: A Case Study

Roch Molléro[1], Dominik Neumann[3,4], Marc-Michel Rohé[1], Manasi Datar[3],
Hervé Lombaert[1], Nicholas Ayache[1], Dorin Comaniciu[2], Olivier Ecabert[3],
Marcello Chinali[5], Gabriele Rinelli[5], Xavier Pennec[1], Maxime Sermesant[1(✉)],
and Tommaso Mansi[2]

[1] Inria, Asclepios Research Project, Sophia Antipolis, France
{roch-philippe.mollero,maxime.sermesant}@inria.fr
[2] Siemens Corporate Technology, Imaging and Computer Vision, Princeton, NJ, US
[3] Siemens Corporate Technology, Imaging and Computer Vision, Erlangen, Germany
[4] Pattern Recognition Lab, Friedrich-Alexander-Universität,
Erlangen-nürnberg, Germany
[5] Ospedale Pediatrico Bambino Gesù, Rome, Italy

Abstract. Computer models of the heart are of increasing interest for
clinical applications due to their discriminative and predictive power.
However the personalisation step to go from a generic model to a patient-
specific one is still a scientific challenge. In particular it is still difficult
to quantify the uncertainty on the estimated parameters and predicted
values. In this manuscript we present a new pipeline to evaluate the
impact of fibre uncertainty on the personalisation of an electromechanical model of the heart from ECG and medical images. We detail how
we estimated the variability of the fibre architecture among a given population and how the uncertainty generated by this variability impacts
the following personalisation. We first show the variability of the personalised simulations, with respect to the principal variations of the fibres.
Then discussed how the variations in this (small) healthy population of
fibres impact the parameters of the personalised simulations.

1 Introduction

Cardiac modelling aims at understanding cardiac diseases (such as heart failure,
dissynchrony or tachycardia), helping diagnosis and predicting cardiac response
to therapy (e.g. cardiac resynchronization therapy, or radiofrequency ablation).
In order to impact clinical practice, generic models have to be adjusted to a given
patient, which is personalisation [1–4]. This is still a challenging part, and often
computationally demanding, therefore most of the approaches are deterministic.
However there are several sources of uncertainty, both due to the data and the
models [5,6]. In this work we present the propagation of the uncertainty coming
from the lack of knowledge on cardiac fibres for a given patient. Indeed, it is still
difficult to obtain measurements on the fibre architecture for a given patient

© Springer International Publishing Switzerland 2015
H. van Assen et al. (Eds.): FIMH 2015, LNCS 9126, pp. 448–456, 2015.
DOI: 10.1007/978-3-319-20309-6_51

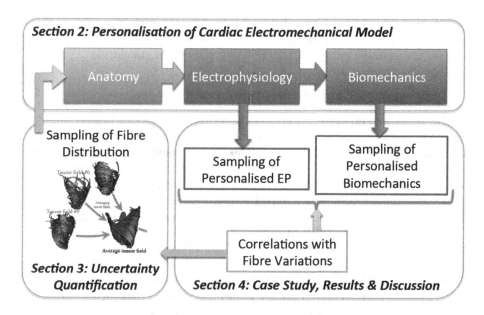

Fig. 1. Global scheme of fibre variability propagation along personalisation pipeline.

in-vivo, therefore we have to rely on prior knowledge. In order to propagate this uncertainty, it has first to be quantified. This was done by computing statistics on a small population of healthy hearts (details in Sect. 3). Then the personalisation pipeline has to be efficient enough so that a sampling of this uncertainty can be propagated. Finally we obtained a sampling of the distribution of the parameters and personalised simulations (see Fig. 1).

We illustrated this method on a paediatric dilated cardiomyopathy case (details in Sect. 4.1). From a clinical standpoint, it is very difficult to predict the dramatic evolution of such rapidly-evolving case, even with advanced imaging. The aim of the project is to test if parameters derived from biophysical models could help predicting the outcome of such cases.

2   Personalisation of Cardiac Electromechanical Model

2.1   Robust Segmentation of Myocardium from MRI

Patient-specific heart morphology is obtained from short-axis cine magnetic resonance images (MRI). To that end, a robust, data-driven machine learning approach is employed [7] to estimate surface meshes of the left endocardium, left outflow tract, left epicardium, right endocardium, right outflow tract and right inflow tract. Each surface is estimated using marginal space learning and probabilistic boosting trees, constrained by a shape model learned from a database of hundreds of cases, thus ensuring inter-patient point correspondence. Next, each surface is tracked over the entire cine sequence using a combination of tracking by detection and tracking by registration. Finally, the surface meshes at mid-diastole are selected to generate a closed surface of the biventricular myocardium, which is transformed into a tetrahedral volume mesh for simulation[1].

2.2 Personalised Cardiac Electrophysiology Model

Cardiac electrophysiology (EP) is modelled using the approach presented in [4]. Cardiac transmembrane potentials are calculated according to the mono-domain Mitchell-Schaeffer (MS) model as it offers a good compromise between model observability and fidelity. In this study, we are mostly interested in two parameters: the time during which the ion channels are closed τ_{close}, which captures action potential duration and is directly linked to the QT duration; and tissue diffusivity c, which determines the speed of the electrical wave propagation and is directly linked to the QRS duration. We model fast regional diffusivity for the left c_{LV} and right c_{RV} endocardium to mimic the fast conducting Purkinje network, and slower diffusivity $c_{\text{myo}} \leq c_{\text{LV}}, c_{\text{myo}} \leq c_{\text{RV}}$, for the myocardium. Transmembrane potentials are calculated using LBM-EP, a Lattice-Boltzmann method, which is coupled to a boundary element method approach to calculate the 12-lead cardiac electrocardiogram (ECG) resulting from the cardiac potentials [4]. The model is finally personalised like in [8,9]. BOBYQA, a constrained gradient-free optimization method is used to estimate tissue diffusivity and τ_{close} such that computed QRS duration, QRS electrical axis (EA) and QT duration match the measurements.

2.3 Personalised Cardiac Mechanical Model

The cardiac mechanical model is based on the Bestel-Clement-Sorine (BCS) model [10]. This model describes the heart as a Mooney Rivlin material, and model the stress along the cardiac fibres according to microscopic scale phenomena. Particularly, this model is compatible with the laws of thermodynamics and is able to model the Starling Effect. In this pipeline, it integrates a circulation model representing the 4 phases of the cardiac cycle (aortic pressure modelled by a 4-parameter Windkessel model), and takes the depolarization times and the action potential durations in each point of the mesh as an input to compute the mechanical contraction and relaxation of the myocardium.

As in [3], we only personalize the most influential and independent parameters which are the *maximal contraction* σ, the *viscosity coefficient* μ, the *Bulk Modulus K* and the *Aortic peripheral resistance Rp*. The calibration is performed following [3]: after performing 9 simulations using some specific parameter values that lie in a distribution of acceptable physiological values, the Unscented Transform Algorithm finds in one iteration the set of parameters that best fit the observations within this distribution. In our case, the observations are the minimal LV volume and the time between the two moments the LV is at 50 % of its contraction volume, both calculated from the cine MRI.

---

[1] http://www.cgal.org – computational geometry algorithms library.

3 Population-Based Uncertainty Quantification of Fibres

3.1 Variability Estimation in Atlas Space

One often characterize the variability of a random vector by its mean and covariance matrix since these two first moments completely characterize the Gaussian distribution. However, in more than a few dimensions, the covariance matrix is too large to be computed robustly from only a few data observations. An alternative is to draw just a few samples from the population distribution, either by choosing randomly a number of points from the data observations, or more rationally by selecting a few points that describe the main subspace of variation in the data, for instance through *Principal Component Analysis* (PCA). Within this subspace one could describe the variability using a minimal number of points thanks to the so-called sigma-points at the vertices of a minimal simplex, originally designed for the Unscented-Kalman Filter [11]. However, it is often empirical observed that using symmetric points on all axes is significantly more accurate for underlying symmetric distributions. This is the approach we took in this study to quantify the variability of the fibre architecture.

We used $N = 10$ ex-vivo DTI acquisitions of healthy human hearts, registered in the atlas space [12]. Both left and right ventricles images were generated with this atlas but due to the lower resolution of the right ventricle we chose to use this atlas only for the left ventricle part. On the right ventricle, we instead use a single DTI heart acquisition with high resolution done by Johns Hopkins University (JHU) [13]. Therefore we have no variability estimation of the fibres for the right ventricle.

To compute the mean DTI over the population and quantify the variability, we work in the Log-Euclidean space [14] rather than the standard Euclidean space. The mean DTI is $\bar{D} = \exp\left(\frac{1}{N}\sum_{i=1}^{N}\log(D^{(i)})\right)$ and the data matrix of centred observations is $X = [vect(\log D^{(1)} - \log \bar{D})\ldots vect(\log D^{(N)} - \log \bar{D})]$. The PCA is obtained by diagonalizing the large covariance matrix $\Sigma = XX^{\top}/(N-1)$ or more efficiently we chose to compute the singular-value decomposition (SVD) of the data matrix $X = U\Lambda V^{T}$, where the $N \times N$ diagonal matrix Λ encloses the square root of the eigenvalues of Σ. We choose to only study the first 3 eigenmodes $U_{i=1,2,3}$ which explain 59 % of the variation of the log-tensors seen in the population. Also we noted that the higher modes being increasingly affected by the noise of the DTI acquisition, they increasingly describe the variability due to the noise and not the *intrisic* variability. For each mode, we compute two symmetric images representing the range of variation along the mode at plus or minus one standard deviation as: $M_{i,\pm}(x) = \exp\left(\log(\bar{D}(x)) \pm \sigma_i U_i(x)\right)$.

3.2 From Atlas to Patient Space

In order to relate the atlas space to the geometry of our target patient, we register the mesh of our patient to the mask of both the atlas (for the LV) and the JHU heart (for the RV) with a three-steps framework. First, the mask of the patient is aligned with the mask using a rigid landmark based registration method. Correspondences between the atlas and the target heart are manually checked. Secondly, we perform a similarity registration with five coarse levels and one fine level, each of which are composed of 10 iterations. Finally, we perform a diffeomorphic registration using diffeomorphic demons algorithm with $15 \times 10 \times 5 \times 5$ iterations (from coarsest to finest levels), a Gaussian smoothing factor of 2 in the regularization phase, and an interpolation for the moving image done with B-splines [15]. We then get the full diffeomorphic transformation for each one of our two initial atlases to the target patient mask.

We apply the transformation found in the previous step to the mesh of the patient. For each of the vertices, if the correspondence lies within the RV we use the JHU DTI-image whereas we use the mean or the sampled images of the Lombaert atlas if it lies within the LV. We take the mean (in the Log-Euclidean space) of the tensors of the 5-nearest voxels. The tensor value is then reoriented using the Finite Strain method, and the fibre orientation is taken as its first eigenvector [16]. The results of the fibres personalisation are 7 sets of fibres shown in Fig. 2.

4   Propagation of Fibre Uncertainty on a Case Study

4.1   Clinical Background

The patient is a 16 years old male who had no family history of cardiac disease. After being admitted at the hospital for chest pain, evidence of *reduced ejection fraction* and *dilated left ventricle* led to a first diagnosis of myocarditis. A detailed echocardiographic examination performed 3 month later showed evidence of markedly increased trabeculae of the left ventricular apical and lateral walls, possibly suggesting the presence of *left ventricular non-compaction*. The MRI study did not confirm this diagnosis but only the *idiopathic dilated cardiomyomathy*. After 9 month of follow-up in the clinic, the patient was put on the national heart transplant list due to worsening conditions. The patient is now doing well at follow-up after transplant, and the pathology and histology testing at the hospital confirmed the diagnosis of *idiopathic dilated cardiomyopathy*.

4.2   Goodness of Fit and Variability After Personalization

For each of the 7 tested fibre architectures we personalised EP and EM parameters as described in Sect. 2.2. High goodness of fit between observations and simulations were achieved for all instances : for the ECG, the maximum obtained errors after personalization are 0.2 ms for QRS, 2.9 ms for QT and 0.3° for EA, which is well below 1 % of the measured values for QRS (96 ms) and QT (413 ms), and below 1 % of the maximum possible error (180°) for EA (5°), respectively. Similary in terms of mechanics, the error between simulated and measured minimal volume and the time at 50 % contraction are below 3 %.

Mean	Mode 1	Mode 2	Mode 3

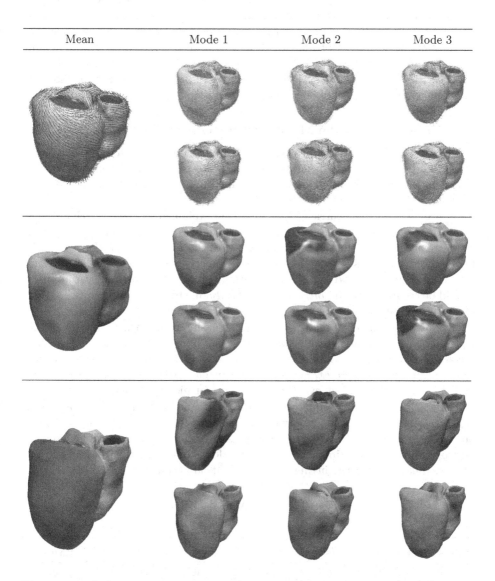

Fig. 2. Top left: Mean fibres. **Top right**: modes of variation plus (top) / minus (bottom) σ coloured by angular variation w.r.t. the mean (from $0°$ blue to $20°$ red). **Middle left**: Mean depolarisation times after EP personalization; from blue (early) to red (late$\approx$100 ms). **Middle right**: Variation from mean depolarisation times colouring from blue (-10 ms) to grey (0 ms) to red ($+10$ ms). **Bottom left**: Local strain at end-systole range from blue (high) to red (low). **Bottom right**: Variation for each different fibre modes after mechanical personalisation (blue: more contraction, red: less than on mean fibre) (Color figure online).

After this step, we can observe the spatial variability of EP depolarisation times and end-diastolic strain between modes in Fig. 2. Although the main features of the ECG are the same, variations in local depolarization times can be up to 10ms from one set of fibre to the mean fibre set due to the difference in current propagation. Interestingly we can notice for some of the sets a correlation between the peaks and zones of the variations of fibre orientation, with those of the depolarisation times, which would be interesting to investigate deeper.

4.3   Uncertainty on the Model Parameters and Discussion

Table 1 shows the values of the parameters after calibration for the mean fibre model, and their relative variation for each fibre set (Mx and Px are the two fibre sets representing the mode x as described in Sect. 3, for x = 1, 2 or 3).

For the EP parameters, we first note that c_{RV} varies the most although the fibres are fixed on this ventricle. This might be explained by the large changes in *direction of depolarisation* on the LV due to changes in fibre orientation, which would require the conductivity of the RV to vary as well to *match the same EA*. Logically, τ_{close} varies very little, since it is directly linked to the QT duration, that is not much affected by fibre orientation. Finally, the observed variabilities should be compared to the *intrinsic uncertainty* due to the parameter estimation process (quantified as high as 45 % for c_{RV} in some cases) [4].

About the mechanical parameters, we can easily explain the variations of the Rp and the σ. It's indeed well known that the fibre architecture has a strong influence on the stroke volume and when we fix all the parameters, we see that the ejection fraction is *maximal for the mean fibre*, with the *largest variations along the mode 2*. To achieve the same level of ejection fraction with a less efficient set of fibre, the peripheral resistance must be lowered and the maximum constraction increased, which is what we observe for all the modes, (and in a larger range for the mode 2). The variations of K and μ are more challenging to interpret directly. They impact directly the slopes of contraction and relaxation phases, thus ensuring the fitting of the time at 50 % contraction.

Table 1. Variability in estimated EP and EM model parameters after personalisation.

Parameter	c_{myo}	c_{LV}	c_{RV}	τ_{close}	σ	μ	K	Rp
Unit	mm^2/s	mm^2/s	mm^2/s	ms	Pa	Pa s	Pa	$Pa\,m^3\,s$
Mean	1.21e3	4.70e3	1.83e4	2.09e2	4.74e6	2.29e5	2.01e7	1.9e7
P1	−17.2 %	−14 %	−14.7 %	−0.48 %	+2.71 %	−5.84 %	−5.21 %	−26.7 %
M1	+11.5 %	+10.6 %	+6.06 %	+0.14 %	+4.2 %	−8.54 %	−11.4 %	−30.9 %
P2	−0.82 %	+2.34 %	+24.2 %	−0.14 %	+2.4 %	−20.1 %	+4.86 %	−54.6 %
M2	+3.28 %	+2.55 %	−30.9 %	−0.14 %	+0.86 %	−2.02 %	+10.6 %	−35.8 %
P3	+3.28 %	+7.02 %	−11.4 %	−0.14 %	+3.75 %	−2.87 %	−7.64 %	−34.4 %
M3	−0.82 %	−7.66 %	+9.06 %	−0.57 %	+0.75 %	−12.6 %	−0.98 %	−20.7 %

5   Conclusion

In this manuscript we detailed how a quantified uncertainty on myocardial fibres could be propagated along an efficient model personalisation pipeline. We presented the need to comprehensively quantify the influence of the parameters on the final output, and reversely to quantify their uncertainty when personalising models in order to fit clinical data. Atlases with mean and principal modes of variations are a good way to hierarchically represent the main directions of variability on quantities with many parameters such as vector or scalar fields. We used that method for the uncertainty on local fibre orientation in each point of the heart, and assesed the variations of personalized parameters according to those uncertainty. Interestingly, if we have prior knowledge on some parameters of the heart, this method could reciprocally give us information on the fibre set with the higest probability.

Finally, several aspects of this pipeline could be further improved for a more general assessment of the uncertainty, in particular with a more advanced personalisation from clinical data (evolution of regional volumes, the whole flow curves..) and an extension of the us of atlases to regional parameters such as conductivity or stiffness maps.

Ackowledgements. This work has been partially funded by the EU FP7-funded project MD-Paedigree (Grant Agreement 600932)

References

1. Xi, J., Lamata, P., Niederer, S., Land, S., Shi, W., Zhuang, X., Ourselin, S., Duckett, S.G., Shetty, A.K., Rinaldi, C.A., et al.: The estimation of patient-specific cardiac diastolic functions from clinical measurements. Med. Image Anal. **17**(2), 133–146 (2013)
2. Krishnamurthy, A., Villongco, C.T., Chuang, J., Frank, L.R., Nigam, V., Belezzuoli, E., Stark, P., Krummen, D.E., Narayan, S., Omens, J.H., et al.: Patient-specific models of cardiac biomechanics. J. Comput. Phys. **244**, 4–21 (2013)
3. Marchesseau, S., Delingette, H., Sermesant, M., Ayache, N.: Fast parameter calibration of a cardiac electromechanical model from medical images based on the unscented transform. Biomech. Model. Mechanobiol. **12**(4), 815–831 (2013)
4. Zettinig, O., Mansi, T., Neumann, D., Georgescu, B., Rapaka, S., Seegerer, P., Kayvanpour, E., Sedaghat-Hamedani, F., Amr, A., Haas, J., Steen, H., Katus, H., Meder, B., Navab, N., Kamen, A., Comaniciu, D.: Data-driven estimation of cardiac electrical diffusivity from 12-lead ECG signals. Med. Image Anal. **18**(8), 1361–1376 (2014)
5. Neumann, D., Mansi, T., Georgescu, B., Kamen, A., Kayvanpour, E., Amr, A., Sedaghat-Hamedani, F., Haas, J., Katus, H., Meder, B., Hornegger, J., Comaniciu, D.: Robust image-based estimation of cardiac tissue parameters and their uncertainty from noisy data. In: Golland, P., Hata, N., Barillot, C., Hornegger, J., Howe, R. (eds.) MICCAI 2014, Part II. LNCS, vol. 8674, pp. 9–16. Springer, Heidelberg (2014)

6. Konukoglu, E., Relan, J., Cilingir, U., Menze, B., Chinchapatnam, P., Jadidi, A., Cochet, H., Hocini, M., Delingette, H., Jaïs, P., Haïssaguerre, M., Ayache, N., Sermesant, M.: Efficient probabilistic model personalization integrating uncertainty on data and parameters: application to eikonal-diffusion models in cardiac electrophysiology. Prog. Biophys. Mol. Biol. **107**(1), 134–146 (2011)

7. Wang, Y., Georgescu, B., Chen, T., Wu, W., Wang, P., Lu, X., Lonasec, R., Zheng, Y., Comaniciu, D.: Learning-based detection and tracking in medical imaging: a probabilistic approach. In: Hidalgo, M.G., Torres, A.M., Gómez, J.V. (eds.) Deformation Models. LNVCB, pp. 209–235. Springer, Dordrecht (2013)

8. Neumann, D., Mansi, T., Grbic, S., Voigt, I., Georgescu, B., Kayvanpour, E., Amr, A., Sedaghat-Hamedani, F., Haas, J., Katus, H., et al.: Automatic image-to-model framework for patient-specific electromechanical modeling of the heart. In: 2014 IEEE 11th International Symposium on Biomedical Imaging (ISBI), pp. 935–938. IEEE (2014)

9. Seegerer, P., Mansi, T., Jolly, M.-P., Neumann, D., Georgescu, B., Kamen, A., Kayvanpour, E., Amr, A., Sedaghat-Hamedani, F., Haas, J., Katus, H., Meder, B., Comaniciu, D.: Estimation of regional electrical properties of the heart from 12-lead ECG and images. In: Camara, O., Mansi, T., Pop, M., Rhode, K., Sermesant, M., Young, A. (eds.) STACOM 2014. LNCS, vol. 8896, pp. 204–212. Springer, Heidelberg (2015)

10. Chapelle, D., Le Tallec, P., Moireau, P., Sorine, M.: Energy-preserving muscle tissue model: formulation and compatible discretizations. Int. J. Multiscale Comput. Eng. **10**(2), 189–211 (2012)

11. Julier, S.J., Uhlmann, J.K.: A new extension of the kalman filter to nonlinear systems. In: International Symposium on Aerospace/Defense Sensing, Simulation and Controls, Orlando, FL, vol. 3, pp. 182–193 (1997)

12. Lombaert, H., Peyrat, J.-M., Croisille, P., Rapacchi, S., Fanton, L., Clarysse, P., Delingette, H., Ayache, N.: Statistical analysis of the human cardiac fiber architecture from DT-MRI. In: Metaxas, D.N., Axel, L. (eds.) FIMH 2011. LNCS, vol. 6666, pp. 171–179. Springer, Heidelberg (2011)

13. Helm, P.A., Tseng, H.J., Younes, L., McVeigh, E.R., Winslow, R.L.: Ex vivo 3d diffusion tensor imaging and quantification of cardiac laminar structure. Magn. Reson. Med. **54**, 850–859 (2005)

14. Arsigny, V., Commowick, O., Ayache, N., Pennec, X.: A fast and log-Euclidean polyaffine framework for locally linear registration. J. Math. Imaging Vis. **33**(2), 222–238 (2009)

15. Vercauteren, T., Pennec, X., Perchant, A., Ayache, N.: Diffeomorphic demons: Efficient non-parametric image registration. NeuroImage **45**(1, Supp. 1), S61–S72 (2009)

16. Peyrat, J.M., Sermesant, M., Pennec, X., Delingette, H., Xu, C., McVeigh, E.R., Ayache, N.: A computational framework for the statistical analysis of cardiac diffusion tensors: application to a small database of canine hearts. IEEE Transa. Med. Imaging **26**(11), 1500–1514 (2007)

Issues in Modeling Cardiac Optical Mapping Measurements

Gwladys Ravon[1,2,3]([✉]), Yves Coudière[1,2,3], Angelo Iollo[1,3], Oliver Bernus[2], and Richard D. Walton[2]

[1] Inria Bordeaux Sud-Ouest, Bordeaux, France
gwladys.ravon@inria.fr
[2] LIRYC, L'Institut de Rythmologie et Modélisation Cardiaque, Université de Bordeaux, 33000 Bordeaux, France
[3] Institut de Mathématiques de Bordeaux, Bordeaux, France

Abstract. Optical mapping allows to visualize cardiac action potentials (AP) on cardiac tissue surfaces by fluorescence using voltage-sensitive dyes. So far, the surface measurements are directly related to surface AP. In a previous study was developed a method to reconstruct three-dimensional depolarization front: the main idea was to solve an inverse problem using the experimental measures on the surfaces. Although the method was very accurate on *in silico* data, it showed difficulties to recover real optical mapping measurements. Here we describe the different directions we followed to improve the results.

Keywords: Mathematical modeling · Optical mapping · Inverse problem

1 Introduction

Optical mapping is an important tool for the understanding of cardiac arrhythmias [1]. It provides surface optical recordings that are linked to surface AP [5]. Although the photons are known to interact with the tissue up to a few mm in depth, it remains very challenging to actually retrieve 3D information from optical recordings. In [2], Khait *et al.* derive a formula to determine the depth of some fixed electrical sources in a phantom. Instead, we propose to solve an inverse problem, so as to recover more accurate and complete information on the electrical sources. The approach is also expected to apply to more general experimental conditions. In a previous study we presented a first attempt to obtain a complete 3D reconstruction following this approach. Although we obtained excellent agreement between the actual location of the source and the location found by the inverse method, on *in silico* data, the results on the experimental data provided by the authors of [2] were disappointing. Actually, we observed a large mismatch between the experimental measures and the measures computed from the optical mapping model and the known experimental locations. Here we explore the model and its optical parameters as a cause of this mismatch.

© Springer International Publishing Switzerland 2015
H. van Assen et al. (Eds.): FIMH 2015, LNCS 9126, pp. 457–465, 2015.
DOI: 10.1007/978-3-319-20309-6_52

2   Optical Mapping

For optical mapping of AP, fluorescent voltage-sensitive dyes that attach to the cells' membrane are injected into a slab of tissue. The tissue is put in a bath, and cameras and lights are placed on both sides of the preparation (epicardium and endocardium, Fig. 1). A filter is placed in front of each camera, that allows to choose the wavelength to be recorded. The epi- and endocardium are alternatively illuminated. The dyes then emit a fluorescent light assumed to be proportional to both the incident light and the transmembrane potential (TMP). Optical images are the surface recordings of this light by the cameras [3]. Images are recorded on the illuminated surface (reflexion mode) and on the opposite one (transillumination mode). For each time step, optical mapping hence produces four images.

The medium has a natural fluorescence (called background F_0) recorded when the tissue is at rest. Fluxes captured during an AP are denoted by F. The signal due only to the AP itself is $F - F_0$. We shall rather use the usual renormalization:

$$g^\star = \overline{F_0} \, \max\left(0, \frac{F - F_0}{F_0}\right). \tag{1}$$

Indeed the max(...) amounts to ignore negative, physically irrelevant, optical signals (due to noise). The multiplication by the average $\overline{F_0}$ of the background signal is a way to retrieve the correct amplitude of the signal. *Our main goal is to reconstruct the 3D front of the AP from these 2D optical data.*

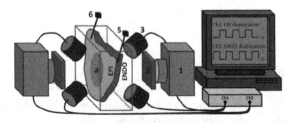

Fig. 1. The optical imaging setup: (1) CCD camera, (2) emission filter, (3) LED illumination, (4) tissue sample, (5) ECG electrode, (6) bipolar stimulating electrode.

3   Model

3.1   Forward Problem

In order to write the mathematical model of these observations, we assume the following: the cameras record photon fluxes through the surfaces, the light interacts with the tissue material in the diffusive regime [2], and a Robin boundary condition can be used to model the interaction between the tissue and its environment. Hence the illumination light is described by its photon density ϕ_0 that solves the diffusion equation

$$\begin{cases} -D_0\Delta\phi_0 + \mu_0\phi_0 = 0 & \text{in } \Omega, \\ \phi_0 + d_0\frac{\partial\phi_0}{\partial n} = 0 & \text{on } \partial\Omega\backslash\Gamma, \quad \text{and } \phi_0 = \frac{I_0\delta_0}{D_0} \text{ on } \Gamma, \end{cases} \tag{2}$$

where $\Omega \subset \mathbb{R}^3$ represents the slab of tissue, Γ is the illuminated surface, and n is the unit normal to $\partial\Omega$, outward of Ω. The fluorescent light is assumed to be proportional to the TMP and the illumination light (multiplicative factor $\beta > 0$). Its photon density solves:

$$\begin{cases} -D\Delta\phi + \mu\phi = \beta(V_m - V_0)\phi_0 & \text{in } \Omega, \\ \phi + d\frac{\partial\phi}{\partial n} = 0 & \text{on } \partial\Omega. \end{cases} \tag{3}$$

In both equations, the optical parameters D, μ, d stand respectively for diffusion coefficient, absorption coefficient and extrapolation distance. The attenuation length is the parameter δ_0 defined by $\delta_0 = \sqrt{\frac{D_0}{\mu_0}}$. The intensity of the illumination, assumed uniform, is the parameter I_0. The multiplicative factor $\beta > 0$ is known for the dye used during the experiments. The dyes are assumed to be uniformly distributed in the tissue. Finally the fluxes measured through the surfaces are given by Fick's law:

$$g = -D\frac{\partial\phi}{\partial n} \quad \text{on the epi or endocardium.} \tag{4}$$

Remark that the experimental flux g^* given by (1) does not satisfy Eq. (3), because of the renormalization. The quantity $F - F_0$ does. However we shall consider g^* as a good approximation of g, following the recommendations of the experimenters.

Since we consider a rectangular slab of tissue, we may have used structured meshes. We choose to work with unstructured meshes in order to allow more general geometries. This is necessary to study data from heart tissues. The diffusion equations are solved with P1-Lagrange finite elements method using the solver *FreeFem++* [4].

3.2   Inverse Problem

The problem of retrieving the 3D spatial distribution of the TMP, denoted by $V_m(t, \mathbf{x})$ from the 2D optical signals at time $t > 0$ is under-determined. Hereafter, $\mathbf{x} = (x, y, z)$ denotes a point in Ω with Cartesian coordinates (x, y, z). Instead of finding the complete distribution $V_m(t, \mathbf{x})$, we look for a depolarization front at each time. Specifically, we assume that a surface $\mathcal{S}(t) = \{\mathbf{x} \in \Omega : f(t, \mathbf{x}) = 0\}$ defined as the level 0 of the function f splits the domain Ω into the region $\Omega_r = \{\mathbf{x} \in \Omega : f(t, \mathbf{x}) > 0\}$ of tissue at rest, and the region $\Omega_p = \{\mathbf{x} \in \Omega : f(t, \mathbf{x}) < 0\}$ of excited tissue. It follows that $V_m(t, \mathbf{x}) = V_p$ if $\mathbf{x} \in \Omega_p$, and $V_m(t, \mathbf{x}) = V_0$ if $\mathbf{x} \in \Omega_r$. We consider simple depolarization fronts $\mathcal{S}$ modeled

- *either by the sphere* centered in $\mathbf{x}_0 \in \Omega$ and expanding with the velocity $c > 0$ after the given time $t = t_0 \geq 0$, defined by the level-set function $f(t, \mathbf{x}) = |\mathbf{x} - \mathbf{x}_0| - c(t - t_0)$,

– *or by the fixed ellipsoid* centered in $\mathbf{x}_0 \in \Omega$ and with radiuses r_x, r_y, $r_z > 0$, defined by the level-set function $f(t, \mathbf{x}) = \frac{(x-x_0)^2}{r_x^2} + \frac{(y-y_0)^2}{r_y^2} + \frac{(z-z_0)^2}{r_z^2} - 1$.

This level-set approach generalizes to more complex AP, once these simple cases are completely understood. In both cases, the inverse problem reduces to the identification of a small parameters set $\mathcal{P} = (\mathbf{x}, c, t_0) \subset \mathbb{R}^5$ (sphere) or $\mathcal{P} = (\mathbf{x}, r_x, r_y, r_z) \subset \mathbb{R}^6$ (ellipsoid). In order to identify these parameters, we minimize the least squares difference $e(\mathcal{P})$ between the actual measurements and the measurements computed from Eqs. (2)–(4) with a TMP as above:

$$e(\mathcal{P}) = \frac{1}{2} \sum_{i=1}^{4} \|g_{\mathcal{P}}^i - g^{\star,i}\|_{L^2(\mathbb{S}_i)}^2, \tag{5}$$

where the functions $g^{\star,i}$ are the data. Here i refers to one of the four images ($i \in \{1, 2, 3, 4\}$), and the surface $\mathbb{S}_i$ is either the epicardium or the endocardium, as detailed in Table 1. Although this is the natural way to define the cost function, the value I_0 of the illumination in Eq. (2) is unknown, while the optical parameters are. Consequently, and since Eqs. (2)–(4) are linear, the density ϕ_0, or ϕ, can only be computed up to a multiplicative constant. The mapping $I_0 \mapsto g^i$ is also linear, the measurement in-silico g^i is consequently proportional to I_0, and we can change the cost function to account for this unknown value. A first idea is to identify the intensity I_0, and consider the following modified cost function:

Table 1. References of the measures

#	Illuminated surface	Measured surface
1	Epicardium	$\mathbb{S}_1 = $ epicardium
2	Epicardium	$\mathbb{S}_2 = $ endocardium
3	Endocardium	$\mathbb{S}_3 = $ endocardium
4	Endocardium	$\mathbb{S}_4 = $ epicardium

$$e(\mathcal{P}) = \frac{1}{2} \sum_{i=1}^{2} \sum_{j=1}^{2} \|I_0^i g_{\mathcal{P}}^{ij} - g^{\star,ij}\|_{L^2(\mathbb{S}_j)}^2, \tag{6}$$

where $i, j \in \{1, 2\}$, i stands for the illuminated surface while j stands for the measured surface ($i = j$ for the reflexion mode, and $i \neq j$ for the transillumination mode). Since the intensities are different, we have two additional parameters to retrieve, I_0^1 and I_0^2. A second idea consists in comparing the normalized fluxes:

$$e(\mathcal{P}) = \frac{1}{2} \sum_{i=1}^{4} \left\| \frac{g^i}{\|g^i\|} - \frac{g^{\star,i}}{\|g^{\star,i}\|} \right\|_{L^2(\mathbb{S}_i)}^2, \tag{7}$$

where the inner norms $\|\cdot\|$ are also L^2 norms on the surface $\mathbb{S}_i$. In this case there is no additional parameter to be identified, but the problem becomes nonlinear with respect to V_m.

In all cases, a fixed-step gradient method followed by the BFGS algorithm is used to solve the inverse problem. We need the gradient of all the cost functions e with respect to the unknown parameters $\mathcal{P}$, which is computed with the adjoint method.

4   Summary of the Results and Issues

4.1   Summary of Previous Results

In this part we quickly recall the first results we obtained. We first compared the formula derived in [2] and our method for the expanding sphere and the cost function (5) on data *in silico*.

Figure 2 shows the depth of the source computed from Khait's formula (diamonds) and from our method (squares) as a function of time. Results are presented for inclusions at four different locations, excitation time $t_0 = 0$ and velocity $c = 0.5\,\text{ms}^{-1}$. The last example was carried out on a cylinder, in order to illustrate the case of a complex geometry. For all cases we retrieved the complete unknown location $\mathbf{x}_0$ of the source, at any time, and even after the breakthrough, with an accuracy up to machine precision. Additionally, the velocity c could be recovered from any time-sequence of data. These were very good results, based on data *in-silico*.

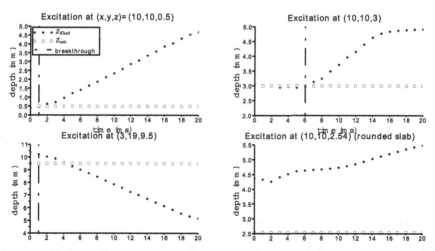

Fig. 2. Comparison between with Khait's formula (diamonds) and our method (squares). The vertical lines mark the breakthrough of the wave on the observed surface.

The method was then tested on the experimental data from the optical phantom experiments set up by Khait and coworkers [2]. In this case, fixed ellipsoidal sources were considered and the intensity I_0 of the illumination was unknown. We chose to work with the normalized cost function (7), so that we had to identify six parameters.

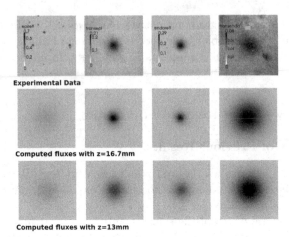

Experimental Data

Computed fluxes with z=16.7mm

Computed fluxes with z=13mm

Fig. 3. Results for one set of experimental data. Domain size: $40 \times 40 \times 20$ mm. Columns 1 and 2: records on the epicardium. Columns 3 and 4: records on the endocardium.

Figure 3 shows the results obtained for a phantom located at a depth $z_0 = 13$ mm. Although the reconstruction of the photon fluxes looks qualitatively correct (first and second rows of images), the reconstructed depth is $z^* = 16.7$ mm.

In order to understand this large error (3 mm over a total depth of 20 mm), we computed the theoretical observations associated to the exact experimental location of phantom source: from $\mathcal{P}$ known, we compute the TMP distribution $V_m(\mathbf{x})$, solve Eqs. (3) and (4), and finally compute the observation g with Eq. (4). These normalized fluxes (third row of images on Fig. 3) clearly have different amplitudes than the experimental signal. These fluxes are also more diffuse than the experimental ones. We observe the same behavior with seven other experimental phantoms. We tried to replace the measure g^* (Eq. (1)) by the simpler difference $\max(F - F_0, 0)$ and obtained similar errors on the location. *We deduce from these results that the optical model (2) and (3) is questionable. In the next section, we identify and study in depth several possible sources of error.*

4.2   Model Improvements to Fix the Mismatch

Uniform Illumination. First, we addressed the approximation of uniform distribution of the illumination I_0. We tested the effect of several spatially distributed illumination intensities $I_0(x, y)$: constant, narrow Gaussian, diffuse Gaussian, supposed to mimic the experimental lights. They read

$$I_0(x, y) = A \exp \left[- \left(\frac{(x - x_0)^2}{\sigma_x^2} + \frac{(y - y_0)^2}{\sigma_y^2} + \frac{(x - x_0)(y - y_0)}{2\sigma_{xy}} \right) \right]. \quad (8)$$

We observed the same fluxes g for the three cases. Indeed, the source term $(V_m(\mathbf{x}) - V_0)\phi_0(\mathbf{x})$ in the fluorescence equation is the same in all cases because

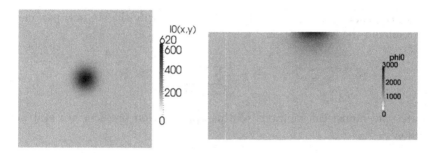

Fig. 4. Narrow Gaussian example for an epi-illumination. Domain size: $40 \times 40 \times 20$ mm. On the left: spatial distribution of I_0, on the right: photon density ϕ_0, cut plane in z-direction.

the photon density ϕ_0 becomes quickly constant inside the tissue (see Fig. 4). Finally it is not a limitation to consider an uniform illumination.

Optical Parameters. Another possibility is that the values of the optical parameters are not correct. In order to understand the role of these parameters, we derive a dimensionless version of the equations. Consider a space scale $L > 0$ and let rescale $\mathbf{x}$ as : $\mathbf{x}' = \frac{\mathbf{x}}{L}$. Equations (2) and (3) rewritten

$$\begin{cases} -\frac{D_0}{L^2}\Delta\phi_0 + \mu_0\phi_0 = 0 & \text{in } \Omega, \\ \phi_0 + \frac{d_0}{L}\frac{\partial\phi_0}{\partial n} = 0 & \text{on } \partial\Omega\backslash\Gamma, \\ \phi_0 = \frac{I_0\delta_0}{D_0} & \text{on } \Gamma, \end{cases} \qquad \begin{cases} -\frac{D}{L^2}\Delta\phi + \mu\phi = \beta(V_m - V_0)\phi_0 & \text{in } \Omega, \\ \phi + \frac{d}{L}\frac{\partial\phi}{\partial n} = 0 & \text{on } \partial\Omega. \end{cases}$$

We define $\gamma = \frac{I_0\delta_0}{D_0}$ and rescale the density as $\phi_0' = \frac{\phi_0}{\gamma}$ and $\phi' = \frac{\phi}{\gamma}$, in such a way that:

$$\begin{cases} -\frac{D_0}{L^2}\Delta\phi_0' + \mu_0\phi_0' = 0 & \text{in } \Omega, \\ \phi_0' + \frac{d_0}{L}\frac{\partial\phi_0'}{\partial n} = 0 & \text{on } \partial\Omega\backslash\Gamma, \\ \phi_0' = 1 & \text{on } \Gamma, \end{cases} \qquad \begin{cases} -\frac{D}{L^2}\Delta\phi' + \mu\phi' = \beta(V_m - V_0)\phi_0' & \text{in } \Omega, \\ \phi' + \frac{d}{L}\frac{\partial\phi'}{\partial n} = 0 & \text{on } \partial\Omega. \end{cases}$$

Dividing the diffusion equation by μ_0 (resp. μ) the dimensionless system reads:

$$\begin{cases} -\overline{\delta_0}^2\Delta\phi_0' + \phi_0' = 0 & \text{in } \Omega, \\ \phi_0' + \overline{d_0}\frac{\partial\phi_0'}{\partial n} = 0 & \text{on } \partial\Omega\backslash\Gamma, \\ \phi_0' = 1 & \text{on } \Gamma, \end{cases} \qquad \begin{cases} -\overline{\delta}^2\Delta\phi' + \phi' = (V_m - V_0)\phi_0' & \text{in } \Omega, \\ \phi' + \overline{d}\frac{\partial\phi'}{\partial n} = 0 & \text{on } \partial\Omega. \end{cases}$$

where $\overline{\delta_0}^2 = \frac{D_0}{\mu_0 L^2}$, $\overline{d_0} = \frac{d_0}{L}$, $\overline{\delta}^2 = \frac{D}{\mu L^2}$ and $\overline{d} = \frac{d}{L}$ are dimensionless optical parameters. Finally the fluxes are given by $g' = \phi'$ on the epi or endocardium. The other parameters, I_0 and β are hidden in a dimensionless number γ and we compare the experimental fluxes $g^\star$ to $\gamma g'$. The optical system is characterized by the four optical parameters. Adding the two terms γ (one for each illumination), we end up with six parameters. We tried to solve a second inverse problem:

knowing the characteristics $\mathcal{P}$ of the inclusion, identify the six new parameters by minimizing the cost function

$$e_1(\overline{\delta_0}^2, \overline{d_0}, \overline{\delta}^2, \overline{d}, \gamma^1, \gamma^2) = \frac{1}{2} \sum_{i=1}^{2} \sum_{j=1}^{2} \|\gamma^i g'^{ij} - g^{\star,ij}\|_{L^2(\mathbb{S}_j)}^2. \qquad (9)$$

To date, the numerical solutions to this optimization problem are still being computed.

Other Possibilities. In our model we do not consider the distance between the preparation and the camera. We could consider the diffusion of the photon density in the air by ensuring the continuity of the fluxes at the border medium/air. Instead, we impose a Robin condition on all the surfaces. We consider that the fluxes are recorded directly through the tissue surfaces, and not through the Plexiglas, because the Plexiglas has a negligible absorption coefficient.

5   Discussion

Modeling the optical measurements by the diffusion Eqs. (2) and (3) is widely used in cardiac optical mapping. But when we confront the measurements obtained with this model to the experimental ones in a well controlled setup, we observe an important mismatch (Fig. 3). We studied several ideas that might explain the differences, always with a negative result. To our opinion, this is likely to suggest that the diffusive regime is a too coarse approximation of the interaction between light and matter in the cardiac context. If this is confirmed by further experiments, the complete radiative transfer equation (RTE) might be used to model the measurements.

6   Conclusion

The aim of this study was to detail our investigations concerning the current model of cardiac optical mapping measurements. Having recalled previous results we described our attempts to improve our mathematical model. The most likely assumption was that the illumination on the tissue was not constant. Few tests showed that it was not the key. We eliminated several other sources of error, but there remains some more. We keep on working on the dimensionless problem in order to identify its parameters. Otherwise the RTE might be used.

Acknowledgments. This work is partially supported by the grant number ANR-10-IAHU-04 from the French government, and by the *Conseil Régional Aquitaine*.

References

1. Rosenbaum, D.S., Jalife, J.: Optical Mapping of Cardiac Excitation and Arrhythmias. Wiley-Blackwell, Chichester (2001)
2. Khait, V.D., Bernus, O., Mironov, S.F., Pertsov, A.M.: Method for 3-dimensional localization of intramyocardial excitation centers using optical imaging. J. Biomed. Opt. **11**, 34007 (2006)
3. Walton, R.D., Lawrence Xavier, C.D., Tachtsidis, I., Bernus, O.: Experimental validation of alternating transillumination for imaging intramural wave propagation. Conf. Proc. IEEE Eng. Med. Biol. Soc. **2011**, 1676–1679 (2011)
4. Hecht, F.: New development in FreeFem++. J. Numer. Math. **20**, 251–265 (2012)
5. Bishop, M.J., Rodriguez, B., Eason, J., Whiteley, J.P., Trayanova, N., Gavaghan, D.J.: Synthesis of voltage-sensitive optical signals: application to panoramic optical mapping. Biophys. J. **90**(8), 2938–2945 (2006)

Data-Driven Model Reduction for Fast, High Fidelity Atrial Electrophysiology Computations

Huanhuan Yang, Tiziano Passerini[✉], Tommaso Mansi, and Dorin Comaniciu

Siemens Corporation, Corporate Technology, Imaging and Computer Vision,
Princeton, NJ, USA
tiziano.passerini@siemens.com

Abstract. Understanding and predicting atrial electrophysiology, for diagnosis and therapy planning purposes, calls for methods able to accurately represent the complex patterns of atrial electrical activity, and to produce very fast predictions to be suitable for use in the clinical practice. We apply a data-driven approach for the model reduction of an atrial cellular model. The reduced model predicts cellular action potentials (AP) in a simple form but is effective in capturing the physiological complexity of the original model. The model construction starts from an AP manifold learning which reduces the AP manifold dimension to 15, and continues with a regression model learning to predict the 15 components in the reduced AP manifold. The regression model has the potential to drastically improve the performance of atrial tissue-level electrophysiology (EP) modeling, enabling a 75 % reduction of the computational cost with the same time step and up to two order of magnitudes smaller computational time with larger time steps. The model is also capable of describing the restitution properties of the AP, as demonstrated in tests with varying diastolic intervals. This model has great potential use for real-time personalized atrial EP modeling, and the same modeling technique can be extended to the study of other excitable myocardial tissues.

1 Introduction

Computational modeling of healthy and diseased atrial electrophysiology (EP) has a great potential for use in clinical practice. It can provide non-invasive, cost effective and personalized assessment of the state of the atria; furthermore, it can support the planning and guidance of atrial therapies (such as the ablation therapy for atrial fibrillation) by predicting the response of the patient. To make them suitable for use in the clinic, EP models ought to be (1) computationally efficient, (2) reliable in capturing detailed cardiac biology, and (3) easy to be personalized, directly or statistically, from clinical data.

Recent EP models are capable of describing more and more complex cellular mechanisms, which are crucial for the detailed description of the organ electrical activity. However, these models are computationally demanding due to the many and coupled algebraic and ordinary differential equations accounting for different

© Springer International Publishing Switzerland 2015
H. van Assen et al. (Eds.): FIMH 2015, LNCS 9126, pp. 466–474, 2015.
DOI: 10.1007/978-3-319-20309-6_53

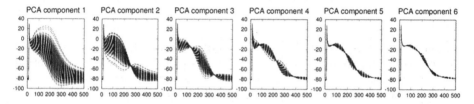

Fig. 1. Dimensional reduction of the AP manifold generated by the CRN model. The first six PCA modes are plotted.

ionic channels and gating variables. As an example, the Courtemanche-Ramirez-Nattel (CRN) human atrial cell model [4] features 35 static parameters and 21 ordinary differential equations to describe 12 ionic channels, the corresponding gating variables and ionic concentrations. Another challenge is the availability of robust and efficient methods to personalize EP models, especially ones featuring a large number of parameters. Several simplified phenomenological models have been proposed as computationally efficient surrogates, such as the FitzHugh-Nagumo (FHN) model [5]. However, these models usually lack the capability of describing important physiological properties. Moreover, to the best of our knowledge no simplified model is currently available for human atria-specific cellular EP.

The topic of reduced-order modeling for cardiac electrophysiology has been explored in the literature. Most relevant to this work, an approach based on a Galerkin method for the solution of the bidomain equations, combined with proper orthogonal decomposition, has recently been proposed in [3]. Motivated by recent progresses in meta-modeling [9], we apply a data-driven approach to the reduction of state-of-the-art cellular models used for atria simulation in literature. The reduced model learned by regression keeps the ability to capture the complex dynamics of the original biophysically detailed model, while in simple form and depending on a smaller number of parameters. This makes the model efficient and suitable for use for large scale simulations at the organ level. To the best of our knowledge, this represents the first example of application of a model reduction technique based on statistical learning to the multiscale modeling of cardiac EP. In this work, we focus on the CRN atrial cell model. We first use the principal component analysis (PCA) to reduce the dimensionality of the AP manifold (Sect. 2.1). We then learn a regression model of AP dynamics at the cell level, given a set of CRN model parameters (Sect. 2.2). Finally we use this reduced cellular model for tissue-level EP modeling (Sect. 2.3). As reported in Sect. 3, AP manifold dimension can be reduced to 15 despite being the output of a non-linear system. Our regression model demonstrates the ability of capturing the physiological complexity of cardiac AP, and it drastically improves the performance of atrial tissue-level electrophysiology modeling by achieving a 75 % reduction of the computational cost with the same computational time step and two order of magnitudes smaller computational time with larger time steps (in the order of seconds to compute one heart cycle of electrical activation in a patient-specific atrial anatomy, using a regular workstation).

2   Methods

Although the methods described in the following do not depend on the specific choice of the cellular model, we focus here on a model suitable for the description of atrial electrophysiology. The CRN cell model was developed based on human atrial cell data and has been validated for use in both tissue [2] and organ [1] level simulations. The dominating equation is

$$\frac{dv}{dt} = -\frac{I_{ion} + I_{stim}}{C_m},$$

(1)

where I_{ion} is the total current flowing through 12 ion channels, I_{stim} is a transient stimulus current added to simulate electrical pacing, and C_m is the membrane capacitance per unit volume.

For modeling electrophysiology at the tissue level, we resort to the monodomain model coupled with the CRN cell model. The model equation has the form

$$\frac{\partial v}{\partial t} = \frac{1}{C_m} \nabla \cdot D \nabla v - \frac{I_{ion} + I_{stim}}{C_m},$$

(2)

where v is the transmembrane potential, I_{ion} is the ionic current from the coupled cell model, and D is the anisotropic diffusion tensor.

2.1   AP Manifold Learning for Dimensionality Reduction

The solution of Eq. (1) is the time-dependent transmembrane action potential (AP) $v(t)$. We use manifold learning techniques to reduce the dimensionality of the manifold Ω_{AP} to which $v(t)$ belongs. That is, we analyze the number of intrinsic parameters q that are necessary to capture the observed AP data $v(t)$. Let n be the number of observations. For each observation i, we choose a unique set of model parameters $\boldsymbol{\theta}^i \in \mathbb{R}^{1 \times p}$ and compute the AP in m time snapshots. The results are gathered in the observation vector $\mathbf{v}^i = [v^i(t_1), \cdots, v^i(t_m)] \in \mathbb{R}^{1 \times m}$. The $n \times m$ observation matrix $\mathbf{Y} = [(\mathbf{v}^1)^T, \cdots, (\mathbf{v}^n)^T]^T$ represents a sampling of the AP manifold. Before dimension reduction, the AP matrix is zscored. To uncover the intrinsic structure of the AP manifold, we adopt the principal component analysis (PCA) [7].

2.2   Regression Model

We then learn a regression model of action potential dynamics given a sample space of the CRN model parameters $\boldsymbol{\theta} \in \mathbb{R}^{1 \times p}$ and their corresponding AP outputs $\mathbf{v} \in \mathbb{R}^{1 \times m}$.

Model Input. The CRN model parameters $\boldsymbol{\theta}$ are the regression inputs.

Model Output. Instead of the time-series of action potential $v(t)$, we consider the reduced AP representation computed in Sect. 2.1 as output: the regression model predicts the q embedding coordinates $\mathbf{v}_{emb}$, which are then used to reconstruct the AP time frames $\mathbf{v}$ with the basis provided by PCA.

Model Construction. The data-driven model writes $\mathbf{v}_{emb} = \mathbf{f}(\boldsymbol{\theta})$, $\mathbf{f} : \mathbb{R}^p \to \mathbb{R}^q$. Rather than directly regressing $\mathbf{v}_{emb}$ from $\boldsymbol{\theta}$, we propose a two-step model construction process based on statistical learning. We first regress some phenomenological features of the AP, and then use them as additional inputs for the second regression step to increase the accuracy of the overall prediction. The rationale behind this choice is that many phenomenological features of AP can be regressed with high accuracy from $\boldsymbol{\theta}$, as shown in Sect. 3, and adding them as features helps further constraining the second regression problem.

More precisely, we characterize the action potential by different properties: peak voltage (V_{peak}), resting membrane potential (V_{rest}), action potential duration (APD). We learn a prediction model for these quantities $\mathbf{f}_1 : \boldsymbol{\theta} \mapsto [V_{peak}, V_{rest}, APD]$ using the projection pursuit regression (PPR) method [6]. In this work, we use the R function ppr. The second step of the full statistical learning procedure is to regress the embedding coordinates $\mathbf{v}_{emb}$ by including predicted values of those AP quantities as inputs together with the CRN model parameters $\boldsymbol{\theta}$. This step estimates a model $\mathbf{f}_2 : [\boldsymbol{\theta}, \mathbf{f}_1(\boldsymbol{\theta})] \mapsto \mathbf{v}_{emb}$ using the PPR method as in the first step.

AP Prediction. Given a new set of parameters $\hat{\boldsymbol{\theta}}$, the AP prediction is as follows

$$\hat{\boldsymbol{\theta}} \to \hat{\mathbf{v}}_{emb} = \mathbf{f}_2(\hat{\boldsymbol{\theta}}, \mathbf{f}_1(\hat{\boldsymbol{\theta}})) \to \hat{\mathbf{v}} \text{ reconstruction from } \hat{\mathbf{v}}_{emb}.$$

2.3   Application to Tissue-Level Atrial EP Modeling

We present a framework for efficient patient-specific simulation of cardiac transmembrane potential.

Computational Domain Preparation from Medical Images. Starting from atria images (e.g., from CT), the left and right atrium are automatically segmented using a machine learning approach [15]. Regional atrial wall thickness can be extracted from high resolution images, but this may be challenging to obtain for the whole atria. We follow a simpler approach, applying a uniform mesh thickening based on thresholding a level-set representation of the atrial surface on an isotropic Cartesian grid. Grid nodes lying at the sino-atrial node region are manually annotated. If available, information on the fibers orientation can be readily included in the computational domain, as described in [14] for the case of ventricular tissue. In this work we do not model the presence of fibers in the atrial tissue. Since tissue anisotropy plays a role in the monodomain equation, but not in the cell model, the proposed model reduction strategy is not affected by this choice.

Lattice-Boltzmann Method for Cardiac EP. We solve the monodomain tissue model on a Cartesian grid by applying the Lattice-Boltzmann method introduced in [11] to the discretization of the model equation. The method uses a 7-connectivity topology (6 connections + central position) and Neumann boundary conditions.

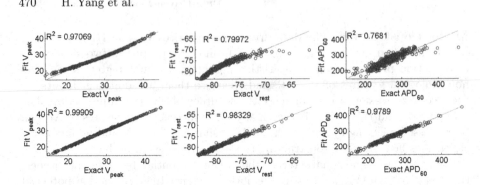

Fig. 2. Regression on V_{peak}, V_{rest}, and APD_{60} by two methods: PLSR (row 1), PPR (row 2)

Registration of the Regression Cellular Model. The regression model we designed is discrete. Given a set of model parameters, the regression model predicts the full time sequence $\mathbf{v}_{ref}$ of AP, in m time snapshots (as for the observations in the training data set). To obtain the potential value at each time in the heart cycle, we align $\mathbf{v}_{ref}$ at the time $t_{upstroke}$ of AP upstroke or depolarization; then we extract the proper snapshot from $\mathbf{v}_{ref}$. To monitor $t_{upstroke}$ in each cell, we propose to use the simple Mitchell-Schaeffer (MS) model [10] (having only one gating variable). For the sake of clarity, we assume that the time step dt used by the monodomain solver is a multiple of the time step used to sample $\mathbf{v}_{ref}$. After the upstroke, at time $t^i = t_{upstroke} + i \cdot dt$, the last term in Eq. (2) is computed as follows

$$-\frac{I_{ion}+I_{stim}}{C_m}\left(\mathbf{x}, t^i\right) = \frac{\mathbf{v}_{ref}(i)-v(\mathbf{x}, t^{i-1})}{dt} \ ,$$

with $i \in \mathbb{N}$ corresponding to the selected time snapshot of $\mathbf{v}_{ref}$. After AP repolarization, we switch to use the MS model to monitor the next heart beat.

3   Experiments and Results

Sampling. In this work, we considered as model input a set of 12 parameters with each controlling an ionic channel in the CRN model. The chosen parameters represent maximal ionic conductances and maximal channel currents. A sample was generated by scaling the baseline values listed in [4], and scaling factors were chosen randomly from a log-normal distribution with mean value 1 and standard deviation 0.3. In non-reported experiments we verified that using these 12 parameters rather than the entire set of 35 parameters in the original model can capture a large variety of AP patterns, comparable with the output of the original model. Different samples can be obtained with alternative choices of the parameters to be used as model input. For instance, the parameters selection could be guided by a sensitivity analysis of relevant features of the cell model (e.g., the APD). This would allow the definition of different samples for different applications. In an extreme example such as the study of atrial fibrillation, characterized by a significantly shortened APD, a specialized sample could be constructed by perturbing the parameters after proper adjustment of the reference values, as shown in [8].

For the generation of the database, an electrical stimulus was applied with amplitude $-20\,\mathrm{pA/pF}$ starting at $t = 10\,\mathrm{ms}$ and lasting for $1\,\mathrm{ms}$. In the first $500\,\mathrm{ms}$, $v(t)$ was recorded in 1000 time snapshots. In our first experiment, the training set consisted of 1000 observations and the testing set consisted of 500 different observations. The goodness of fit of the predicted output is measured in different ways (and always on the original data before zscore or dimension reduction): R^2 value, Maximum Amplitude Difference (MAD, in mV) defined by $MAD(y_i, y_j) = |\max(y_i(t)) - \max(y_j(t))|$, and absolute differences between Areas Under the Curves (AUC). The resting potential V_{rest} is defined as the potential recorded at the end of the simulation, i.e. at $t = 500\,\mathrm{ms}$. The action potential duration APD_{20} (APD_{40}, APD_{60}) is defined as the time from AP onset ($\arg_t \max \frac{dv}{dt}$) to the $-20\,\mathrm{mV}$ ($-40\,\mathrm{mV}$, $-60\,\mathrm{mV}$) repolarization moment.

Dimension reduction of the AP manifold was performed with PCA, and 15 PCA components were found to be sufficient to accurately capture the AP dynamics ($R^2 > 0.99$ in reconstructing APD in our training data). Figure 1 illustrates the modes of AP variations estimated by PCA components. In particular, the first mode captures the AP amplitude, the following modes capture variations of curve curvatures in different phases of the heart beat.

Regression Model Construction. First of all, two regression methods were tested to predict $[V_{peak}, V_{rest}, APD_{60}, APD_{40}, APD_{20}]$: partial least squares regression (PLSR) and PPR. We report their R^2 values in Fig. 2. We see that the most widely used method PLSR can accurately predict V_{peak}, but fails in regressing V_{rest}. This demonstrates the non-linear complexity of the CRN atrial model, compared with sample variations of a ventricle model which can be easily predicted by PLSR with R^2 value 0.98 [12]. This results is consistent with the mode variations represented by PCA. PPR performs very well in this test, with all the R^2 values above 0.97.

In the second step of model construction, we use extra parameters predicted from step one: $\mathbf{f}_1 = [V_{peak}, V_{rest}, APD_{60}, APD_{40}, APD_{20}]$. To predict $\mathbf{v}_{emb}$, i.e. the embedding coordinates in the 15-dimensional reduced space Ω_{AP}^{PCA} of Ω_{AP}, we tested two approaches: $(\mathcal{M}_1)$ use $[\boldsymbol{\theta}, \mathbf{f}_1]$ as model input; $(\mathcal{M}_2)$ use $[\boldsymbol{\theta}, \mathbf{f}_1]$ as model input, and iteratively regress components of $\mathbf{v}_{emb}$, i.e. use $[\boldsymbol{\theta}, \mathbf{f}_1, \mathbf{v}_{emb}(1), \cdots \mathbf{v}_{emb}(i)]$ as model input to predict $\mathbf{v}_{emb}(i+1)$. We report in Table 1 the fitting errors by PPR prediction relative to mean values of the training data. The errors in MAD and V_{rest} measure the AP amplitudes, errors in APD measure the variation of curve curvatures (AP patterns), and AUC measures the errors in a global way. As one can see, errors in APD and V_{rest} are small with both approaches, while iteratively regressing the PCA components significantly decreased the MAD error. Overall, method $\mathcal{M}_2$ produced the best prediction (Fig. 3).

Application to Tissue-Level EP Modeling. We tested the application of this regression model to 3D tissue-level EP modeling. We used as reference a

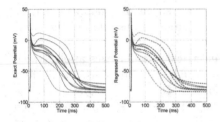

Fig. 3. AP regression by PPR. Left: exact, Right: $\mathcal{M}_2$.

Table 1. Errors of the regressed AP (by PPR) compared to the original AP.

Relative Error in (90-percentile, Mean, SD)						
Method	$\mathcal{M}_1$			$\mathcal{M}_2$		
$MAD(\%)$	45.72	23.25	16.50	2.098	0.985	1.465
$V_{rest}(\%)$	1.64	0.74	0.85	1.42	0.67	0.80
$APD_{60}(\%)$	6.12	2.66	3.12	6.12	2.73	3.31
$APD_{40}(\%)$	5.98	2.84	3.05	5.61	2.76	2.84
$APD_{20}(\%)$	8.62	4.05	4.94	7.87	3.51	7.47
$AUC(\%)$	2.99	1.39	1.55	2.70	1.26	1.58

given time series $\mathbf{v}_{ref}$ of cellular AP, with resting potential $-80.83\,\mathrm{mV}$ and AP amplitude $8\,\mathrm{mV}$. To monitor upstroke, we used the MS model with parameters as in [10], except for choosing the change-over voltage parameter v_{gate} to be $0.46 = (-40 + 80.83)/(8+80.83)$, consistent with the I_{Na} channel upstroke $-40\,\mathrm{mV}$ in the original CRN model. We simulate the potential propagation on a patient-specific geometry of left and right atrium reconstructed from medical images, with stimulus applied on the sino-atrial node. Cell model registration started at the moment $v = -10\,\mathrm{mV}$ right after AP depolarization, and stopped while $v = -75\,\mathrm{mV}$ after AP repolarization. The time step dt was $0.05\,\mathrm{ms}$. Using registration of the proposed regression model reduced the computational time by $75\,\%$ (from 84 seconds of computation time for solving one heart cycle with the original CRN model, to $23\,\mathrm{s}$ with the reduced model). The computation speed was dramatically improved as we increased the time step for the regression model registration, which is prohibited by stability requirements of numerical solvers for the original CRN model. For $dt = 1\,\mathrm{ms}$, the computational cost was reduced to $1.5\,\%$ and the simulation was almost real-time (in the order of one second per heart cycle). All computations were performed on a standard workstation. Some snapshots of the simulation are reported in Fig. 4.

Fig. 4. Simulation on the atria by the regression model registration. From left to right: t $=10\,\mathrm{ms}$; t $= 130\,\mathrm{ms}$; t$=250\,\mathrm{ms}$; t $= 370\,\mathrm{ms}$.

Restitution Study by Diastolic Interval Change. The reduced model can be extended to recover the restitution properties of AP with varying diastolic interval (DI). DI is defined as $CL - APD_{60}$, with CL being the cycle length of heart beat. In a preliminary experiment, we added DI to the set of model parameters $\boldsymbol{\theta}$; the baseline value of DI was $250\,\mathrm{ms}$, and it was randomly sampled to generate the observation matrix, as explained above. Random scale factors for DI were in log-normal distribution with mean value 1 and standard deviation 0.3. We increased the number of training data to be 1500 since the AP dynamics is more complex. In this test the prediction fit of AP was not as good as in previous ones, but the overall fitting errors were in an acceptable range (in the order of 5 %). This result suggests that the regression model has potential use in modeling complex pathological patterns in diseases such as atrial fibrillation.

4   Discussion and Conclusion

In this paper, we use a data-driven approach for the CRN cellular model reduction. The regression model we designed is able to capture the AP dynamics of the original CRN model, while in a very simple form. Before the model construction, the AP manifold dimension was reduced 15 parameters using a manifold learning technique (PCA). In model construction, different AP quantities (APD for instance) were accurately regressed using the PPR method. The embedding coordinates of the action potential in its reduced space were then regressed by PPR using the above quantities as extra parameters and an iterative regression strategy on the outputs was applied. This approach returns accurate results even with sample parameters having standard deviation 0.3. It's also interesting to see that this approach can still be used to study the restitution properties of AP with changing diastolic intervals. Finally and most significantly, the application of this regression model to tissue-level EP modeling dramatically improves the computational efficiency: it decreases the computational time up to two orders of magnitude as compared to using the original non-reduced model and enables almost real-time computations (order of seconds for computing a heart cycle on a standard workstation).

The statistical learning approach is general enough, that it can be applied to the reduction of any other cardiac cell model. Moreover, with a proper design of the training data set the reduced model can be tailored to the specific application, for the prediction of the most relevant features (for instance, a specific ionic current or a channel blocker).

In future work, we will focus on more precise ways for monitoring AP upstroke, such as the use of the eikonal equation for the depolarization time, or a regression approach on the sodium ionic channel which controls the AP depolarization phase. We also intend to study the AP restitution properties with a more sophisticated control of the action potential duration alternans, as done in [13], with the aim of modeling complex pathological patterns such as atrial fibrillation.

References

1. Aslanidi, O., Colman, M., Stott, J., Dobrzynski, H., Boyett, M., Holden, A., Zhang, H.: 3D Virtual Human Atria: a computational platform for studying clinical atrial fibrillation. Prog Biophys. Mol. Biol. **107**, 156–168 (2011)
2. Atienza, F., Almendral, J., Moreno, J., Vaidyanathan, R., Talkachou, A., Kalifa, J., Arenal, A., Villacastin, J., Torrecilla, E., Sanchez, A., Ploutz-Snyder, R., Jalife, J., Berenfeld, O.: Activation of inward rectifier potassium channels accelerates atrial fibrillation in humans: evidence for a reentrant mechanism. Circulation **114**, 2434–2442 (2006)

3. Boulakia, M., Schenone, E., Gerbeau, J.F.: Reduced-order modeling for cardiac electrophysiology. application to parameter identification. Int. J. Num. Meth. Biomed. Eng. **28**(6–7), 727–744 (2012)

4. Courtemanche, M., Ramirez, R., Nattel, S.: Ionic mechanisms underlying human atrial action potential properties: insights from a mathematical model. Am. J. Physiol. **275**, H301–H321 (1998)

5. FitzHugh, R.: Impulses and physiological states in theoretical models of nerve membrane. Biophys. J. **1**, 445–466 (1961)

6. Friedman, J.H., Stuetzle, W.: Projection pursuit regression. J. Am. Stat. Assoc. **76**, 817–823 (1981)

7. Friedman, J., Hastie, T., Tibshirani, R.: The Elements of Statistical Learning: Data Mining, Inference, and Prediction. Springer, New York (2009)

8. Krummen, D., Bayer, J., Ho, J., Ho, G., Smetak, M., Clopton, P., Trayanova, N., Narayan, S.: Mechanisms of human atrial fibrillation initiation: clinical and computational studies of repolarization restitution and activation latency. Circ. Arrhythm. Electrophysiol. **5**(6), 1149–1159 (2012)

9. Mansi, T., Georgescu, B., Hussan, J., Hunter, P.J., Kamen, A., Comaniciu, D.: Data-driven reduction of a cardiac myofilament model. In: Ourselin, S., Rueckert, D., Smith, N. (eds.) FIMH 2013. LNCS, vol. 7945, pp. 232–240. Springer, Heidelberg (2013)

10. Mitchell, C., Schaeffer, D.: A two-current model for the dynamics of cardiac membrane. Bull. Math. Biol. **65**(5), 767–793 (2003)

11. Rapaka, S., Mansi, T., Georgescu, B., Pop, M., Wright, G., Kamen, A., Comaniciu, D.: LBM-EP: Lattice-Boltzmann method for fast cardiac electrophysiology simulation from 3D images. Med. Image Comput. Comput Assist. Interv. **15**(2), 33–40 (2012)

12. Sobie, E.: Parameter sensitivity analysis in electrophysiological models using multivariable regression. Biophys. J. **96**(4), 1264–1274 (2009)

13. Kanu, U., Iravanian, S., Gilmour, R., Christini, D.: Control of action potential duration alternans in canine cardiac ventricular tissue. IEEE Trans. Biomed. Eng. **58**(4), 894–904 (2011)

14. Zettinig, O., Mansi, T., Neumann, D., Georgescu, B., Rapaka, S., et al.: Data-driven estimation of cardiac electrical diffusivity from 12-lead ECG signals. Med. Image Anal. **18**(8), 1361–1376 (2014)

15. Zheng, Y., Barbu, A., Georgescu, B., Scheuering, M., Comaniciu, D.: Four-chamber heart modeling and automatic segmentation for 3-D cardiac CT volumes using marginal space learning and steerable features. IEEE Trans. Med. Imaging **27**(11), 1668–1681 (2008)

Sensitivity of the Electrocardiography Inverse Solution to the Torso Conductivity Uncertainties

N. Zemzemi[1]($^{\boxtimes}$), R. Aboulaich[2], N. Fikal[2], and E. El Guarmah[3]

[1] Inria Bordeaux Sud-Ouest Carmen Project, Le Chesnay, France
nejib.zemzemi@inria.fr
[2] LERMA Mohammadia Engineering School, Agdal, Morocco
[3] BEFRA, Royal Air Force School, Dakhla, Morocco

Abstract. Electrocardiography imaging (ECGI) is a new non invasive technology used for heart diagnosis. It allows to construct the electrical potential on the heart surface only from measurement on the body surface and some geometrical informations of the torso. The purpose of this work is twofold: First, we propose a new formulation to calculate the distribution of the electric potential on the heart, from measurements on the torso surface. Second, we study the influence of the errors and uncertainties on the conductivity parameters, on the ECGI solution. We use an optimal control formulation for the mathematical formulation of the problem with a stochastic diffusion equation as a constraint. The descretization is done using stochastic Galerkin method allowing to separate random and deterministic variables. The optimal control problem is solved using a conjugate gradient method where the gradient of the cost function is computed with an adjoint technique. The efficiency of this approach to solve the inverse problem and the usability to quantify the effect of conductivity uncertainties in the torso are demonstrated through a number of numerical simulations on a 2D geometrical model. Our results show that adding $\pm 50\,\%$ uncertainties in the fat conductivity does not alter the inverse solution, whereas adding $\pm 50\,\%$ uncertainties in the lung conductivity affects the reconstructed heart potential by almost $50\,\%$.

1 Introduction

The ECGI procedure helps medical doctors to target some triggers of cardiac arrhythmia and consequently plan a much more accurate surgical interventions [1]. The mathematical problem behind the ECGI solution is known to be ill posed [2]. Therefore, many regularization techniques have been used in order to solve the problem [3–7]. Thus, the obtained solution depends on the regularization method and parameters [8]. Furthermore, the inverse solution does not depend only on the regularization method but also to the physical parameters and the geometry of the patient. These dependencies have not been taken into account in most (if not all) of the studies. In particular, in most of the studies in the literature the torso is supposed to be a homogenous. Moreover, when the conductive heterogeneities are considered, they are fixed according to data obtained from

© Springer International Publishing Switzerland 2015
H. van Assen et al. (Eds.): FIMH 2015, LNCS 9126, pp. 475–483, 2015.
DOI: 10.1007/978-3-319-20309-6_54

textbooks. The problem is that these data are different from a paper to another [9–11]. Their variability is mainly due to the difference between the experiment environments and other factors related to the measurement tools. Only few works have been performed in order to study the effect of conductivities uncertainty in the propagation of the electrical potential in the torso [12–14]. In [13], authors use the stochastic finite elements method (SFEM) to evaluate the effect of lungs muscles and fat conductivities on the forward problem. In [14], authors used a principal component approach to predict the effect of conductivities variation on the body surface potential.

However, to the best of our knowledge, no work in the literature has treated the influence of conductivity uncertainties of the ECGI inverse solution. In this work, we propose to use an optimal control approach to solve the inverse problem and to compute the potential value on the heart, we use an energy functional [15]. with SPDE constraints as proposed in [13]. We are interested in assessing the effect of the variability of tissue conductivity on the solution of the ECGI problem. The resolution of optimal control will be made using an iterative procedure based on the conjugate gradient method and the numerical approximation will be performed using the SFEM.

2   Methods

2.1   Solving Stochastic Forward Problem of Electrocardiography

Following [13], we represent the stochastic characteristics of the forward solution of the Laplace equation by the generalized chaos polynomial. For the space domain we use simplified analytical 2D model representing a cross-section of the torso (see Fig. 1) in which the conductivities vary stochastically. Let us first denote by D the spatial domain and Ω the probability space. By supposing that the conductivity parameter depends on the space and on the stochastic variable $\sigma(x, \xi)$, the solution of the Laplace equation does also depend on space and the stochastic variable $u(x, \xi))$. The stochastic forward problem of electrocardiography can be written as follows:

$$\begin{cases} \nabla.(\sigma(x, \xi) \nabla u(x, \xi)) = 0 & \text{in } D \times \Omega, \\ u(x, \xi) = u_0 & \text{on } \Gamma_{\text{int}} \times \Omega, \\ \sigma(x, \xi)\frac{\partial u(x, \xi)}{\partial n} = 0 & \text{on } \Gamma_{\text{ext}} \times \Omega, \end{cases} \tag{1}$$

where, Γ_{int} and Γ_{ext} are the epicardial and torso boundaries respectively, $\xi \in \Omega$ is the stochastic variable (it could also be a vector) and u_0 is the potential at the epicardial boundary.

2.2   Numerical Descretization of the Stochastic Forward Problem

We use the stochastic Galerkin (SG) method in order to solve Eq. (1). A stochastic process $X(\xi)$ of a parameter or a variable X is represented by weighted sum of orthogonal polynomials known as probability density functions PDFs

$\{\Psi_i(\xi)\}$. More details about the different choices of PDFs could be found in [16]. In our case we use the Legendre polynomials which are more suitable for uniform probability density.

$$X(\xi) = \sum_{i=0}^{p} \hat{X}_i \Psi_i(\xi)$$

where $\hat{X}_i$ are the projections of the random process on the stochastic basis $\{\Psi_i(\xi)\}_{i=1}^{p}$. The mean value and the standard deviation of X over Ω are then computed as follows

$$\mathrm{E}(X) = \int_{\Omega} \sum_{i=0}^{p} \hat{X}_i \Psi_i(\xi) = \hat{X}_0, \quad \mathrm{stdev}[X] = (\sum_{i=1}^{p} \hat{X}_i^2 \int_{\Omega} \Psi_i(\xi)^2)^{\frac{1}{2}}$$

Since in our study we would like to evaluate the effect of the conductivity randomness of the different torso organs on the electrical potential. Both of σ and u are now expressed in the Galerkin space as follows:

$$\sigma(x, \xi) = \sum_{i=0}^{p} \hat{\sigma}_i(x) \Psi_i(\xi), \qquad u(x, \xi) = \sum_{j=0}^{p} \hat{u}_j(x) \Psi_i(\xi)$$

By substituting (2), (3) into the elliptic Eq. (1) and by projecting the result on the polynomial basis $\{\Psi_k(\xi)\}_{k=1}^{p}$ we obtain the following system: For $k = 0, ..., p$:

$$\begin{cases} \sum_{i=0}^{p} \sum_{j=0}^{p} T_{ijk} \nabla.(\hat{\sigma}_i(x) \nabla) \hat{u}_j(x)) = 0 \text{ in } D, \\ \qquad \hat{u}_0(x) = u_0(x) \text{ on } \Gamma_{\text{int}}, \\ \qquad \hat{u}_j(x) = 0 \text{ on } \Gamma_{\text{int}} \quad \forall j = 1, ...p, \\ \qquad \hat{\sigma}_i(x) \dfrac{\partial \hat{u}_j(x)}{\partial n} = 0 \text{ on } \Gamma_{\text{ext}} \ \forall i, j = 0, ...p, \end{cases} \qquad (2)$$

where $T_{ijk} = E[\Psi_i(\xi), \Psi_j(\xi), \Psi_k(\xi)]$. By applying the standard finite elements variational formlation and Galerkin projections we obtain a linear system of size $(p \times dof)$, where dof is the number of the degrees of freedom for the Laplace equation in the deterministic framework.

2.3   Stochastic Inverse Problem Problem

In this section we propose to build an optimal control problem taking into account the variability of the tissue conductivities in the torso. Let us first propose a gold standard solution representing the true electrical potential in the torso. Supposing that the true conductivity distribution in the torso is σ_T, at a given time the true electrical potential in the torso u_T is governed by the

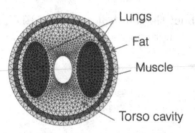

Fig. 1. 2D computational mesh of the torso geometry showing the different regions of the torso considered in this study (fat, muscle, lungs, torso cavity).

Laplace equation, it depends on the extracellular potential at the epicardium u_{ex} as follows:

$$\begin{cases} \nabla.(\sigma_T \nabla u_T) = 0 & \text{in } D, \\ u_T = u_{ex} & \text{on } \Gamma_{\text{int}}, \\ \sigma_T \frac{\partial u_T}{\partial n} = 0 & \text{on } \Gamma_{\text{ext}}. \end{cases} \tag{3}$$

From the deterministic forward problem solution u_T, we extract the electrical potential at the external boundary and we denote it by $f = u_{T/\Gamma_{\text{ext}}}$. For the inverse problem formulation, we write a cost function that takes into account the uncertainties in the torso conductivities. We then use an energy cost function as used in [15,17] constrained by the stochastic conductivity Laplace formulation as presented in the previous paragraph. We look for the current density and the value of the potential on the epicardial boundary $(\eta, \tau) \in L^{-\frac{1}{2}}(\Gamma_{\text{int}}) \times L^{\frac{1}{2}}(\Gamma_{\text{int}})$ minimizing the following cost function

$$\begin{cases} J(\eta, \tau) = \frac{1}{2}E\Big(\|v(x,\xi) - f\|^2_{L^2(\Gamma_{\text{ext}})} + \frac{1}{2}\Big\|\sigma(x,\xi)\frac{\partial u(x,\xi)}{\partial n} - \eta\Big\|^2_{L^2(\Gamma_{\text{int}})}\Big) \\ \text{with } v(x,\xi) \text{ solution of:} \\ \nabla.(\sigma(x,\xi) \nabla v(x,\xi)) = 0 & \text{in } D \times \Omega, \\ v(x,\xi) = \tau & \text{on } \Gamma_{\text{int}} \times \Omega, \\ \sigma(x,\xi)\frac{\partial v(x,\xi)}{\partial n} = 0 & \text{on } \Gamma_{\text{ext}} \times \Omega. \end{cases} \tag{4}$$

In order to solve this minimization problem, we use a conjugate gradient method as used in [15] where the components of the gradient of the cost function are computed using an adjoint method. The gradient of the functional J is given by:

$$\begin{cases} < \frac{\partial J(\eta,\tau)}{\partial \eta}.\phi > = -E[\int_{\Gamma_{\text{int}}} (\sigma\frac{\partial v}{\partial n} - \eta)\phi d\Gamma_{\text{int}}] & \forall \phi \in L^2(\Gamma_{\text{int}}), \\ < \frac{\partial J(\eta,\tau)}{\partial \tau}.h > = E[\int_{\Gamma_{\text{int}}} \sigma\frac{\partial \lambda}{\partial n} h d\Gamma_{\text{int}}] & \forall h \in L^2(\Gamma_{\text{int}}), \\ \text{with } \lambda \text{ solution of:} \\ \nabla.(\sigma(x,\xi)\nabla\lambda(x,\xi)) = 0 & \text{on } D \times \Omega, \\ \lambda(x,\xi) = \sigma(x,\xi)\frac{\partial v(x,\xi)}{\partial n} - \eta & \text{on } \Gamma_{\text{int}} \times \Omega, \\ \sigma(x,\xi)\frac{\partial \lambda(x,\xi)}{\partial n} = -(v - f) & \text{on } \Gamma_{\text{ext}} \times \Omega. \end{cases} \tag{5}$$

2.4   The Conjugate Gradient Algorithm

The stochastic Cauchy problem has been reformulated as a minimization one in the previous section. For the numerical solution, we use a conjugate gradient optimization procedure. The algorithm is then as follows,

Step 1. Given $f \in L^2(\Gamma_{\text{ext}})$, choose an arbitrary initial guess

$$(\varphi_p, t_p) \in L^2(\Gamma_{\text{int}}) \times L^2(\Gamma_{\text{int}})$$

Step 1.1. Solve the well-posed stochastic forward problem:

$$\begin{cases} \nabla.(\sigma(x,\xi)\nabla v^p(x,\xi)) = 0 & \text{in } D \times \Omega, \\ \sigma(x,\xi)\frac{\partial v^p(x,\xi)}{\partial n} = 0 & \text{on } \Gamma_{\text{ext}} \times \Omega, \\ v^p(x,\xi) = t_p & \text{on } \Gamma_{\text{int}} \times \Omega. \end{cases} \tag{6}$$

Step 1.2. Solve the stochastic adjoint problem:

$$\begin{cases} \nabla.(\sigma(x,\xi)\nabla \lambda^p(x,\xi)) = 0 & \text{in } D \times \Omega, \\ \lambda^p(x,\xi) = \sigma(x,\xi)\frac{\partial v(x,\xi)}{\partial n} - \varphi_p & \text{on } \Gamma_{\text{int}} \times \Omega, \\ \sigma(x,\xi)\frac{\partial \lambda^p(x,\xi)}{\partial n} = -(v(x,\xi) - f) & \text{on } \Gamma_{\text{ext}} \times \Omega. \end{cases} \tag{7}$$

step 1.3. We evaluate the gradient of the cost function:

$$\nabla J(\varphi_p, t_p) = \left(E\left[\varphi^p - \sigma(x,\xi)\frac{\partial v^p(x,\xi)}{\partial n}\right], E\left[\sigma(x,\xi)\frac{\partial \lambda^p(x,\xi)}{\partial n}\right] \right) \tag{8}$$

Step 1.4. Determine the descent direction d_p as follows:

$$\begin{cases} \gamma_{p-1} = \frac{\|\nabla J(\varphi_p, t_p)\|^2}{\|\nabla J(\varphi_{p-1}, t_{p-1})\|^2} \\ d_p := (d_1^p, d_2^p) = -\nabla J(\varphi_p, t_p) + \gamma_{p-1} d_{p-1} \end{cases} \tag{9}$$

We compute: $(\varphi_{p+1}, t_{p+1}) = (\varphi_p, t_p) + \alpha_p d_p$ where the scalar α_p is obtained through a linear search by:

$$\alpha_p = -\frac{E[\int_{\Gamma_{\text{ext}}} z^p(v^p - f)d\Gamma_{\text{ext}}] + E[\int_{\Gamma_{\text{int}}} (\sigma\frac{\partial z^p}{\partial n} - d_1^p)(\sigma\frac{\partial v^p}{\partial n} - \varphi_p)d\Gamma_{\text{int}}]}{E[\int_{\Gamma_{\text{ext}}} (z^p)^2 d\Gamma_{\text{ext}}] + E[\int_{\Gamma_{\text{int}}} (\sigma\frac{\partial z^p}{\partial n} - d_1^p)^2 d\Gamma_{\text{int}}]} \tag{10}$$

where z^p is the solution of the following stochastic problem:

$$\begin{cases} \nabla.(\sigma(x,\xi)\nabla z^p(x,\xi)) = 0 & \text{in } D \times \Omega \\ z^p(x,\xi) = d_2^p & \text{on } \Gamma_{\text{int}} \times \Omega \\ \sigma(x,\xi)\frac{\partial z^p(x,\xi)}{\partial n} = 0 & \text{on } \Gamma_{\text{ext}} \times \Omega \end{cases} \tag{11}$$

Step 2. Having obtained (φ_p, t_p) for $p \geq 0$, set $p = p + 1$ and repeat from step 1.1 until the prescribed stopping criterion is satisfied. As a stopping criterion we choose $\|E[v^p(x,\xi)] - f\|_{L^2(\Gamma_{\text{ext}})} \leq \epsilon$ or $\| \nabla J(\varphi_p, t_p) \| \leq \epsilon$. In practice, we take $\epsilon = 0.001$.

3   Results

In this section we present the numerical results of the stochastic inverse problem. In order to assess the effect of the conductivity uncertainties of each of lungs and fat conductivities on the reconstructed electrical potential on the heart boundary, we start by generating our ground truth solution by solving the Eq. (3). For the sake of simplicity we take a harmonic function on the heart boundary: where the exact extracellular potential $u_{ex}(x, y) = \exp(x) \sin(y)$.

Since we assume that the uncertainty of the conductivity value follows a uniform probability density, as probability density functions $\{\Psi_i\}$ we use the Legendre polynomials defined on the interval $\Omega = [-1, 1]$. We also suppose that the true conductivity uncertainty interval is centered by σ_T, the true conductivity. We propose to study three cases. In the first case we suppose that there is no uncertainties. In the second (respectively, third) case we suppose that all the conductivities values are known but the conductivity of fat (respectively, lungs) where we gradually increase the uncertainty from zero to $\pm 50\%$ of the true conductivity value σ_T. We solve the stochastic inverse problem following the algorithm described in the previous section. We measure the effect of the uncertainties using relative error (RE) and the correlation coefficient (CC). In Table 1, we show the performance of the stochastic method. We used different level of uncertainties: 0%, $\pm 3\%$, $\pm 10\%$, $\pm 20\%$, $\pm 30\%$ and $\pm 50\%$. We find that the relative error of the inverse solution has been barely affected by the uncertainties of the fat conductivity even for high uncertainty levels. In fact the RE (respectively, CC) is 0.1245 (respectively, 0.993) when there is no uncertainties and for $\pm 50\%$ of uncertainties the RE (respectively, CC) is 0.1276 (respectively, 0.991). On the contrary the effect of the lung conductivity uncertainties is high: The RE increases from 0.1245 when we don't consider the uncertainties to 0.4852 when we have $\pm 50\%$ of uncertainties on the lung conductivity.

The effect of the uncertainty on the correlation coefficient could also be qualitatively seen in Fig. 2, where the pattern of the mean value of the stochastic inverse solution looks the same in Fig. 2(a) (no uncertainties) and (b) ($\pm 50\%$ of uncertainties on the fat conductivity) and different in Fig. 2(c) ($\pm 50\%$ of uncertainties on lungs conductivity). Similarly the effect of uncertainties on the relative error could be qualitatively seen in Fig. 3. As shown in Table 1, the error does not change too much from no uncertainties (Fig. 3(a)) to $\pm 50\%$ of fat conductivity uncertainty (Fig. 3(b)). Whereas the error is high for $\pm 50\%$

Table 1. Relative error and correlation coefficient of the stochastic inverse solution for different levels of uncertainty on the fat and lungs conductivities.

	Conductivity uncertainties	0 %	±3 %	±10 %	±20 %	±30 %	±50 %
Relative error	Fat	0.1245	0.1245	0.1248	0.1248	0.1251	0.1276
	Lungs	0.1245	0.1263	0.1439	0.2208	0.3333	0.4852
Corr coeff	Fat	0.9930	0.9930	0.9945	0.9943	0.9980	0.991
	Lungs	0.9930	0.9933	0.9899	0.9767	0.9660	0.885

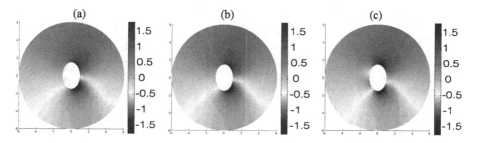

Fig. 2. Mean value of the SFEM inverse solution: No uncertainties (a), ±50 % uncertainty in the fat conductivity (b) and ±50 % uncertainty in lungs conductivity (c)

of lungs conductivity uncertainty (Fig. 3(c)). The propagation of uncertainties from the conductivities to the the inverse problem solution is reflected in the deviation of the stochastic inverse solution from the ground truth presented in Fig. 4. We remark that the error is concentrated in the heart boundary Γ_{int}, it reaches 0.8 for ±50 % of lungs conductivity uncertainty and 0.25 for ±50 % of fat conductivity uncertainty.

4   Discussion

Solving the inverse problem in electrocardiography imaging based on an optimal control problem allowed us to quantify the effect of the torso organs conductivity uncertainties using a stochastic finite element method. This work is the first mathematical study of the effect of the conductivity uncertainties on the inverse problem solution. Our results show that increasing the level of the fat conductivity uncertainty from zero to ±50 % of its original value does not alter the

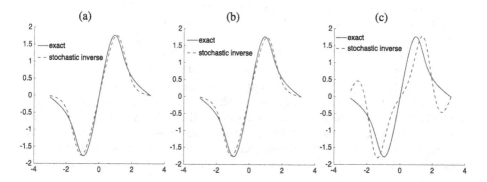

Fig. 3. Comparison of the SFEM solution to the ground truth on the heart boundary Γ_{int}: (a) no uncertainties, (b) ±50 % uncertainties in the fat conductivity, (c) ±50 % uncertainties in lungs conductivity. Exact solution (blue continuous line). Stochastic inverse solution mean value (red dashed line). X-axis polar coordinate angle from $-\pi$ to π. Y-axis value of the electrical potential on the boundary Γ_{int}.

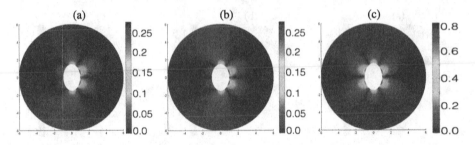

Fig. 4. Deviation of the SFEM inverse solution from the ground truth solution: No uncertainties (a), $\pm 50\,\%$ uncertainty in the fat conductivity (b) and $\pm 50\,\%$ uncertainty in lungs conductivity (c).

quality of the reconstructed potential. This is in line with the results presented in [13] for the forward problem when introducing $\pm 50\,\%$ uncertainties in the fat conductivity. On the contrary, the results that we obtained for the uncertainties on the lungs conductivity show an important effect on the ECGI solution. In fact the relative error is about $50\,\%$ when introducing $\pm 50\,\%$ of uncertainty and the CC is significantly altered. This result is different from he results presented in [13] for the forward solution with $\pm 50\,\%$ uncertainties in the lungs conductivity where the standard deviation does not exceed $3\,\%$ of the mean value.

5   Conclusions

In this work we presented a novel approach to study sensitivity to parameters values in data completion inverse problem and that could have application in a wide range of bioelectric and biomedical inverse problems resolution. We used a stochastic finite element method in order to take into account the variability of the conductivity values in the ECGI inverse problem formulated in an optimal control manner. We used a conjugate gradient method to solve this problem where the gradient of the cost function was computed using an adjoint method. We have described the different steps of the algorithm used to solve this stochastic inverse problem. The numerical simulation that we conducted in a simple 2D case showed that there is an important sensitivity of the solution to the lungs conductivity, whereas the uncertainties on the fat conductivity did not affect too much the inverse solution. It is still not clear if the difference in the sensitivity to the uncertainties is due to volume of the organs and/or their positions and proximity to the heart and/or the magnitude of their mean conductivities. This question would be a subject of a mathematical and numerical investigation as a continuation of this work.

Acknowledgments. We would like to thank the LIRIMA Laboratory which financially supported the teams ANO and EPICARD to perform this work.

References

1. Shah, A.J., Lim, H.S., Yamashita, S., Zellerhoff, S., Berte, B., Mahida, S., Hooks, D., Aljefairi, N., Derval, N., Denis, A., et al.: Non invasive ecg mapping to guide catheter ablation. JAFIB: J. Atrial Fibrillation 7(3) (2014)
2. Hadamard, J.: Lectures on Cauchy's Problem in Linear Partial Differential Equations. Yale University Press, New Haven (1923)
3. Ghosh, S., Rudy, Y.: Application of l1-norm regularization to epicardial potential solution of the inverse electrocardiography problem. Ann. Biomed. Eng. 37(5), 902–912 (2009)
4. Zakharov, E., Kalinin, A.: Algorithms and numerical analysis of dc fields in a piecewise-homogeneous medium by the boundary integral equation method. Comput. Math. Model. 20(3), 247–257 (2009)
5. Li, G., He, B.: Localization of the site of origin of cardiac activation by means of a heart-model-based electrocardiographic imaging approach. Biomed. Eng. IEEE Trans. 48(6), 660–669 (2001)
6. Doessel, O., Jiang, Y., Schulze, W.H.: Localization of the origin of premature beats using an integral method. Int. J. Bioelectromagnetism 13, 178–183 (2011)
7. Zemzemi, N., Dubois, R., Coudiere, Y., Bernus, O., Haissaguerre, M.: A machine learning regularization of the inverse problem in electrocardiography imaging. In: Computing in Cardiology Conference (CinC) 2013, 1135–1138 (2013)
8. Hansen, P.C., O'Leary, D.P.: The use of the l-curve in the regularization of discrete ill-posed problems. SIAM J. Sci. Comput. 14(6), 1487–1503 (1993)
9. Foster, K.R., Schwan, H.P.: Dielectric properties of tissues and biological materials: a critical review. Crit. Rev. Biomed. Eng. 17(1), 25–104 (1988)
10. Faes, T., Van Der Meij, H., De Munck, J., Heethaar, R.: The electric resistivity of human tissues (100 hz-10 mhz): a meta-analysis of review studies. Physiol. Meas. 20(4), R1 (1999)
11. Gabriel, S., Lau, R., Gabriel, C.: The dielectric properties of biological tissues: Ii. measurements in the frequency range 10 hz to 20 ghz. Phys. Med. Biol. 41(11), 2251 (1996)
12. Van Oosterom, A., Huiskamp, G.: The effect of torso inhomogeneities on body surface potentials quantified using tailored geometry. J. Electrocardiol. 22(1), 53–72 (1989)
13. Geneser, S.E., Kirby, R.M., MacLeod, R.S.: Application of stochastic finite element methods to study the sensitivity of ecg forward modeling to organ conductivity. Biomed. Eng. IEEE Trans. 55(1), 31–40 (2008)
14. Weber, F.M., Keller, D.U., Bauer, S., Seemann, G., Lorenz, C., Dossel, O.: Predicting tissue conductivity influences on body surface potentialsan efficient approach based on principal component analysis. Biomed. Eng. IEEE Trans. 58(2), 265–273 (2011)
15. Aboulaïch, R., Abda, A.B., Kallel, M., Bal, G., Jollivet, A., Bresson, X., Chan, T.F., Flenner, A., Hewer, G.A., Kenney, C.S., et al.: Missing boundary data reconstruction via an approximate optimal control. Inverse Prob. Imaging 2(4), 411–426 (2008)
16. Le Maître, O.P., Reagan, M.T., Najm, H.N., Ghanem, R.G., Knio, O.M.: A stochastic projection method for fluid flow: Ii. random process. J. Comput. Phys. 181(1), 9–44 (2002)
17. Andrieux, S., Baranger, T., Abda, A.B.: Solving cauchy problems by minimizing an energy-like functional. Inverse Probl. 22(1), 115 (2006)

Author Index

Printed in the United States
By Bookmasters